T0305572

Flight Mechanics Modeling and Analysis

Flight Mechanics Modeling and Analysis comprehensively covers flight mechanics and flight dynamics using a systems approach.

This book focuses on applied mathematics and control theory in its discussion of flight mechanics to build a strong foundation for solving design and control problems in the areas of flight simulation and flight data analysis. The second edition has been expanded to include two new chapters and coverage of aeroservoelastic topics and engineering mechanics, presenting more concepts of flight control and aircraft parameter estimation.

This book is intended for senior undergraduate aerospace students taking Aircraft Mechanics, Flight Dynamics & Controls, and Flight Mechanics courses. It will also be of interest to research students and R&D project-scientists of the same disciplines.

Including end-of-chapter exercises and illustrative examples with a MATLAB®-based approach, this book also includes a Solutions Manual and Figure Slides for adopting instructors.

Features:

- Covers flight mechanics, flight simulation, flight testing, flight control, and aeroservoelasticity.
- Features artificial neural network- and fuzzy logic-based aspects in modeling and analysis of flight mechanics systems: aircraft parameter estimation and reconfiguration of control.
- Focuses on a systems-based approach.
- Includes two new chapters, numerical simulation examples with MATLAB®-based implementations, and end-of-chapter exercises.
- Includes a Solutions Manual and Figure Slides for adopting instructors.

Flight Mechanics Modeling and Analysis

Second Edition

Jitendra R. Raol
Jatinder Singh

CRC Press is an imprint of the
Taylor & Francis Group, an **informa** business

Second edition published 2023
by CRC Press
6000 Broken Sound Parkway NW, Suite 300, Boca Raton, FL 33487-2742

and by CRC Press
4 Park Square, Milton Park, Abingdon, Oxon, OX14 4RN

CRC Press is an imprint of Taylor & Francis Group, LLC

First edition published by CRC Press 2008

Library of Congress Cataloging-in-Publication Data
Names: Raol, J. R. (Jitendra R.), 1947- author. | Singh, Jatinder, 1964- author.
Title: Flight mechanics modeling and analysis / Jitendra R. Raol, Jatinder Singh.
Description: Second edition. | Boca Raton, FL : CRC Press, [2023] |
Includes bibliographical references and index. |
Identifiers: LCCN 2022038871 (print) | LCCN 2022038872 (ebook) |
ISBN 9781032276090 (hbk) | ISBN 9781032276106 (pbk) | ISBN 9781003293514 (ebk)
Subjects: LCSH: Flight control–Simulation methods. | Aerodynamics–Mathematics. |
Airplanes–Control systems–Mathematical models.
Classification: LCC TL589.4 .R36 2023 (print) | LCC TL589.4 (ebook) |
DDC 629.132/3011–dc23/eng/20221018
LC record available at https://lccn.loc.gov/2022038871
LC ebook record available at https://lccn.loc.gov/2022038872

ISBN: 9781032276090 (hbk)
ISBN: 9781032276106 (pbk)
ISBN: 9781003293514 (ebk)

DOI: 10.1201/9781003293514

Typeset in Times
by codeMantra

Access the Support Material: www.routledge.com/9781032276090

This edition is dedicated

to Dr. S. Balakrishna

who had pioneered the activity in the late 60's in the country, and for his immense contributions to mainly the experimental (as well as theoretical) flight mechanics and control, one of the most thriving, formidable and difficult disciplines in India;

And

in loving memory to

Dr. S. Srinathkumar

who dedicated his life time in pursuit of flight control, and was a main architect for the LCA Control Laws.

Contents

Preface

Today the subject of Flight Mechanics: Modeling and Analysis assumes greater importance in design, development, analysis, and evaluation via flight tests, of aerospace vehicles, especially in fly-by-wire (FBW) aircraft development programs. Study of flight mechanics is important because the understanding of flight dynamics of aerospace vehicles presupposes good understanding of flight mechanics. This study is undertaken via control theoretic-systems approach without undermining the importance of the related aerodynamics that inherently provides the basis of flight mechanics modeling and analysis; this approach and related methods provide natural synergy with flight mechanics applications to design and development of flight control laws, flight simulation, aircraft system identification, aircraft parameter estimation, and handling qualities evaluation for a piloted vehicle. Interestingly, several such aspects also apply to unmanned (uninhibited) aerial vehicles (UAVs), micro air vehicles (MAVs), and rotorcrafts. Collectively, all the foregoing topics are generally studied and researched under the banner of flight dynamics, but we like to use the catchy terminology of flight mechanics and control, and hence, the flight mechanics modeling and analysis as the name of the present volume, which is *a much-revised version* of the previous CRC book by the same name.

In this book, we study some of the foregoing topics and emphasize the use of flight mechanics-knowledge via modeling and analysis, the latter are the control-systems-theoretic approaches. We illustrate several concepts and methods of flight mechanics as applied to system identification, aircraft parameter estimation, flight simulation/control, and handling qualities evaluations; where appropriate, we use several numerical simulation examples coded (by the authors in MATLAB, which is the trade mark of MathWorks Ltd. USA); the user should have an access to PC-based MATLAB software and its other toolboxes: signal processing, control system, system identification, neural networks, fuzzy logic, and aerospace.

This edition is the revised version in the following sense: (i) the aerodynamics principles and fundamental are given an enhanced treatment; (ii) a new chapter on engineering dynamics is introduced; (iii) additional material for the chapter on equations of motion is included; (iv) the transfer function analysis methods is unified with the material on mathematical model building, and the additional aspects of stability and control are presented; (v) flight simulation material is expanded with additional concepts on desktop simulator and hardware-in-the-loop simulation (HILS); (vi) flight control chapter is expanded with the basic aspects of (a) control system concepts, (b) stability augmentation systems, (c) autopilot design, (d) some flight control design examples, and (e) some additional material on design of flight control; (vii) the aircraft parameter estimation is expanded with the newer material on nonlinear estimation, and neural network-based techniques, related examples, and Lyapunov stability analysis; (viii) a new chapter is introduced on aero-servo-elastic concepts with emphasis on modeling, design, and parameter estimation; and (ix) in appendixes B and C, the concepts of artificial neural networks and fuzzy logic that are used for aircraft parameter estimation and flight control, respectively, are briefly given.

There are several good books on flight mechanics/dynamics but these either do not approach the subject from systems' point of view or the treatment of such aspects as outlined above is somewhat limited or specialized. The systems approach inherently borrows the ideas and concepts from applied mathematics; mathematical model building, especially based on empirical methods, control theory, and signal/data analysis methods; also, concepts from linear algebra and matrix computation are greatly used. Collective study of these concepts – aerodynamics, engineering dynamics, and control system methods – lends itself to a synergy that can be described in short as flight mechanics modeling and analysis. The new-generation paradigms of artificial neural network- and fuzzy logic-based modeling are gradually making their way into flight dynamic modeling and associated parameter estimation, simulation, and control technologies; hence, these topics are also integrated in here, perhaps, for the first time in a book on flight mechanics.

The end users of this integrated technology of Flight Mechanics Modeling and Analysis will be aero-systems-educational institutions, aerospace R&D laboratories, aerospace industries, flight test agencies, and transportation/automotive industry. Interestingly, some other industrial and mechanical engineering centers might be able to derive a good benefit from certain material of this book.

Acknowledgements

The subject of flight mechanics has been studied for nearly 100 years, in some ways. The advances in the theory and approaches to understanding of various aspects of flight mechanics have been of varied nature and types, and several researchers and engineers all over the world have made contributions to this specialized field, which has emerged as an independent discipline, from the basic aerodynamics and the applied mechanics.

We are very grateful to Dr. S. Balakrishna and Dr. S. Srinathkumar (late) who initiated the research and a number of project activities in flight mechanics and control in CSIR-NAL six decades ago. Balakrishan built, in-house, the first 3DOF motion-based simulator in 1976 in NAL for research on human operator modeling, and proposed mathematical models based on time-series approach. The major impetus to the activity came from the country's indigenous fighter aircraft and missile development programs, which supported the sponsored activities in the areas of modeling and identification, flight simulation, dynamic wind tunnel-based experiments, and flight control. Srinathkumar lead the control law team for a decade on the indigenous fighter aircraft development program of the country, and the light combat aircraft (LCA) made a successful maiden flight, with the control law designed and synthesized by him and his team, in Bangalore on 4th January 2001, this was a historical moment for the country. We are hence grateful to several aeronautical and flight-testing agencies in India and a few overseas (e.g., CALSPAN, USA; DLR, Germany; and UK, via cooperative exchange programs) that have, in some ways, supported the research and development in these areas in CSIR-NAL.

We appreciate constant technical support from several colleagues, of the flight mechanics and control discipline/division, who have greatly, tirelessly, and very ably furthered the cause of the flight mechanics and control research and application from the CSIR-NAL's platform. Certain contributions in the area of flight simulation and aircraft parameter estimation from: Niranjan, T. (late); Srikanth, K., Thomas, M., Pashilkar, A. A., and Madhuranath, P.; Parameswaran, V., Basappa, K., Sudesh, K. K., Shantakumar, N., and Girija, G.; Mohan Ram V.S.; Sheikh, I.; Singh, G. K.; Savanur, Shobha; Sara Mohan George, and Selvi, S. S. are gratefully appreciated.

The first author is very grateful to Prof. G. N. V. Rao and Prof. M. R. Ananthasayanam, Dept. of Aerospace Engg., IISc. (Indian Institute of Science), Bangalore, for teaching him some basics of flight mechanics; and to Dr. Ranjit. C. Desai (M. S. University of Baroda, Vadodara) and (late) Prof. Naresh Kumar Sinha (McMaster University, Hamilton, Canada) for teaching him control and system identification theories. We are also grateful to Kyra Lindholm, Kendall Bartels, and others from CRC Press who have been involved in this new project of the revision of this book for their continual support and attention. The first edition of this book was initiated by one of the great editors of the CRC press Mr. Jonathan Plant in early 2007.

We are as ever, very grateful to our spouses and children for their understanding, endurance, care, and affection. We are grateful to Mayur, J. Raol, and Mrs. Amina Khan Mayur for the suggestions of the design of the cover for this new edition.

Acknowledgements

Authors

Jitendra R. Raol had received B. E. and M. E. degrees in electrical engineering from M. S. University (MSU) of Baroda, Vadodara, in 1971 and 1973, respectively, and Ph.D. (in electrical & computer engineering) from McMaster University, Hamilton, Canada, in 1986, and at both the places, he was also a postgraduate research and teaching assistant. He had joined the National Aeronautical Laboratory (NAL) in 1975. At CSIR-NAL, he was involved in the activities on human pilot modeling in fix- and motion-based research flight simulators. He re-joined NAL in 1986 and retired in July 2007 as Scientist-G (and Head, flight mechanics, and control division at CSIR-NAL).

He had visited Syria, Germany, The United Kingdom, Canada, China, the United States of America, and South Africa on deputation/fellowships to (i) work on research problems on system identification, neural networks, parameter estimation, multi-sensor data fusion, and robotics; (ii) present technical papers at international conferences; and/or (iii) deliver guest lectures at some of these places. He had given several guest lectures at many Indian colleges and universities, and Honeywell (HTSL, Bangalore).

He was a Fellow of the IEE/IET (United Kingdom) and a senior member of the IEEE (United States). He is a life-fellow of the Aeronautical Society of India and a life member of the Systems Society of India. During his studies at the MSU, he had received Suba Rao memorial prize and M. C. Ghia charitable fellowship. In 1976, he had won K. F. Antia Memorial Prize of the Institution of Engineers (India) for his research paper on nonlinear filtering. He was awarded a certificate of merit by the Institution of Engineers (India) for his paper on parameter estimation of unstable systems. He had received the best poster paper award from the National Conference on Sensor Technology (New Delhi) for a paper on sensor data fusion. He had also received a gold medal and a certificate for a paper related to target tracking (from the Institute of Electronics and Telecommunications Engineers, India). He is also one of the (5) recipients of the CSIR (Council of Scientific and Industrial Research, India) prestigious technology shield for the year 2003 for the leadership and contributions to the development of Integrated Flight Mechanics and Control Technology for Aerospace Vehicles in the country; the shield was associated with a plaque, a certificate, and a project-grant-prize of INRs. 3,000,000 for the project work. He was one of the five recipients of the Chellaram Foundation Diabetes Research Award-2018 for the best paper (presented at the 2nd International Diabetes Summit, March 2018, Pune, India, which carried a prize of 100,000 INRs.). He has received Sir Thomas Ward memorial prize of the Institution of Engineers (India) in 2019 (jointly) for the paper on Image Centroid Tracking with Fuzzy Logic…, and it carried a gold medal and a certificate. He is featured in the list of the Stanford University (USA) as one of the top 2% scientists/researchers of the world for the year 2019.

He has published nearly 150 research papers and numerous technical reports. He had Guest-edited two special issues of Sadhana (an engineering journal published by the Indian Academy of Sciences, Bangalore) on (i) advances in modeling, system identification, and parameter estimation (jointly with Late Prof. Dr. Naresh Kumar Sinha) and (ii) multi-source, multi-sensor information fusion. He had also Guest-edited two special issues of the Defense Science Journal (New Delhi, India) on (i) mobile intelligent autonomous systems (jointly with Dr. Ajith K. Gopal, CSIR-SA), and (ii) aerospace avionics and allied technologies (jointly with Prof. A. Ramachandran, MSRIT).

He has co-authored an IEE/IET (London, UK) Control Series book Modeling and Parameter Estimation of Dynamic Systems (2004), a CRC Press (Florida, USA) book Flight Mechanics Modeling and Analysis (2009), a CRC Press book Nonlinear Filtering: Concepts and Engineering Applications (2017), and a CRC Press book Control Systems: Classical, Modern, and AI based Approaches (2019). He has also authored CRC Press books Multi-sensor Data Fusion with MATLAB (2010), and Data Fusion Mathematics–Theory and Practice (2015). He has edited (with Ajith K. Gopal) a CRC press book Mobile Intelligent Autonomous Systems (2012).

He has served as a member/chairman of numerous advisory-, technical project review-, and doctoral examination committees. He has also conducted sponsored research and worked on several projects from industry as well as other R&D organizations to CSIR-NAL with substantial budget. Under his technical guidance, eleven doctoral and eight master research scholars have had received their degrees successfully. He is a reviewer of a dozen national/international journals, and has evaluated several M. Tech./Doctoral theses (from India and overseas). He had been with MSRIT (M. S. Ramaiah Institute of Technology, Bengaluru) as emeritus professor for five years; with the Govt. College of Engineering, Kunnur (Kerala) as a senior research advisor; and with the Department of Aerospace Engineering (IISc., Bangalore) as a consultant on modeling and parameter estimation for the Type I diabetes patients' data for a period of three months.

His main research interests have been and are data fusion, system identification, state/parameter estimation, flight mechanics-flight data analysis, H-infinity filtering, nonlinear filtering, artificial neural networks, fuzzy logic systems, genetic algorithms, and soft technologies for robotics.

He has also authored a few books as the collection of his 320 (free-) verses on various facets closely related to science, philosophy, evolution, and life itself. He has also contributed 62 articles and 830 'bites' (long quotes) on matrubharti.com (#1 Indian Content Community) and 7 ebooks on Amazon KDP (kindle direct publishing) covering social, philosophical, science, and human life-related aspects and issues. He is one of the most downloaded one hundred English authors of the matrubharti.com for the year 2021.

His new area of study and research is data-systems analytics (DaSyA).

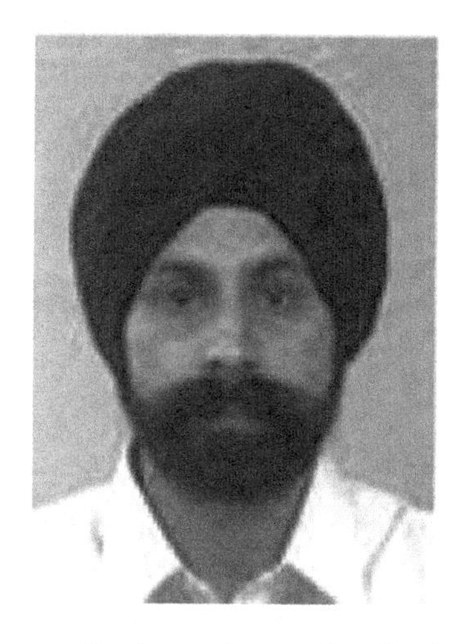

Jatinder Singh joined as a senior scientist in the flight mechanics and control division of NAL in 1998. He has a PhD from IIT, Kanpur. He was a recipient of the Alexander von Humboldt fellowship and has worked as a guest scientist at the DLR Institute for Flight Systems, Braunschweig, Germany, before joining NAL. He is a senior member of the AIAA, and a life member of Aeronautical Society of India and the Systems Society of India. He is recipient of the NAL foundation day award, and was also a member of the team, under the leadership of Dr. J. R. Raol, that won the CSIR technology shield for the year 2003 that was awarded to the flight mechanics and control division of NAL for their contribution to the integrated flight mechanics and control technology for aerospace vehicles. He has several conference and journal papers to his credit. He has also co-authored the book *Modelling and Parameter Estimation for Dynamic Systems* published by the IEE/IET, London, UK, in 2004. He has been extensively involved in aircraft modeling and parameter estimation work and has made significant contribution to the LCA (light combat aircraft) program. His expertise is in the area of flight mechanics-flight data analysis, system identification, aerodynamic modeling for fixed and rotary wing aircraft, Kalman filtering, and artificial neural networks. Some of the major projects he has worked on include: (i) Aero database validation and update of LCA (TEJAS), (ii) flight identification of intermediate Jet Trainer HJT-36 Aircraft, (iii) online flight path reconstruction and parameter estimation, (iv) development of neural network software for flush air data system, (v) stall data analysis, (vi) flight data analysis for SARAS-LTA (Light Transport Aircraft), (vii) EC 135 & ALH (Advanced Light Helicopter) rotorcraft system identification, and (viii) FOC upgradation of Mirage 2000 Aircraft.

Presently, he is a chief scientist and heads the division of flight mechanics and control at the CSIR-NAL, Bangalore.

Introduction

Flight mechanics is a bit difficult subject to understand, especially for non-aerospace scientists and engineers; even for aerospace scientists and engineers with background in pure aerodynamics, propulsion, and structures, it would be difficult to appreciate various aspects of flight mechanics, especially when applied to flight simulation, flight control, and aircraft parameter estimation. It is strongly felt, based on our own experience, that an approach based on the control-system theory would be very useful for better understanding of flight mechanics: its analysis and applications. The approach taken here is via simple concepts of mathematical model building, introduction of flight mechanics, and mathematical representations of its concepts. This leads to the equations of motion of an aerospace vehicle and the simplifications of these equations to arrive at linear models, which are often used for aircraft parameter estimation, design of flight control laws, and handling qualities (HQ) evaluation. Subsequently, the original non-linear (or sometimes extended linear models, like piece-wise models) can be used for further studies and interpretations.

The problems in aircraft dynamics are, often but not always, related to simulation, flight control, and system identification. Application of these to an aerospace vehicle centers on the knowledge of flight mechanics of the vehicle. The main point of this book is to strengthen the base in flight mechanics by using system-theoretic concepts of mathematical model building, system identification, aircraft parameter estimation, simulation, and control that provide the defining principles and techniques. Often, flight mechanics and flight dynamics are used interchangeably; however, study of flight dynamics encompasses flight mechanics analysis. In this book, the emphasis is on flight mechanics with associated studies in model building and their use in various applications that broaden the scope to flight dynamics; however, for the historical reasons, we like and prefer the catchy name that is the title of this book.

Flight mechanics of an aerospace vehicle is approached from the fundamental aspects of study of a dynamic system. The dynamics of aircraft flight are described by its equations of motion [1–3]. The main forces that act on the aircraft are inertial, gravitational, aerodynamic, and propulsive. The aerodynamic forces have a major contribution in the flight of an atmospheric vehicle.

In fact, aircraft stability and control should be studied from the point of view of applied science because it utilizes information, data and analytical tools from several disciplines: (i) applied mathematics, (ii) control and systems dynamics, (iii) aerodynamics and computational fluid dynamics (CFD), (iv) wind-tunnel testing, (v) flight testing, (vi) aeroservoelasticity, (vii) flying/ HQs, (viii) flight dynamics simulation, and (ix) flight control [4]. Thus, the field of aircraft stability and control has evolved as a mature discipline in the design development, and certification of atmospheric vehicles, and mainly airplanes.

Figure I.1 depicts the interconnection of several disciplines that form the input/output (I/O) to the study of flight mechanics.

DOI: 10.1201/9781003293514-1

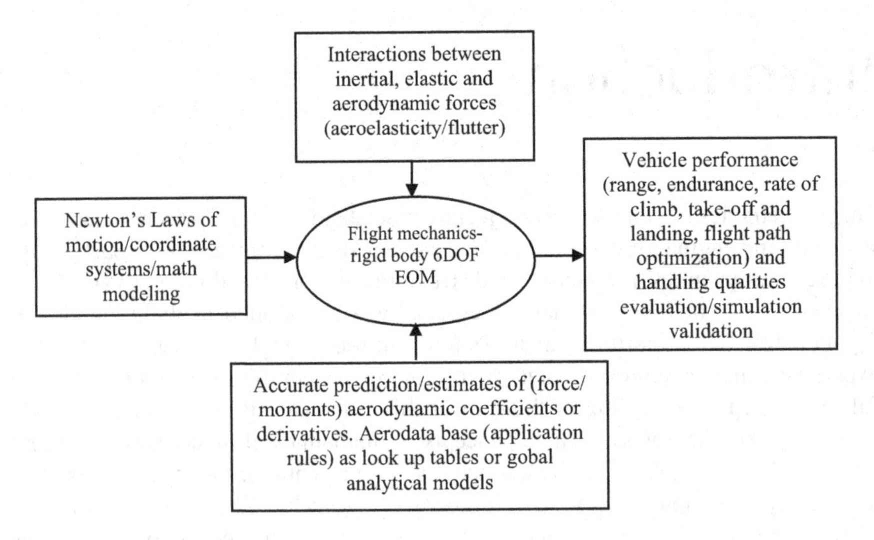

FIGURE I.1 Interaction of several disciplines in the study of flight mechanics (collectively often called flight dynamics).

I.1 MODELING

The modeling of aircraft dynamics involves mainly the characterization of the aerodynamic forces and moments. Modeling, identification, parameter estimation, and simulation play a very significant role in the present-day system analysis of complex dynamic systems including aerospace vehicles. The information obtained from the application of these techniques can be routinely used for flight simulation, design and development of flight control laws, and prediction (including simulation) of dynamic phenomena of these vehicles.

Flight motions of an aircraft can be described by non-linear coupled differential equations based on Newton's laws of motion of a rigid body, with several associated constants and parameters that are mass and geometry-related coefficients, aerodynamic coefficients, and aerodynamic stability and control derivatives. Simplification can be made by using relevant assumptions. An aircraft is considered as a dynamic system that can be studied directly by doing some experiments with it or by utilizing its mathematical model. For aircraft, both are possible: the latter is done in the design and development stage, and the former is done when the vehicle is ready for flight tests. Flight mechanics analysis, flight simulation, flight control design, and partly the handling quality analysis (HQA) are carried out using the mathematical models. Subsequently, the flight data are obtained from flight tests on the aircraft at several flight conditions (defined by altitude and Mach number). These data are pre-processed and analyzed by system identification/parameter estimation procedures [5] and the mathematical models (i.e., their parameters) are updated, if deemed necessary.

For system identification and estimation of the aerodynamic derivatives from flight test data, one needs to conduct appropriate flight tests and acquire these data.

For this task, certain special maneuvers are performed by the test pilot to excite the modes of the test aircraft. Then, one uses parameter/state estimation techniques. The aerodynamic derivatives (stability and control derivatives) that form parameters in the mathematical model of an aircraft are required for one or more of the following reasons: (i) they explain aerodynamic, stability, and control behavior of the vehicle, thereby describing its static/dynamic behavior; (ii) the mathematical models (and the associated parameters) are required for the design of flight control systems; and (iii) high-fidelity simulators need accurate mathematical models of aircraft. Three main approaches for the estimation of these derivatives are (i) analytical methods, like DATACOM, CFD; (ii) wind-tunnel testing of scaled models of aircraft; and (iii) flight testing and subsequent flight data analysis.

Main focus here is on the third approach. The subject of flight tests is very vast, and we would concentrate only on the flight tests techniques/experiments that are just necessary to generate dynamic responses of aircraft with a view to further analyzing these data in order to extract aerodynamic derivatives of aircraft. Similar methods with certain special or degenerate experiments are applicable for other atmospheric or space vehicles. For projectiles, aeroballistic test range facilities are often used. Specifically, some important principles and techniques of system identification, and state/parameter estimation are covered in this book. The applications of these techniques to kinematic consistency checking and estimation of aerodynamic derivatives from flight data are discussed. The determination of aircraft performance (drag polars) and HQs from flight test data is also of great importance. The drag polars can be successfully determined from the dynamic maneuvers data using parameter estimation methods. Thus, several important concepts and aspects of flight mechanics analysis from modeling, and simulation point of view are presented. Here, modeling is approached from the system identification and parameter estimation point of view.

I.2 FLIGHT SIMULATION

This is the subject of system's analysis, like you know the mathematical model of the dynamic system, e.g., aircraft dynamics, and you want its responses to a given input, here the pilot command signals, and the responses are the changes in angle of attack, speed, altitude, pitch attitude, and roll angle, or the yaw angle, etc. The flight simulation of an aircraft provides a virtual scenario on the ground itself (in the simulation laboratory/facility) of the flight conditions in air, and this helps in understanding the flight dynamics, and the behavior of the aircraft during the take-off and the landing conditions. The flight simulation helps in design and validation of flight control laws, the HQ evaluation of the aircraft, and understanding the effects of the faults in flying methods, and the faults in various subsystems of the aircraft, e.g., avionics, engine, aircraft control surfaces, etc., by simulation of many such conditions in the laboratory, so a lot of risk is avoided. There are mainly two types of flight simulators: (i) for training the pilots who would fly an airline aircraft, or a fighter aircraft, and (ii) for research, mainly for the design and development of aircraft, and its control laws, like hardware-in-the-loop simulator. For each category, there are again two major systems: (i) fixed base and (ii) motion based; of course within these two categories, there are a few variations, like with full visual cues, limited degree of freedom, etc.

I.3 FLIGHT CONTROL

Many modern fighter aircraft are designed with unstable configurations, meaning thereby that they do not possess natural or inherent stability, i.e., they have relaxed static stability. Some of the merits of such design are (i) improvement in performance (i.e., increased lift, *L*/*D*, drag ratio) and (ii) improvement in maneuverability. Essentially, the stability task has been partially transferred to the control task and hence the emphasis on flight control. Flight control systems for aerospace vehicles are basically extended applications of the classical control methods and/or the so-called modern control approaches to the problem of design and development of accurate/sophisticated control laws with one or more of the following purposes: (i) to improve the otherwise poor static stability or low damping in dynamics in a given axis, (ii) to provide stability to inherently unstable aircraft/dynamics, (iii) to improve the HQs of the aircraft and pilot-aircraft interactions/coupling, (iv) to improve the safety and reliability of the aircraft's functions, e.g., in the presence of certain kind of failures, and (v) to reduce the workload of the pilot in handling the complex flight missions and tasks, e.g., autopilot. Recently, there is upward surge of applications of artificial neural networks (ANNs) and fuzzy logic/system (FL/S) to aid the flight control/autopilot systems.

Figure I.2 depicts a synergy of the development process of flight control and flight simulation for an aircraft [6].

I.4 ARTIFICIAL NEURAL NETWORKS (ANNs) IN CONTROL

ANNs are emerging paradigms for solving complex problems in science and engineering [7–10]. The ANNs have the following features: (i) they mimic some simple behavior of a human brain; (ii) they have massively parallel architecture/topology; (iii) they can be represented by adaptive circuits with input channel, weights (parameters/coefficients), one or two hidden layers, and output channel with some non-linearities; (iv) the weights can be tuned to obtain optimum performance of the neural network in modeling of a dynamic system or non-linear curve fitting; (v) they require training algorithms to determine the weights; (vi) they can have feedback-type arrangement within the neuronal structure leading to recurrent neural networks (RRNs); (vii) the trained network can be used for predicting the behavior of the dynamic system, and also for parameter estimation; (viii) they can be easily coded and validated using standard software procedures; (ix) optimally structured neural network architectures can be hard-wired and embedded into a chip for practical applications – this will be the generalization of the erstwhile analog circuits-cum-computers; and (x) then, the neural network-based system can be truly termed as a new generation powerful/parallel computer.

The aerodynamic model (used for design of flight control laws) could be highly non-linear and dependent on many physical variables. The difference between the mathematical model and the real system may cause performance degradation. To overcome this drawback, ANN can be used and its weights adjusted to compensate for the effect of the modeling errors. There are several ways the ANNs can be used for control augmentation: (i) conventional control can be aided by ANNs controllers

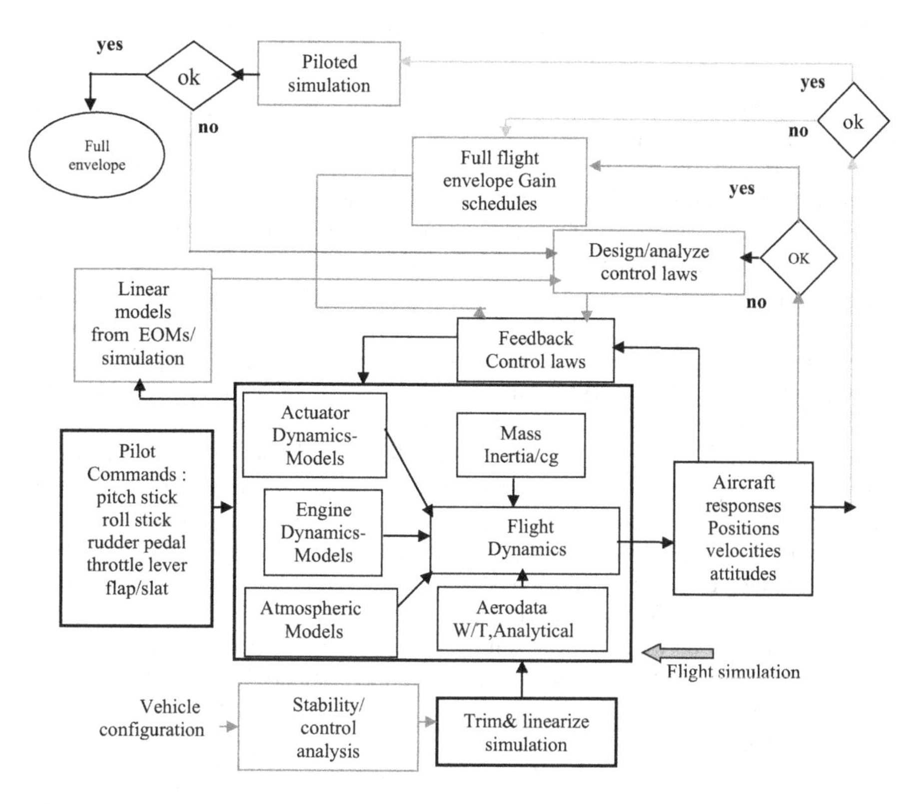

FIGURE I.2 System synergy between flight simulation and control. (From Madhuranath, P. Introduction to flight simulation. In *Aircraft Flight Control and Simulation* (edrs. Chetty S., and Madhuranath, P.), NAL Special Publication, SP-9717, National Aerospace Laboratories, Bangalore, August, 1997.)

for online learning to represent the local inverse dynamics of an aircraft, (ii) attempt to compensate for uncertainty without explicitly identifying changes in the aircraft model, (iii) the ANN's non-linearity can be made adaptive and used in the desired dynamics block of the flight controller, (iv) the learning ability can be incorporated into the gain scheduling process, and (v) sensor/actuator failure detection and management. Some of the benefits would be: (i) the controller becomes more robust and more insensitive to the plant parameter variations, and (ii) the online learning ability would be useful in handling certain unexpected behavior, of course, in a limited way, e.g., fault diagnosis and reconfiguration.

In this book, the ANNs are used for aircraft parameter estimation from realistically simulated flight data.

I.5 FUZZY LOGIC-BASED CONTROL

The fuzzy logic is also emerging as a new paradigm for non-linear modeling especially to represent certain kind of uncertainty, called vagueness, with more rigor. The fuzzy logic-based systems have the following features: (i) they are based on multi-valued logic as against the bi-valued (crisp) logic, (ii) they do not have any specific architecture like neural networks, (iii) they are based on certain rules that need to be a priori specified, (iv) fuzzy logic is a machine intelligent approach in which desired behavior can be specified by the rules in which an expert's (or a design engineer's) experience can be captured, (v) fuzzy logic system deals with approximate reasoning in uncertain situations where truth is a matter of degree, and (vi) fuzzy system is based on the computational mechanism (algorithm) with which decisions can be inferred despite incomplete knowledge. This is the process of inference engine.

Fuzzy logic-based control is suited to multivariable and non-linear processes. The measured plant variables are first fuzzified. Then, the inference engine is invoked. Finally, the results are defuzzified to convert the composite membership function of the output into a single crisp value. This specifies the desired control action. The heuristic fuzzy control does not require deep knowledge of the to-be-controlled process. The heuristic knowledge of the control policy should be known a priori. There are several ways fuzzy logic can be used to augment the flight control system: (i) FL will approximately duplicate some of the ways a pilot might respond to an aircraft that is not behaving as expected due to a damage or failure, (ii) to incorporate the complex non-linear strategies based on pilot's or system design engineer's experience and intelligence within the control law, (iii) adaptive fuzzy gain scheduling (AGS) using the fuzzy relationships between the scheduling variables and controller parameters, and (iv) fuzzy logic-based adaptive tuning of Kalman filter for adaptive estimation/control.

In this book, FL is used for reconfiguration of control law in the strategy of fault identification and management.

I.6 EVALUATION OF AIRCRAFT CONTROL-PILOT INTERACTIONS

The HQA and prediction for a fighter aircraft are important in the design and development of a flight control system [11]. Traditionally pilot's opinion ratings have been on 1- to 10-point scale, called Cooper-Harper scale. Each point gives a qualitative opinion of the pilot's view of the aircraft's overall behavior for a specified task/mission. This looks somewhat like multi-valued description. The ratings/qualitative description is also specified in the three levels: rating 1–3.5 → Level 1, rating 3.5–6.5 → Level 2, and rating 6.5–10 → Level 3. The description of the behavior of the aircraft with pilot in the loop sounds somewhat like rule-based logical enunciation of the pilot's assessment of the aircraft's performance. The Levels 1–3 have gradation more coarse than the pilot's rating scale. To quantify this assessment, several HQ criteria have been evolved that are largely based, in one way or the other, on the fundamental tenets/concepts of the (conventional) control theory, e.g., bandwidth, rise time, settling time, gain, and phase margin (GPMs), etc. [12]. It might be, perhaps, feasible to establish some connectivity between multi-valued logic-fuzzy system, H-infinity concept, and HQ criteria: the FL/S are useful for representing, in some concrete form, the uncertainty in a system's model – this uncertainty plays an important role in evaluating the robustness of a flight control system – the stability and related aspects of which are evaluated using the HQ criteria.

A possible synergy of many such aspects presented in this book is highlighted in Figure I.3 [13]. Further, system-oriented synergy is presented in Figure I.4.

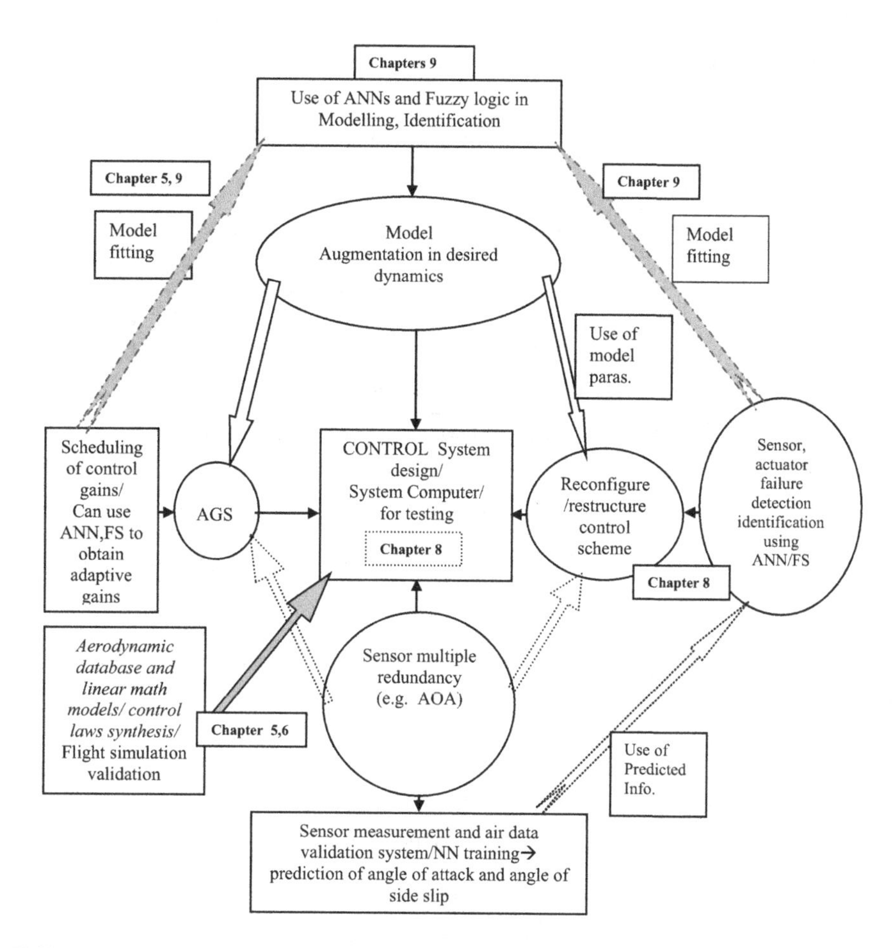

FIGURE I.3 System synergy of aircraft control system aided by artificial neural networks (ANN), fuzzy system (FS), system identification (SID), and restructuring schemes (AGS: adaptive gain scheduling). (From: Raol, J. R. Intelligent and allied technologies for flight control - a brief review. *ARA Journal*, Vol. 2001–2002, No. 25–27, 2002.)

Thus, the important features of flight mechanics-dynamics are measurements, representations, and predictions, i.e., modeling, and analysis of aerodynamic forces, and evaluation of HQs. Related main problems in engineering, especially for most atmospheric vehicles, are [2,6,14–16] (i) stability in motion; (ii) responses of the vehicle to propulsive and control input changes; (iii) responses to atmospheric gust and turbulence; (iv) aeroelastic oscillations; and (v) performance in terms of speed, altitude, range, and fuel assessed from flight maneuvers and testing. Reference [17] is a recent volumetric treatise on flight dynamics. Also, the field of 'soft' computing [18] is gaining importance in the aerospace field, and hence, some aspects are dealt with in this book. There is an immense scope of extending soft computing to flight simulation (like ANN-polynomial models for aero database representation), and human operator modeling while performing tasks in a flight simulator, etc.

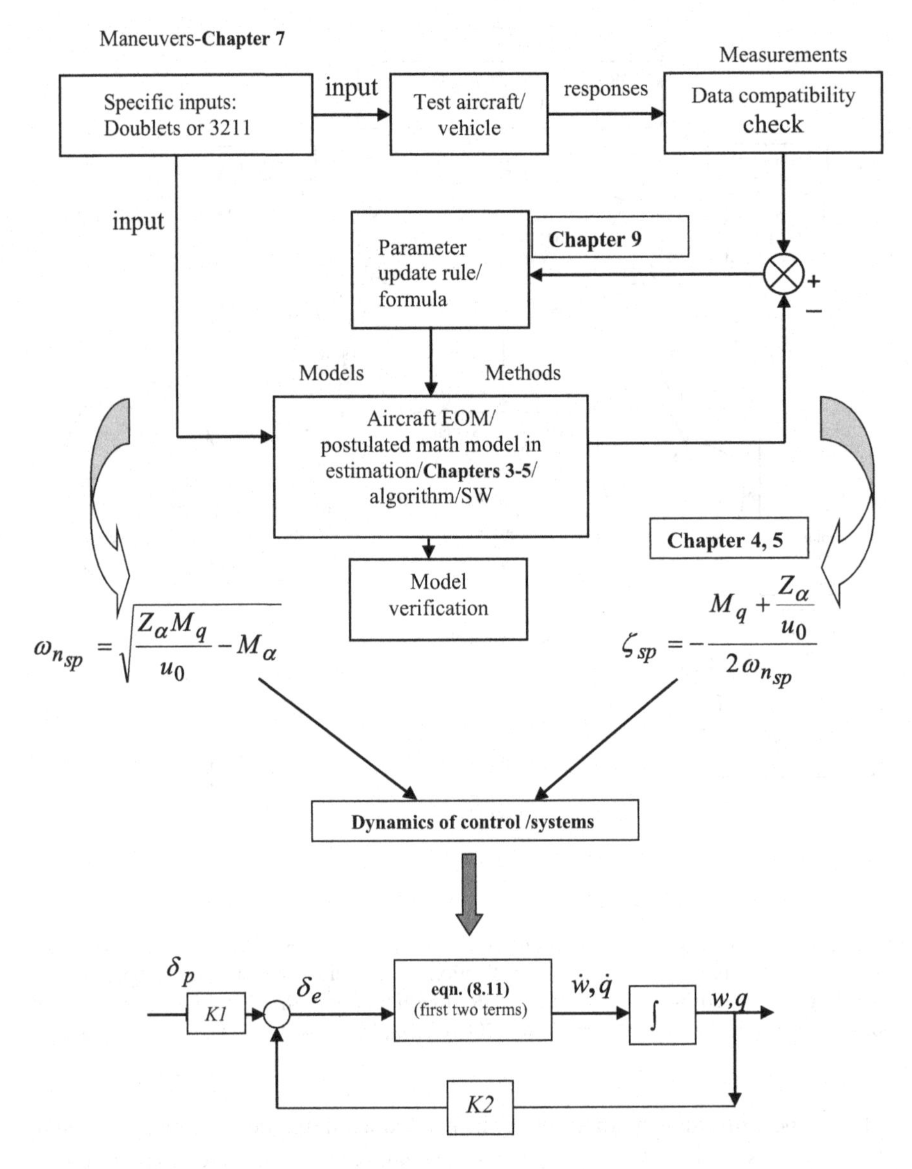

FIGURE 1.4 System synergy of maneuvers, parameter estimation, aerodynamic effects, and aircraft and control dynamics.

1.7 CHAPTER HIGHLIGHTS

In Chapter 1, we introduce several important fundamental and basic aspects of aerodynamics that are required to understand the behavior of an aircraft when it is flying in the atmosphere. In the flight mechanics modeling and analysis, these aspects play an implicit role, and hence, some basic knowledge and information are just sufficient.

Chapter 2 deals with some fundamental aspects of engineering mechanics/dynamics, since an aircraft is an engineering system, and it has its own dynamics; and all the topics briefly discussed here are very relevant to understanding of flight mechanic of an atmospheric vehicle; and when combined with the concepts of Chapter 1, the flight mechanics stands on a firmer ground.

Chapter 3 deals with equations of motion of aerospace vehicles, especially for aircraft, helicopter, and missile. The solution of these equations provides the dynamic responses of the vehicle. The aerodynamic derivatives are the kernel of the equations of motion. Together, these models capture the dynamic behavior of the vehicle in motion. In Chapter 4, we discuss the fundamental concepts of aerodynamic derivatives and models. The aerodynamic derivatives as the constituents of the aerodynamic model building are explained in sufficient details to render better appreciation of the material of the later chapters.

In Chapter 5, we resort to simplification of the equations of motion, since inherently these could be complicated and non-linear. Before that, we introduce the concepts and methods of mathematical structures and model building for dynamic systems: the conventional transfer function (TF), time-series, and state-space modeling aspects are treated. The delta operator TF theory is also highlighted. The simplified EOM (equations of motion) with embedded aerodynamic derivatives provide easier and better understanding of the behavior of the vehicle. It is here that some of the concepts and methods of Chapter 3 will be useful.

In Chapter 6, the concepts and approaches to simulation of aircraft dynamics are discussed. Depending on the availability of the detailed models of the subsystems of the aircraft, the simulation could be made simpler or more sophisticated. Its use in understanding of dynamics of a flying vehicle, in validation of flight control laws, and trying out parameter estimation exercises/algorithms need not be overemphasized. In Chapter 7, we discuss types of the input excitation signals (pilot commands) and flight test maneuvers that are an integral part of flight test exercises of an aircraft under certification or other special test trials.

In Chapter 8, we introduce some important fundamental concepts of control theory, requirements of flight control, stability augmentation systems, basic autopilot systems, and principles of reconfiguration-fuzzy logic control. Flight control examples are given. Some modeling and analysis aspects of aircraft fault detection, and identification are discussed. Several important features of modeling and design processes are outlined.

Chapter 9 deals with concepts and methods of system identification and parameter estimation as applied to real or simulated flight test data. Some important methods are discussed, and the results of several practical case studies are presented. Recent aspects of analytical global modeling, and neural network/fuzzy logic-based parameter estimation approaches are also treated. Aircraft parameter estimation results using Gaussian sum extended Kalman filter and information filters are discussed, and analytical conditions for the convergence of these algorithms using Lyapunov method are derived.

In Chapter 10, concepts of HQA of aircraft are treated. The HQA helps in the evaluation of aircraft design (though in a limited way), its dynamic behavior and flight control laws at simulation stage as well as after the flight tests are conducted.

For fighter and large transport aircraft and rotorcrafts, evaluation of pilot-vehicle interactions via HQA is very important in the early new vehicle development/modification programs. Aspects of HQA and pilot-induced oscillations, better known as pilot-vehicle interactions, are also discussed.

Chapter 11 deals with basic concepts of aeroservoelasticity: modeling, design, and estimation for flexible aircraft.

It would be, of course, necessary to have a reasonably good background in the basics of linear control systems and linear algebra.

In Appendixes A, B, C, D, and E, we compile several important aspects related to atmospheric models, ANNs, fuzzy logic, and systems/signal to support the material of the various chapters.

Disclaimer: Although enough care has been taken in working out the solutions of examples/exercises and presentation of various theories and case study results in this book, any practical applications of these should be made with proper care and caution; any such endeavors would be the readers'/users' own responsibility. Some MATLAB programs developed for the illustration of various concepts via examples in this book would be accessible to the readers from the book's URL of the CRC Press.

The end users of this integrated technology will be educational institutions, aerospace R&D laboratories, aerospace industries, flight test agencies, and transportation/automotive industry. Interestingly, some other industrial and engineering centers might be able to derive a good benefit from certain material of this book.

REFERENCES

1. McRuer, D. T., Ashkenas, I., and Graham, D. *Aircraft Dynamics and Automatic Control.* Princeton University Press, Princeton, NJ, 1973.
2. Nelson, R. C. *Flight Stability and Automatic Control*, 2nd Edn. McGraw Hill International Editions, New York, 1998.
3. Cook, M. V. *Flight Dynamics Principles.* Arnold, London, 1997.
4. Roskam, J. Evolution of airplane stability and control: A designer's viewpoint. *Journal of Guidance, Control, and Dynamics*, 14(3), 481–487, 1991.
5. Raol, J. R., Girija, G., and Singh, J. Modelling and parameter of dynamic systems. *IEE Control Series*, Vol. 65, IEE, London, UK, 2004.
6. Madhuranath, P. Introduction to flight simulation. In *Aircraft Flight Control and Simulation* (edrs. Chetty S., and Madhuranath, P.), NAL Special Publication, SP-9717, National Aerospace Laboratories, Bangalore, 1997.
7. Zurada, J. M. *Introduction to ArtificialNneural System.* West Publishing Company, New York, 1992.
8. Haykin, S. *Neural Networks-A Comprehensive Foundation.* IEEE, New York, 1994.
9. Kosko, B. *Neural Networks and FuzzySystems-A Dynamical Systems Approach to Machine Intelligence.* Prentice-Hall, Englewood Cliffs, NJ, 1995.
10. King, R. E. *Computational Intelligence in ControlEngineering.* Marcel Dekker, New York, 1999.
11. Mooij, H. A. *Criteria for Low Speed Longitudinal Handling Qualities (of Transport Aircraft with Closed-Loop Flight Control Systems).* Martinus Nijhoff Publishers, the Netherlands, 1984.
12. Sinha, N. K. *Control Systems.* Holt, Rinehart and Winston, Inc., New York, 1988.

13. Raol, J. R. Intelligent and allied technologies for flight control - a brief review. *ARA Journal*, 2001–2002, 25–27, 2002.
14. Etkin, B., and Reid, L. D. *Dynamics of Flight – Stability and Control*, 3rd Edn. John Wiley, New York, 1996.
15. Yechout, T. R., Morris, S. L., Bossert, D. E., and Hallgren, W. F. *Introduction to Aircraft Flight Mechanics-Performance, Static Stability, Dynamic Stability, and Classical Feedback Control.* AIAA Education Series, Reston, VA, 2003.
16. Ranjan, V. *Flight Dynamics, Simulation, and Control-For Rigid and Flexible Aircraft.* CRC Press, Boca Raton, FL, 2015.
17. Stengal, R. F. *Flight Dynamics.* Princeton University Press. Prenceton, NJ, 2004.
18. Hajela, P. Soft computing in multidisciplinary aerospace design – new directions for research. *Progress in Aerospace Science*, 38, 1–21, 2002.

1 Aerodynamic Principles and Fundamentals

1.1 AERODYNAMIC CONCEPTS AND RELATIONSHIPS

An atmospheric vehicle moves in the flow field of (free) air. Air pressure (P), air temperature (T), air density (ρ), and air velocity (V) are very important considerations in the study of such flight vehicles, and a knowledge of (P, ρ, T, V) at a point in the air defines a flow field; here, e.g., $P=P(x, y, z)$, and so on. In any flight mechanics-cum-dynamics (FMD) study and analysis, the three aspects are very important: (i) aircraft stability and control, (ii) aircraft performance, and (iii) vehicle's dynamic trajectory analysis. Aircraft stability and control aspects are the direct outcomes of the application of the control theory of dynamic systems to aircraft, and here, the aircraft is considered as a dynamic system. The aircraft performance mainly depends on the aircraft engine being used for propelling the vehicle through the atmosphere. The aircraft dynamic trajectory is the time history of its flight path and of many components of this flight path through the atmosphere. All these three major studies and study of flight mechanics constitute the main body of the flight dynamics.

1.1.1 Air Pressure

Pressure is defined as the force acting on a unit area, and it is due to the time rate of change of momentum (mass times velocity of the air parcels) of the gas/air molecules imparting on that surface; $P = \lim_{dA \to 0}\left(\frac{dF}{dA}\right)$, pressure at a point $\left(\mathrm{N/m^2};\mathrm{atm}\right)$; $1\ \mathrm{atm} = 1.01 \times 10^5\ \mathrm{N/m^2} = 2016\ \mathrm{lb/ft^2} = 2016/(12)^2\ \mathrm{lb/in^2} = 14$ PSI. For an aircraft, it is usually measured by using a pitot-static system of the kind shown in Figure 1.1 that measures the total and static air pressures. The pressure tube is a concentric with center tube measuring the total pressure and the outer tube surrounding it measuring the static pressure. As the aircraft moves forward, the airflow comes to rest at the mouth of the pitot tube. This is known as the stagnation point, and the pressure at this point is given by

$$P_T = P_s + \frac{1}{2}\rho V^2 \tag{1.1}$$

Here, P_T is the total pressure, which is often measured at the nose boom. P_s is the static pressure upstream away from the body, and V is the free stream velocity. The $\frac{1}{2}\rho V^2$ is termed as the dynamic pressure and is generally denoted by $\bar{q}$, and ρ 'rho' is the air density. Equation (1.1) holds good only for low-speed flights where the compressibility effects can be ignored. Also, note that both the dynamic and static

DOI: 10.1201/9781003293514-2

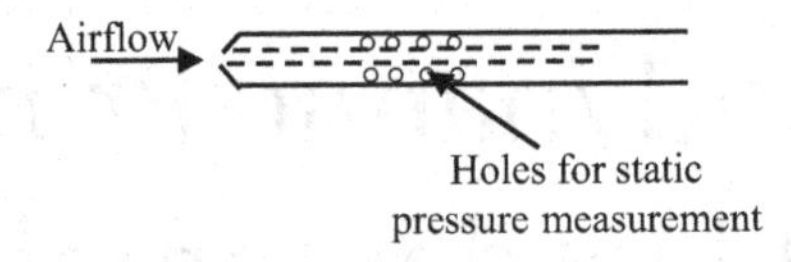

FIGURE 1.1 Pitot-static tube.

pressure act on the inner tube, while only static pressure acts on the outer tube. The difference between the two is the dynamic pressure, which is used to determine the velocity V of the aircraft, and the dynamic pressure is the direct consequence of the aircraft's velocity. At high speeds (Mach number, $M > 0.3$), the pitot measurements must be corrected for compressibility effects. The corrected dynamic pressure $\bar{q}_c$ can be written as [1]:

$$\bar{q}_c = P_T - P_s = P_s\left[\left(1+\frac{\gamma-1}{2}M^2\right)^{\frac{\gamma}{\gamma-1}} - 1\right] \tag{1.2}$$

Here, γ is the ratio of specific heats (heat capacity with constant pressure to heat capacity with constant volume) and is taken as 1.4; and it is also known as adiabatic index, and Laplace's coefficient. The Mach number is defined as the ratio of the speed of the aircraft to the speed of sound. It is also worth mentioning here that the static pressure is subject to errors called position errors. It is a known fact that for any given static source location, the position error varies with a change in speed and altitude. The normal atmospheric pressure at sea level is 101 kN/m^2. It is often expressed in terms of bars: 1 bar = 100 kN/m^2. Thus, one millibar = 100 N/m^2 = Pa (Pascal). Thus, when an aircraft lies in the atmospheric air, the air pressure acts on all its external body parts, fuselage, wings, and engine. The air with its dynamic pressure provides lift to the aircraft as well as opposes its forward motion (i.e., it provides the drag force), and the balance of this with the weight of the whole aircraft and the propulsive force supplied by aircraft engine keeps the aircraft in its steady-state motion in the air. So, the aircraft flight is the balance of various forces acting on it!

1.1.2 Air Density

In general, density of any material is defined as its mass per its own unit volume:

$$\rho = \text{Mass / Volume} = \lim_{dv\to 0}\left(\frac{dm}{dv}\right); \left(\text{kg / m}^3\right) \tag{1.3}$$

It signifies the fact how much material is compacted in so much volume. The relation between pressure, density, and temperature, according to gas laws for a perfect gas, is given by:

$$P = \rho RT \tag{1.4}$$

Here, R is the (universal) gas constant (287 J/kg-K; J is Joules, and K is the temperature in Kelvin; =8.31432 J/Kelvin-mole). A perfect gas is one in which the intermolecular forces (which are the ramifications of the complex interaction of the electromagnetic properties of the electrons and nucleus) are negligible; air at standard conditions can be approximated by a perfect gas. Another factor related to density that occurs in many formulae is the ratio of the air density ρ to the sea-level density ρ_0 (1.225 kg/m^3):

$$\sigma = \frac{\rho}{\rho_0} \tag{1.5}$$

The related quantity, the specific weight, is defined as: Wt./volume = m g/volume; here, m is the mass and g is the acceleration due to gravity, 9.81 m/s^2. We have seen that the density enters in the formula for dynamic pressure. This means that the aircraft will have higher dynamic pressure in the dense air. At higher altitudes where the air density is low, the dynamic pressure acting on the aircraft will be also low!

1.1.3 Air Temperature

The particles in a gas are in a constant motion, and each particle has kinetic energy; and hence, the temperature T of the gas is thus directly proportional to the average molecular kinetic energy, KE = (3/2) kT; k is the Boltzmann constant (1.38×10^{-23} Joules/Kelvin). The pressure and temperature are also related via gas law. The temperature of the atmosphere varies with altitude and can be represented by a linear relation:

$$T = T_1 + lh \tag{1.6}$$

Here, l is the lapse rate that indicates the rate of change in temperature with altitude. In an aircraft, a temperature probe is used to measure the total temperature of the air. Assuming adiabatic conditions (no loss of heat), the total temperature T_t in terms of ambient temperature T is given by:

$$T_t = T\left[1 + \frac{\gamma - 1}{2} M^2\right] \tag{1.7}$$

1.1.4 Altitudes

In aeronautics and aviation flight, we talk about several types of altitude, the primary among them being the absolute, geometric, and geo-potential altitude. Geometric altitude is the altitude-height from mean sea level. It does not vary with temperature or with the change of gravity. The absolute altitude is measured from the center of the earth. If H_a denotes the absolute altitude, H_{GM} denotes the geometric altitude, and R_E is the radius of earth, then

$$H_a = H_{GM} + R_E \tag{1.8}$$

If g_0 is the acceleration due to gravity at sea level, then the 'g' varies as [1,2]:

$$\frac{g}{g_0} = \left[\frac{R_E}{R_E + H_{GM}}\right]^2 \tag{1.9}$$

The 'g' is the local gravitational acceleration at a given absolute height, H_a.

The geo-potential altitude H_{GP} is in fact a fictitious altitude defined based on the reference gravity at sea level ($g_0 = 9.81$ m/sec^2). The relation between the geometric and geo-potential altitude is given by:

$$H_{GP} = \left[\frac{R_E}{R_E + H_{GM}}\right] H_{GM} \tag{1.10}$$

The difference between H_{GM} and H_{GP} is small for lower altitudes, but it grows as the altitude increases.

Other altitudes are the pressure altitude, density altitude, and temperature altitude. We know that the static pressure varies with altitude. A sensor called altimeter, calibrated using the standard atmosphere, measures the static pressure and relates it to altitude. The altitude indicated by the altimeter is therefore the pressure altitude. The density and the temperature altitudes, likewise, are the altitudes in the standard atmosphere corresponding to the measured density and temperature. The density altitude is often used for piston engine aircraft since their power is generally proportional to air density. At a certain geometric altitude, there is a set of P, T, and the density values for each of these, one can read-off the altitudes from the SATM table; e.g., at a certain altitude, the pressure $P = 4.72\ 10^4$ N/m$^2 \rightarrow$ 6000 m (altitude/ht.); and $T = 255.7$ K$\rightarrow$Temperature altitude is 5 km; and density $= P/(RT) = 0.641$ kg/m$^3 \rightarrow$ density altitude is 6.25 km.

Note: Some call the geometric altitude as the system altitude. Some define an absolute altitude as the height above the surface of the earth at any given surface location. If the aircraft is above the mountain, then the absolute altitude is the vertical clearance between the aircraft and the surface of the mountain. Also, the true altitude is defined as the height above the sea level. The pressure altitude is defined as the height from the reference where the pressure is 29.921 In. (760 mm) of Hg (mercury). Other related definitions are as follows: the Geoid height is the height of the actual surface of the earth from the surface of the ideal spheroid of the Earth, the elevation is the height from the actual surface, and the geodetic height is from the spheroid (of the Earth).

1.1.5 Airspeeds-IAS, CAS, EAS, TAS

The basic speed of sound is given by:

$$a = \sqrt{\gamma RT} \tag{1.11}$$

For sea-level conditions (T is 288 K), the speed of sound is computed to be nearly 340 m/s. It has already been mentioned that the pitot-static tube can be used to measure the airspeed. The pressure difference, $(P_T - P_s)$, measured by the pitot-static tube is passed on to an indicator that is calibrated to standard sea-level conditions. The airspeed read from this instrument is called the indicated airspeed (IAS). The IAS will be affected by instrument and position errors. Instrument errors could arise from mechanical inaccuracies, while position errors are caused by location of the pitot-static tube in the flow field that is distorted because of the interference from fuselage or wing. Correction for these errors in IAS will yield the calibrated airspeed (CAS). Modern aircraft have airspeed indicators which directly read CAS [1,2]. True airspeed (TAS) is the actual airspeed of the aircraft relative to the air. The relationship between the TAS and the ground speed is given by:

$$\text{TAS} = V_g + V_w \tag{1.12}$$

Here, V_g is the speed w.r.t the ground and V_w is the wind speed. Equivalent airspeed (EAS) is defined as the speed at standard sea-level conditions which produce the same dynamic pressure as the TAS:

$$\frac{1}{2}\rho_0(\text{EAS})^2 = \frac{1}{2}\rho(\text{TAS})^2 \tag{1.13}$$

or

$$\text{TAS} = \frac{\text{EAS}}{\sqrt{\sigma}} \tag{1.14}$$

where $\sigma = \dfrac{\rho}{\rho_0}$

Thus, TAS results when EAS is corrected for density altitude. Knowing TAS, Mach number can be computed using the relation:

$$M = \frac{\text{TAS}}{a} = \frac{\text{TAS}}{\sqrt{\gamma RT}} \tag{1.15}$$

In terms of total pressure P_T and static pressure P at a given flight altitude, TAS and EAS for the subsonic flight ($M < 0.3$) can be obtained using the relation:

$$\text{TAS} = \sqrt{\frac{2(P_T - P)}{\rho}} \tag{1.16}$$

and

$$\text{EAS} = \sqrt{\frac{2(P_T - P)}{\rho_0}} \tag{1.17}$$

Interestingly, Mach number can be viewed as the ratio of the inertia force to elastic force. For high subsonic flight ($M > 0.3$), TAS and CAS are given by:

$$\text{TAS} = \sqrt{\frac{2a^2}{\gamma - 1}\left[\left(\frac{P_T - P}{P} + 1\right)^{\frac{\gamma-1}{\gamma}} - 1\right]} \tag{1.18}$$

and

$$\text{CAS} = \sqrt{\frac{2a_0^2}{\gamma - 1}\left[\left(\frac{P_T - P}{P_0} + 1\right)^{\frac{\gamma-1}{\gamma}} - 1\right]} \tag{1.19}$$

The 'a_0' is the speed of sound at sea-level condition and P_0 is the static pressure at sea level ($1.01325 \times 10^5 \text{N/m}^2$). In other words, the relationship between TAS and CAS can also be expressed as:

$$\text{TAS} = \text{CAS}\sqrt{\frac{a}{a_0}} \tag{1.20}$$

Aircraft speeds are classified as: subsonic $M < 1$; transonic $0.8 \le M \le 1.2$; supersonic $1 < M < 5$; and hypersonic $M \ge 5$.

1.1.6 Bernoulli's Continuity Equations

Continuity equation relates density, velocity, and area at one section of the flow to the same quantities at another section. In Figure 1.2, the flow is shown to be bounded by two streamlines. The mass flow rates at sections 1 and 2 are given by [3]:

$$\dot{m}_1 = \rho_1 A_1 V_1; \ \dot{m}_2 = \rho_2 A_2 V_2 \tag{1.21}$$

Since, mass can neither be created nor be destroyed (in a general sense, otherwise, we have to use $E = mc^2$):

$$\dot{m}_1 = \dot{m}_2 \text{ or } \rho_1 A_1 V_1 = \rho_2 A_2 V_2 \tag{1.22}$$

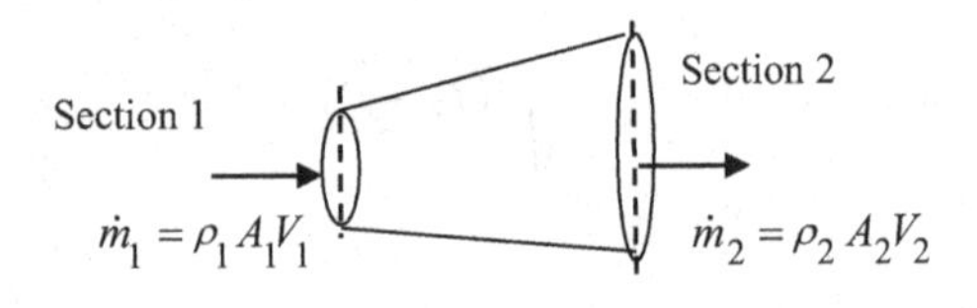

FIGURE 1.2 Stream tube.

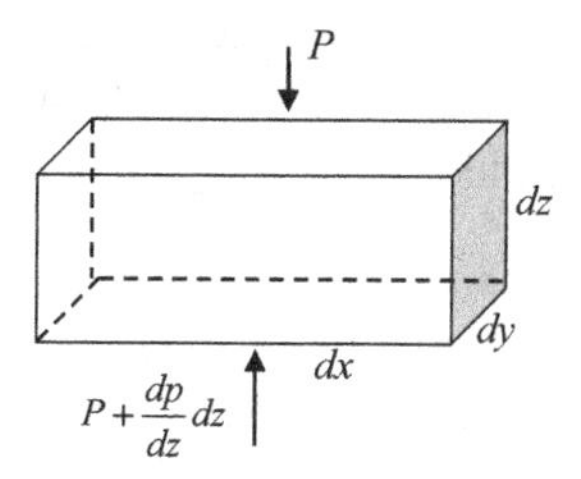

FIGURE 1.3 Forces on a fluid element for developing Bernoulli's equation.

This is the continuity equation for steady flow. Another popular equation that relates pressure and velocity at one point in the flow to another point is the Bernoulli's equation. It essentially follows from the Newton's second law, which states that Force = mass × acceleration. In Figure 1.3, if P is the pressure on face 1 of the fluid element (shown by a cuboid), the net force is given by:

$$F = Pdxdy - \left(P + \frac{dP}{dz}dz\right)dxdy \tag{1.23}$$

or

$$F = -\frac{dP}{dz}(dxdydz) \tag{1.24}$$

Since mass is density multiplied by volume, we have:

$$m = \rho(dxdydz) \tag{1.25}$$

Also, along the z-axis, acceleration is given by the rate of change of velocity:

$$a = \frac{dV}{dt} = \frac{dV}{dz}\frac{dz}{dt} = \frac{dV}{dz}V \tag{1.26}$$

From Newton's second law

$$F = ma \tag{1.27}$$

Substituting for F and a (here 'a' is acceleration) from eqn. (1.26), we have:

$$F = -\frac{dP}{dz}(dxdydz) = \rho(dxdydz)\frac{dV}{dz}V \tag{1.28}$$

or

$$dP = -\rho VdV \tag{1.29}$$

Equation (1.29) is called the momentum equation for steady, inviscid flow. Considering point 1 with quantities P_1 and V_1, and point 2 with P_2 and V_2 in a flow, the momentum equation can be integrated to yield:

$$P_1 + \frac{1}{2}\rho V_1^2 = P_2 + \frac{1}{2}\rho V_2^2 \tag{1.30}$$

This is called the Bernoulli's equation for inviscid incompressible flows with gravity terms neglected. As speed increases, compressibility effects need to be accounted for and density can no longer be assumed constant. Bernoulli's equation for compressible flows and with gravity term included is given by [1]:

$$\frac{\gamma}{\gamma-1}\frac{P}{\rho} + \frac{1}{2}V^2 + gz = \text{constant} \tag{1.31}$$

1.1.7 Mach Number

The Mach number is a very important similarity parameter, which comes out of the fluid dynamic equations of motion and captures the compressibility effect. It is defined as:

$$M = \frac{\text{TAS}}{a} = \frac{\text{TAS}}{\sqrt{\gamma RT}} \tag{1.32}$$

An expression for Mach number can also be obtained in terms of P_T and P_s, by integrating Bernoulli's equation [2]:

$$M = \frac{2}{\gamma-1}\left[\left(\frac{P_T}{P_s}\right)^{\frac{\gamma-1}{\gamma}} - 1\right]^{0.5} \tag{1.33}$$

Equation (1.33) can be used to compute Mach number at subsonic speeds, i.e., for $M < 1$.

1.1.8 Reynold's Number

Reynold's number is the ratio of inertia force to the viscous force and is defined as (in the form of units' analysis):

$$\text{Re} = \frac{\text{mass} \times \text{acceration}}{(-)\mu \times \text{Area} \times (dV/dx)}$$

$$= \frac{\text{density} \times d^3 \times \text{Velocity/time}}{(-)\mu V d} = \frac{\text{density} \times d^2 \times (d/\text{time}) \times V}{(-)\mu V d}$$

$$\text{Re} = \frac{\rho V d}{\mu} \text{ as a dimensionless number} \tag{1.34}$$

Here, d is the characteristic length (usually the mean aerodynamic chord, c-bar), V is the true airspeed, and μ is the coefficient of viscosity of the fluid/air. If during a wind tunnel test on an aircraft model, the Reynolds and Mach numbers are same as the full-scale flight vehicle, then the flow about the model and the full-scale vehicle will be identical.

1.1.9 Viscosity

Frictional force in a flowing fluid is termed as viscosity of the fluid defined as the ratio of the shear force to the rate of shear deformation. Higher the friction, higher the viscosity. In that sense, liquids are more viscous than gases. Mathematically, if τ represents the frictional force per unit area (also called shear stress/pressure) and dV/dx represents the velocity gradient, then the coefficient of viscosity can be obtained from the relation:

$$\text{Viscous Force} = \mu \times \text{Area} \times \frac{dV}{dx} \tag{1.35}$$

The viscosity of gases increases with an increase in temperature as per the relation (Rayleigh's formula):

$$\mu \propto T^{3/4} \tag{1.36}$$

The ratio of absolute viscosity μ to density ρ is called the kinematic viscosity υ, i.e.,

$$\upsilon = \frac{\mu}{\rho} \tag{1.37}$$

1.2 AIRCRAFT FORCE PARAMETERS

The relative wind acting on the aircraft produces the so-called total aerodynamic force – the components of which are briefly explained now. Interestingly, the four forces are defined with respect to three different coordinate systems: the lift and drag are relative to the wind, weight is relative to Earth, and thrust is defined relative to the aircraft's orientation. In a normal cruise flight, the weight (downward) is balanced by the lift (upward), and the drag (backward) is balanced by the thrust (forward).

1.2.1 Lift

It is easier to understand the concept of aircraft lift with Newton's laws rather than Bernoulli's. The Bernoulli principle/approach is the one that is generally used to describe lift [4]. This explanation focuses on the shape of the wing. It does not satisfactorily explain the inverted flight and ground effect. The description of lift based on Newton's laws can explain power curve, high-speed stalls, and ground effect. It is conventionally believed and explained that the wing produces the lift due to the fact that air travels faster over the top of the wing, and this creates a lower pressure than the bottom of the wing, and this differential pressure produces the upward lift. It is not clear here why the air moves faster on the top. As per the Newton's laws, the wing changes air's momentum, which generates the force on the wings. Thus, the wing continuously diverts lot of air downward. Thus, the lift of a wing is directly proportional to the quantity of the air diverted down and the downward speed of the mass of air. This downward velocity is called 'downwash'. As the wing moves along, the air is diverted down at the rare as well as the air is pulled up at the leading edge leading to 'upwash'. This combination of the upwash and the downwash, in general, contributes to the overall lift. Thus, as the air is bent around the top, it is pulled from above and this pulling causes the pressure to become lower above the wing. The acceleration of the air above the wing in the downward direction is responsible for giving the wing lift. The air is moving up at the leading edge, and at the trailing edge, it is diverted downward, with the top air accelerating towards the trailing edge. The air being fluid when it is moving and comes in the contact with a curved surface, it will try to follow that surface. This is known as the Coanda effect. Here, Newton's two laws are useful: for the air (any fluid) to bend, there must be a force acting on it, and the air must put an equal and opposite force on the wing surface that caused it to bend. The air follows a curved path due to its viscosity, which is a property of any fluid – the resistance to flow, a property of stickiness of the fluid. The relative velocity of the air molecules nearest to the surface (with respect to the surface) is zero, and it progressively increases as we move away from the surface. Since the air near the surface has a change in velocity, the air flow is bent towards the surface. The volume of the air around the wing seems attached to the surface, and this phenomenon is known as the boundary layer. Effectively, the wing is forcing the air down, or pulling the air downward from the above. The wing that is producing the lift would be at a certain AOA (angle of attack) that determines the lift (Figure 1.4). The AOA is adjusted to get certain lift for the speed and the load. The lift begins to decrease at a certain AOA, and it could be 12–15 deg. The force necessary to bend the air to such a steep angle is much greater than that could be supported by the viscosity of the air. This is the cause of the separation of the air from the wing and leads to the stall of the wing, the reduction/loss of the lift.

We have seen that the lift of the wing is proportional to the amount of the air diverted down times the downward velocity of that air. As the aircraft's speed increases, the more air is diverted. The AOA is reduced to maintain a constant lift. At higher altitudes, the air density is lesser and lesser, and hence, the air diverted for

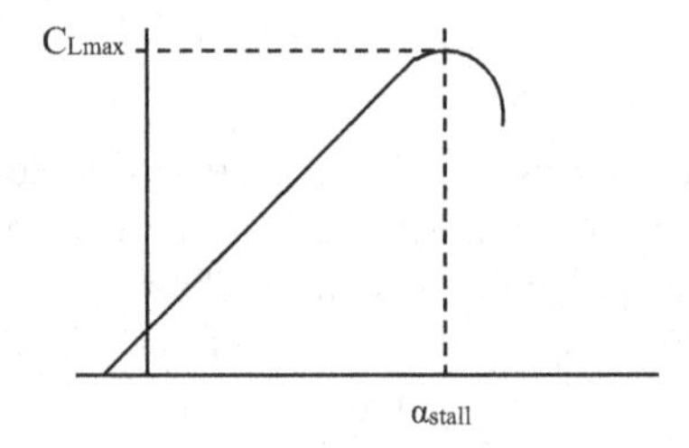

FIGURE 1.4 Lift curve and stall.

the same speed is less. This requires an increase in AOA. The air is left in motion after the aircraft passes ahead, the still air ends up with a downward velocity, and thus, the air has been imparted with energy. The lift thus requires power (being supplied by the engine or by gravity). This power is proportional to the amount of the air diverted down times the square of the velocity of the diverted air. We know that the lift of wing is proportional to the amount of the air downward times the downward velocity of the air. Hence, the power required to lift the aircraft is proportional to the load (weight) times the vertical velocity of the air. If the aircraft speed is doubled, the amount of air diverted doubles. The AOA should be reduced to reduce the vertical velocity to half for the same lift. Thus, the power required for the lift has been reduced to half. The power to create lift is, in fact, inversely proportional to the speed of the aircraft.

The amount of the air in the downwash also changes along the wing. The downwash comes off the wing in the form of a sheet. The wing at the root diverts more air (than at the tip); the net effect is that the downwash sheet begins to curl outward around itself, producing the wing vortex, tightness of the curling of the vortex being proportional to the rate of change of the lift along the wing. At the wing tip, the lift is nearly zero, and the tip vortices are tightest.

Near the ground, the upwash is reduced due to the fact that the ground reduces the circulation of the air under the wing, requiring reduced downwash of the air (to provide the lift). The AOA is reduced and hence the induced power. This makes the wing more efficient due to the ground effect. Thus, the ground effect, the improved efficiency of the wing, is more correctly explained rather than being the result of the air compressed between the wing and the ground. Normally, the lift has vertical as well as horizontal components, and it is usually nearly equal and opposite to the weight (of the aircraft) time the load factor.

1.2.2 Weight

This is the weight of an aircraft, and it is directed downward from the C. G. of the aircraft toward the center of the Earth, $W=$mass of the aircraft times the acceleration due to gravity (g).

1.2.3 Thrust

This force is produced by the engine of the aircraft and is directed in the forward direction along the axis of the engine and is normally parallel to the x-axis of the aircraft. In general, the thrust can have horizontal and vertical components. The thrust is usually nearly equal and opposite to the drag force. Ironically, in a low-speed, high-powered climb, the lift is less than the weight; here, the thrust is supporting the weight.

1.2.4 Drag

Drag is the opposing force experienced by an aircraft moving forward with TAS V. It is a retarding force that acts parallel to and in opposite direction to V. The drag force

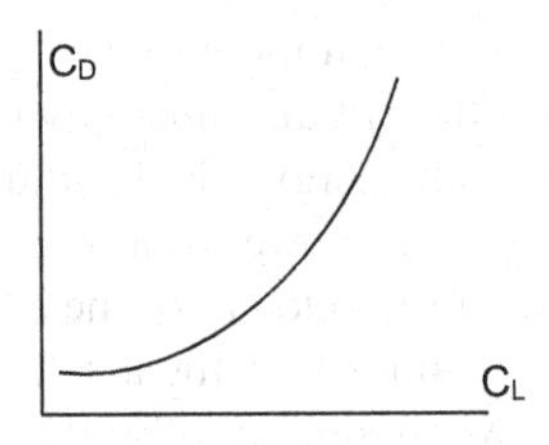

FIGURE 1.5 Typical drag polar curve.

can act parallel to the surface (friction drag along the surface), and the pressure drag against the surface. The friction drag is proportional to viscosity, i.e., the stickiness of the fluid. The pressure drag mainly depends on the mass density of the air. A part of the pressure drag that a wing produces depends on the amount of lift, and this is called induced drag, and the remaining drag force is called the parasite drag. A part of the parasite drag that is not due to friction is called 'form' drag, because it is sensitive to the form and shape of the airplane.

The drag (D) can also have a vertical component. Ironically, in a low-power, high-speed descent, the lift is less than the weight – here, the drag is supporting the part of the weight. Mathematically, D is expressed as:

$$D = C_D \bar{q} S \tag{1.38}$$

Here, C_D is the drag coefficient (Figure 1.5), $\bar{q}$ is the dynamic pressure, and S is the reference wing area. The total drag coefficient at subsonic speeds is given by:

$$C_D = C_{D_p} + C_{D_i} \tag{1.39}$$

Here, C_{D_p} is the profile drag coefficient, which is made up of skin friction drag and the pressure drag due to separation; C_{D_i} is the induced drag coefficient, which varies as the square of lift coefficient and is given by:

$$C_{D_i} = \frac{C_L^2}{\pi A e} \tag{1.40}$$

Here, A denotes the wing aspect ratio and e is the span efficiency factor, which generally varies from 0.8 to 0.95. The drag coefficient is also expressed in the form:

$$C_D = C_{D0} + \frac{C_L^2}{\pi A e} \tag{1.41}$$

Here, C_{D0} is the zero-lift drag coefficient. In addition to these drag components, another component is the wave drag that occurs at supersonic flows and is the result of the pressure increase across the shock wave. Wave drag can be minimized by keeping the supersonic airfoils thin with sharp leading edges.

1.2.5 Load Factor

Load factor is defined as the ratio of lift to weight:

$$n = \frac{L}{W} \tag{1.42}$$

In straight and level flight, since $L = W$, the load factor would be unity. For a turning flight, $n > 1$. The load factor is also referred to in terms of g's, e.g., during a turn, if the lift is 4 times that of the aircraft weight, the aircraft is said to be making a 4g turn. The higher the load factor, the smaller will be the radius of turn. The load factor gradient is defined as: $n_\alpha = \frac{\partial L}{\partial \alpha}\frac{1}{W}$. This quantity is very much used in HQ specifications.

1.2.6 Drag Polars

The C_L vs C_D plot is called drag polar. This is typically plotted for specified Mach number and altitude and for a given location of C. G. Systematic evaluation of the drag polars of an aircraft using dynamic maneuvers can be carried out over the full AOA range of the aircraft. Reference [5] describes several parameter/state estimation methods for the determination of drag polars from flight data.

1.3 AERODYNAMIC DERIVATIVES – PRELIMINARY DETERMINATION

The preliminary estimates of the derivatives are generally obtained by using methods like DATCOM (data compendium/handbook methods) and CFD (computational fluid dynamics). These methods are valid for low AOA and subsonic/supersonic Mach regions. At high AOA, the effects of flow separation, boundary-layer flow, vortex flow, and shock intensity and location at transonic Mach complicate the accurate determination of these derivatives. For these conditions, the data from the wind tunnel experiments and flights of the similar aircraft are used in conjunction with DATCOM. Subsequently or concurrently, the derivatives are determined from suitable wind tunnel experiments. Further refinements can be made by using advanced CFD codes. The estimates thus obtained are employed for studies related to the selection of the base line configurations for aircraft and missiles. These estimates are generally known as predicted values (or initial reference values) of the aerodynamic derivatives or simply the 'predictions'. After conducting several WT tests, the preliminary estimates are refined and the vehicle configurations are optimized for obtaining the desired performance and controllability and used for studies on load estimation. Subsequently, the prototype vehicle is built, and preliminary flight tests are conducted. One of the purposes of these flight tests is to estimate the aerodynamic derivatives from the flight data generated by conducting certain specific maneuvers on the vehicle (usually aircraft and rotorcrafts); these aspects are discussed in Chapters 7 and 9. Mathematical models (specifically the aerodynamic derivatives or stability and control derivatives which occur in the equations of motion) of aircraft dynamics are generally available from wind tunnel experiments and analytical methods prior to flight tests. Due to extensive wind tunnel testing and progress in aerodynamics and system technologies, reasonably accurate mathematical representations (models) are available. As a result of these, satisfactory characteristics, validated through extensive flight simulation, could be designed into the aircraft prior to flight. This has given more weightage to model verification exercise to be performed along with pilot's assessment. The vehicle's dynamical characteristics are described by equations of motion and parameters, which have physical meaning. These parameters are to be estimated from flight test data. These estimated parameters are compared with those obtained from wind tunnel experiments and analytical methods (e.g., computational fluid dynamics/DATCOM – all put together as prediction methods). The flight-determined stability and control derivatives (FDD) are also used in handling quality criteria (Chapter 10) to assess the overall pilot-aircraft interactions and performance. Hence, Taylor series (expansion) of aerodynamic coefficients is generally found very useful in representing the stability and control derivatives.

1.4 AIRCRAFT PROPULSION AND ITS PERFORMANCE

A few important aspects of flight vehicle performance are mentioned here. The drag force acting on an aircraft is typically a function of the lift coefficient and Mach number.

$$C_D = f(C_L, M) \tag{1.43}$$

The aircraft drag polar, eqn. (1.43), can be expressed as:

$$C_D = C_{D_0} + KC_L^2 \tag{1.44}$$

Here, K can be estimated from flight data using parameter estimation techniques. The theoretical value for the drag-due-to-lift factor K at subsonic speeds is given by $1/\pi Ae$, while its value at transonic and supersonic speeds is close to $1/C_{L\alpha}$. Assuming thrust T is inclined at angle σ_T with the flight path direction, the equations of motion for a level ($\theta = 0$) un-accelerated flight can be expressed as:

$$T\cos\sigma_T = D; T\sin\sigma_T + L = W \tag{1.45}$$

For small values of σ_T, eqn. (1.45) becomes:

$$T = D; L = W \tag{1.46}$$

Rearranging eqn. (1.46), the required thrust for an un-accelerated level flight is given by:

$$T = \frac{W}{L/D} \tag{1.47}$$

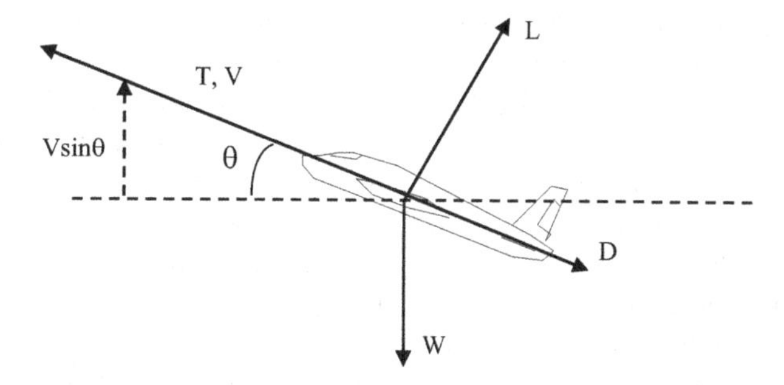

FIGURE 1.6 Airplane in a steady climb.

Thus, minimum thrust is required where *L/D* is maximum. If *V* is the velocity of the aircraft during level un-accelerated flight, and lift is equal to weight, with *P* as the required power:

$$C_L \frac{1}{2}\rho V^2 S = W \text{ or } V = \sqrt{\frac{2W}{\rho S C_L}}; P = TV = \frac{W}{C_L/C_D}\sqrt{\frac{2W}{\rho S C_L}} = \sqrt{\frac{2W^3 C_D^2}{\rho S C_L^3}} \quad (1.48)$$

Another important parameter in aircraft performance is the rate of climb. Compared to the level un-accelerated flight, the thrust in this case, in addition to overcoming drag, will also be required to compensate for the component of weight (Figure 1.6):

$$T = D + W \sin\theta \quad (1.49)$$

The vertical velocity component gives the rate of climb, ROC = $V \sin\theta$.

1.5 AIRCRAFT SENSORS-INSTRUMENTATION SYSTEMS

Aircraft sensors have several instruments that aid the pilot in flying; these can be categorized according to their use for: flight, engine, navigational aid, environmental instruments, and electrical system. Some of the instruments that fall in the category of flight instruments are discussed in Refs. [1,6]. The sensors/transducers are used to measure control positions, pressures, temperatures, and loads. The electrical signal from each transducer is routed thru an instrumentation wiring to a signal conditioning circuit or device in the airplane. The signal conditioning would include multiplexing, commutating, digitizing, ADC/DAC conversions, time code generation, and pulse code modulation. Optical head-up display allows the pilot to observe the flight instrument information presented in the head-up form as well as the out of window view of the world.

1.5.1 Air-Data Instruments

These instruments utilize a pitot-static system to measure the ambient atmospheric pressure. The ambient or static pressure is measured through a set of holes provided on the side of a tube projecting into the free stream, typically on both sides of the fuselage (Figure 1.1). To minimize the errors in the measured static pressure, the sensor is placed clear of the disturbed airflow. The pitot or the total pressure is sensed by open end of the pitot-static tube that is directly facing the oncoming airflow. The pitot pressure is the sum of static and dynamic pressure, the latter caused by the forward movement of the aircraft. Nose boom allows measurements of pressure and flow angles far in front of the fuselage. For flow angles, the differential pressures are measured in both the axes: horizontal and vertical relative to the aircraft. These differential pressures along with the dynamic pressure (at the aircraft nose boom) are used and the flow angles are computed. In many situations, five-hole probes are employed to measure 3-D flows. These probes can determine pitch and yaw angles also. Interestingly, a simple relation exists between yaw coefficient and measured pressures:$C_{\text{yaw}} = \dfrac{P_l - P_r}{P\min_{\max}}$. From this coefficient, yaw angle can be computed. However, in reality, the algorithms to accurately compute various such quantities would be complicated.

1.5.2 Pressure Altimeter

This is used to measure the aircraft altitude. It is basically a pressure transducer calibrated to read height above a specific pressure datum. It measures the ambient pressure and indicates the value of the altitude on the instrument dial. The altimeter readings can become inaccurate at higher altitudes where the pressure change with change in height is small.

1.5.3 Air Speed Indicator

The air speed indicator measures the dynamic pressure, which is computed as the difference between the pitot (total) and static pressure, and converts it to airspeed. Since the airspeed indicator is usually calibrated to standard sea-level conditions, the speed it indicates is called the indicated airspeed (IAS). The IAS corrected for the instrumentation errors and compressibility effects gives the EAS. The TAS can be obtained from the EAS by accounting for the change in the air density from standard sea-level conditions, eqn. (1.17). The compressibility and density corrections are normally included in the navigation computer from which CAS and TAS can be found.

1.5.4 Mach Meter

This instrument comprises both the pressure altimeter and the air speed indicator. Mach number is defined as the ratio of TAS to local speed of sound. In Mach meter, the altimeter and ASI capsules are linked to the pointer of the Mach meter through links and gears. The pointer rotates on a dial indicating the airspeed in terms of Mach number.

1.5.5 Vertical Speed Indicator

This is also known as the rate of climb/descent indicator. It indicates the rate of change of height per minute as the aircraft is climbing or descending. It essentially senses the pressure difference between the pressure in the instrument casing and the static pressure inside the capsule, due to change in height. The capsule is connected to a pointer which moves on the dial. In level flight, the pressure in the instrument casing is equal to the static pressure in the capsule and the pointer will indicate zero.

1.5.6 Accelerometers

These measure the body axis accelerations, excluding the component of acceleration due to gravity or other inertial forces. It measures the externally applied force on its case. Thus, it will indicate 1g of upward acceleration (due to the lift) for an aircraft in steady-level flight, whereas the actual acceleration in the steady-level flight is zero (1g upward due to lift, and 1g downward due the weight from gravity).

1.6 ENERGY AWARENESS, CONVERGENCE, AND MANAGEMENT

The four types of energy very crucial in/for flight management of an airplane are (i) potential energy (PE) due to the altitude of an aircraft, (ii) kinetic energy proportional to the square of the airspeed (→dynamic pressure), (iii) the chemical energy in the fuel, and (iv) the energy left behind the aircraft that disturbs the air and leaves it warmer. There are seven aspects of energy conversion in the flight of an aircraft: (i) in the gliding phase, the altitude is gradually lost and the drag helps to hold the aircraft, the airspeed does not change, and the fuel is not burnt; (ii) in the climb, the fuel is spent to handle the opposing drag force, and the altitude is gained; (iii) in the cruise flight, the fuel is burnt to handle the drag, and the altitude and airspeed are not changing much; (iv) if the stick is pulled, the aircraft is slowed down and ascend, and if done quickly, the drag will not consume much energy, and the engine will not have enough time to convert much fuel; (v) if the stick is pushed, the aircraft will speed up, and start descending, and if done quickly, the drag and engine power will not affect the balance of energy much in this push-over maneuver; (vi) in the early phase of the takeoff (-roll), drag is very small, and the altitude is not changed much, and hence, all the engine power is used for gaining the speed; and (vii) at the end of every flight, one may be able to maintain the altitude without using much of the engine power by gradually losing the airspeed for the drag.

The airspeed and altitude can be converted back and forth. If the engine produces more power, then the aircraft will descent at a lesser rate (rate of descent), or even it will start ascending. Throttle controls the power, and the latter can be used: (i) to overcome the drag, (ii) to climb, and (iii) to accelerate the aircraft. In the aircraft's trimmed conditions, when the control stick is pulled, the new energy is converted into the altitude. During the takeoff (defeating the trim, it is not effective), the energy from the engine is used to gain speed, more than the altitude.

1.7 ANGLE OF ATTACK AWARENESS AND MANAGEMENT

The AOA is the angle at which the air (its velocity vector) hits the wing. An airplane if left alone, is trimmed for a definite AOA, by its very structure. To make changes in the AOA, adjust the pitch attitude/angle (using pressure on the stick), then trim to remove the pressure. It is not advisable to trim the aircraft for pitch attitude, and for rate of climb; but can trim for airspeed, i.e., to trim for the AOA. Since it is difficult to perceive the AOA directly (unless there is an instrument displaying it), we have three things: (i) pitch attitude, the angle that the fuselage (aircraft's nose) makes relative to the horizontal (axis); (ii) angle of climb, the angle between the flight path and the horizontal; and (iii) angle of incidence, the angle at which the wings are attached to the fuselage; we have the following relationship: pitch attitude (this angle is changed by pulling the stick) + incidence (changed by extending the flaps) = angel of climb + AOA. In straight and level flight, one can control AOA by controlling pitch attitude. There is a relationship between the airspeed (IAS, V) and the AOA as follows: Coefficient of lift (C_L depends on AOA) = (weight of the aircraft × load factor)/ ($0.5\,\rho V^2 S$); thus if the speed goes down, the coefficient of life should increase and hence the AOA, up to the just beginning of the stall.

APPENDIX 1A AIRFLOW

The air passes over to an airfoil (e.g., wing of an aircraft) upward in the front, and it is called upwash, and at the aft of the wing downward and this is called downwash. We note here that pressure is related to velocity, as can be seen from the Bernoulli's principle, which is derived from the law of conservation of energy that involves the kinetic energy of air and the PE stored in the 'springiness' of the air, the energy is stored in the pressurized air:

Pressure (P) = Force/area = F × distance/(area × distance) = work (energy)/volume is the PE per unit volume. The kinetic energy of the moving air is given as:

KE per unit volume = (0.5) 'rho' (velocity)^2; hence, Mechanical energy per volume = PE + KE.

Thus, by neglecting the chemical and heat energies, and in the absence of any pumped energy, one can conclude that a given air parcel has a constant mechanical energy as it flows past the wing; hence, when the velocity is increased, the pressure decreases, and vice versa; this is an essence of the Bernoulli's principle.

The most practical result of the flow of air, the flow field over an object (e.g., wing of an aircraft) is that the object experiences a force and is normally called an aerodynamic force: (i) due to air pressure distribution on the surface (normal/perpendicular to the surface) and (ii) air friction (air rubbing the surface, shear stress; tangential to the surface).

Air is compressible, meaning its density changes in the response of the applied pressure, although the density is proportional to the pressure, and yet the pressure change is more important because the lift depends on the pressure difference between the top and bottom of the wing, and the pressure drag depends on the pressure difference. Of course, the flight depends directly on the total density, but it does not depend directly on the total pressure, but depends on the pressure difference. The compressibility signifies to the first order how density depends on the pressure.

At the subatomic level, air consists of particles: molecules of nitrogen, oxygen, water, and various other substances. There are some other aspects about the air hitting the airfoil (wing): (i) air pressure on the top of the wing is only a few percent lower than that of on the bottom, (ii) the shape of the top of the wing is crucial, and (iii) the wing creates a pressure field that strongly deflects even the far-away packets of the fluid.

A simple expression for the lift is given as: Lift = airspeed (V) × circulation × density × wing span. Here, the circulation is proportional to the coefficient of lift (C_L) and the airspeed; it is proportional to AOA. In a way, the lift requires circulation and vortices; and a vortex is a bunch of air circulating around itself, i.e., around the vortex line. The circulation required to generate the lift is attributed to a bound (-ed) vortex line; this binds to the wing and travels with the airplane, and spills off each wingtip; each wing forms a trailing (-edge) vortex (wake vortex) which extends for miles behind the aircraft; if a small aircraft flies in the wake of a large aircraft, then it may get easily flipped. The circulatory motion in a vortex involves nontrivial amount of kinetic energy, and this generates an induced drag.

The wing produces circulation in proportion to its AOA and the airspeed; this means that air above the wing moves faster, and this in turn produces a low pressure in accordance with the principle of Bernoulli; this low pressure pulls up on the wing and pulls down on the air as per the Newton's laws.

1A.1 Boundary Layer

The velocity of the fluid air next to the wing is zero; next to the surface, there is a thin layer, which is called the boundary layer. The aircraft wing works very well when the airflow is attached to it by a simple boundary layer, and the opposite of this is a separated flow. For the attached flow, there is no pressure change from the wing surface out to the full-speed flow. The separated airflow does not follow the contour of the wing.

In the laminar flow, every small parcel of the air has a definite velocity, and the velocity varies smoothly from place to place. In the turbulent flow, the velocity fluctuates as a function of time at any given point; and at any given time, the velocity changes rapidly from point to point. The attached turbulent flow produces mixed flow (up, down, left, right, faster, slower); and for the separated laminar flow, there would be some reverse flow, yet the pattern in the space will be much smoother than the turbulent flow, and the flow will not fluctuate in time.

The small objects moving slowly through viscous fluids have low Reynolds numbers (less than 10 will have a laminar flow everywhere), and large objects like an aircraft moving through thin fluids, like air, have high Reynolds numbers; the ones with >10 are expected to create at least some turbulence.

APPENDIX 1B AIRCRAFT ENGINES

The piston engines with propellers still rule the roost at low-speed flights, while gas turbine engines are used for jet propulsion at higher speeds. The turbo jets are equipped with a compressor, a combustion chamber, turbine, and exhaust nozzle. Thrust is provided by reaction of the exhaust gases thrown backward through the nozzle. Another type is the turbo fan or the bypass engine which has a large fan that accelerates the air ahead of the compressor, thus resulting in more thrust with higher efficiency. While turbojet and turbofan engines have rotating parts, there is another type called the ramjet that uses the ram effect generated from the forward speed to pass the compressed air to the combustion chamber and out of the exhaust nozzle at very high speeds. As is obvious, ramjet will give no thrust at all at zero forward speeds.

The most of the reciprocating engines (also known as an internal-combustion, IC engines) use liquid fuel, and all these engines need the systems for: (i) ignition in the combustion chamber, (ii) cooling, and (iii) lubrication [7]. The reciprocating piston movement is converted to rotary motion to turn the propeller by the connecting rod and the crankshaft. The opening and closing of the two valves (to let in a mixture of fuel and to let out the burnt gases/exhaust) are mechanized by a cam geared to the crankshaft.

The other engines used for aircraft propulsion are type of turbine engines, wherein the vanes fitted in the wheel are struck by the force of a moving fluid, e.g., air. For an aircraft turbine engine, the force of the striking hot flowing gases is used. Some of these engines are geared to propellers. The vibration stress is much reduced due to rotary rather than reciprocating mechanism; one lever controls the speed and power. The continuous ignition system of a reciprocating engine is not needed.

A turbojet uses a series of fan-like compressor blades to suck the air into the engine and then compress it; there is a series of rotor and stator blades; the former gather and push air backward into the engine, and the latter serve to straighten the flow of this air; and it continues to travel from the low-compression set to the high-compression set. The combustion chamber receives the high-pressure air, mixes fuel with it, and then burns this mixture; to produce the hot and very high-velocity gases, which strike the blades of the turbine and cause it to spin rapidly; and the turbine is mounted on a shaft that is connected to the compressor; and the spinning causes the compression sections to function. The hot and highly accelerated gases go into the engine's exhaust section, which is designed to give additional acceleration to the gases and thereby increase the thrust.

In a turbofan engine, one or more rows of compressor blades extend beyond the usual compressor blades, and a fan burner allows the burning of additional fuel in the fan airstream. With the burner on, the thrust is doubled and it can operate on high speeds and high altitudes. This engine has greater thrust for takeoff, climbing and cruising for the same quantity of the fuel compared to the turbojet engine.

A turboprop engine combines the feature of turbojet and propeller engines. The former is more efficient at high speed and high altitudes; and the latter is more efficient at speeds under 400 mph and below 30,000 ft. The engine uses a gas turbine to turn a propeller and depends on the propeller for the thrust, rather than on the high-velocity gases going out of the exhaust; and in a real sense, it is not a 'jet engine'.

1B.1 Engine Thrust Computations

The maximum propeller efficiency is nearly 90%. The thrust horsepower provided by the propeller is given by:

$$\text{thp} = \frac{TV}{550} = (\text{bhp})\eta_p \tag{1B.1}$$

Here, T is thrust in lbs, V is velocity in ft/s, bhp is brake horsepower, 550 is the conversion factor, and η_p is the propeller efficiency.

1B.2 For Turbojet-Type Engine

The jet engine thrust is given by:

$$\text{Thrust} = \frac{\text{weight of air in pounds per sec.} \times \text{velocity}}{g\left(\text{in ft/s}^2\right)} \tag{1B.2}$$

As a rule of thumb, at 375 miles per hour (mph), one pound of thrust equals one hp. Net thrust T_n = gross thrust T_g – momentum drag

$$= \frac{WV_{\text{js}}}{g} + \left(p_s - p_a\right) S_{\text{na}} - \frac{WV_a}{g} \tag{1B.3}$$

$$\frac{W}{g}\left(V_{\text{js}} - V_a\right) + \left(p_s - p_a\right) S_{\text{na}} \tag{1B.4}$$

Here, W is the weight flow rate of the air passing through the engine, V_{js} is the jet stream velocity, V_a is the speed of the aircraft, p_s is the static pressure across the propeller nozzle, p_a is the atmospheric pressure, and S_{na} is the propeller nozzle area.

EPILOGUE

In this chapter, we have briefly captured several fundamental concepts and principles of aerodynamics, and sensors/instrumentation aspects. In fact, for major topics of this chapter, there are a number of books in the literature on aerodynamics, sensors, and aircraft engines. We feel that the presented information would be just sufficient to begin with the further study of flight mechanics modeling and analysis concepts and even some advanced engineering methods that are dealt with in the remaining chapters of this revised edition.

REFERENCES

1. Nelson, R. C. *Flight Stability and Automatic Control*, 2nd Edn. McGraw-Hill International Editions, New York, 1998.
2. Olson, W. M. *Aircraft Performance Flight Testing*, Technical Information Handbook, AFFTC-TIH-99-01, September 2000.
3. Anderson, J. D. Jr. *Introduction to Flight*, 3rd Edn. McGraw-Hill, New York, 1989.
4. Anon. How airplanes fly. http://www.allstar.fiu.edu/aero/, accessed 2007.
5. Girija, G., Basappa, M., Raol, J. R., and Madhuranath, P. Evaluation of methods for determination of drag polars of unstable/augmented aircraft, AIAA Paper 2000-0501, USA.
6. Harris, D. *Ground Studies for Pilots – Flight Instruments & Automatic Flight Control Systems*, 6th Edn. Blackwell Science Ltd., Hoboken, NJ, 2004.
7. Anon. Aeronautics-Aircraft Propulsion-Level 2. http://www.allstar.fiu.edu/aero/flight60, -65.htm, accessed February 2022.

2 Engineering Dynamics

2.1 INTRODUCTION

This chapter gives the reader an insight into the various aspects of engineering dynamics. The subject of engineering dynamics is a blend of physics, applied mathematics, basic logic, and computational methods. An airplane is considered as a dynamic system, and hence, many principles and concepts of engineering dynamics are directly applicable to the study of aircraft flight mechanics–flight dynamics. Important aspects here are the use of the tools that describe the motion and solve the equations of motion of the vehicle, wherein the kinematics of a particle or a rigid body supports the laws of dynamics. The concerned scientists and engineers are interested in studying the movements of such bodies on Earth, in atmosphere, and in space and are also interested in predicting their behavior for longer time spans. Such an understanding would help them design, develop, modify, and use these vehicles for their intended purposes. The basic study actually starts with the vector mechanics algebra; however, it is presupposed to have been studied by the readers at the level of first two semesters' course level in mathematics. The basic vector calculus can be easily used to derive the formulae for velocity and acceleration. The Newtonian mechanics plays very important role in the study of flight mechanics as dealt in Chapter 3.

The kinematical description can be understood by using either extrinsic or intrinsic coordinates: The former means that the description is extrinsic to the knowledge of the path followed by the point (rectangular Cartesian coordinates), and the latter means that the description uses the knowledge of the path. In the Cartesian coordinates, the position is known in terms of the distances measured along the three mutually orthogonal straight lines that represent reference directions. In the intrinsic coordinates, the unit vectors are defined in terms of the properties of the path, and these coordinates are known as path variables; alternatively, this description is also known as tangent and normal components. The fundamental variable for a denoted path is the arc length 's' along a curve, measured from some starting point A to the point of interest, B. We also know that because 's' changes with time, the position is an implicit function of time; this also means that the position is a vector function of the arc length. The velocity and acceleration are oriented parallel to the straight path. However, acceleration will not be parallel to the velocity for a smooth curvilinear path, unless $\dot{s} = 0$.

To begin with, some basic techniques are discussed to understand the kinematics of a particle and rigid body motion; primarily, an aircraft is treated as a rigid body [1,2]. Displacement of points relative to various reference frames is discussed that helps one to correlate position, velocity, and acceleration from fixed and moving reference frames. Constraint conditions, when a body moves with restrictions imposed on it by the other bodies, are discussed. Basic laws governing the relationship between

DOI: 10.1201/9781003293514-3

rigid body motion and the forces that act on the body are derived. The fundamental concepts of linear and angular momentum and the impact of their variation on rigid body motion are explained. These are further used to derive the Newton–Euler equations of motion of a rigid body. Associated concepts pertaining to dynamic virtual work, virtual displacement, generalized coordinates, and velocities are discussed. To further enhance understanding of the basic concepts of analytical mechanics, alternative formulations to derive equations of motion, based on Hamilton's principle, are discussed [3]. The chapter ends with a simplified explanation of the physics behind the working of gyroscope.

2.2 KINEMATICS

The kinematics is concerned with the static aspects of mechanics wherein the inherent cause of the motion is not taken into account.

2.2.1 Rectangular Cartesian Coordinates

Rectangular Cartesian coordinates are associated with orthogonal *xyz* axes that are right-handed by convention. Such coordinates are generally used to define position, velocity, and force vectors. For instance, position vector can be defined in terms of its components *x*, *y*, and *z*. Figure 2.1 shows the components of the position vector $\vec{R}$. Assuming that the components of $\vec{R}$ vary with time, the position vector $\vec{R}$ can be expressed as

$$\vec{R} = x(t)\hat{i} + y(t)\hat{j} + z(t)\hat{k} \tag{2.1}$$

where $\hat{i}$, $\hat{j}$, and $\hat{k}$ are unit vectors. It is easy to obtain the velocity vector by differentiating (2.1) for position vector.

$$\vec{V} = u(t)\hat{i} + v(t)\hat{j} + w(t)\hat{k} \tag{2.2}$$

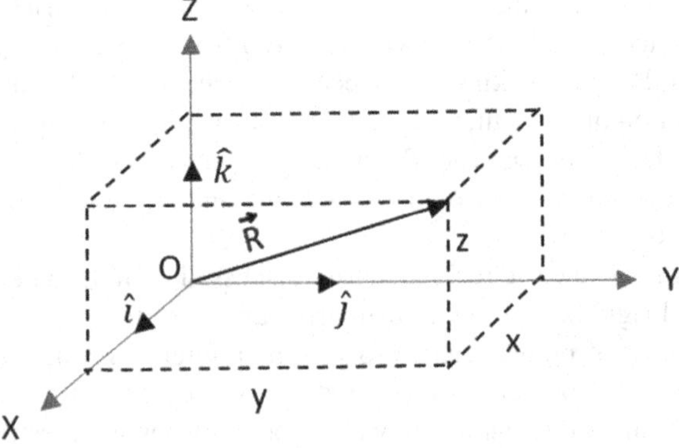

FIGURE 2.1 Rectangular Cartesian coordinates.

Further differentiation of the equation for $\vec{V}$ shall provide the acceleration

$$\vec{a} = a_x(t)\hat{i} + a_y\hat{j} + a_z\hat{k} \tag{2.3}$$

Here, $u = \dot{x}$, $v = \dot{y}$, $w = \dot{z}$; and $a_x = \ddot{x}$, $a_y = \ddot{y}$, $a_z = \ddot{z}$. Although, the rectangular Cartesian coordinate system is simple and easy to apply, it has limitations, particularly when one wishes to describe the motion more accurately, for example, projectile motion over a long range. For such cases, curvilinear coordinates are found to be more suitable.

2.2.2 Curvilinear Coordinates

These use three parameters (α, β, γ) to describe the location of a point. However, the unit vectors associated with these parameters are allowed to be variable. The general relation between the rectangular and Cartesian coordinates is expressed as

$$\alpha = \alpha(x, y, z);\ \beta = \beta(x, y, z);\ \gamma = \gamma(x, y, z) \tag{2.4}$$

Holding two of the parameters constant and varying the third over a range of values shall give a family of curves. More such curves forming a spatial mesh can be generated by varying each parameter at a time while holding the other two constant. If the curves intersect orthogonally, as is the case in most of the problems, then the parameters (α, β, γ) are said to be orthogonal curvilinear coordinates. Cylindrical (Figure 2.2) and spherical coordinates (Figure 2.3) are considered to be special cases of curvilinear coordinates. Here, z represents the reference axis, R is the transverse line, and θ is the azimuth angle traced by the xy plane about the reference z axis. R, θ, and z are cylindrical coordinates. The transformation from (R, θ, z) to (x, y, z) can be expressed as

$$x = R\cos\theta;\ y = R\sin\theta;\ z = z \tag{2.5}$$

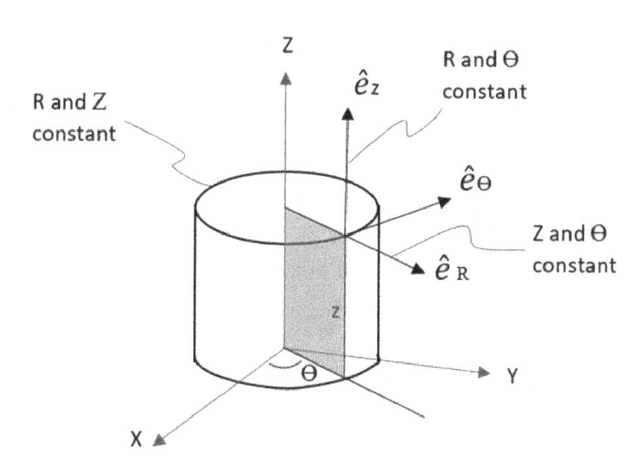

FIGURE 2.2 Cylindrical coordinates.

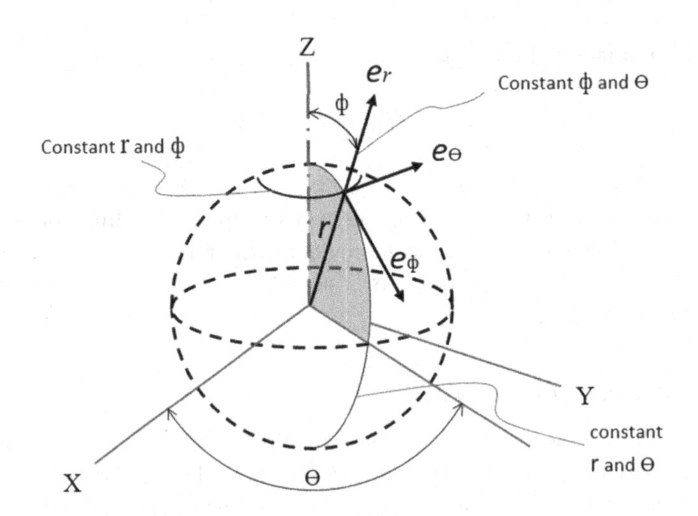

FIGURE 2.3 Spherical coordinates.

Here, R and θ are also known as polar coordinates. In Figure 2.2, $\hat{e}_R$, $\hat{e}_\theta$, and $\hat{e}_z$ are the unit vectors in the transverse, azimuthal, and axial directions. Projecting them onto the xy plane yields

$$\hat{e}_R = \cos\theta\ \hat{i} + \sin\theta\ \hat{j};\ \hat{e}_\theta = -\sin\theta\hat{i} + \cos\theta\ \hat{j};\ \hat{e}_z = \hat{k} \tag{2.6}$$

Another form of coordinates used to locate a point is the spherical coordinates (Figure 2.3) and is defined by the radial distance r from a fixed point (which could be the origin) to the point of interest, the azimuthal angle θ, and the polar angle φ. The transformation from (r,θ,φ) to (x,y,z) is given by

$$x = r\sin\varphi\ \cos\theta;\ y = r\sin\varphi\ \sin\theta;\ z = r\ \cos\varphi \tag{2.7}$$

The spherical coordinate unit vectors $\hat{e}_r$, $\hat{e}_\varphi$, and $\hat{e}_\theta$ are tangent to the respective coordinate curves and are mutually orthogonal, such that $\hat{e}_r \times \hat{e}_\varphi = \hat{e}_\theta$.

2.3 RELATIVE MOTION

This motion is concerned with movement of a body with respect to a fixed or moving point.

2.3.1 Displacement and Time Derivatives

Consider Figure 2.4, in which the absolute position of point P is observed from a fixed reference frame XYZ having origin at O, and its relative position is observed from a moving reference frame xyz with origin at O'. The absolute position $\vec{r}_{P/O}$ of P

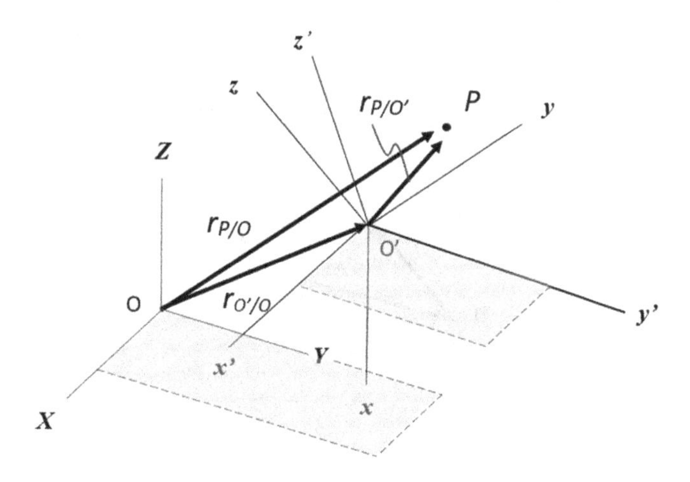

FIGURE 2.4 Position of P observed from fixed and moving reference frames.

in XYZ coordinate system has components X_P, Y_P, and Z_P. The relative position of P in moving xyz coordinate system is $\vec{r}_{P/O'}$ having components x_P, y_P, and z_P. Evidently,

$$\vec{r}_{P/O} = \vec{r}_{O/O'} + \vec{r}_{P/O'} \tag{2.8}$$

However, to evaluate the sum of components of the vector in (2.8), it is necessary to have a common set of unit vectors. We, therefore, introduce an axis frame x', y', z' which is parallel to XYZ but whose origin coincides with O'. The coordinates of P w.r.t. (x', y', z') are (x'_P, y'_P, z'_P), while the coordinates of O' w.r.t. origin O in XYZ frame are (X_O, Y_O, Z_O). Since the unit vectors $(\hat{I}, \hat{J}, \hat{K})$ and $(\hat{i}', \hat{j}', \hat{k}')$ are in the same direction, we can define the position of P as

$$X_P = X'_O + x'_P;\ Y_P = Y'_O + y'_P;\ Z_P = Z'_O + z'_P \tag{2.9}$$

Equation (2.9) provides the position of point P w.r.t the moving reference frame (x', y', z'). However, determining the coordinates of point P w.r.t (xyz) frame requires the use of rotation transformation $[R]$ between the frames (xyz) and (x', y', z') to get

$$\begin{aligned} &\begin{bmatrix} X_P & Y_P & Z_P \end{bmatrix}^T \\ &= \begin{bmatrix} X'_O & Y'_O & Z'_O \end{bmatrix}^T + [R]^T \begin{bmatrix} x_P & y_P & z_P \end{bmatrix}^T \end{aligned} \tag{2.10}$$

where $\begin{bmatrix} \hat{i}' & \hat{j}' & \hat{k}' \end{bmatrix}^T = [R] \begin{bmatrix} \hat{i} & \hat{j} & \hat{k} \end{bmatrix}^T$ and

$$[R] = \begin{bmatrix} d_{x'x} & d_{x'y} & d_{x'z} \\ d_{y'x} & d_{y'x} & d_{y'x} \\ d_{z'x} & d_{z'x} & d_{z'x} \end{bmatrix} \tag{2.11}$$

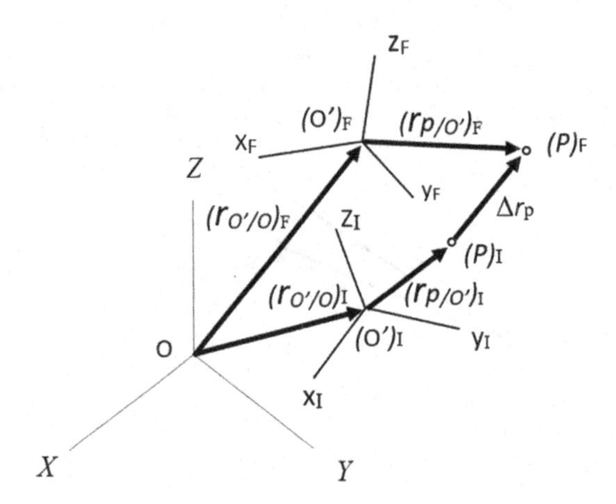

FIGURE 2.5 Displacement of P viewed from moving reference frames.

where $d_{i'j}$ defines the cosine of the angle between axes i' and j, with i and j representing x, y, or z axis. $d_{i'j}$ are also called direction cosines. Also, note that $\begin{bmatrix}\hat{i} & \hat{j} & \hat{k}\end{bmatrix}^T = [R']\begin{bmatrix}\hat{i}' & \hat{j}' & \hat{k}'\end{bmatrix}^T$ where $[R'] = [R]^{-1} = [R]^T$. Thus, knowing the transformation matrix $[R]$ enables us to know the coordinates of point P with respect to either the fixed XYZ or the moving (xyz).

Next, we consider a situation where the point P shifts from an initial position P_I to final position P_F, as seen in Figure 2.5. The displacement vector $\Delta\vec{r}_P$ from the initial to the final position, in the XYZ coordinate system is given by

$$\Delta\vec{r}_P = \left(\Delta\vec{r}_{P/O}\right)_F - \left(\Delta\vec{r}_{P/O}\right)_I \tag{2.12}$$

This displacement vector, in terms of its components, can be written as

$$\begin{bmatrix}\Delta r_{PX}\\ \Delta r_{PY}\\ \Delta r_{PZ}\end{bmatrix} \equiv \begin{bmatrix}\Delta X_P\\ \Delta Y_P\\ \Delta Z_P\end{bmatrix} = \begin{bmatrix}(\Delta X_P)_F\\ (\Delta Y_P)_F\\ (\Delta Z_P)_F\end{bmatrix} - \begin{bmatrix}(\Delta X_P)_I\\ (\Delta Y_P)_I\\ (\Delta Z_P)_I\end{bmatrix} \tag{2.13}$$

Using (2.12) into (2.13), we can define the *absolute displacement* of point P as

$$\begin{bmatrix}\Delta r_{PX}\\ \Delta r_{PY}\\ \Delta r_{PZ}\end{bmatrix} = \begin{bmatrix}\Delta r_{O'X}\\ \Delta r_{O'Y}\\ \Delta r_{O'Z}\end{bmatrix} + [R]_F^T \begin{bmatrix}(x_P)_F\\ (y_P)_F\\ (z_P)_F\end{bmatrix} - [R]_I^T \begin{bmatrix}(x_P)_I\\ (y_P)_I\\ (z_P)_I\end{bmatrix} \tag{2.14}$$

On the other hand, the *relative displacement* $\left(\Delta\vec{r}_P\right)_{xyz}$ is defined only by the change in the values of x_P, y_P and z_P, as given in

$$(\Delta\vec{r}_P)_{xyz} = \left[\left((x_P)_F - (x_P)_I\right)\hat{i}\right] + \left[\left((y_P)_F - (y_P)_I\right)\hat{j}\right] + \left[\left((z_P)_F - (z_P)_I\right)\hat{k}\right] \quad (2.15)$$

Equation (2.15) can be rearranged as

$$\begin{bmatrix} (x_P)_F \\ (y_P)_F \\ (z_P)_F \end{bmatrix} = \begin{bmatrix} (x_P)_I \\ (y_P)_I \\ (z_P)_I \end{bmatrix} + \begin{bmatrix} (\Delta\vec{r}_P)_{xyz} \ \hat{i} \\ (\Delta\vec{r}_P)_{xyz} \ \hat{j} \\ (\Delta\vec{r}_P)_{xyz} \ \hat{k} \end{bmatrix} \quad (2.16)$$

Substituting eqn. (2.16) into eqn. (2.14), we have

$$\begin{bmatrix} \Delta r_{PX} \\ \Delta r_{PY} \\ \Delta r_{PZ} \end{bmatrix} = \begin{bmatrix} \Delta r_{O'X} \\ \Delta r_{O'Y} \\ \Delta r_{O'Z} \end{bmatrix} + \left\{[R]_F^T - [R]_I^T\right\} \begin{bmatrix} (x_P)_I \\ (y_P)_I \\ (z_P)_I \end{bmatrix} + [R]_F^T \begin{bmatrix} (\Delta\vec{r}_P)_{xyz} \ \hat{i} \\ (\Delta\vec{r}_P)_{xyz} \ \hat{j} \\ (\Delta\vec{r}_P)_{xyz} \ \hat{k} \end{bmatrix} \quad (2.17)$$

The left-hand side of equation can be expressed in terms of its unit vector components as

$$\begin{bmatrix} \Delta\vec{r}_P \cdot \hat{I} \\ \Delta\vec{r}_P \cdot \hat{J} \\ \Delta\vec{r}_P \cdot \hat{K} \end{bmatrix} = \begin{bmatrix} \Delta\vec{r}_{O'} \cdot \hat{I} \\ \Delta\vec{r}_{O'} \cdot \hat{J} \\ \Delta\vec{r}_{O'} \cdot \hat{K} \end{bmatrix} + \left\{[R]_F^T - [R]_I^T\right\} \begin{bmatrix} (x_P)_I \\ (y_P)_I \\ (z_P)_I \end{bmatrix}$$

$$+ [R]_F^T \begin{bmatrix} (\Delta\vec{r}_P)_{xyz} \ \hat{i} \\ (\Delta\vec{r}_P)_{xyz} \ \hat{j} \\ (\Delta\vec{r}_P)_{xyz} \ \hat{k} \end{bmatrix} \quad (2.18)$$

We can multiply (2.18) by $[R]_F$ to eliminate initial coordinates and have all the components in the final (xyz) orientation.

$$\begin{bmatrix} \Delta\vec{r}_P \cdot \hat{i} \\ \Delta\vec{r}_P \cdot \hat{j} \\ \Delta\vec{r}_P \cdot \hat{k} \end{bmatrix} = \begin{bmatrix} \Delta\vec{r}_{O'} \cdot \hat{i} \\ \Delta\vec{r}_{O'} \cdot \hat{j} \\ \Delta\vec{r}_{O'} \cdot \hat{k} \end{bmatrix} + \left\{[I] - [R]_F [R]_I^T\right\} \begin{bmatrix} (x_P)_F \\ (y_P)_F \\ (z_P)_F \end{bmatrix}$$

$$+ [R]_F [R]_I^T \begin{bmatrix} (\Delta\vec{r}_P)_{xyz} \ \hat{i} \\ (\Delta\vec{r}_P)_{xyz} \ \hat{j} \\ (\Delta\vec{r}_P)_{xyz} \ \hat{k} \end{bmatrix} \quad (2.19)$$

Having explained displacement, velocity can now be defined as the derivative of infinitesimal displacement over infinitesimal time. If $d\vec{r}_P$ and $d\vec{r}_{O'}$ represent the differential quantities of the absolute displacement of origin O' and point P, $(d\vec{r}_P)_{xyz}$ represents the differential quantity of the relative displacement, and $\overline{d\theta}$ represents the infinitesimal rotation that allows the coordinate system (xyz) to move to its final orientation, then

$$d\vec{r}_P = d\vec{r}_{O'} + (d\vec{r}_P)_{xyz} + \overline{d\theta} \times \vec{r}_{P/o'} \tag{2.20}$$

Dividing eqn. (2.20) by dt that defines the time interval over which the displacement of point P takes place, the velocity of P is given by

$$\vec{v}_P = \vec{v}_{O'} + (\vec{v}_P)_{xyz} + \vec{\omega} \times \vec{r}_{P/o'} \tag{2.21}$$

where $\vec{\omega}$ is the angular velocity of frame xyz and $(\vec{v}_P)_{xyz}$ is the velocity relative to xyz. From Figure 2.3, we have already seen that

$$\vec{r}_{P/O} = \vec{r}_{O/O'} + \vec{r}_{P/O'} \tag{2.22}$$

Differentiating (2.20) w.r.t time gives

$$\vec{v}_P = \vec{v}_{O'} + \frac{d}{dt}\vec{r}_{P/o'} \tag{2.23}$$

Comparing eqns. (2.22) and (2.23), we have

$$\frac{d}{dt}\vec{r}_{P/o'} = (\vec{v}_P)_{xyz} + \vec{\omega} \times \vec{r}_{P/o'} \tag{2.24}$$

Since $(\vec{v}_P)_{xyz} = \dfrac{\partial(\vec{r}_{P/o'})}{\partial t}$, the analogy between $\vec{r}_{P/o'}$ and any other vector, say $\vec{A}$, leads to the inference that

$$\dot{\vec{A}} = \frac{\partial \vec{A}}{\partial t} + \vec{\omega} \times \vec{A} \tag{2.25}$$

2.3.2 Angular Velocity and Acceleration

We have seen that calculation of velocity using a moving reference frame requires the value of angular velocity $\vec{\omega}$, which can be described as the sum of infinitesimal rotations about various axes. If $x_n y_n z_n$ represents one such auxiliary axis having angular velocity $\omega_n \hat{e}_n$, where $\hat{e}_n$ is the unit vector along the axis of rotation, then angular velocity is given by

$$\vec{\omega} = \sum_n \omega_n \hat{e}_n \tag{2.26}$$

The angular acceleration $\vec{\alpha}$ is defined as the rate of change of angular velocity $\vec{\omega}$.

$$\vec{\alpha} \equiv \frac{d\vec{\omega}}{dt} \tag{2.27}$$

If $\vec{\Omega}_n$ denotes the angular velocity of auxiliary frame $x_n y_n z_n$, applying the standard rule for time derivative yields

$$\frac{d\hat{e}_n}{dt} = \vec{\Omega}_n \times \hat{e}_n \tag{2.28}$$

The angular acceleration $\vec{\alpha}$ can then be expressed in the following form

$$\vec{\alpha} = \frac{d\vec{\omega}}{dt} = \sum_n \left(\dot{\omega}_n \hat{e}_n + \vec{\Omega}_n \times \omega_n \hat{e}_n \right) \tag{2.29}$$

2.3.3 Velocity and Acceleration Using a Moving Reference Frame

The components of $\left(\vec{v}_P\right)_{xyz}$ are the rates of change of the coordinates x, y, and z, such that

$$\left(\vec{v}_P\right)_{xyz} = \dot{x}_P \hat{i} + \dot{y}_P \hat{j} + \dot{z}_P \hat{k} \tag{2.30}$$

The relative acceleration can then be obtained by taking the derivative of above equation

$$\left(\vec{a}_P\right)_{xyz} = \ddot{x}_P \hat{i} + \ddot{y}_P \hat{j} + \ddot{z}_P \hat{k} \tag{2.31}$$

$$\frac{d}{dt}\left(\vec{v}_P\right)_{xyz} = \left(\vec{a}_P\right)_{xyz} + \vec{\omega} \times \left(\vec{v}_P\right)_{xyz} \tag{2.32}$$

Consider the term $\vec{\omega} \times \vec{r}_{P/o'}$. Using the expression for $\frac{d}{dt}\vec{r}_{P/o'}$ from (2.24), the time derivative of $\vec{\omega} \times \vec{r}_{P/o'}$ yields

$$\frac{d}{dt}\left(\vec{\omega} \times \vec{r}_{P/o'}\right) = \left(\vec{\alpha} \times \vec{r}_{P/o'}\right) + \vec{\omega} \times \left[\left(\vec{v}_P\right)_{xyz} + \vec{\omega} \times \vec{r}_{P/o'} \right] \tag{2.33}$$

We also have the following expression for acceleration

$$\vec{a}_P = \vec{a}_{O'} + (\vec{a}_P)_{xyz} + (\vec{\alpha} \times \vec{r}_{P/o'}) + \vec{\omega} \times (\vec{\omega} \times \vec{r}_{P/o'}) + 2\left[\vec{\omega} \times (\vec{v}_P)_{xyz}\right] \quad (2.34)$$

Equations (2.30) and (2.34) define the absolute velocity and acceleration of point P using the moving reference frame xyz.

2.4 KINEMATICS OF CONSTRAINT RIGID BODIES

2.4.1 General Equations

Consider a rigid body with coordinate system xyz forming the body-fixed axis system and the global axis XYZ system with origin O as the reference (Figure 2.6). Since the relative distance between the points O', A, and B in the rigid body is fixed, they satisfy the following relations for velocity and acceleration.

$$(\vec{v}_A)_{xyz} = 0.; \; (\vec{a}_A)_{xyz} = 0 \quad (2.35)$$

$$\vec{v}_A = \vec{v}_{O'} + \vec{\omega} \times \vec{r}_{A/o'} \quad (2.36)$$

and

$$\vec{a}_A = \vec{a}_{O'} + (\vec{\alpha} \times \vec{r}_{A/o'}) + \vec{\omega} \times (\vec{\omega} \times \vec{r}_{A/o'}) \quad (2.37)$$

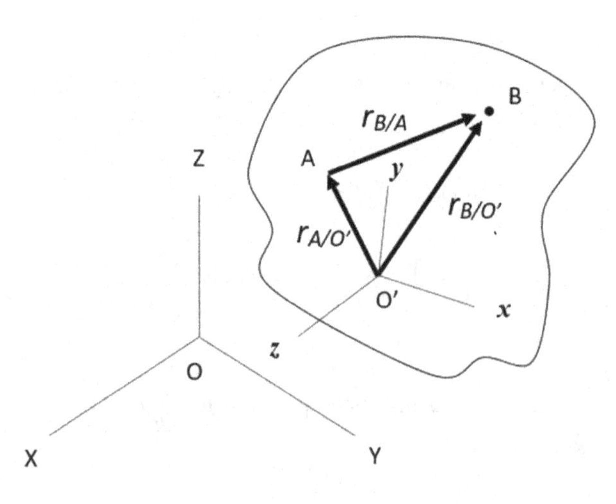

FIGURE 2.6 Motion of points O', A, and B in a rigid body.

Similarly, the velocity of point B is given by

$$\vec{v}_B = \vec{v}_{O'} + \vec{\omega} \times \vec{r}_{B/o'}$$

$$\vec{v}_B = \vec{v}_A + \vec{\omega} \times \vec{r}_{B/A}; \ \vec{r}_{B/A} = \vec{r}_{B/o'} - \vec{r}_{A/o'} \tag{2.38}$$

Thus, to analyze the motion of a rigid body, one needs to solve for the equations of displacement, velocity, and acceleration.

2.4.2 Eulerian Angles

The Euler angles are the three angles that describe the orientation of a body-fixed coordinate system xyz with respect to a global coordinate system XYZ (Figure 2.7). These are very useful in defining the aircraft attitude with respect to Earth. The three Euler angles used to define the rotations that transform Earth-fixed axis to body axis, at any particular instant of time, are the heading angle ψ, the pitch angle θ, and the bank or roll angle ϕ. The order in which the rotations are made is also important since angles do not obey the commutative law. To begin with, we consider the body-fixed axis frame xyz to be aligned with the absolute reference frame XYZ. The first rotation by angle ψ is considered about the Z axis. Consider the positive ψ rotation clockwise. The new orientation after the first rotation is denoted as x_1, y_1, z_1, where z_1 is identical to Z. Next, we rotate about axis y_1 by angle θ to arrive at the new orientation x_2, y_2, z_2,

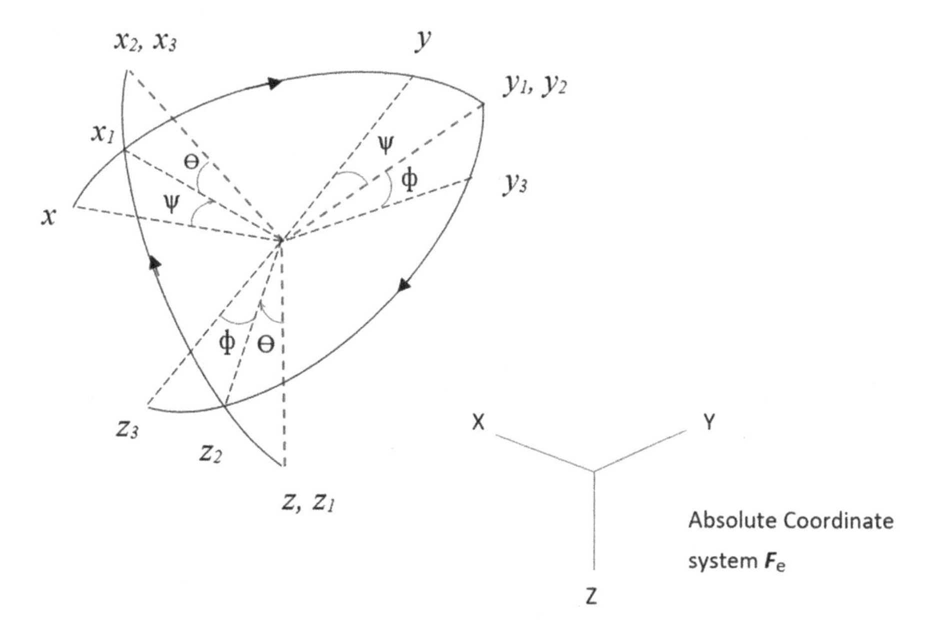

FIGURE 2.7 Eulerian angles.

where y_2 is identical to y_1. Pitch up is considered positive. Finally, the third rotation by angle ϕ is about the axis x_2 to arrive at the orientation x_3, y_3, z_3, where x_3 is identical to x_2. The transformation from Earth-fixed coordinate frame $\mathcal{F}_e$ to body-fixed coordinate frame $\mathcal{F}_b$ is given by

$$\mathcal{F}_b = \begin{bmatrix} 1 & 0 & 0 \\ 0 & \cos\phi & \sin\phi \\ 0 & -\sin\phi & \cos\phi \end{bmatrix} \begin{bmatrix} \cos\theta & 0 & -\sin\theta \\ 0 & 1 & 0 \\ \sin\theta & 0 & \cos\theta \end{bmatrix} \times \begin{bmatrix} \cos\psi & \sin\psi & 0 \\ -\sin\psi & \cos\psi & 0 \\ 0 & 0 & 1 \end{bmatrix} \mathcal{F}_e \tag{2.39}$$

The inverse transformation is

$$\mathcal{F}_e = \begin{bmatrix} \cos\psi & -\sin\psi & 0 \\ \sin\psi & \cos\psi & 0 \\ 0 & 0 & 1 \end{bmatrix} \begin{bmatrix} \cos\theta & 0 & \sin\theta \\ 0 & 1 & 0 \\ -\sin\theta & 0 & \cos\theta \end{bmatrix} \times \begin{bmatrix} 1 & 0 & 0 \\ 0 & \cos\phi & -\sin\phi \\ 0 & \sin\phi & \cos\phi \end{bmatrix} \mathcal{F}_b \tag{2.40}$$

2.5 INERTIAL EFFECTS

Everybody with mass has inertia also; hence, these effects should be taken into account.

2.5.1 Linear and Angular Momentum

If we consider external forces $\vec{F}_A$ and $\vec{F}_B$, and internal forces $\vec{f}_A$ and $\vec{f}_B$ acting on a system of two particles A and B (Figure 2.8), then the internal forces which the particles exert on each other will be equal and opposite in magnitude, as dictated by Newton's third law.

$$\vec{f}_A = -\vec{f}_B \tag{2.41}$$

Since the internal forces are collinear, the moment of these forces about the origin O shall also be equal and opposite in magnitude.

$$\vec{r}_A \times \vec{f}_A = -\vec{r}_B \times \vec{f}_B \tag{2.42}$$

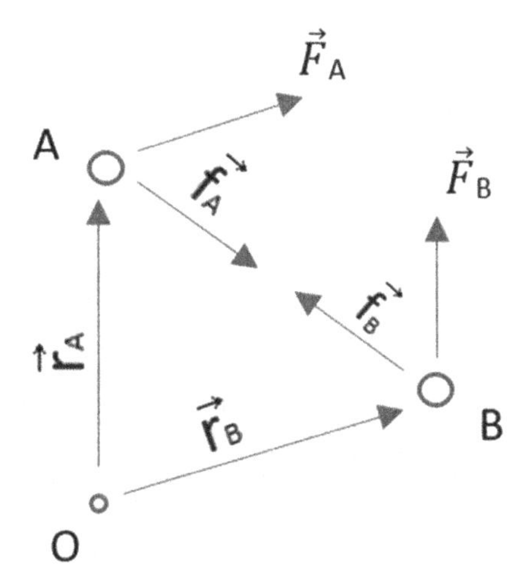

FIGURE 2.8 Interaction between forces on two particle system.

Since the resultant forces is the sum of internal and external forces, according to Newton's second law, we have

$$\begin{aligned}\vec{F}_A + \vec{f}_A &= m_A \times \vec{a}_A \\ \vec{F}_B + \vec{f}_B &= m_B \times \vec{a}_B\end{aligned} \tag{2.43}$$

When computing the resultant force $\sum \vec{F}$, the internal forces being equal and opposite will cancel out each other, and we have

$$\sum \vec{F} = \vec{F}_A + \vec{F}_B = m_A \vec{a}_A + m_B \vec{a}_B \tag{2.44}$$

Likewise, the resultant moment about origin O can be defined as

$$\sum \vec{M} = \vec{r}_A \times \left(\vec{F}_A + \vec{f}_A\right) + \vec{r}_B \times \left(\vec{F}_B + \vec{f}_B\right) = \vec{r}_A \times m_A \vec{a}_A + \vec{r}_B \times m_B \vec{a}_B \tag{2.45}$$

Since internal forces $\vec{f}_A$ and $\vec{f}_B$ being equal and opposite in magnitude cancel out each other, the contribution to total moment is only from the external forces

$$\sum \vec{M} = \vec{r}_A \times \vec{F}_A + \vec{r}_B \times \vec{F}_B = \vec{r}_A \times m_A \vec{a}_A + \vec{r}_B \times m_B \vec{a}_B \tag{2.46}$$

Considering more particles of the system in above analysis does not alter the fact that the total moment is generated only by the sum of external forces acting on the body.

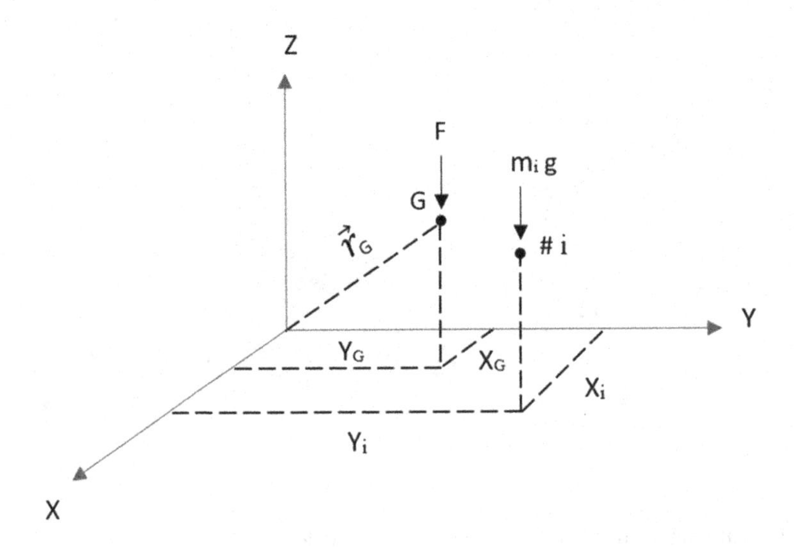

FIGURE 2.9 System of N particles ($i = 1,\ldots,N$).

Thus, if there are N particles in the system (Figure 2.9), the total force and moment acting on the body is the sum of all external forces and moments.

$$\sum \vec{F} = \sum_{i=1}^{N} \vec{F}_i = \sum_{i=1}^{N} m_i \vec{a}_i \tag{2.47}$$

$$\sum \vec{M} = \sum_{i=1}^{N} \vec{r}_i \times \vec{F}_i = \sum_{i=1}^{N} \vec{r}_i \times m_i \vec{a}_i \tag{2.48}$$

Replacing the acceleration by the second derivative of position, the above expression for total force can be written as

$$\sum \vec{F} = \sum_{i=1}^{N} \vec{F}_i = \sum_{i=1}^{N} m_i \frac{d^2}{dt^2} \vec{r}_i = \frac{d^2}{dt^2} \sum_{i=1}^{N} (m_i \ \vec{r}_i) \tag{2.49}$$

If m_{sys} is the mass of the rigid body and its centre of mass G is located at $\vec{r}_G$ from the origin O, then

$$m_{sys} = \sum_{i=1}^{N} m_i \tag{2.50}$$

$$\sum \vec{F} = \frac{d^2}{dt^2} \sum_{i=1}^{N} m_i \ \vec{r}_G = \frac{d^2}{dt^2} \left(m_{sys} \vec{r}_G \right) = m_{sys} \vec{a}_G \tag{2.51}$$

If the origin O of the XYZ frame in Figure 2.9 is moving with an acceleration $\vec{a}_O$, then the resultant moment of all the particles about point O can be expressed as

$$\sum \vec{M} = \sum_{i=1}^{N} \vec{r}_i \times m_i \left(\vec{a}_O + \vec{a}_i\right) \tag{2.52}$$

In view of eqns. (2.50) and (2.51), the expression (2.52) becomes

$$\begin{aligned}\sum \vec{M} &= m_{\text{sys}}\, \vec{r}_G \times \vec{a}_O + \sum_{i=1}^{N} \left[\vec{r}_i \times m_i \vec{a}_i\right] \\ &= m_{\text{sys}}\, \vec{r}_G \times \vec{a}_O + \sum_{i=1}^{N} \vec{r}_i \times m_i \frac{d}{dt}\vec{v}_i \\ &= m_{\text{sys}}\, \vec{r}_G \times \vec{a}_O + \sum_{i=1}^{N} \left[\frac{d}{dt}\left(\vec{r}_i \times m_i \vec{v}_i\right) - \frac{d}{dt}\left(\vec{r}_i\right) \times m_i \vec{v}_i\right]\end{aligned} \tag{2.53}$$

Note that $m_i \vec{v}_i$ is the linear momentum of particle i in the moving reference frame xyz. Since $\frac{d}{dt}\left(\vec{r}_i\right) = \vec{v}_i$, the cross product in the last term of the above expression goes to zero. The expression for total moment thus becomes

$$\sum \vec{M} = m_{\text{sys}}\, \vec{r}_G \times \vec{a}_O + \frac{d}{dt}\vec{H} \tag{2.54}$$

where $\vec{H} = \sum_{i=1}^{N} \left(\vec{r}_i \times m_i \vec{v}_i\right)$. If point O happens to be the centre of mass, then $\vec{r}_G \equiv 0$, and eqn. (2.54) reduces to

$$\sum \vec{M} = \frac{d}{dt}\vec{H} \tag{2.55}$$

In the above relation, $\vec{H}$ is the moment of momentum, also known as the angular momentum, about the origin O of xyz reference frame. Thus, eqns. (2.51) and (2.54) determine the translation and rotation of a rigid body. If ω is the rigid body angular velocity, then the velocity of particle $\vec{v}_i$, w.r.t origin O, can be expressed as $\vec{v}_i = \vec{\omega} \times \vec{r}_i$. The angular momentum can, therefore, also be expressed as

$$\vec{H} = \sum_{i=1}^{N} m_i \left\{\vec{r}_i \left(\vec{\omega} \times \vec{r}_i\right)\right\} \tag{2.56}$$

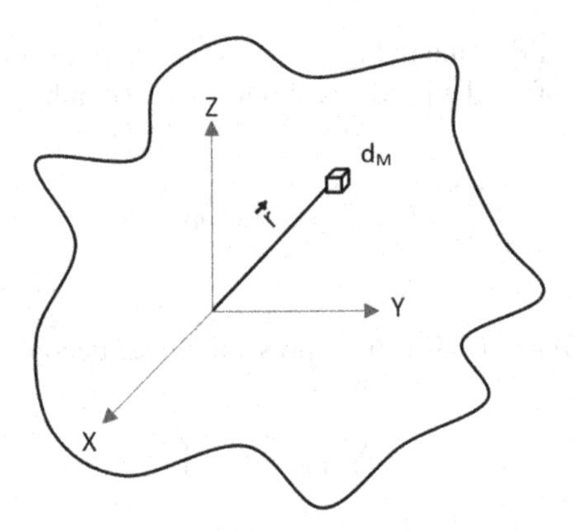

FIGURE 2.10 Differential mass dm in the body fixed frame.

2.5.2 Inertial Properties

Consider the differential mass dm in body-fixed frame xyz (Figure 2.10). Let the angular velocity and position vector of the differential mass be

$$\vec{\omega} = \omega_x\vec{i} + \omega_y\vec{j} + \omega_z\vec{k} \text{ and } \vec{r} = x\vec{i} + y\vec{j} + z\vec{k} \tag{2.57}$$

Substituting these into eqn. (2.56) and converting the summation to integral yields

$$\vec{H} = \iiint\left(x\vec{i} + y\vec{j} + z\vec{k}\right)\left[\left(\omega_x\vec{i} + \omega_y\vec{j} + \omega_z\vec{k}\right)\times\left(\vec{i} + y\vec{j} + z\vec{k}\right)\right];$$
$$\vec{H} = H_x\vec{i} + H_x\vec{j} + H_x\vec{k} \tag{2.58}$$

where

$$H_x = \omega_x I_x - \omega_y I_{xy} - \omega_z I_{xz}$$
$$H_y = -\omega_x I_{xy} + \omega_y I_y - \omega_z I_{yz} \tag{2.59}$$
$$H_z = -\omega_x I_{xz} - \omega_y I_{yz} + \omega_z I_z$$

$$I_x = \iiint\left(y^2 + z^2\right)dm, I_y = \iiint\left(x^2 + z^2\right)dm, I_z = \iiint\left(x^2 + y^2\right)dm$$
$$I_{xy} = \iiint xy\ dm, I_{yz} = \iiint yz\ dm, I_{xz} = \iiint xz\ dm \tag{2.60}$$

The angular momentum vector $\vec{H}$ in eqn. (2.58) may be expressed in the matrix form as

$$[\vec{H}] = [I]\{\vec{\omega}\} \tag{2.61}$$

where $[I]$ is the inertia matrix given by

$$I = \begin{bmatrix} I_x & -I_{xy} & -I_{xz} \\ -I_{xy} & I_y & -I_{yz} \\ -I_{xz} & -I_{yz} & I_z \end{bmatrix} \tag{2.62}$$

2.5.3 Rate of Change of Angular Momentum

If the xyz frame is rotating at an angular velocity $\vec{\Omega}$, and the body's angular velocity is $\vec{\omega}$ which is different from the reference frame angular velocity, then the orientation of the body w.r.t frame will be changing and the inertia matrix may also become time-dependent. The time derivative of eqn. (2.58) will be

$$\frac{d}{dt}\left(\vec{H}\right) = \frac{\partial \vec{H}}{\partial t} + \vec{\Omega} \times \vec{H} \tag{2.63}$$

Equation (2.63) requires us to define time derivative of $\vec{\omega}$. In a reference frame rotating at an angular velocity $\vec{\Omega}$, this time derivative is given by

$$\frac{d\vec{\omega}}{dt} = \frac{\partial \vec{\omega}}{\partial t} + \vec{\Omega} \times \vec{\omega} \tag{2.64}$$

Normally, to avoid having the moments and product of inertia varying with time, we consider xyz to be a body fixed reference frame. This sets $\vec{\Omega} = \vec{\omega}$ and the last term in eqn. (2.64) vanishes. Combining eqns. (2.61)–(2.64), rate of angular momentum can be expressed in the following form

$$\left[\vec{H}\right] = [I]\left\{\frac{\partial \vec{\omega}}{\partial t}\right\} + \vec{\Omega} \times \vec{H}; \ \left[\vec{H}\right] = [I]\left\{\frac{d\vec{\omega}}{dt}\right\} + \{\vec{\omega}\} \times ([I]\{\omega\}) \tag{2.65}$$

2.6 NEWTON–EULER EQUATIONS OF MOTION

2.6.1 Fundamental Equations and Planar Motion

The basic laws that govern the rigid body motion are eqn. (2.51) for the resultant force and eqn. (2.55) for the resultant moment, which define the time rate of change of linear and angular momentum, respectively:

$$\begin{aligned} \sum \vec{F} &= m\vec{a}_G \\ \sum \vec{M} &= \frac{d}{dt}\vec{H} \end{aligned} \tag{2.66}$$

In several cases, the resultant force component and the acceleration of the centre of mass are considered in an inertial reference frame. The vector equations in eqn. (2.66) can be written in scalar form in terms of the components of forces and moments along the x, y, and z axes of the global coordinate system.

$$\sum \vec{F}\cdot\vec{i} = m\left(\vec{a}_G\cdot\vec{i}\right), \sum \vec{F}\cdot\vec{j} = m\left(\vec{a}_G\cdot\vec{j}\right), \sum \vec{F}\cdot\vec{k} = m\left(\vec{a}_G\cdot\vec{k}\right)$$

$$\sum \vec{M}\cdot\vec{i} = \left(\frac{\partial \vec{H}}{\partial t} + \vec{\omega}\times\vec{H}\right)\cdot\vec{i}, \sum \vec{M}\cdot\vec{j} = \left(\frac{\partial \vec{H}}{\partial t} + \vec{\omega}\times\vec{H}\right)\cdot\vec{j}, \tag{2.67}$$

$$\sum \vec{M}\cdot\vec{k} = \left(\frac{\partial \vec{H}}{\partial t} + \vec{\omega}\times\vec{H}\right)\cdot\vec{k}$$

Making use of eqns. (2.63) and (2.67) can be expressed in a compact matrix form as

$$\begin{Bmatrix} \sum \vec{F}\cdot\vec{i} \\ \sum \vec{F}\cdot\vec{j} \\ \sum \vec{F}\cdot\vec{k} \end{Bmatrix} = m\begin{Bmatrix} \left(\vec{a}_G\cdot\vec{i}\right) \\ \left(\vec{a}_G\cdot\vec{j}\right) \\ \left(\vec{a}_G\cdot\vec{k}\right) \end{Bmatrix} \tag{2.68}$$

$$\begin{Bmatrix} \sum \vec{M}\cdot\vec{i} \\ \sum \vec{M}\cdot\vec{j} \\ \sum \vec{M}\cdot\vec{k} \end{Bmatrix} = [I]\begin{Bmatrix} \dot{\vec{\omega}}_x \\ \dot{\vec{\omega}}_y \\ \dot{\vec{\omega}}_z \end{Bmatrix} + \begin{Bmatrix} \vec{\omega}_x \\ \vec{\omega}_y \\ \vec{\omega}_z \end{Bmatrix} \times \left[I \begin{Bmatrix} \vec{\omega}_x \\ \vec{\omega}_y \\ \vec{\omega}_z \end{Bmatrix} \right] \tag{2.69}$$

Equations (2.68) and (2.69) for moment can be simplified further by assuming xyz to be principal axes, thereby making the products of inertia $I_{xy} = I_{yz} = I_{zx} = 0$. With this assumption, eqn. (2.69) can be written as

$$\sum \vec{M}\cdot\vec{i} = I_x\dot{\vec{\omega}}_x + \vec{\omega}_y\vec{\omega}_z\left(I_z - I_y\right)$$

$$\sum \vec{M}\cdot\vec{j} = I_y\dot{\vec{\omega}}_y + \vec{\omega}_z\vec{\omega}_x\left(I_x - I_z\right) \tag{2.70}$$

$$\sum \vec{M}\cdot\vec{k} = I_z\dot{\vec{\omega}}_z + \vec{\omega}_x\vec{\omega}_y\left(I_y - I_x\right)$$

The expressions in eqn. (2.70) are also referred to as Euler's equations of motion.

2.6.2 N-E Equations for a System

As in the case of a system of particles constrained to form a rigid body, a similar approach can lead to a system of rigid bodies to be treated as a unified system. Consider a pair of bodies A and B, such that the body A exerts a force $\vec{f}_{BA}$ and a moment $\vec{M}_{BA}$ on body B. Likewise, as shown in Figure 2.11, body B exerts a force $\vec{f}_{AB}$ and a moment $\vec{M}_{AB}$ on body A. These internal forces and moments acting between the two bodies follow the Newton's third law, which states that

$$\vec{f}_{AB} = \vec{f}_{BA} \text{ and } \vec{M}_{AB} = \vec{M}_{BA} \tag{2.71}$$

These forces and moments acting on bodies A and B can be equivalently represented by a force $m\vec{a}_G$ and a couple $\frac{d\vec{H}}{dt}$, acting on the centre of mass of the bodies A and B, as shown in Figure 2.5. The same can be extended to a system composed of N number of bodies. Since the system shown in Figure 2.11 is equivalent to the system in Figure 2.12, it follows that summation of $m\vec{a}_G$, for N number of bodies, shall be equal to the resultant forces acting on the system. Likewise, summation of their moment

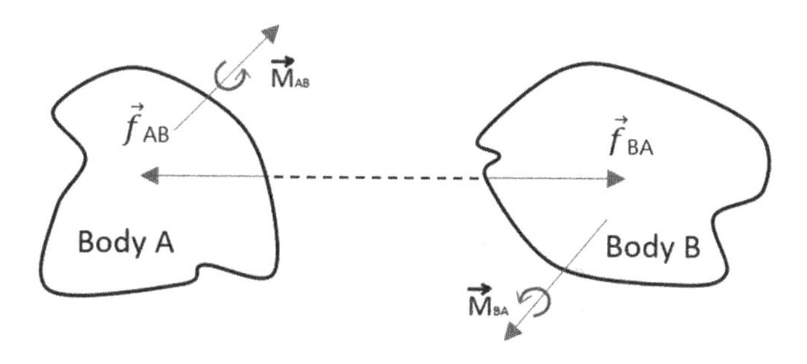

FIGURE 2.11 Forces acting on a pair of rigid bodies.

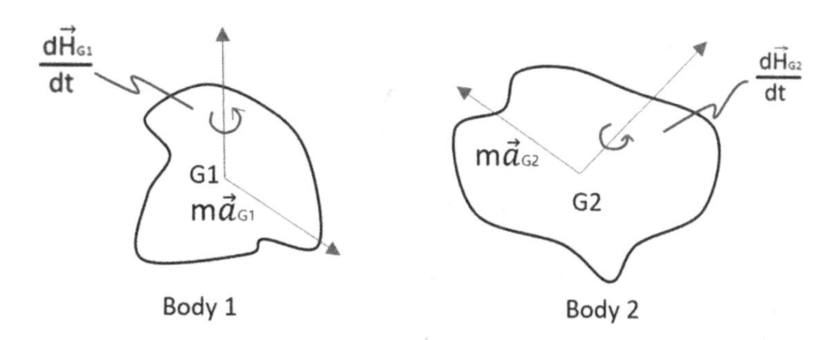

FIGURE 2.12 Force–couple system equivalent to the system in Figure 2.11.

about any point D shall yield the total moment $\vec{M}_D$ about that point. So the Newton–Euler equations for the system can be expressed as

$$\sum \vec{F} = \sum_{i=1}^{N} m_i \vec{a}_{G_i}$$
$$\sum \vec{M}_D = \sum_{i=1}^{N} \frac{d\vec{H}_i}{dt} + \sum_{i=1}^{N} \vec{r}_{G_i/D} \times m_i \vec{a}_{G_i} \tag{2.72}$$

In eqn. (2.72), following Newton's third law, we need not consider the forces and moments that are internal to the system. Therefore, while 12 scalar equations are required to describe the motion of each rigid body, only six equations consisting of three force and three moment equations will be required when two bodies are considered as single system.

2.6.3 Principles of Momentum and Energy

We have already discussed that external force and moment acting on a rigid body are described by the rate of change of linear momentum $\vec{P}$ and angular momentum $\vec{H}$, respectively.

$$\sum \vec{F} = \frac{d}{dt}\vec{P} \quad \text{where} \quad \vec{P} = m\vec{v}_G \tag{2.73}$$

where $\vec{v}_G$ is the velocity of centre of mass of the rigid body. Further, we have

$$\sum \vec{M} = \frac{d}{dt}\vec{H} \quad \text{where} \quad \vec{H} = I\vec{\omega} \tag{2.74}$$

Integrating eqn. (2.74) between time instant t_1 to t_2, and rearranging to get the final values of linear and angular momentum, leads to:

$$\vec{P}_2 = \vec{P}_1 + \int_{t_1}^{t_2} \sum \vec{F}\, dt$$
$$\vec{H}_2 = \vec{H}_1 + \int_{t_1}^{t_2} \sum \vec{M}\, dt \tag{2.75}$$

Equation (2.75) represents the impulse–momentum relation, which can be used to compute the linear and angular velocities of the rigid body. The difficulty lies in the evaluation of the integral terms because the force and moment are often not known apriori. Further, the orientation of the unit vectors and the corresponding components, which vary with time, should also be known for computing the integral terms. A simplified form of eqn. (2.75) can be used to study the action of impulsive forces, wherein velocity is assumed to change instantly from time t_i^- to t_i^+ in time Δt, while the position remains unchanged:

$$\vec{P}\left(t_i^+\right) = \vec{P}\left(t_i^-\right) + \sum \vec{F}_{\text{imp}} \Delta t$$
$$\vec{H}\left(t_i^+\right) = \vec{H}\left(t_i^-\right) + \sum \vec{M}_{\text{imp}} \Delta t \tag{2.76}$$

Combining the Newton–Euler relations given in eqns. (2.72) and (2.75) yields the following equations of impulse–momentum principle

$$\int_{t_1}^{t_2} \sum \vec{F}\, dt = \sum_{i=1}^{N} m_i \left(v_{Gi}\right)_2 - \sum_{i=1}^{N} m_i \left(v_{Gi}\right)_1$$
$$\int_{t_1}^{t_2} \sum \vec{M}_D\, dt = \left\{ \sum_{i=1}^{N} \left(\vec{H}_{Gi}\right)_2 + \sum_{i=1}^{N} \left(\vec{r}_{G_i/D} \times m_i v_{Gi/D}\right)_2 \right\} - \left\{ \sum_{i=1}^{N} \left(\vec{H}_{Gi}\right)_1 + \sum_{i=1}^{N} \left(\vec{r}_{G_i/D} \times m_i v_{Gi/D}\right)_1 \right\} \tag{2.77}$$

Equation (2.77) applies only if point D is either stationary or it is the centre of mass of the system. If the impulse component vanishes, then

$$\sum_{i=1}^{N} m_i \left(v_{Gi}\right)_2 = \sum_{i=1}^{N} m_i \left(v_{Gi}\right)_1$$
$$\left\{ \sum_{i=1}^{N} \left(\vec{H}_{Gi}\right)_2 + \sum_{i=1}^{N} \left(\vec{r}_{G_i/D} \times m_i v_{Gi/D}\right)_2 \right\} = \left\{ \sum_{i=1}^{N} \left(\vec{H}_{Gi}\right)_1 + \sum_{i=1}^{N} \left(\vec{r}_{G_i/D} \times m_i v_{Gi/D}\right)_1 \right\} \tag{2.78}$$

Equation (2.78) refers to the conservation principle which states that the system's linear momentum and angular momentum are constant.

2.7 ANALYTICS MECHANICS

2.7.1 Generalized Coordinates and Kinematical Constraints

Generalized coordinates are the variables that uniquely define the current location of the system. For any system, a minimum number of generalized coordinates are required which also defines the number of degrees of freedom of that system. For example, consider the system shown in Figure 2.13 made up of linkages AB and BC. It is assumed that the point A is fixed while the point C can move along X-axis. The system, thus, has only one degree of freedom. The relative position of the linkages can be defined using θ. It is to be noted that the relative position of C with respect

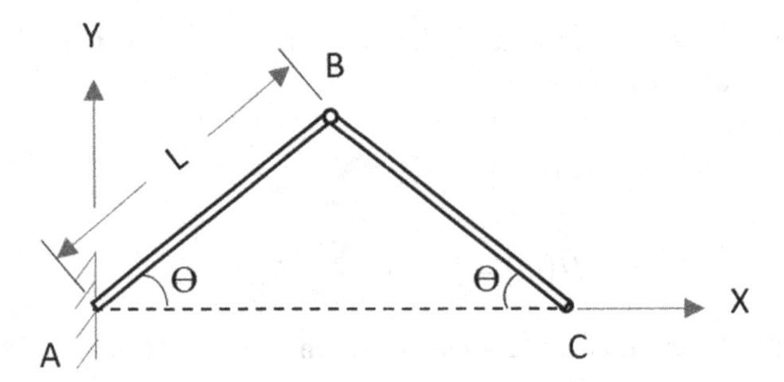

FIGURE 2.13 Frame structure in static equilibrium.

to A can also be defined by using the horizontal distance l between the two, where $l = 2L\cos\theta$. Thus, the choice of generalized coordinates to define the position of a system need not be unique. Also, it is generally possible to define one set of generalized coordinates in terms of the other. A rigid body moving is space, like an airplane, has six degrees of freedom. For such a system, the generalized coordinates comprising of three position coordinates of the center of mass and the three Eulerian angles relative to a fixed reference frame can define its position and orientation. A body constrained to move in a plane has only three degrees of freedom that can be defined by using the two position coordinates and the rotation about the axis perpendicular to the plane. Generalized coordinates can be either constrained or unconstrained; if the number of generalized coordinates equals the number of degrees of freedom of the system, these are called unconstrained or independent coordinates. If they exceed the number of degrees of freedom, they must satisfy an additional set of kinematical equations, called the constraint equations. For such cases, the number of constraint equations equals the number of generalized coordinates minus the degrees of freedom.

2.7.2 Virtual Displacements

Virtual denotes something that does not really occur. Virtual movement or virtual rotation of any point in a body refers to an infinitesimal displacement of the point. Normally, the symbol δ is used instead of d to imply that the displacement is virtual and not incremental. For static equilibrium, the sum of all forces and moments about a point B in a rigid body should be zero.

$$\sum \vec{F} = 0 \text{ and } \sum \vec{M} = 0 \tag{2.79}$$

If δr_B and $\delta\theta$ denote the virtual movement and rotation of point B, then the virtual work δW done by the resultant force and moment would also vanish:

$$\sum \vec{F} \cdot \vec{\delta r_B} = 0 \text{ and } \sum \vec{M} \cdot \vec{\delta\theta} = 0 \tag{2.80}$$

Extending this further, the work done by a set of forces and moments on a system in static equilibrium vanishes:

$$\delta W = \sum_{i=1}^{n} \vec{F}_i \cdot \delta \vec{r}_B + \sum_{j=1}^{n} \vec{M}_j \cdot \delta \vec{\theta} = 0 \tag{2.81}$$

Equation (2.81) is also referred to as principal of virtual work. Another perspective of virtual displacement is to define it as the difference between the system's actual position at any given point of time and the position at the same instant when the forces acting on the system get altered. In most of the cases, the position of any point B in a system, with respect to a known reference point O, can be defined in terms of the generalized coordinates q_n and the time t:

$$\delta \vec{r}_B = \delta \vec{r}_{B/O}\left(q_n, t\right) \tag{2.82}$$

For calculating the virtual displacement, it is assumed that time t is held constant while the generalized coordinates q_n undergo an incremental change. Using partial differentiation chain rule, the absolute position of point B is given by

$$\delta \vec{r}_B = \sum_{i=1}^{N} \left\{ \frac{\partial}{\partial q_i} \delta \vec{r}_{B/O}\left(q_n, t\right) \right\} \delta q_i \tag{2.83}$$

where N is the total number of generalized coordinates. Since time is not constant, chain rule for differential displacement can be applied to obtain the true differential displacement of point B in terms of the generalized coordinates q_i and their increments dq_i over an infinitesimal time interval are given by

$$d\vec{r}_B = \sum_{i=1}^{N} \left\{ \frac{\partial}{\partial q_i} \delta \vec{r}_{B/O}\left(q_n, t\right) \right\} dq_i + \frac{\partial}{\partial t} \delta \vec{r}_{B/O}\left(q_n, t\right) dt \tag{2.84}$$

Dividing eqn. (2.83) by dt gives an expression for the velocity of point B as

$$\vec{v}_B = \sum_{i=1}^{N} \left\{ \frac{\partial}{\partial q_i} \delta \vec{r}_{B/O}\left(q_n, t\right) \right\} \frac{dq_i}{dt} + \frac{\partial}{\partial t} \delta \vec{r}_{B/O}\left(q_n, t\right) \tag{2.85}$$

Equations (2.83) and (2.85) show that virtual displacement $\delta \vec{r}_B$ depends linearly on the virtual increments δq_i, while the velocity $\vec{v}_B$ is a linear function of the generalized velocities $\dot{q}_i$. Denoting $\dfrac{\partial \delta \vec{r}_B}{\partial q_i}$ as $\vec{v}_{B_i}$, the displacement and velocity of point B can be expressed as:

$$\begin{aligned}\delta \vec{r}_B &= \sum_{i=1}^{N} \vec{v}_{B_i}(q_n,t)\,\delta q_i \\ \vec{v}_B &= \sum_{i=1}^{N} \vec{v}_{B_i}(q_n,t)\,\dot{q}_i + \vec{v}_{P_t}(q_n,t)\end{aligned} \tag{2.86}$$

2.7.3 Generalized Forces

Velocity and virtual displacement of a system can be expressed in terms of generalized coordinates. The virtual work done to virtually displace a point in a system by $\delta \vec{r}_n$, with forces $\vec{F}_n$ acting on it, is

$$\delta W = \sum_n \vec{F}_n \cdot \delta \vec{r}_n \tag{2.87}$$

Substituting for $\delta \vec{r}_n$ from eqn. (2.86) into eqn. (2.87) above, we can get virtual work in terms of increments in generalized coordinates.

$$\delta W = \sum_n \vec{F}_n \sum_{i=1}^{N} \vec{v}_{n_i}\;\delta q_i \tag{2.88}$$

If Q_i symbolize the generalized forces, where each generalized force represents the net force acting in the direction of the corresponding generalized coordinate q_i, then

$$Q_i = \sum_n \vec{F}_n\;\vec{v}_{n_i}(q_n,t) = \sum_{i=1}^{N} \vec{F}_n\,\frac{\partial \vec{r}_n}{\partial q_i} \tag{2.89}$$

The virtual work in terms of generalized forces can then be expressed as

$$\delta W = \sum_{i=1}^{N} Q_i \delta q_i \tag{2.90}$$

Equation (2.87) can be extended further to include the virtual work resulting from virtual rotation of the body due to couple $\vec{\Upsilon}$

$$\delta W = \sum_n \vec{F}_n \cdot \delta \vec{r}_n + \sum_k \vec{\Upsilon}_k \cdot \delta \vec{\theta}_k \tag{2.91}$$

The virtual rotation $\delta\vec{\theta}_k$ in terms of angular velocity and generalized coordinates q_i can be expressed as

$$\delta\vec{\theta}_k = \sum_{i=1}^{N} \vec{\omega}_{ki}(q_n,t)\,\delta q_i \tag{2.92}$$

The preceding equation for virtual work can now be written as

$$\delta W = \sum_n \vec{F}_n \cdot \sum_{i=1}^{N} \vec{v}_{n_i}(q_n,t)\,\delta q_i + \sum_k \vec{\Upsilon}_k \cdot \sum_{i=1}^{N} \vec{\omega}_{ki}(q_n,t)\,\delta q_i \tag{2.93}$$

Combining eqns. (2.90) and (2.93) leads to the equation for the generalized forces

$$Q_i = \sum_n \vec{F}_n \cdot \sum_{i=1}^{N} \vec{v}_{n_i}(q_n,t) + \sum_k \vec{\Upsilon}_k \cdot \sum_{i=1}^{N} \vec{\omega}_{ki}(q_n,t) \tag{2.94}$$

2.7.4 Lagrange's Equations

If $\vec{r}_n$ and $\vec{v}_n$ denote the position and velocity of a particle n in a system, then the relations expressing their dependence on generalized coordinates can be written as

$$\vec{v}_n = \sum_{i=1}^{N} \left(\frac{\partial \vec{r}_n}{\partial q_i} \right) \dot{q}_i + \frac{\partial \vec{r}_n}{\partial t} \tag{2.95}$$

$$\delta\vec{r}_n = \sum_{i=1}^{N} \left(\frac{\partial \vec{r}_n}{\partial q_i} \right) \delta q_i \tag{2.96}$$

If $\sum \vec{F}_n$ represents the resultant force acting on particle having acceleration a_n, then the principle of dynamic virtual work states that

$$\delta W_n = \sum \vec{F}_n \cdot \delta\vec{r}_n = (m_n\vec{a}_n)\cdot\delta\vec{r}_n;\ (m_n\vec{a}_n)\cdot\delta\vec{r}_n - \delta W_n = 0 \tag{2.97}$$

Substituting for $\delta\vec{r}_n$ from eqn. (2.96) into eqn. (2.97), we get

$$\sum_{i=1}^{N} m_n\vec{a}_n \cdot \left(\frac{\partial \vec{r}_n}{\partial q_i} \right) \delta q_i - \delta W_n = 0 \tag{2.98}$$

Replacing $\vec{a}_n$ by $d\vec{v}_n/dt$ in eqn. (2.98) yields

$$\sum_{i=1}^{N} m_n \frac{d\vec{v}_n}{dt} \cdot \left(\frac{\partial \vec{r}_n}{\partial q_i}\right) \delta q_i - \delta W_n = 0 \tag{2.99}$$

Using the rule of time derivative of a product, eqn. (2.99) can be rewritten as

$$\sum_{i=1}^{N} \left[\frac{d}{dt}\left\{ m_n \vec{v}_n \cdot \left(\frac{\partial \vec{r}_n}{\partial q_i}\right)\right\} - m_n \vec{v}_n \cdot \frac{d}{dt}\left(\frac{\partial \vec{r}_n}{\partial q_i}\right)\right] \delta q_i - \delta W_n = 0 \tag{2.100}$$

In eqn. (2.100), the derivative operation on $\dfrac{\partial \vec{r}_n}{\partial q_i}$ can be performed as

$$\frac{d}{dt}\left(\frac{\partial \vec{r}_n}{\partial q_i}\right) = \frac{\partial}{\partial q_i}\left(\frac{d}{dt}\vec{r}_n\right) = \frac{\partial \vec{v}_n}{\partial q_i} \tag{2.101}$$

From eqn. (2.95), it follows that

$$\frac{\partial \vec{v}_n}{\partial \dot{q}_i} \equiv \frac{\partial \vec{r}_n}{\partial q_i} \tag{2.102}$$

Using (2.102), eqn. (2.100) for virtual work can be expressed as

$$\sum_{i=1}^{N} \left[\frac{d}{dt}\left\{ m_n \vec{v}_n \cdot \left(\frac{\partial \vec{v}_n}{\partial \dot{q}_i}\right)\right\} - m_n \vec{v}_n \cdot \frac{\partial \vec{v}_n}{\partial q_i}\right] \delta q_i - \delta W_n = 0 \tag{2.103}$$

Using the commutative property of the dot product, eqn. (2.103) can be expressed as

$$\sum_{i=1}^{N} \left[\frac{d}{dt}\left\{ m_n \frac{\partial}{\partial \dot{q}_i}\left(\frac{1}{2}\vec{v}_n . \vec{v}_n\right)\right\} - m_n \frac{\partial}{\partial q_i}\left(\frac{1}{2}\vec{v}_n . \vec{v}_n\right)\right] \delta q_i - \delta W_n = 0 \tag{2.104}$$

Bringing m_n inside the parentheses along with $\left(\dfrac{1}{2}\vec{v}_n \cdot \vec{v}_n\right)$ gives the familiar expression of kinetic energy K_E, so that eqn. (2.104) becomes

$$\sum_{i=1}^{N} \left[\frac{d}{dt}\left(\frac{\partial K_E}{\partial \dot{q}_i}\right) - \left(\frac{\partial K_E}{\partial q_i}\right)\right] \delta q_i - \delta W_n = 0 \tag{2.105}$$

If $P_E(q_i, t)$ is the potential energy expressed as function of the generalized coordinates, then virtual work is defined as the amount by which P_E decreases as q_i undergoes an infinitesimal shift to $q_i + \delta q_i$. Thus,

$$\delta W = P_E(q_i, t) - P_E(q_i + \delta q_i, t) \tag{2.106}$$

Using the chain rule of partial differentiation, we have

$$\delta W = \sum_{i=1}^{N} \left[-\frac{\partial}{\partial q_i} P_E(q_i, t) \right] \delta q_i \tag{2.107}$$

Since generalized force, from eqn. (2.90), is the coefficient of δq_i in the expression for δW,

$$\delta W = \left[Q_i - \frac{\partial}{\partial q_i} P_E \right] \delta q_i \tag{2.108}$$

Equation (2.105) can therefore be expressed as

$$\sum_{i=1}^{N} \left[\frac{d}{dt}\left(\frac{\partial K_E}{\partial \dot{q}_i} \right) - \left(\frac{\partial K_E}{\partial q_i} \right) + \left(\frac{\partial P_E}{\partial q_i} \right) - Q_i \right] \delta q_i = 0 \tag{2.109}$$

The generalized forces Q_i can now be expressed as

$$Q_i = \frac{d}{dt}\left(\frac{\partial K_E}{\partial \dot{q}_i} \right) - \frac{\partial K_E}{\partial q_i} + \frac{\partial P_E(q_i, t)}{\partial q_i}, \quad i = 1, \ldots, N \tag{2.110}$$

which are Lagrange's equations. Introducing the Lagrangian function $\mathcal{L} = K_E - P_E$, and considering the fact that potential energy P_E does not depend upon the generalized velocities, we have

$$\frac{\partial \mathcal{L}}{\partial \dot{q}_i} \equiv \frac{\partial K_E}{\partial \dot{q}_i} \tag{2.111}$$

The Lagrange's equations in eqn. (2.110) can now be expressed as

$$Q_i = \frac{d}{dt}\left(\frac{\partial \mathcal{L}}{\partial \dot{q}_i} \right) - \frac{\partial \mathcal{L}}{\partial q_i}, \quad i = 1, \ldots, N \tag{2.112}$$

2.8 CONSTRAINT GENERALIZED COORDINATES

Differential equations of motion, whether derived from Lagrangian or Newton–Euler approach, need to be solved in order to simulate the system's response. The equations of motion generally become complex and cannot always be solved analytically. When generalized coordinates are unconstrained, the number of differential equations of motion is equal to the number of generalized coordinates. However, when generalized coordinates are constrained, solving the equations of motion additionally requires kinematical constraint equations. The process of using of constrained generalized coordinates in formulating the equations of motion can be quite involved and, therefore, mostly requires numerical techniques to solve the equations. Differential algebraic equation solver, in the same manner as one uses Runge–Kutta, is often used for solving unconstrained equations of motion. Another approach to solve for both the dynamic equations and constraint is to use the elimination method. A possible solution could be obtained by using the well-known singular value decomposition (SVD) method.

2.9 ALTERNATIVE FORMULATIONS

Concept of generalized coordinates cannot be extended to flexible bodies which, in addition to kinetic and potential energy, will also store strain energy. This leads to the need to derive models of systems in which relative position of the mass points is not kinematically constrained. Hamilton's principle offers an alternative to Lagrange's equations to formulate equations for relativistic systems.

2.9.1 Hamilton's Principle

In Hamilton's principle, unlike Lagrange's formulation, generalized coordinates are not used to describe position. Consider a single particle for which the principle of dynamic virtual work, from eqn. (2.97), states that

$$\delta W - (m\vec{a})\cdot\delta\vec{r} = 0 \tag{2.113}$$

Using the rule for differentiating a product leads to

$$\delta W - \frac{d}{dt}(m\vec{v}\cdot\delta\vec{r}) + m\vec{v}\,\frac{d}{dt}\delta\vec{r} = 0 \tag{2.114}$$

The last term in eqn. (2.114) can be written as

$$m\vec{v}\,\frac{d}{dt}\delta\vec{r} = m\vec{v}\,\delta\vec{v} = \delta\left(\frac{1}{2}\,m\vec{v}\cdot\vec{v}\right) = \delta K_E \tag{2.115}$$

where K_E is the kinetic energy of the particle. Equation (2.114) can now be written as

$$\delta W + \delta K_E - \frac{d}{dt}(m\vec{v}\cdot\delta\vec{r}) = 0 \tag{2.116}$$

The virtual work, in a general situation, is made up of nonconservative and conservative effects. Further, virtual work done by the conservative forces is the negative of the virtual change in the potential energy. Thus,

$$\begin{aligned}\delta W &= \left[(\delta W)_{\text{nonconservative}} + (\delta W)_{\text{conservative}}\right]\\ &= \left[(\delta W)_{\text{nonconservative}} - \delta P_E\right]\end{aligned} \tag{2.117}$$

Considering eqn. (2.117) and extending eqn. (2.116) to n number of particles, we have

$$\delta W - \delta P_E + \delta K_E - \sum_n \frac{d}{dt}(m\vec{v}_n \cdot \delta\vec{r}_n) = 0 \tag{2.118}$$

In eqn. (2.118), K_E represents the kinetic energy of all particles in the system and δW represents the virtual work done by all forces; the last term contains the particle's momentum $m\vec{v}_n$; now, integrating the equation over time interval t_0 to t_1 yields

$$\int_{t_0}^{t_1} (\delta W - \delta P_E + \delta K_E)dt - \left[\sum_n (m\vec{v}_n \cdot \delta\vec{r}_n)\right]_{t_0}^{t_1} = 0 \tag{2.119}$$

For all particles contained in the system, consider the actual and variational paths the system traces from initial time t_0 to final time t_1, as shown in Figure 2.14. At the final time t_1, the true and the variational path should intersect such that, at $t = t_1$, $\delta\vec{r} = 0$. Under these conditions, eqn. (2.119) reduces to

$$\int_{t_0}^{t_1} (\delta W - \delta P_E + \delta K_E)dt = 0 \tag{2.120}$$

Considering the Lagrangian $\mathcal{L} = K_E - P_E$, equation can be expressed in a more compact form as

$$\int_{t_0}^{t_1} (\delta W + \delta\mathcal{L})dt = 0 \tag{2.121}$$

The foregoing form represents the Hamilton's principle. Since, the variational paths traced are arbitrary and different from true path, except at the initial and final instants, eqn. (2.121) leads to many relations; Hamilton's principle must be satisfied for each of the curve traced.

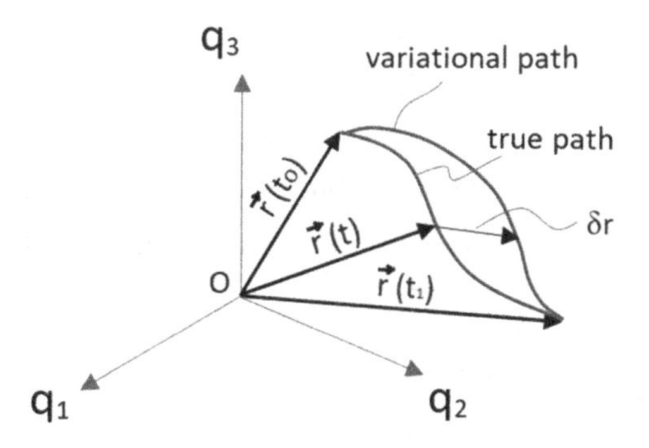

FIGURE 2.14 True and variational paths traced by a system.

2.9.2 Generalized Momentum Principles

For a system having a finite number of degrees of freedom, the relation between momentum and kinetic energy can be obtained using Hamilton's principle. Consider a single particle whose position in the xyz frame of reference is given by $q_1 = x$, $q_2 = y$, and $q_3 = z$. Further, the momentum components are defined by $m\dot{x}$, $m\dot{y}$, and $m\dot{z}$, the kinetic energy $K_E = \frac{1}{2}m\left(\dot{x}^2 + \dot{y}^2 + \dot{z}^2\right)$, such that we have

$$\frac{\partial K_E}{\partial \dot{x}} = m\dot{x}; \quad \frac{\partial K_E}{\partial \dot{y}} = m\dot{y}; \quad \frac{\partial K_E}{\partial \dot{z}} = m\dot{z} \tag{2.122}$$

Considering the kinetic energy of a single particle, differentiation of K_E w.r.t. generalized velocity $\dot{q}_n$ yields

$$\frac{\partial K_E}{\partial \dot{q}_n} = \frac{\partial}{\partial \dot{q}_n}\left(\frac{1}{2}m\vec{v}\cdot\vec{v}\right) = m\vec{v}\cdot\frac{\partial \vec{v}}{\partial \dot{q}_n} \tag{2.123}$$

Using the relation between the particle's velocity and generalized coordinate from eqn. (2.95), we have

$$\vec{v}\left(\dot{q}_i, q_i, t\right) = \sum_{j=1}^{N}\left(\frac{\partial}{\partial q_j}\vec{r}\left(q_i, t\right)\right)\dot{q}_j + \frac{\partial}{\partial t}\vec{r}\left(q_i, t\right) \tag{2.124}$$

From the identity described in eqn. (2.102), we have

$$\frac{\partial}{\partial \dot{q}_n}\vec{v}\left(\dot{q}_i, q_i, t\right) = \frac{\partial}{\partial q_j}\vec{r}\left(q_i, t\right) \tag{2.125}$$

Equation (2.123) now becomes

$$\frac{\partial K_E}{\partial \dot{q}_n} = m\vec{v}.\frac{\partial \vec{r}}{\partial q_n} \tag{2.126}$$

In eqn. (2.126), $\frac{\partial K_E}{\partial \dot{q}_n}$ is the total momentum of a system in the direction of increasing q_n and is referred to as generalized momentum. Since potential energy P_E is not a function of generalized velocities, eqn. (2.126) can also be expressed in terms of Lagrangian $\mathcal{L}$, where $\mathcal{L} = K_E - P_E$.

$$p_n = \frac{\partial K_E}{\partial \dot{q}_n} = \frac{\partial \mathcal{L}}{\partial \dot{q}_n} \tag{2.127}$$

where p_n is the notation generally used to represent generalized momenta. If the mechanical energy $E = K_E + P_E$, then expressing E in terms of $\mathcal{L}$ yields

$$E = 2K_E - \mathcal{L} = \{(m\dot{x})\dot{x} + (m\dot{y})\dot{y} + (m\dot{z})\dot{z}\} - \mathcal{L} \tag{2.128}$$

In eqn. (2.128), K_E has been expressed in terms of momentum components. Introducing a new quantity called the *Hamiltonian* $\mathcal{H}$, and replacing the momentum components with p_n and the velocity components with $\dot{q}_n$, we can express *Hamiltonian* $\mathcal{H}$ as

$$\mathcal{H} = \sum_{j=1}^{N} p_j \dot{q}_j - \mathcal{L} \tag{2.129}$$

Differentiating eqn. (2.129) with time yields

$$\frac{d\mathcal{H}}{dt} = \sum_{j=1}^{N} \left(\dot{p}_j \dot{q}_j - \frac{\partial \mathcal{L}}{\partial q_j} \dot{q}_j \right) - \frac{\partial \mathcal{L}}{\partial t} \tag{2.130}$$

Since $\mathcal{H}$ is a general function of p_j, q_j, and t, $\frac{d\mathcal{H}}{dt}$ can also be expressed as

$$\frac{d\mathcal{H}}{dt} = \sum_{j=1}^{N} \frac{\partial \mathcal{H}}{\partial p_j} \dot{p}_j + \sum_{j=1}^{N} \frac{\partial \mathcal{H}}{\partial q_j} \dot{q}_j + \frac{\partial \mathcal{H}}{\partial t} \tag{2.131}$$

Comparing the two expressions of eqns. (2.130) and (2,131) for $\frac{d\mathcal{H}}{dt}$, we have

$$\dot{q}_j = \frac{\partial \mathcal{H}}{\partial p_j}, \frac{\partial \mathcal{L}}{\partial q_j} = -\frac{\partial \mathcal{H}}{\partial q_j}, \frac{\partial \mathcal{L}}{\partial t} = -\frac{\partial \mathcal{H}}{\partial t} \tag{2.132}$$

From eqn. (2.127), we have

$$\frac{d}{dt} p_j = \frac{d}{dt} \left(\frac{\partial \mathcal{L}}{\partial \dot{q}_n} \right) \tag{2.133}$$

Substituting the expressions from eqns. (2.132) and (2.133) into eqn. (2.112), we have

$$Q_i = \dot{p}_j + \frac{\partial \mathcal{H}}{\partial q_j} \tag{2.134}$$

$$\dot{q}_j = \frac{\partial \mathcal{H}}{\partial p_j} \tag{2.135}$$

The eqns. (2.134) and (2.135) are Hamilton's equations of motion.

2.10 GYROSCOPIC EFFECTS

To understand the physics behind the principal of gyroscope, consider a typical schematic of gyroscope shown in Figure 2.15 [4]. Assume the wheel spins at a constant rate ω_s and the gyroscope precesses at a constant rate of ω_p about the fixed point at the base, with θ being constant. The spinning of the wheel will create forces F_1 and F_2 that act in opposite directions. The gravitational force g on the gyroscope creates a clockwise torque M such that it is able to maintain the angle θ as the gyroscope precesses. This in turn prevents the gyroscope from falling over. This feature of a fast spinning axisymmetric object, that is able to maintain its precession axis, has found immense use in navigation. For this, a spinning wheel is powered by a motor and mounted onto a gimbal (metal frame) to eliminate the effect of any external torque (Figure 2.16). Consequently, the gyroscope itself experiences no change in orientation and can, therefore, provide useful information of roll and pitch in boats, ships, and airplanes.

2.10.1 Free Motion of Axisymmetric Body

For an axisymmetric body experiencing torque free motion, as observed in the inertial frame of reference, the angular momentum vector acts along the precession axis. Consider Figure 2.17 with global axis frame XYZ representing the inertial reference frame and the local axis frame xyz having its origin at the C. G., G, with x-axis aligned with and parallel to X-axis. All the moments, inertia terms, and angular

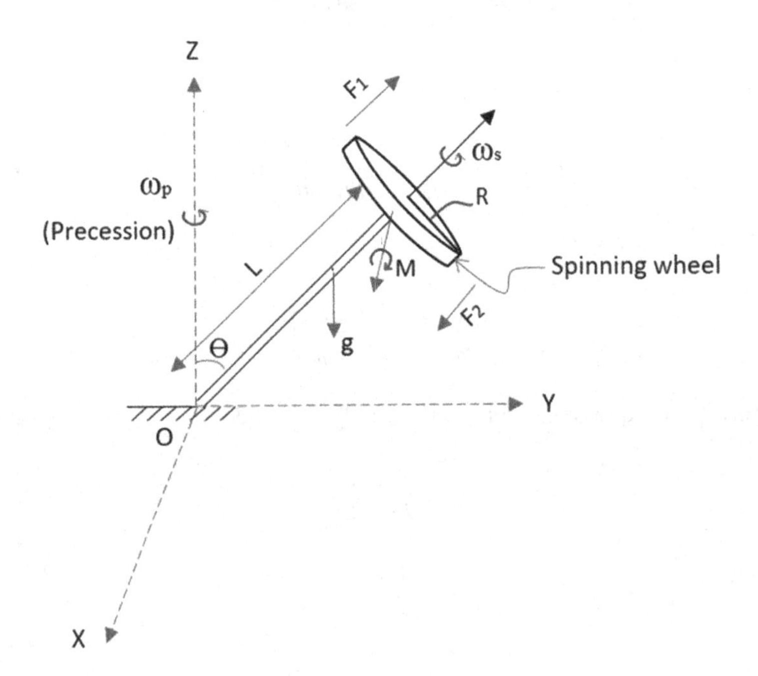

FIGURE 2.15 Principal of gyroscope.

momentum are with respect to G. From Figure 2.16, angle θ is related to angular velocity ω_s and angular momentum vector $\vec{H}_G$ as follows

$$\cos\theta = \frac{\vec{H}_G \cdot \hat{j}}{\left|\vec{H}_G\right|} \tag{2.136}$$

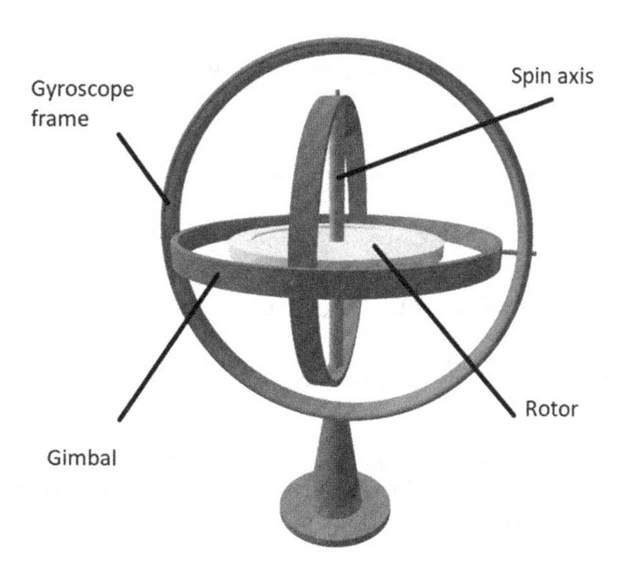

FIGURE 2.16 Gyroscope gimbal unit.

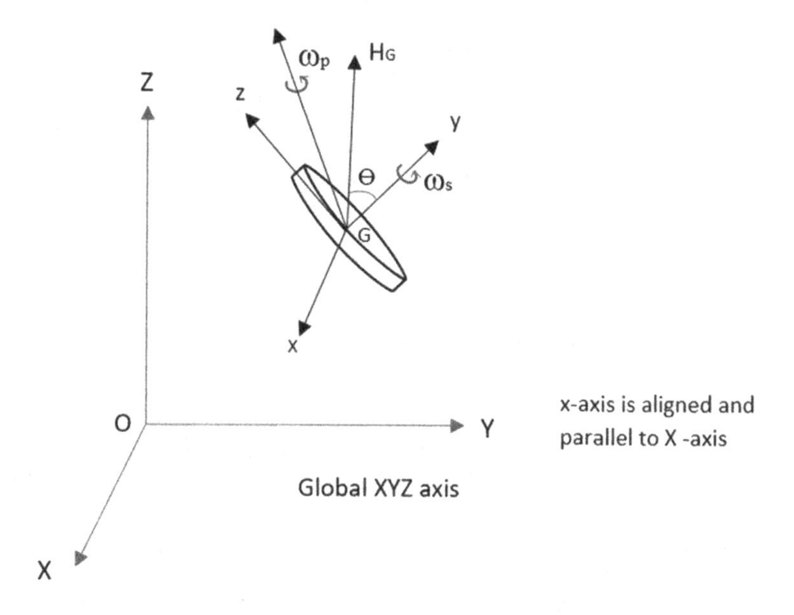

FIGURE 2.17 Torque-free motion of axisymmetric body.

where $\hat{j}$ is the unit vector along the y-axis and $|\vec{H}_G|$ is the magnitude of angular momentum vector $\vec{H}_G$. Differentiating eqn. (2.136) w.r.t time gives

$$-\sin\theta\frac{d\theta}{dt} = \frac{\vec{H}_G}{|\vec{H}_G|}\left(\frac{d\hat{j}}{dt}\right); \ \frac{d\theta}{dt} = -\frac{\vec{H}_G}{\sin\theta|\vec{H}_G|}\left(\frac{d\hat{j}}{dt}\right) \tag{2.137}$$

From the definition of vector derivatives, we know that

$$\left(\frac{d\hat{j}}{dt}\right) = \left(\vec{\omega}\times\hat{j}\right) \tag{2.138}$$

Also,

$$\vec{\omega} = \omega_x\hat{i} + \omega_y\hat{j} + \omega_z\hat{k} \tag{2.139}$$

$$\vec{H}_G = I_x\omega_x\hat{i} + I_y\omega_y\hat{j} + I_z\omega_z\hat{k} \tag{2.140}$$

where ω_x, ω_y, and ω_z are the components of the angular velocity vector and I_x, I_y, and I_z are the principal moments of inertia along the x, y, and z directions. Substituting eqns. (2.139) and (2.140) into eqn. (2.137), we get

$$\frac{d\theta}{dt} = -\frac{(I_z - I_x)\omega_z\omega_x}{\sin\theta|\vec{H}_G|} \tag{2.141}$$

For $I_z = I_x$, $\frac{d\theta}{dt} = 0$, which implies that angle θ is constant. Further, if we choose $\in = 0$, then precession axis coincides with the angular momentum vector $\vec{H}_G$. Since θ is constant, the angular momentum can be expressed as

$$\vec{H}_G = \left(|\vec{H}_G|\cos\theta\right)\hat{j} + \left(|\vec{H}_G|\sin\theta\right)\hat{k} \tag{2.142}$$

Equating the components of $\hat{i}$, $\hat{j}$, and $\hat{k}$ from eqns. (2.10.3) and (2.10.5), and defining $I_z = I_x = I_\omega$, we have

$$\omega_x = 0; \omega_y = \frac{\left(|\vec{H}_G|\cos\theta\right)}{I_y}; \omega_z = \frac{\left(|\vec{H}_G|\sin\theta\right)}{I_\omega} \tag{2.143}$$

Also, from Figure 2.16, we can express the angular velocity components as

$$\omega_x = \frac{d\theta}{dt} = 0; \omega_y = \omega_p\cos\theta + \omega_s; \omega_z = \omega_p\sin\theta \tag{2.144}$$

Solving the above equations for ω_p and ω_s, we get

$$\omega_p = \frac{|\vec{H}_G|}{I_\omega} \text{ and } \omega_s = |\vec{H}_G| \cos\theta \frac{(I_\omega - I_y)}{I_\omega \, I_y} \tag{2.145}$$

With $\vec{H}_G$, I_y, I_ω θ being constant, the angular velocities ω_p and ω_s are also constant.

EPILOGUE

In this chapter, we have given a brief description of the engineering dynamics which when combined with the material of Chapter 1 on aerodynamic principles and fundamentals would help readers to grasp several aspects of equations of motion of the aircraft and aerodynamic behavior as succinctly described by the aerodynamic coefficients and stability and control derivatives. The reference [3] though a big volume is a very comprehensive treatise on the subject. Reference [5] is another very useful volume on the subject of engineering mechanics-cum-dynamics.

EXERCISES

2.1 What is the main difference between kinematics and dynamics of a system?
2.2 Where the curvilinear coordinates are useful?
2.3 Why constrained equations are required?
2.4 Where and why the concept of virtual work is useful?
2.5 What does Hamiltonian actually signify?

REFERENCES

1. Raol, J. R., and Singh, J. *Flight Mechanics Modelling and Analysis*. CRC Press, Taylor & Francis, Boca Raton, FL, 2010.
2. Raol, J. R., Girija, G., and Singh, J. Modelling and parameter estimation of dynamic systems. *IET/IEE Control Series*, London, UK, Vol. 65, August 2004.
3. Ginsberg, J. *Engineering Dynamics*. Cambridge University Press, Cambridge, UK, 2008.
4. Anon. https://www.real-world-physics-problems.com/gyroscope-physics.html, accessed May 2022.
5. Palanichamy, M. S., Nagan, S., and Elango, P. *Engineering Mechanics: Dynamics*. Tata McGraw-Hill Publishing Company Limited, New Delhi, 1998.

3 Equations of Motion

3.1 INTRODUCTION

A dynamic system is defined as a set of components or activities that interact together to achieve a desired task. To study a system, be it a chemical plant or a flight vehicle, the first thing required is a mathematical model (MM, as a reality with reduced complexity) that closely represents the system under study. In other words, a model is a simplified representation of a system that helps to understand, predict, and possibly control the behavior of the system. It presents knowledge of the system in usable form and can help simulate the behavior of the system.

A flight vehicle is a complex dynamic system that would have a number of models for the various (dynamic) components that make up the system (Figure 3.1). In this chapter, we focus on describing the model form that is a reasonable representation of the aircraft dynamics. When the aircraft flies, it experiences various forces arising from the flow of the air over the aircraft frame, gravity forces, propulsive forces, and inertial forces. To achieve a steady unaccelerated flight, these forces must balance out one another. The upward force due to lift should be in equilibrium with the downward force due to the weight of the aircraft, so it does not experience upward or downward motion (unintended one). Similarly, the forward thrust force should be in equilibrium with the opposing drag force and so that the aircraft does not accelerate and, hence, is in a steady motion. An aircraft satisfying this requirement is said to be in a state of equilibrium or flying at *trim* condition. Normally at trim, the translational and angular accelerations are zero. The body axis linear accelerations of any point on the aircraft are the projections of the acceleration vector of the point (with respect to the inertial space) on the body axes. The accelerations, unlike the angular rates, are different for different point on the rigid body aircraft.

To describe the aircraft motion in flight would require the (i) equations of motion and (ii) models of the aerodynamic force and moment coefficients to be postulated (Figure 3.2). The aircraft equations of motion comprise of the aircraft motion variables (aircraft states) such as the airspeed, flow angles (angle-of-attack, AOA and sideslip angles, AOSS), angular velocities/rates (pitch, roll, and yaw), and attitude angles (bank, pitch, and heading) formulated as differential equations. The collection of these aircraft motion variables is called the state vector that helps describe the motion of the flight vehicle at any given point of time. The aircraft equations of motion are described in various text books and reports [1–6]. The core issue in flight mechanics is to evaluate the performance and dynamic characteristics of the aircraft. The laws that govern the motion of an aircraft are those of Newton, provided the aircraft is considered as a rigid body. It is well known that, at high velocities, the aircraft dynamics can and will get affected by structural elasticity. However, for the purpose of understanding the dynamics of flight, it is a convenient to consider the aircraft to be a rigid body with its mass (assumed to be) concentrated at its center of gravity (C. G). There are ways of adding on the aeroelastic effects to the rigid formulation.

DOI: 10.1201/9781003293514-4

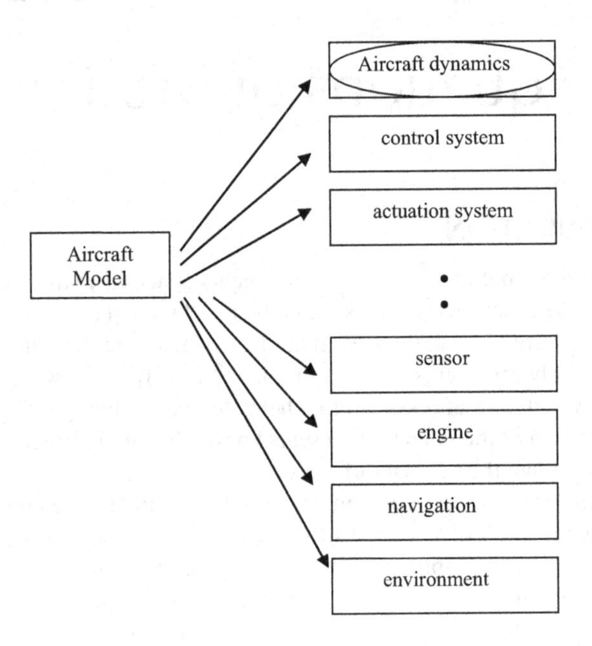

FIGURE 3.1 Aircraft model components.

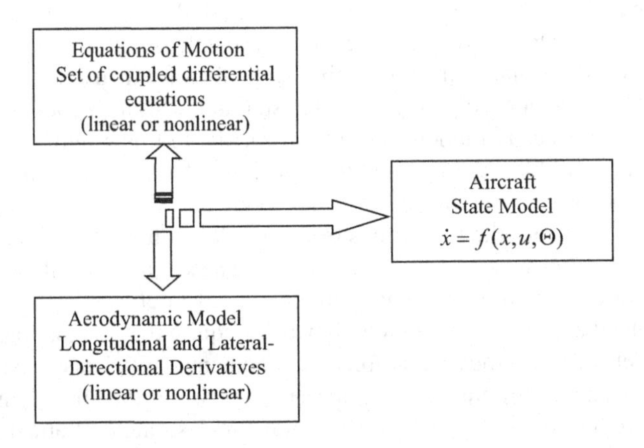

FIGURE 3.2 Schematic of aircraft dynamic modeling.

With this important assumption of the aircraft as a rigid body, we can elaborate on the forces and moments experienced by the aircraft in flight, which we referred to in the definition of equilibrium of aircraft flight.

3.2 RIGID BODY EQUATIONS OF MOTION (EOM)

When describing aircraft flight path, it is necessary to mention the reference axis system of the aircraft and its relation to the earth to which one constantly relates. Thus, a proper understanding of the various axes system used and the associated

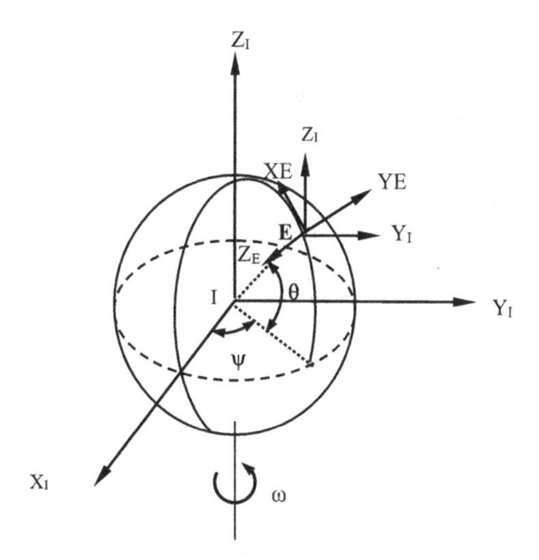

FIGURE 3.3 Inertial axes system.

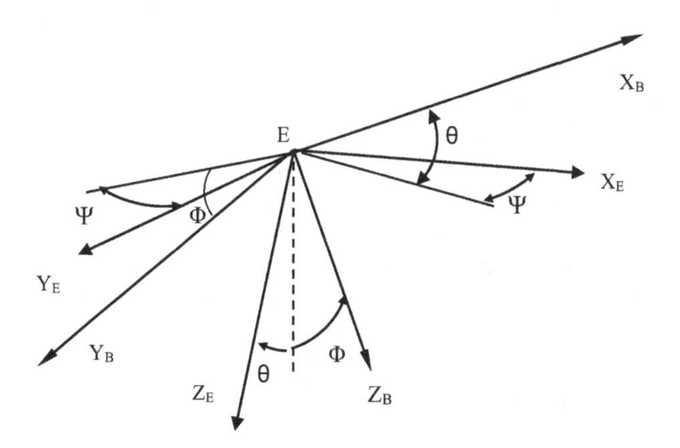

FIGURE 3.4 Earth axes system.

transformation assume some importance. The often-used axes system are the inertial, earth, body, wind, and stability axes, with definitions and assumptions as set next.

Inertial Axes: The earth's center is assumed to be the origin of this axis system. This reference frame is assumed to be stationary with the *Z* axis pointing to the north pole and the *X* axis points to the zero-degree longitude (Figure 3.3). The *Y*-axis, perpendicular to the *XZ* plane, completes the right-handed orthogonal axis system.

Earth Axes: The origin of the earth axes system lies on the surface of the earth, with the *X* and *Y* axes pointing to the north (N) and east (E) direction. The *Z* axis points down (D) toward the center of the earth. This reference frame is fixed to the earth and, at times, referred to as the NED frame (Figure 3.4). For most aircraft modeling and simulation problems, it is assumed to be the inertial frame of reference where Newton's laws are valid.

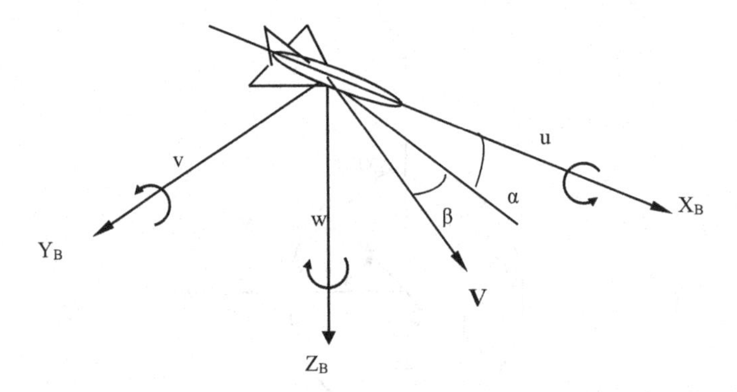

FIGURE 3.5 Body axes system.

Body Axes: This reference frame is fixed to the aircraft with its origin at the aircraft C.G. and rotates with the aircraft and moves in space. The *X* axis points forward through the aircraft nose and is called the longitudinal axis. The positive direction of *Y* axis is along the right wing and named as the lateral axis. Thus, *X*- and *Y*-axes are aligned with some geometric features of the aircraft. The *Z* axis is perpendicular to the *XY* plane and positive downward (Figure 3.5). It will point toward the earth when the aircraft is in straight and level flight. In an alternative principal-body axes, the *X*-axis is aligned with the principal inertia axis. This makes the cross-product of inertia I_{zx} zero. This would simplify the lateral equations of motion. Interestingly, the aerodynamic-body axes (are sometimes known as stability axes in the USA and wind axes in the UK) have the *X*-axis aligned with the projection of the velocity vector, in a datum flight condition, on to the plane of symmetry. The aerodynamic-body axes are at an angle α, AOA relative the geometric body axis. Since, these axes are "natural" axes system, they are used for defining the basic aerodynamic forces, lift and drag. Basically, the body axis angular rates and Euler angles are most naturally measured quantities. This system is consistent in the use of the force coefficients in the body axis. However, it yields nonlinear observation equations for flow angles and velocity and also has singularity at 90 deg. of pitch angle. Interestingly, the pilot is much aware of the rotations of the aircraft about its C.G. rather than any other axis fixed in space. The pilot "feels" the accelerating movement of the aircraft because her/his alignment is in the body axis system.

Wind Axes: Here, the origin is at the aircraft C. G. with the *X*-axis pointing forward and aligned to the velocity vector. The positive *Y*-axis passes through the right wing and the *Z* axis points downward. The aircraft flow angles (angle-of-attack and sideslip) define the relation between the wind and the body axes system (Figure 3.5). Since the direction of the aircraft velocity vector keep changing during the flight, the wind axes system, unlike the body-axes system, is not fixed.

Stability Axes: This axes system is similar to that of body axes. The origin is at the aircraft C. G., and the *Y* stability axis is aligned with the *Y* body axis that passes out through the right wing. The *X* stability axis is inclined to *X* body axis at an angle AOA (Figure 3.6). The aerodynamic data are generated in a wind tunnel in body axes or in stability axes. Thus, a clear understanding of the various axes system used and

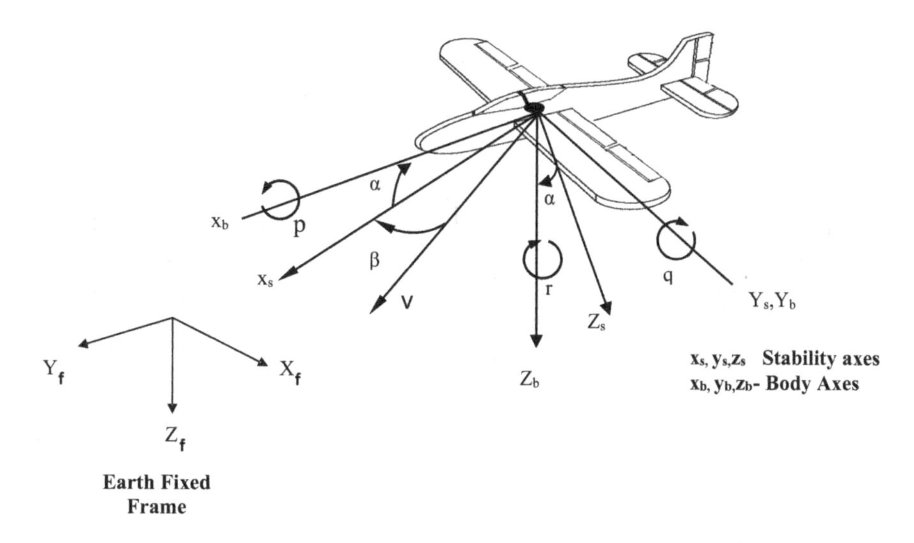

FIGURE 3.6 Stability axes system.

the ability to transform from one to the other is essential to the understanding of aircraft dynamics. The axes system are aligned together by the direction cosine matrix that is a function of the Euler angles ϕ,θ,ψ.

To evaluate aircraft performance (by using the indices such as range, rate of climb, take-off, and landing distance), the aircraft is generally treated as a point mass model and the lift, drag, and thrust forces acting on it are identified. Although, the point mass model is useful for quick check to describe the aircraft dynamics properly, the complete six-degree-of-freedom (6DOF) nonlinear equations would be required. In between, these two extremes lie the linearized models that are useful for control design applications (Figure 3.7). Considering the aircraft to behave like a dynamic system in flight, an input given by the pilot will cause the forces and moments acting on the aircraft to interact with its inherent natural characteristics thereby generating responses, which contain the natural dynamic behavior of the aircraft that can be described by a set of equations called EOM with certain important assumptions.

i. Aircraft is a rigid body. This assumption, though strictly not valid, is sufficiently accurate for the purpose of describing aircraft motion in flight. It allows the aircraft motion to be described by the translation of its C.G. and the rotation about it. In reality, various components of aircraft might move in relation to each other: engine, rotorcraft disc, control devices, bending of wings (this would bring in the aspects of aero-elasticity in the flight dynamics problems) due to air loads.
ii. The earth axes system can be used as the inertial frame of reference in which Newton's laws are valid, and the earth is assumed to be fixed in the space. The inertial reference frame is fixed or moves at a constant velocity relative to earth. This description is reasonably accurate for short-term guidance and control.

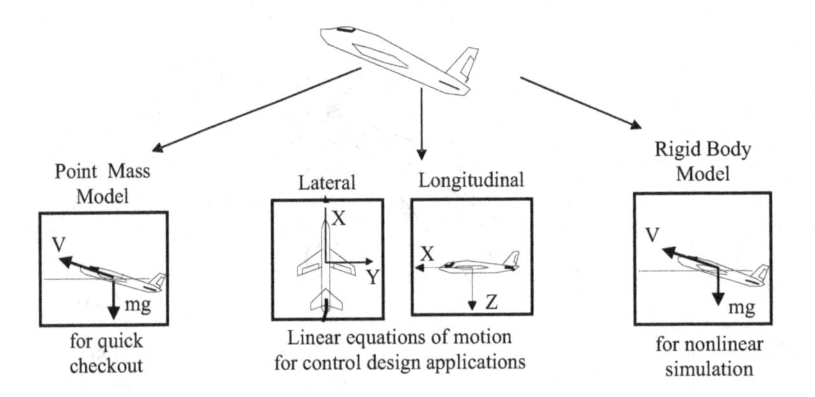

FIGURE 3.7 Methods of modelling body dynamics.

iii. Aircraft mass is constant; its distribution is assumed to be invariant, though not always true.
iv. The earth is flat for the period of time of dynamic analysis. The curvature effects would be required to be taken into account for long-range flights (like for missiles, spacecrafts).
v. Aircraft XZ plane is a plane of symmetry.
vi. The steady flight is assumed to be undisturbed; this will simplify several equations. If flow angles are assumed to be small, then sine and cosine terms would be simplified. This would render the applicability of the EOMs to the conditions where small perturbation could occur.

These basic assumptions are appropriately used in formulating the EOMs and their simplification as the case might be. The EOM of the aircraft can be written by considering its linear and angular momentum. Let p be the linear momentum vector (mV, mass times velocity) and H the angular momentum vector (inertia times angular velocity or rate), each measured in the inertial reference frame. Newton's second law states that the summation of all the external forces acting on a body is equal to the time rate of change of momentum of the body:

$$\frac{dp}{dt} = \frac{d(mV)}{dt} = \sum F \tag{3.1}$$

Similarly, the summation of all external moments acting on the body is equal to the time rate of change of the angular momentum (moment of momentum):

$$\frac{dH}{dt} = \sum M \tag{3.2}$$

Rewriting the vector form of eqns. (3.1) and (3.2) into scalar form by resolving the force and angular momentum into components along the axes the x, y, and z axes (of the inertial reference frame that is a right-handed system of the Cartesian axes), the EOM can be written as

$$F_x = \frac{d}{dt}(mu); \;\; F_y = \frac{d}{dt}(mv); \;\; F_z = \frac{d}{dt}(mw) \tag{3.3}$$

These equations describe the motion of the airframe C.G. as seen by an observer in the x, y, z frame. The moment equations can be expressed in the scalar form as:

$$L = \frac{d}{dt}H_x; \;\; M = \frac{d}{dt}H_y; \;\; N = \frac{d}{dt}H_z \tag{3.4}$$

The angular momentum equation is given by

$$H = I\omega; \;\; I_{x,y,z} = \begin{bmatrix} I_x & -I_{xy} & -I_{xz} \\ -I_{xy} & I_y & -I_{yz} \\ -I_{zx} & -I_{zy} & I_z \end{bmatrix} \tag{3.5}$$

Here, H is the angular momentum, I is the inertia tensor (the rotary analog of the mass), and ω is the angular velocity of the aircraft. The derivative of eqn. (3.5) gives

$$\frac{d}{dt}H = I \cdot \frac{d\omega}{dt} + \frac{dI}{dt}\omega \tag{3.6}$$

Referring to the inertia and angular velocities in the space fixed inertial frame, the inertia tensor will change as the aircraft rotates about the axes. This would contribute to dH/dt through dI/dt, and the resulting equations would have time varying parameters. To overcome this, we consider the body axis system that is fixed to the aircraft, so that the rotary inertial properties are constant (dI/dt's are zero). This implies that the vector quantities V and H in eqns. (3.1) and (3.2) need to be determined in the body axis. A vector in a body axis frame, rotating at an angular speed of ω, can be represented by the equation:

$$\left.\frac{d(.)}{dt}\right|_I = \left.\frac{d(.)}{dt}\right|_B + \omega \times (.) \tag{3.7}$$

Here, subscripts I and B refer to the inertial and body axis frame of references. Using this expression, the forces and moments in the inertial axis system can now be represented in the body axis system as follows:

$$F = m\left.\frac{dV}{dt}\right|_I = m\left.\frac{dV}{dt}\right|_B + m(\omega \times V) \tag{3.8}$$

$$M = \left.\frac{dH}{dt}\right|_I = \left.\frac{dH}{dt}\right|_B + \omega \times H \tag{3.9}$$

It must be emphasized here that the mass is assumed to be constant. This would not be true for long-range aircraft flights, missiles, and spacecrafts because of expending of fuels/shedding away certain extended components/stages. The inertial forces and moments can now be resolved in the body axis system.

3.3 RESOLUTION OF INERTIAL FORCES AND MOMENTS

The right-hand side of eqn. (3.8) contains the terms $\dot{V}$ (dV/dt), which represents the time derivative of velocity with respect to the body axis system. The cross-product terms ($\omega \times V$) arise from the centripetal acceleration along any given axis due to angular velocities about the remaining two axes. In component form we have:

$$V = u\hat{i} + v\hat{j} + w\hat{k}$$

$$\omega = p\hat{i} + q\hat{j} + r\hat{k} \tag{3.10}$$

$$H = H_x\hat{i} + H_y\hat{j} + H_z\hat{k}$$

Using eqn. (3.10) in eqn. (3.8), we have

$$F = m\{[\dot{u}\hat{i} + \dot{v}\hat{j} + \dot{w}\hat{k}] + [(p\hat{i} + q\hat{j} + r\hat{k}) \times (u\hat{i} + v\hat{j} + w\hat{k})]\}; \times \text{ is cross product}$$

$$F = m[\dot{u} + qw - rv]\hat{i} + m[\dot{v} + ru - pw]\hat{j} + m[\dot{w} + pv - qu]\hat{k}$$

Thus, the force components acting along the X, Y, and Z body axis are as follows:

$$F_x = m(\dot{u} + qw - rv)$$

$$F_y = m(\dot{v} + ru - pw) \tag{3.11}$$

$$F_z = m(\dot{w} + pv - qu)$$

Example 3.1

Assume two-dimensional accelerated flight of an aircraft. In this situation, the $U + \Delta U$ and $V + \Delta V$ would have rotated by an incremental angle $R\Delta t$. Of course, the aircraft would have also rotated through the same angle. Here, U and V are the total velocities of the aircraft in forward and lateral directions. Derive the equation for the rectilinear longitudinal acceleration.

Solution 3.1

$$a_X = \lim(\Delta t -> 0) \frac{(U + \Delta U)\cos(R\Delta t) - U - (V + \Delta V)\sin(R\Delta t)}{\Delta t}$$

Due to the limit $\frac{\sin(R\Delta t)}{\Delta t} = R$ and for small changes/perturbation, we get $a_X = \dot{U} - VR$.

Example 3.2

Assume two-dimensional accelerated flight of an aircraft. Derive the equation for the corresponding rectilinear lateral acceleration.

Solution 3.2

We have for this case the following formula:

$a_X = \lim(\Delta t -> 0) \dfrac{(U+\Delta U)\sin(R\Delta t) + (V+\Delta V)\cos(R\Delta t) - V}{\Delta t}$. With the same argument as the previous one, we get $a_Y = \dot{V} + UR$.

Such small perturbation terms appear in eqn. (3.11) as can be clearly seen. From eqn. (3.9), we have $M = \dot{H} + (\omega \times H)$. The total moment about a given axis is due to both direct angular acceleration about that axis and those arising from linear acceleration gradients due to combined rotations about all axes. The components of angular momentum about the three axes are given by

$$
\begin{aligned}
H_x &= pI_x - qI_{xy} - rI_{xz} \\
H_y &= -pI_{xy} + qI_y - rI_{yz} \\
H_z &= -pI_{zx} - qI_{zy} + rI_z
\end{aligned}
\tag{3.12}
$$

Even though there could be variation in the mass and its distribution during a flight due to consumption of the fuel, use of external stores, the assumption that the mass and mass distribution of the air vehicle is constant is reasonable and simplifies the expression for the rate of change of angular momentum. The time derivative of the angular momentum components expressed in body axis are given by

$$
\begin{aligned}
\dot{H}_x &= \dot{p}I_x - \dot{q}I_{xy} - \dot{r}I_{xz} \\
\dot{H}_y &= -\dot{p}I_{xy} + \dot{q}I_y - \dot{r}I_{yz} \\
\dot{H}_z &= -\dot{p}I_{zx} - \dot{q}I_{zy} + \dot{r}I_z
\end{aligned}
\tag{3.13}
$$

To get the total time rate of change of angular momentum and thus the scaler components of the moments, we must add to these the contributions arising due to steady rotations $(\omega \times H)$.

$$
\begin{aligned}
L &= \dot{H}_x - rH_y + qH_z \\
M &= rH_x + \dot{H}_y - pH_z \\
N &= -qH_x + pH_y + \dot{H}_z
\end{aligned}
\tag{3.14}
$$

Substituting for angular momentum components and regrouping results, the moment equations can now be written (with respect to the body axes system) as

$$
\begin{aligned}
L &= \dot{p}I_x - \dot{q}I_{xy} - \dot{r}I_{xz} + qr(I_z - I_y) + (r^2 - q^2)I_{yz} - pqI_{xz} + rpI_{xy} \\
M &= -\dot{p}I_{xy} + \dot{q}I_y - \dot{r}I_{yz} + rp(I_x - I_z) + (p^2 - r^2)I_{xz} - qrI_{xy} + pqI_{yz} \\
N &= -\dot{p}I_{xz} - \dot{q}I_{yz} + \dot{r}I_z + pq(I_y - I_x) + (q^2 - p^2)I_{xy} - rpI_{yz} + qrI_{xz}
\end{aligned} \tag{3.15}
$$

The total forces and moments acting on the aircraft are made up of contributions from the aerodynamic, gravitational, and aircraft thrust.

3.4 RESOLUTION OF AERODYNAMIC, GRAVITY, AND THRUST FORCES

The most significant contribution to the external forces and moments comes from the aerodynamics. The three aerodynamic forces and moments in terms of nondimensional coefficients are as follows:

$$
\begin{aligned}
F_{xA} &= \overline{q}SC_x;\ F_{yA} = \overline{q}SC_y;\ F_{zA} = \overline{q}SC_z \\
L &= C_l\,\overline{q}Sb;\ M = C_m\,\overline{q}S\overline{c};\ N = C_n\,\overline{q}Sb
\end{aligned} \tag{3.16}
$$

Here, the coefficients C_x, C_y, C_z, C_l, C_m and C_n are the nondimensional body-axis force and moment coefficients, S represents the reference wing area, $\overline{c}$ is the mean aerodynamic chord, b is the wing span, and $\overline{q}$ is the dynamic pressure. Further theory of aerodynamic coefficients is discussed in Chapter 4. In particular, the coefficients would be dependent on Mach number, flow angles, altitude, control surface deflections, and propulsion system. For an example, a deflection of a control surface would mean and an effective change of the camber of a wing. This will affect the lift, drag, and moments and hence the aerodynamic coefficients. In fact, these are "static" coefficients and are generally determined experimentally from various wind tunnel tests. Hundreds of such experiments are required to be done for various scaled-down models of the given aircraft. The aerodynamic effects when the aircraft does maneuvering flights are captured in the so-called aerodynamic derivatives (Chapter 4). Here, a brief on these aerodynamic coefficients [7] is given to connect the equations of motion with the aerodynamics. Typical values of these coefficients are given in Figure 3.8a and b.

Lift Coefficient: This coefficient, C_L, is related to C_z. Its contributions come from wings, fuselage, and horizontal tail. It varies with alpha and Mach number. With an increase in lift, always increase in drag is associated. If the camber of the wings is not much, then the lift is obtained from the higher dynamic pressure. Generally, in an aircraft, there would be some mechanisms to increase an effective camber of the wings. These are the additional small surfaces like flaps, the deployment of which would provide required lift at low dynamic pressures. The dependence of the lift coefficient on the AOSS and altitude is usually small and hence neglected. However, the effect of the thrust and the Mach number would be significant on the lift coefficient.

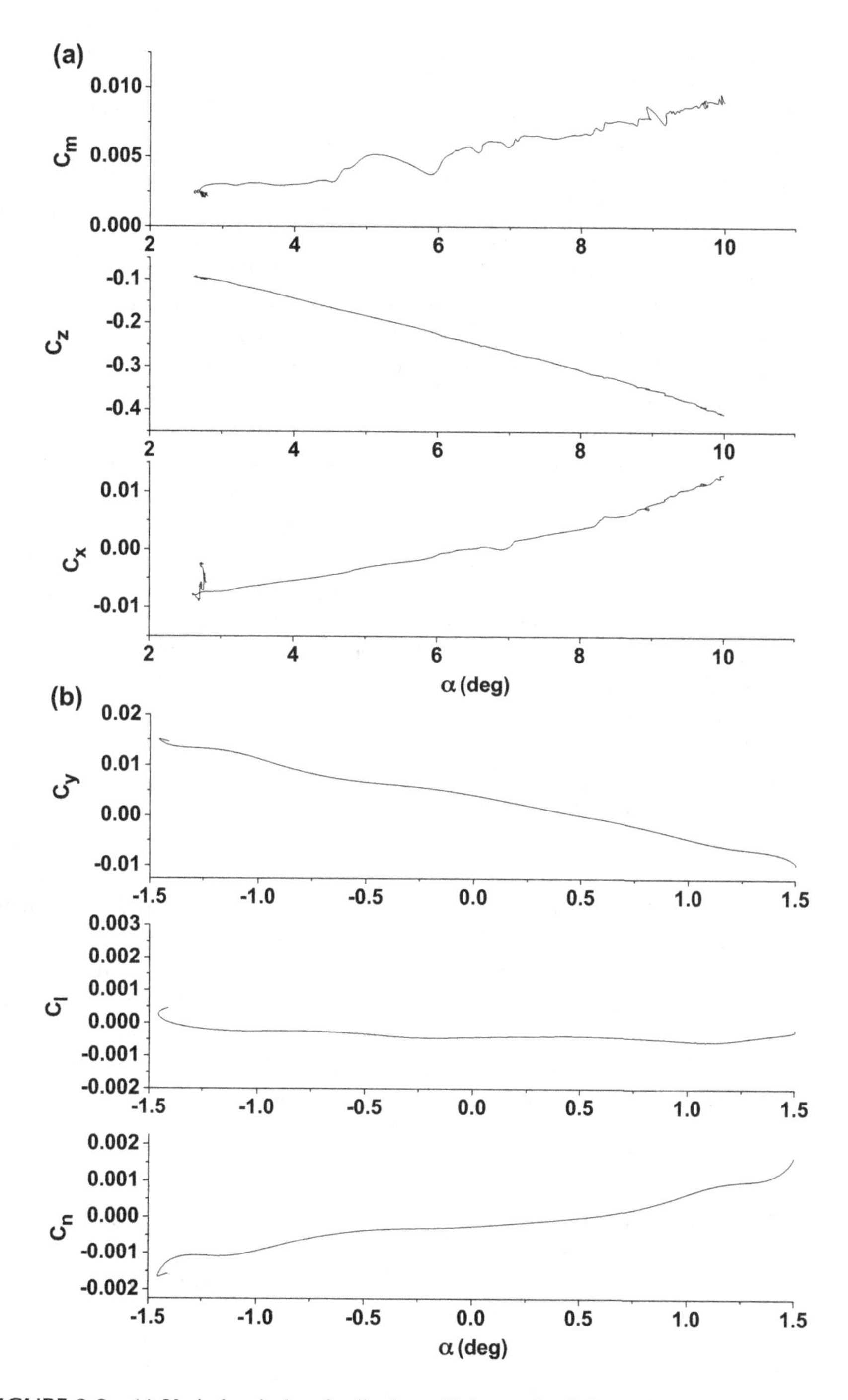

FIGURE 3.8 (a) Variation in longitudinal coefficients of a fighter aircraft during wind-up turn [$H = 9$ km, $M = 0.9$]. (b) Variation in lateral coefficients of a fighter aircraft during Dutch-roll motion [$H = 2$ km, $M = 0.6$].

With higher thrust, the peak of the lift curve (the lift coefficient plotted v/s alpha) could increase, and the peak would also shift to a higher angle of attack. The effect of Mach number is such that the slope of the curve would increase up to a certain Mach number and then start decreasing. For supersonic speeds, the peak would decrease.

Drag Coefficient: This coefficient, C_D, is related to C_y, C_N (the normal force coefficient), and C_x in general; the latter is called the axial force coefficient. This is an important design parameter for the aircraft. One would always like to have minimal drag, no resistance to the aircraft and the maximum possible lift (so that the aircraft can carry very large loads) for an ideal situation. This intuitive requirement defines the lift/drag (*L/D*) ratio, and it is a performance parameter. The drag force is not necessarily a linear, combination of several components: friction, form, induced, and wave drag. The individual contributions of these effects on the total drag would change with the flight conditions.

Side Force Coefficient: This force is the result of the aircraft having AOSS different from zero. In general, the side slip force depends on alpha, beta, Mach number, rudder control surface deflection, roll rate, and yaw rate.

Pitching Moment Coefficient: This is very important parameter for the stability of the aircraft, especially in its derivative form, i.e. $C_{m\alpha}$, which is supposed to be negative in general. However, when it is positive, the aircraft does not have the required static stability. This aspect is discussed in Chapter 4. If the slope of the moment curve is very high, then the aircraft is said to have the pitch "stiffness." Thus, a trade-off is required between the static stability and the maneuverability. The reduced/relaxed static-stability airframes are specially designed to obtain high *L/D* ratio and good maneuverability. The statically unstable configurations for obtaining such benefits, however, need to be artificially stabilized by feedback control filters/transfer functions/algorithms. The moment coefficient, primarily depends on alpha, Mach number, altitude, pitch rate, elevator (or elevon for delta wing aircraft) control surface deflection, inertial damping, aerodynamic damping term $C_{m\dot{\alpha}}$, and flap displacement.

Rolling Moment and Yawing Moment Coefficients: These coefficients depend primarily on alpha, beta, Mach number, roll rate, yaw rate, aileron control surface displacement, and rudder control surface deflection. These coefficients are better understood in the form of their components that are discussed in Chapter 4.

Next, to resolve the gravity force acting on the aircraft into components along the body axes, Figure 3.9, the relative orientation of the gravity vector to the body

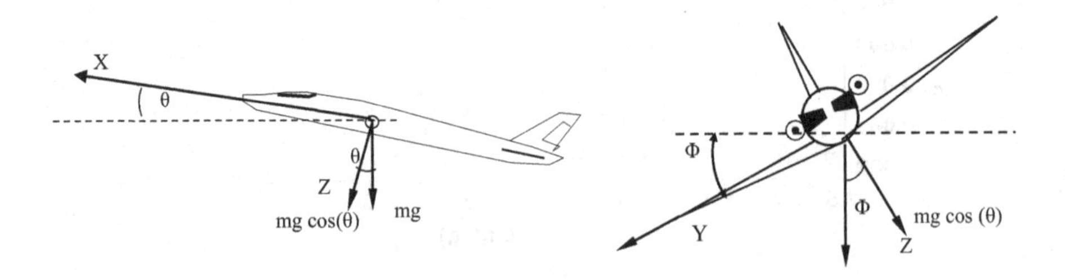

FIGURE 3.9 Gravity force components.

axes system is required. If ϕ,θ and ψ are the angular orientations of the gravity vector with respect to body x, y, and z axes, respectively, the components can be written as

$$
\begin{aligned}
F_{xG} &= -mg\sin\theta \\
F_{yG} &= mg\sin\phi\cos\theta \\
F_{zG} &= mg\cos\phi\cos\theta
\end{aligned}
\tag{3.17}
$$

The third force comes from the engine thrust. Assuming the engine tilt angle to be σ_T with respect to the body X axis, the total applied forces, accounting for the contribution from aerodynamic gravity and thrust, can be expressed as:

$$
\begin{aligned}
F_x &= \bar{q}SC_x - mg\sin\theta + T\cos\sigma_T \\
F_y &= \bar{q}SC_y + mg\sin\phi\cos\theta \\
F_z &= \bar{q}SC_z + mg\cos\phi\cos\theta - T\sin\sigma_T
\end{aligned}
\tag{3.18}
$$

At times, contribution from the gyroscopic moment from the propulsive system is added to the moment equations. This contribution is, however, neglected in the present model postulates. Although the forces and moments expressed here have been derived for the body axis frame of reference, the orientation of the aircraft has to be described with respect to fixed reference frame. Assuming the body axis and the fixed frame to be parallel, three angular rotations in the following specific order are applied to determine the Euler angles (Figure 3.10):

i. Rotate the fixed frame through yaw angle ψ to attain frame position x_1, y_1, z_1
ii. Rotate x_1, y_1, z_1 frame through pitch angle θ to attain the position x_2, y_2, z_2
iii. Rotate x_2, y_2, z_2 frame through roll angle ϕ to attain the position x_3, y_3, z_3

The frame x_3, y_3, z_3 is the final orientation of the body axis to the fixed axis. The relationship between the body axis angular rates p,q, and r and the Euler rates $\dot{\psi}$,$\dot{\theta}$, and $\dot{\phi}$, derived from Figure 3.10, can be expressed as:

$$
\begin{aligned}
\dot{\phi} &= p + q\tan\theta\sin\phi + r\tan\theta\cos\phi \\
\dot{\theta} &= q\cos\phi - r\sin\phi \\
\dot{\psi} &= r\cos\phi\sec\theta + q\sin\phi\sec\theta
\end{aligned}
\quad ; \quad
\begin{bmatrix} p \\ q \\ r \end{bmatrix} =
\begin{bmatrix} 1 & 0 & -\sin\theta \\ 0 & \cos\phi & \sin\phi\cos\theta \\ 0 & -\sin\phi & \cos\phi\cos\theta \end{bmatrix}
\begin{bmatrix} \dot{\phi} \\ \dot{\theta} \\ \dot{\psi} \end{bmatrix}
\tag{3.19}
$$

By integrating eqn. (3.19), one can determine the Euler angles.

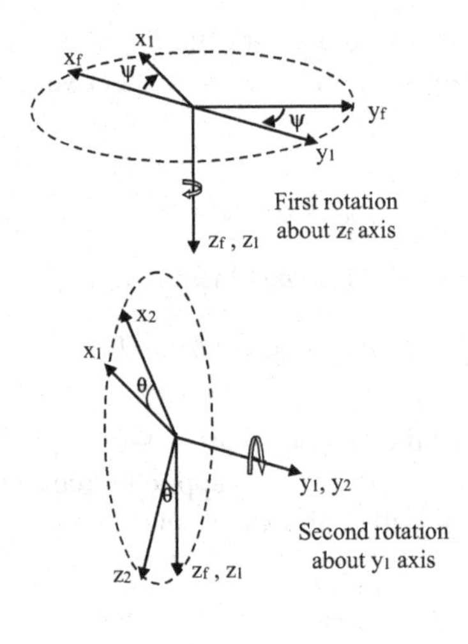

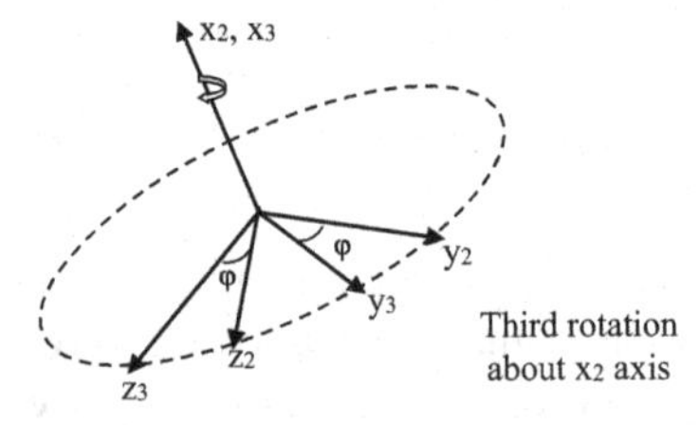

FIGURE 3.10 Body axis to inertial axis transformation.

Example 3.3

Express the inertial velocity components $\dot{x}, \dot{y}, \dot{z}$ in terms of the velocity components *u*, *v*, *w* in body axis.

Solution 3.3

In Figure 3.10, rotation of frame through ψ to position x_1, y_1, z_1 gives

$$\dot{x} = u_1 \cos\psi - v_1 \sin\psi$$

$$\dot{y} = u_1 \sin\psi + v_1 \cos\psi$$

$$\dot{z} = w_1 \quad \text{(the rotation is about } z \text{ axis)}$$

Next, rotation through θ to position x_2, y_2, z_2 gives

$$u_1 = u_2 \cos\theta + w_2 \sin\theta$$

$$v_1 = v_2 \text{ (the rotation is about } y_1 \text{ axis)}$$

$$w_1 = w_2 \cos\theta - u_2 \sin\theta$$

Final rotation through ϕ to position x_3, y_3, z_3 gives

$$u_2 = u \text{ (the rotation is about } y_1 \text{ axis)}$$

$$v_2 = v\cos\phi - w\sin\phi$$

$$w_2 = v\sin\phi + w\cos\phi$$

Substituting u_2, v_2, w_2 into equations for u_1, v_1, w_1, we obtain

$$u_1 = u\cos\theta + [v\sin\phi + w\cos\phi]\sin\theta$$

$$v_1 = v\cos\phi - w\sin\phi$$

$$w_1 = [v\sin\phi + w\cos\phi]\cos\theta - u\sin\theta$$

Substituting u_1, v_1, w_1 into equation for $\dot{x}, \dot{y}, \dot{z}$, we obtain

$$\dot{x} = [u\cos\theta + (v\sin\phi + w\cos\phi)\sin\theta]\cos\psi - [v\cos\phi - w\sin\phi]\sin\psi$$

$$\dot{y} = [u\cos\theta + (v\sin\phi + w\cos\phi)\sin\theta]\sin\psi + [v\cos\phi - w\sin\phi]\cos\psi$$

$$\dot{z} = [v\sin\phi + w\cos\phi]\cos\theta - u\sin\theta$$

In matrix form

$$\begin{bmatrix} \dot{x} \\ \dot{y} \\ \dot{z}, \dot{H} \end{bmatrix}
= \begin{bmatrix} \cos\theta\cos\psi & \sin\theta\sin\phi\cos\psi - \cos\phi\sin\psi & \sin\theta\cos\phi\cos\psi + \sin\phi\sin\psi \\ \cos\theta\sin\psi & \sin\theta\sin\phi\sin\psi + \cos\phi\cos\psi & \sin\theta\cos\phi\sin\psi - \sin\phi\cos\psi \\ -\sin\theta & \cos\theta\sin\phi & \cos\theta\cos\phi \end{bmatrix}
\times \begin{bmatrix} u \\ v \\ w \end{bmatrix} + \begin{bmatrix} W_x \\ W_y \\ W_{z,v} \end{bmatrix}$$

The last vector contains: wind component toward the north, the east, and vertical, respectively.

Note that the matrix is the same as the transformation matrix given in eqn. (3A.9).

A common quantity related to the spatial position is the flight path angle given as

$$\sin\gamma = \frac{\dot{H}}{\sqrt{\dot{x}^2 + \dot{y}^2 + \dot{z}^2}}$$

3.5 COMPLETE SETS OF EOM

Using eqn. (3.11) and eqns. (3.15)–(3.19), the following nine coupled differential equations describe the aircraft motion in flight.

Force Equations:

$$\begin{aligned} m(\dot{u} + qw - rv) &= \bar{q}SC_x - mg\sin\theta + T\cos\sigma_T \\ m(\dot{v} + ru - pw) &= \bar{q}SC_y + mg\sin\phi\cos\theta \\ m(\dot{w} + pv - qu) &= \bar{q}SC_z + mg\cos\phi\cos\theta - T\sin\sigma_T \end{aligned} \tag{3.20}$$

Moment Equations

$$\begin{aligned} \bar{q}SbC_l &= \dot{p}I_x - \dot{q}I_{xy} - \dot{r}I_{xz} + qr(I_z - I_y) + (r^2 - q^2)I_{yz} - pqI_{xz} + rpI_{xy} \\ \bar{q}ScC_m &= -\dot{p}I_{xy} + \dot{q}I_y - \dot{r}I_{yz} + rp(I_x - I_z) + (p^2 - r^2)I_{xz} - qrI_{xy} + pqI_{yz} \\ \bar{q}SbC_n &= -\dot{p}I_{xz} - \dot{q}I_{yz} + \dot{r}I_z + pq(I_y - I_x) + (q^2 - p^2)I_{xy} - rpI_{yz} + qrI_{xz} \end{aligned} \tag{3.21}$$

Euler Angles and Body Angular Velocities

$$\begin{aligned} \dot{\phi} &= p + q\tan\theta\sin\phi + r\tan\theta\cos\phi \\ \dot{\theta} &= q\cos\phi - r\sin\phi \\ \dot{\psi} &= r\cos\phi\sec\theta + q\sin\phi\sec\theta \end{aligned} \tag{3.22}$$

The set of Eqns. (3.20)–(3.22) represents the required 6DOF EOM and can rearranged and expressed in two equivalent forms.

3.5.1 Rectangular Form

When the velocity is represented in rectangular coordinate system, we have:

$$\dot{u} = -qw + rv - g\sin\theta \quad \boxed{+\frac{\bar{q}S}{m}C_x + \frac{T}{m}\cos\sigma_T}$$

$$\dot{v} = -ru + pw + g\cos\theta\sin\phi + \boxed{\frac{\bar{q}S}{m}C_y}$$

$$\dot{w} = -pv + qu + g\cos\theta\cos\phi + \boxed{\frac{\bar{q}S}{m}C_z - \frac{T}{m}\sin\sigma_T}$$

$$\dot{p} = \frac{1}{I_x I_z - I_{xz}^2}\left\{\bar{q}Sb\left(I_z C_l + I_{xz} C_n\right) - qr\left(I_{xz}^2 + I_z^2 - I_y I_z\right) + pqI_{xz}\left(I_x - I_y + I_z\right)\right\}$$

$$\dot{q} = \frac{1}{I_y}\left\{\bar{q}ScC_m - \left(p^2 - r^2\right)I_{xz} + pr\left(I_z - I_x\right) + T(l_{tx}\sin\sigma_T + l_{tz}\cos\sigma_T)\right\} \quad (3.23)$$

$$\dot{r} = \frac{1}{I_x I_z - I_{xz}^2}\left\{\bar{q}Sb\left(I_x C_n + I_{xz} C_l\right) - qrI_{xz}\left(I_x - I_y + I_z\right) + pq\left(I_{xz}^2 + I_x^2 - I_x I_y\right)\right\}$$

$$\dot{\phi} = p + q\tan\theta\sin\phi + r\tan\theta\cos\phi$$

$$\dot{\theta} = q\cos\phi - r\sin\phi$$

$$\dot{\psi} = r\cos\phi\sec\theta + q\sin\phi\sec\theta$$

Note that the terms in the rectangular boxes in the expressions for $\dot{u}, \dot{v}$, and $\dot{w}$ are the linear accelerations. This gives a set of kinematic equations that can be used to check data consistency and estimate calibration (biases and scale factors) errors in the data. The model equations normally used for kinematic consistency check are as follows:

$$\dot{u} = -qw + rv - g\sin\theta \qquad + a_x$$

$$\dot{v} = -ru + pw + g\cos\theta\sin\phi + a_y$$

$$\dot{w} = -pv + qu + g\cos\theta\cos\phi + a_z$$

$$\dot{\phi} = p + q\tan\theta\sin\phi + r\tan\theta\cos\phi \quad (3.24)$$

$$\dot{\theta} = q\cos\phi - r\sin\phi$$

$$\dot{\psi} = r\cos\phi\sec\theta + q\sin\phi\sec\theta$$

$$\dot{h} = u\sin\theta - v\cos\theta\sin\phi - w\cos\theta\cos\phi$$

3.5.2 Polar Form

At times, it is useful to work with equations expressed in polar coordinates:

a. It is more convenient to use equations in polar coordinates because the aerodynamic forces and moments are easier to visualize and express in terms of AOAα, AOSSβ, and *V*.
b. The flow angles α and β can be directly measured and hence would yield simple measurements' equations.
c. The primary demerit is that the system becomes singular at zero velocity, since for the hover condition of a rotorcraft, the angles are not defined. Also singularity occurs for beta = 90 deg.

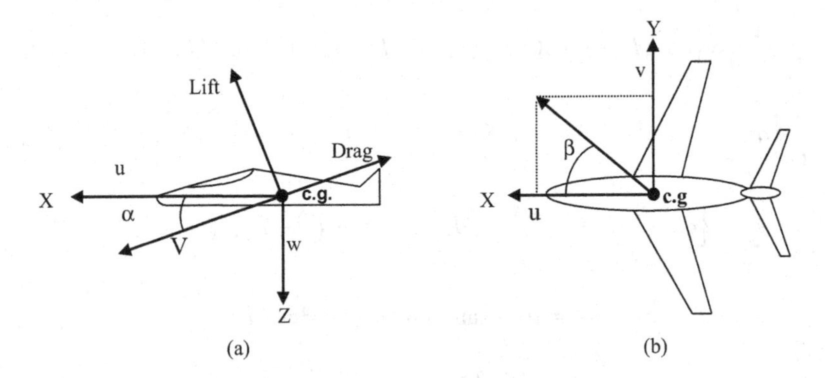

FIGURE 3.11 Flow angles and velocity components. (a) view from side, (b) view from top.

V, α, and β in terms of u, v, and w are given by expressions, Figure 3.11:

$$V = \sqrt{u^2 + v^2 + w^2}; \quad \alpha = \tan^{-1}\frac{w}{u}; \quad \beta = \sin^{-1}\frac{v}{V} \tag{3.25}$$

Differentiating (3.25), we get

$$\dot{V} = \frac{1}{V}(u\dot{u} + v\dot{v} + w\dot{w}); \quad \dot{\alpha} = \frac{u\dot{w} - w\dot{u}}{u^2 + w^2}; \quad \dot{\beta} = \frac{V\dot{v} - v\dot{V}}{V^2\sqrt{\left(1 - v^2/V^2\right)}} \tag{3.26}$$

The velocity components u, v, and w can be expressed in terms of V, α, and β as:

$$u = V\cos\alpha\cos\beta; \quad v = V\sin\beta; \quad w = V\sin\alpha\cos\beta \tag{3.27}$$

Equation (3.27) is obtained from the following transformation:

$\begin{bmatrix} u \\ v \\ w \end{bmatrix} = \begin{bmatrix} \cos\alpha & 0 & -\sin\alpha \\ 0 & 1 & 0 \\ \sin\alpha & 0 & \cos\alpha \end{bmatrix} \begin{bmatrix} \cos\beta & -\sin\beta & 0 \\ \sin\beta & \cos\beta & 0 \\ 0 & 0 & 1 \end{bmatrix} \begin{bmatrix} V \\ 0 \\ 0 \end{bmatrix}$; this is the product of two rotation matrices and the velocity vector in a relative-wind-oriented coordinate system (the wind axes). From (3.25)–(3.27), the V, α, and β form can be formulated:

$$\dot{V} = g(\cos\phi\cos\theta\sin\alpha\cos\beta + \sin\phi\cos\theta\sin\beta - \sin\theta\cos\alpha\cos\beta)$$
$$- \frac{\overline{q}S}{m}C_{D\text{wind}} + \frac{T}{m}\cos(\alpha + \sigma_T)\cos\beta$$

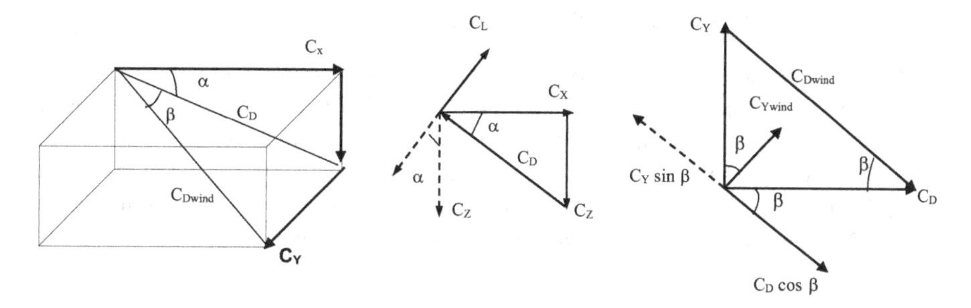

FIGURE 3.12 Wind axis coefficients.

$$\dot{\alpha} = \frac{g}{V\cos\beta}(\cos\phi\cos\theta\cos\alpha + \sin\theta\sin\alpha) + q - \tan\beta(p\cos\alpha + r\sin\alpha)$$

$$-\frac{\bar{q}S}{mV\cos\beta}C_L - \frac{T}{mV\cos\beta}\sin(\alpha + \sigma_T) \tag{3.28}$$

$$\dot{\beta} = \frac{g}{V}(\cos\beta\sin\phi\cos\theta + \sin\beta\cos\alpha\sin\theta - \sin\alpha\cos\phi\cos\theta\sin\beta)$$

$$+ p\sin\alpha - r\cos\alpha + \frac{\bar{q}S}{mV}C_{Y\text{wind}} + \frac{T}{mV}\cos(\alpha + \sigma_T)\sin\beta$$

In (3.28), C_L and C_D are the lift force and drag coefficients in the stability axis. In terms of body-axis coefficients, these are given by Figure 3.12:

$$\begin{aligned} C_L &= -C_Z\cos\alpha + C_X\sin\alpha \\ C_D &= -C_X\cos\alpha - C_Z\sin\alpha \end{aligned} \tag{3.29}$$

Equation (3.29) can further be expressed in terms of C_D and C_Y using the following expressions:

$$\begin{aligned} C_{D\text{wind}} &= C_D\cos\beta - C_Y\sin\beta \\ C_{Y\text{wind}} &= C_Y\cos\beta + C_D\sin\beta \end{aligned} \tag{3.30}$$

Note that there are no small angle approximations used, but equations have singularity at $\theta = 90°$. Also, there are several trigonometric and kinematic nonlinearities in the 6DOF EOM.

Analysis of the full 6DOF equations would require a nonlinear program (algorithm/software). The 6DOF model is useful where nonlinear effects are important and coupling between the longitudinal and lateral modes is strong.

Although it is possible to analyze the 6DOF models, it is always a good idea to work with simplified models (Chapter 5).

3.6 MISSILE DYNAMIC EQUATIONS

Similar to an aircraft, the axis system for missiles for the development of EOM is assumed to be centered at the C.G. and fixed to the body. Figure 3.13 shows the *X*, *Y*, and *Z*, which are also the roll, pitch, and yaw axes, respectively. The *XY* plane represents the yaw plane while the *XZ* plane is the pitch plane; the angle of incidence in the *XZ* plane is α, the AOSS in the *XY* plane is β, and the α_f is the flank angle (Figure 3.13); with a few simple algebraic calculations, we get:

$$\tan\beta = \tan\alpha_f \cos\alpha \tag{3.31}$$

Compared to *v* and *w*, the component *u* is much larger and varies little during the flight. Like the aircraft, the missile too has DOF motion which is represented by three force and three moment equations [8]. The standard set of equations normally used to compute missile dynamics are similar to those in eqns. (3.11) and (3.15):

$$\begin{aligned} F_x &= m(\dot{u} + qw - rv) \\ F_y &= m(\dot{v} + ru) \\ F_z &= m(\dot{w} - qu) \end{aligned} \tag{3.32}$$

$$\begin{aligned} L &= \dot{p}I_x - \dot{q}I_{xy} - \dot{r}I_{xz} + qr(I_z - I_y) + (r^2 - q^2)I_{yz} - pqI_{xz} + rpI_{xy} \\ M &= -\dot{p}I_{xy} + \dot{q}I_y - \dot{r}I_{yz} + rp(I_x - I_z) + (p^2 - r^2)I_{xz} - qrI_{xy} + pqI_{yz} \\ N &= -\dot{p}I_{xz} - \dot{q}I_{yz} + \dot{r}I_z + pq(I_y - I_x) + (q^2 - p^2)I_{xy} - rpI_{yz} + qrI_{xz} \end{aligned} \tag{3.33}$$

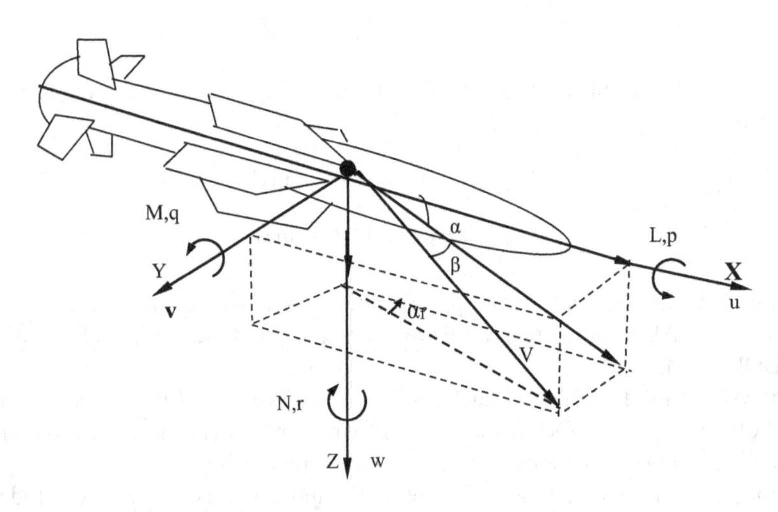

FIGURE 3.13 Axis system for missile.

The *Fx* equation includes the $\dot{u}$ that represents the change in the forward speed. To compute this, we need to know the drag force and the thrust force acting on the missile. Since, w and v are generally small, the contribution of the terms pw and pv in equations for F_y and F_z will be small if and only if the roll rate is small. From another perspective, the rolling motion will cause forces in pitch and yaw plane, a feature that is highly undesirable and should be dealt with at the design stage. The undesirable effects of high roll rate may also be reduced using acceleration feedback from autopilot during the flight. If the roll rate p is assumed to be small and the flight variables q, r, v, and w anyway not being large, the EOM for a missile can be simplified by neglecting the terms involving the products of these variables, e.g. terms like pv, pw, pq, and pr. Also, assuming the missile to be reasonably symmetrical about the XZ and XY planes, the simplified set of equations can be written as:

$$F_x = m(\dot{u} + qw - rv);\ F_y = m(\dot{v} + ru);\ F_z = m(\dot{w} - qu) \tag{3.34}$$

$$L = \dot{p}I_x;\ M = \dot{q}I_y;\ N = \dot{r}I_z \tag{3.35}$$

The rolling moment, eqn. (3.35), shows no coupling between the roll and pitch motions, and the roll and yaw motions. The simplified MMs for a missile are discussed in Chapter 5.

3.7 ROTORCRAFT DYNAMICS

Unlike the fixed-wing aircraft, the lift, propulsion, and control in helicopters are provided by the main rotor, Figure 3.14. The helicopters have the ability to take off vertically, hover, and land even on unprepared terrain. In a helicopter, the upward lift is obtained by keeping the rotor shaft vertical. The main rotor system of the helicopter should be regarded as lift generating mechanism like the wings of an aircraft; the

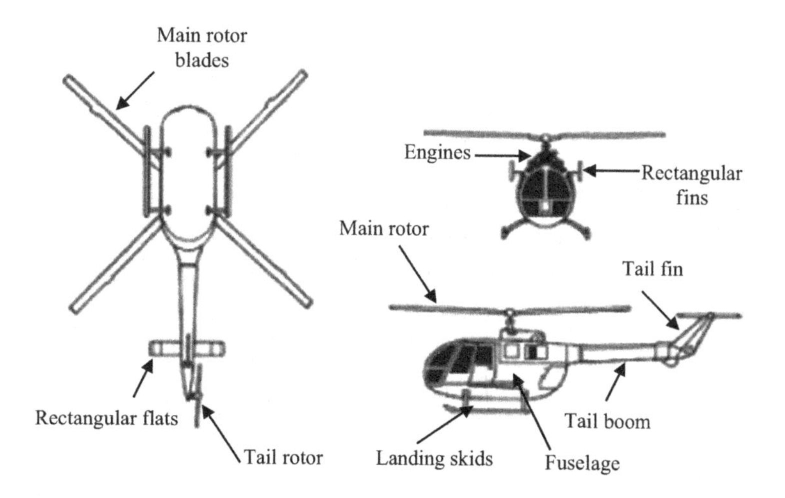

FIGURE 3.14 A 3-view drawing of BO 105.

rotors'-airfoils produce the lift. The rotor system could be: semi-rigid, rigid or fully articulated. The engine provides the power to main rotor and the tail rotor. In a fully articulated rotor, each blade can rotate about the pitch axis that is feathering. This rotation changes the AOA and hence the lift. The blade can move back and forth that is lead/lag movement. The blade can also flap up and down. Generally such system is used for helicopters with more than two blades. A semi-rigid rotor has two blades, and when one blade flaps up the other blade will flap down; there is no lead-lag in this system. The rigid rotor does not provide lead-lag or flapping; some of this aspect is achieved by the blending of the blades.

Aerodynamics of a rotorcraft can be explained by: the momentum theory or the blade element theory; the former provides a basic understanding of the flow through the main rotor and the power required to produce thrust to lift the helicopter, it alone cannot be used to design the rotor blades; and the blade element theory looks into the details of how the thrust is produced by the rotating blades and relates the rotor performance to design parameters.

3.7.1 Momentum Theory

It follows the basic laws of conservation of mass, momentum and energy [9,10]. The action of the air on the rotor blades produces a reaction of the rotor on the air that manifests in the form of thrust at the rotor disk. It is assumed that the rotor disk is composed on infinite number of blades, i.e., the rotor is considered to be an "actuator" disk. It is also assumed that the air is incompressible and frictionless and no rotational energy is imparted to the flow outside the slip stream. Velocity in the slip stream increases from zero at upstream infinity to v at the rotor disk, to w at downstream infinity, Figure 3.15. For the air with density ρ and A being the disc area, the law of conservation of mass gives, from the definition of density:

$$\dot{m} = \rho A v \tag{3.36}$$

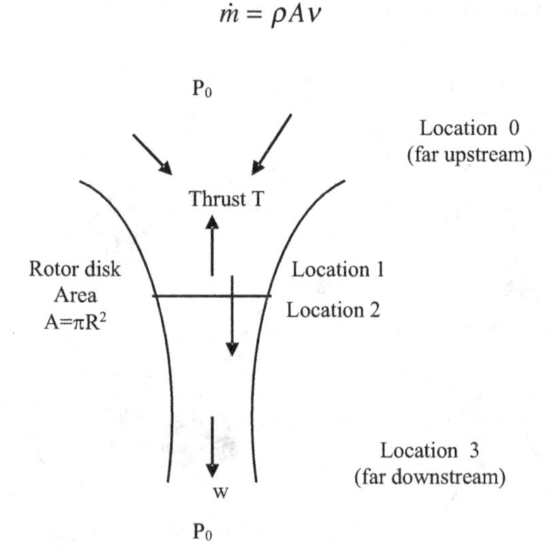

FIGURE 3.15 Flow across the actuator disk.

If T is the thrust at the rotor disk, it will be equal to the rate of change of momentum:

$$T = \dot{m}(w - 0) = \rho A v w \tag{3.37}$$

From the law of conservation of energy, we have:

$$Tv = \frac{1}{2}\dot{m}w^2 \tag{3.38}$$

Substituting for T from the momentum conservation equation:

$$\rho A v w v = \frac{1}{2}\rho A v w^2;\ \text{or } v = \frac{1}{2}w \tag{3.39}$$

Thus, the velocity at downstream infinity is twice the velocity at the rotor disc. From (3.37), and (3.39), the relation between v and thrust T becomes

$$T = \rho A v 2v;\ \text{or } v = \sqrt{\frac{T}{2\rho A}} \tag{3.40}$$

The term 'T/A' is known as disc loading. The induced power (thrust force times distance/time) for the rotor can be written as:

$$P = Tv = T\sqrt{\frac{T}{2\rho A}} = \frac{T^{3/2}}{\sqrt{2\rho A}} \tag{3.41}$$

In non-dimensional form of thrust and power coefficients is given by

$$C_T = \frac{T}{\rho A(\Omega R)^2} \text{ and } C_P = \frac{P}{\rho A(\Omega R)^3} \tag{3.42}$$

Here, Ω is the angular velocity and R is the radius of the rotor. The inflow ratio λ is obtained by using (3.40) and (3.42):

$$\lambda = \frac{v}{\Omega R} = \sqrt{\frac{C_T}{2}} \tag{3.43}$$

The hovering efficiency of the rotor is defined by figure of merit FOM defined as $\text{FOM} = \frac{\text{minumum power required to hover}}{\text{actual power required to hover}}$; the ideal value of which is equal to 1. However, for most rotors, the value lies between 0.75 and 0.8.

3.7.2 Blade-Element Theory

The momentum theory provides estimate of induced power requirements for the rotor, but is not sufficient for designing the rotor blades. The origin of blade element theory can be traced back to the earlier work on marine propellers by Froude.

Subsequently, Drzewiecki took up this study assuming the blade sections to act independently and considering two velocity components ΩR due to rotation and V due to the axial velocity of the rotor. His results indicated correct behavior but were quantitatively in error which was primarily due to the neglecting of the induced velocity at the rotor disc [10]. Several attempts were made later on to account for the induced velocity from momentum theory into the blade element theory. It was only after Prandtl developed the lifting line theory that the influence of the wake velocity at the rotor disc was incorporated. Thus, it was through the wake vortex theory rather than the momentum theory that the induced velocity at the blade section was finally incorporated into the blade element theory. The blade sees the air coming to it due to rotor rotation, as well as due to downward induced velocity. Today, the blade element theory is the foundation for all analysis of helicopter dynamics and aerodynamics. Reference [8] describes in details the blade element theory for a rotor in vertical flight.

3.7.3 Rotorcraft Modeling Formulations

Models of different order may be used depending on what is known about the measurements available, the frequency range of interest and the degree of coupling between the longitudinal and lateral directional motions. Generally, a 6 DOF model is the minimum required for a highly coupled rotorcraft. The six degree of freedom rigid body model for the rotorcraft consists of the usual equations of motion derived from Newton's Law.

Equations for linear accelerations

$$\begin{aligned}
\dot{u} &= \tilde{X} - g\sin\theta \qquad - qw + rv \\
\dot{v} &= \tilde{Y} + g\cos\theta\sin\phi \ - ru + pw \\
\dot{w} &= \tilde{Z} + g\cos\theta\cos\phi - pv + qu
\end{aligned} \tag{3.44}$$

Equations for angular accelerations

$$\begin{aligned}
\dot{p} &= \tilde{L} - \frac{I_{zx}}{I_x I_z - I_{zx}^2}\left(I_y - I_x - I_z\right)pq - \frac{1}{I_x I_z - I_{zx}^2}\left(I_{zx}^2 + I_z^2 - I_y I_z\right)qr \\
\dot{q} &= \tilde{M} - \frac{\left(I_x - I_z\right)}{I_y} rp - \frac{I_{zx}}{I_y}\left(p^2 - r^2\right) \\
\dot{r} &= \tilde{N} - \frac{I_{zx}}{I_x I_z - I_{zx}^2}\left(I_z - I_y + I_x\right)qr - \frac{1}{I_x I_z - I_{zx}^2}\left(I_x I_y - I_{zx}^2 - I_x^2\right)pq
\end{aligned} \tag{3.45}$$

Kinematic equations for Euler rates

$$\dot{\phi} = p + q\sin\phi\tan\theta + r\cos\phi\tan\theta; \ \dot{\theta} = q\cos\phi - r\sin\phi; \ \dot{\psi} = q\frac{\sin\phi}{\cos\theta} + r\frac{\cos\phi}{\cos\theta} \tag{3.46}$$

The equation relating altitude to linear velocities is given by

$$\dot{h} = u\sin\theta - v\cos\theta\sin\phi - w\cos\theta\cos\phi \tag{3.47}$$

In the above equations, the gyroscopic reactions due to rotating elements are neglected and the cross product of inertia I_{xy} and I_{yz} are assumed zero. The presence of gravitational and rotational related terms in eqn. (3.44), and the product of angular rates in eqn. (3.45), gives these equations a nonlinear structure. The terms $\tilde{X}, \tilde{Y}$ and $\tilde{Z}$ in eqn. (3.44) denote the specific forces along the longitudinal, lateral and normal body axis, respectively; the terms $\tilde{L}, \tilde{M}$ and $\tilde{N}$ in eqn. (3.45) represent the specific moments about the longitudinal, lateral and vertical body axis, respectively; these terms basically represent linear and angular acceleration quantities, and are related to resultant aerodynamic force and moment coefficients as follows:

Specific Forces

$$\tilde{X} = X/m; \ \tilde{Y} = Y/m; \ \tilde{Z} = Z/m \tag{3.48}$$

Specific Moments

$$\tilde{L} = \frac{I_z L + I_{zx} N}{I_x I_z - I_{zx}^2}; \ \tilde{M} = \frac{M}{I_y}; \ \tilde{N} = \frac{I_{zx} L + I_x N}{I_x I_z - I_{zx}^2} \tag{3.49}$$

Equation (3.45) can be further simplified by assuming the product of angular rates to be small and neglecting the corresponding terms in the moment equations; so we get:

$$\dot{p} = \tilde{L}; \ \dot{q} = \tilde{M}; \ \dot{r} = \tilde{N} \tag{3.50}$$

The rigid body states described in eqns. (3.44), (3.46), (3.47) and (3.50) are used to characterize the rotorcraft dynamics. Since, the forces and moments are presented as specific quantities in the equations, mass and moment of inertia no longer explicitly appear in the MM.

3.7.4 Limitations of Rigid Body Model

The 6 DOF model involving the rigid body states, is generally adequate to predict rotorcraft dynamics in low and mid frequency range. In the conventional rigid body model, the main rotor dynamics are omitted and the rotor influence is absorbed by the rigid body derivatives. A better prediction at higher frequencies, however, necessitates the inclusion of rotor dynamics in estimation model. One approach to include rotor effects in a 6 DOF model is by introducing equivalent time delays in control inputs [11,12]. These delays can be determined by correlating model response with flight data. This is, however, an inappropriate alternative to modeling the rotor dynamics which are highly complex in nature. Model inversion is required for the feed forward controller in the design of model following control systems and the time delays become time lead on inversion, which means that the future values of the state

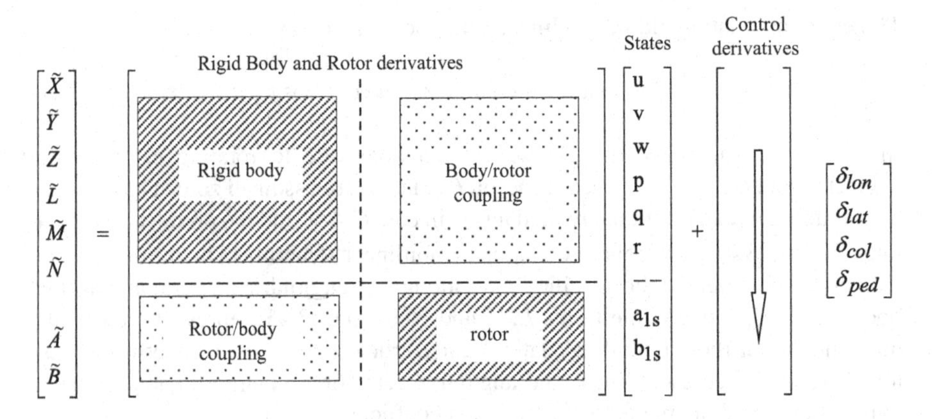

FIGURE 3.16 Extended model with rotor-body coupling for helicopter.

variables are needed in advance. This is unrealistic for an on-line real time process like in-flight simulation. It clearly underlines the need to develop extended models with an explicit representation of rotor dynamic effects. Therefore, the 6 DOF rigid body model is generally extended with additional degrees of freedom representing the longitudinal and lateral rotor flapping.

The basic approach to extend the 6 DOF rigid body model is shown in Figure 3.16. The model structure now includes two additional degrees of freedom representing the longitudinal and lateral flapping [10,11]. The state-space matrix in Figure 3.16 defines the sub matrices pertaining to rigid body, rotor and the cross-coupling matrices for body-to-rotor and rotor-to-body. In a simplified rotor model, the longitudinal and lateral flapping angles can be expressed primarily as a function of body angular rates and control inputs. In Ref. [12], Kaletka and Grünhagen formulated an 8 DOF extended model by redefining the roll and pitch accelerations for rotor/body motion and using them as state variables. Tischler provided a hybrid body/flapping model which coupled the simplified fuselage equations at low frequencies to the simplified rotor equations at high frequencies with equivalent spring terms [13]. The following first order coupled differential equations for longitudinal flapping a_{1s} and lateral flapping b_{1s} can be appended to 6DOF model:

$$\begin{aligned} \dot{a}_{1s} &= \tilde{A} \\ \dot{b}_{1s} &= \tilde{B} \end{aligned} \tag{3.51}$$

Here, $\tilde{A}$ and $\tilde{B}$ are the longitudinal and lateral flapping specific moments.

APPENDIX 3A AIRCRAFT GEOMETRY AND COORDINATE SYSTEMS

The universally accepted aircraft nomenclature representing the translational and rotational motions, the flow angles and the aerodynamic forces and moments acting on the aircraft are mentioned [1,7]. The forward edge of the wing is called the leading edge and the rear edge is called the trailing edge.

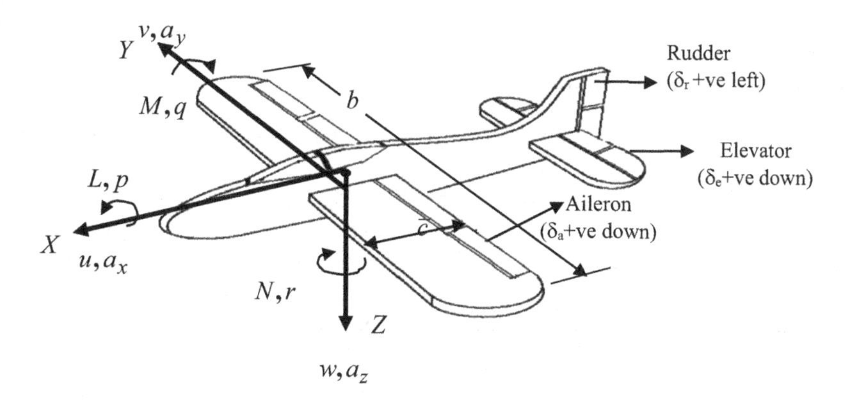

FIGURE 3A.1 Aircraft variables in body- axis/Control surfaces.

3A.1 Aircraft Axis and Notations

The line joining the leading edge to trailing edge is called the chord. If b is the wing span and S represents the area of the wing, then the aspect ratio of the wing is given by $AR = b^2/S$. A reference point of considerable importance is the aerodynamic center. This is the point about which there is no change in the pitching moment as the AOA is increased. Figure 3A.1 shows the aircraft primary control surfaces and the normally accepted sign conventions. All surface positions are angular in deflection. The elevator deflection causes the aircraft to pitch about the Y-axis, aileron deflection causes the aircraft to roll about X-axis, and the rudder deflection causes it to yaw about the Z-axis. A canard is another control surface (not shown in Figure 3A.1) found in fighters, and it is ahead of the wing and is positioned to produce a positive lift. However, it interferes with the smooth flow over the wing and is therefore not found in every fighter aircraft. Other control surfaces are the strakes and trailing edge flaps. Strakes are small fins placed in the horizontal plane on the aircraft forebody projecting into the airflow. This is a passive device, which increases the vortex breakdown AOA. Flaps found on the trailing edge of the wing are called trailing edge flaps. The downward deflection of these flaps causes a significant increase in lift coefficient and lift to drag ratio, i.e., the load factor. The most effective range of downward deflection is from 0 to 20 deg. The notations for the aircraft motion variables body-axis system are put together in Table 3A.1 for better understanding.

3A.2 Axes-Coordinates Transformations and Quaternions

The Euler angles ψ, θ, ϕ are the basic angles of the direction cosine matrix T required to transform a vector from earth to body axes. The relation between the Euler angles and angular rates is given by [8]:

$$\begin{bmatrix} \dot{\psi} \\ \dot{\theta} \\ \dot{\phi} \end{bmatrix} = \frac{1}{\cos\theta} \begin{bmatrix} 0 & \sin\phi & \cos\phi \\ 0 & \cos\theta\cos\phi & -\cos\theta\sin\phi \\ \cos\theta & \sin\theta\sin\phi & \sin\theta\cos\phi \end{bmatrix} \begin{bmatrix} p \\ q \\ r \end{bmatrix} \tag{3A.1}$$

TABLE 3A.1
Aircraft Axis and Notations

	Longitudinal X-axis	Lateral Y-axis	Vertical Z-axis
Velocity components	u	v	w
Angular rates	Roll rate p	Pitch rate q	Yaw rate r
Euler angles	Roll angle ϕ	Pitch angle θ	Heading angle ψ
Accelerations	Longitudinal acceleration a_x	Lateral acceleration a_y	Vertical acceleration a_z
Aerodynamic forces	X	Y	Z
Aerodynamic moments	Rolling moment L	Pitching Moment M	Yawing Moment N
Control deflections	Elevator δ_e	Aileron δ_a	Rudder δ_r
Moment of Inertia	I_{xx}	I_{yy}	I_{zz}

The Euler angles ψ, θ, ϕ can be obtained by integrating the above equations. The matrix T of direction cosines to transform from earth (e) to body (b) axis is given by

$$T = \begin{bmatrix} T_{11} & T_{12} & T_{13} \\ T_{21} & T_{22} & T_{23} \\ T_{31} & T_{32} & T_{33} \end{bmatrix}; \; T_{be} = T \tag{3A.2}$$

Here:

$$\begin{aligned} &T_{11} = \cos\theta\cos\psi && T_{12} = \cos\theta\sin\psi && T_{13} = -\sin\theta \\ &T_{21} = \sin\theta\sin\phi\cos\psi - \cos\phi\sin\psi && T_{22} = \sin\theta\sin\phi\sin\psi + \cos\phi\cos\psi && T_{23} = \cos\theta\sin\phi \\ &T_{31} = \sin\theta\cos\phi\cos\psi + \sin\phi\sin\psi && T_{32} = \sin\theta\cos\phi\sin\psi - \sin\phi\cos\psi && T_{33} = \cos\theta\cos\phi \end{aligned} \tag{3A.3}$$

The singularity encountered as pitch angle θ approaches 90 deg. can be avoided with the use of quaternions. The four components of quaternions are expressed in terms of Euler angles and to the body axis angular rates as follows:

$$\begin{aligned} a_1 &= \cos(\psi/2)\cos(\theta/2)\cos(\phi/2) + \sin(\psi/2)\sin(\theta/2)\sin(\phi/2) \\ a_2 &= \cos(\psi/2)\cos(\theta/2)\sin(\phi/2) - \sin(\psi/2)\sin(\theta/2)\cos(\phi/2) \\ a_3 &= \sin(\psi/2)\cos(\theta/2)\sin(\phi/2) + \cos(\psi/2)\sin(\theta/2)\cos(\phi/2) \\ a_4 &= \sin(\psi/2)\cos(\theta/2)\cos(\phi/2) - \cos(\psi/2)\sin(\theta/2)\sin(\phi/2) \end{aligned} \tag{3A.4}$$

$$\begin{bmatrix} \dot{a}_1 \\ \dot{a}_2 \\ \dot{a}_3 \\ \dot{a}_4 \end{bmatrix} = \begin{bmatrix} 0 & p & q & r \\ -p & 0 & -r & q \\ -q & r & 0 & -p \\ -r & -q & p & 0 \end{bmatrix} \begin{bmatrix} a_1 \\ a_2 \\ a_3 \\ a_4 \end{bmatrix} \tag{3A.5}$$

The four components a_1, a_2, a_3, a_4 satisfy the constraint:

$$a_1^2 + a_2^2 + a_3^2 + a_4^2 = 1 \tag{3A.6}$$

The elements of the transformation matrix T, in terms of the four components of quaternions are:

$$\begin{aligned}
T_{11} &= 2(a_1^2 + a_2^2) - 1 & T_{12} &= 2(a_2 a_3 + a_1 a_4) & T_{13} &= 2(a_2 a_4 - a_1 a_3) \\
T_{21} &= 2(a_2 a_3 - a_1 a_4) & T_{22} &= 2(a_1^2 + a_3^2) - 1 & T_{23} &= 2(a_3 a_4 + a_1 a_2) \\
T_{31} &= 2(a_1 a_3 + a_2 a_4) & T_{32} &= -2(a_1 a_2 + a_3 a_4) & T_{33} &= 2(a_1^2 + a_4^2) - 1
\end{aligned} \tag{3A.7}$$

The Euler angles can be obtained from the quaternions as follows:

$$\psi = \tan^{-1}\left[\frac{T_{12}}{T_{11}}\right]; \ \theta = \tan^{-1}\left[-\frac{T_{13}}{\sqrt{T_{11}^2 + T_{12}^2}}\right]; \ \phi = \tan^{-1}\left[\frac{T_{23}}{T_{33}}\right] \tag{3A.8}$$

The sequence of rotation is important for accurate axes transformation. The first rotation is through angle ψ about the Z axis, then through θ about the Y axis and finally through angle ϕ about the X-axis.

3A.3 Transformation from Body to Earth Axis

To transform from body axis to earth axis, the transformation matrix T_{eb} is given by

$$T_{eb} = \begin{bmatrix} T_{11} & T_{12} & T_{13} \\ T_{21} & T_{22} & T_{23} \\ T_{31} & T_{32} & T_{33} \end{bmatrix}; \ T_{eb} = T'_{be} \tag{3A.9}$$

Since the matrices are orthogonal, the inverse is done by the transpose of the matrix:

$$\begin{aligned}
T_{11} &= \cos\theta\cos\psi & T_{12} &= \sin\theta\sin\phi\cos\psi - \cos\phi\sin\psi & T_{13} &= \sin\theta\cos\phi\cos\psi + \sin\phi\sin\psi \\
T_{21} &= \cos\theta\sin\psi & T_{22} &= \sin\theta\sin\phi\sin\psi + \cos\phi\cos\psi & T_{23} &= \sin\theta\cos\phi\sin\psi - \sin\phi\cos\psi \\
T_{31} &= -\sin\theta & T_{31} &= \cos\theta\sin\phi & T_{33} &= \cos\theta\cos\phi
\end{aligned} \tag{3A.10}$$

3A.4 Transformation from Stability Axis to Body Axis

The origin of the body (b) and stability axes (s) is the same and the only rotation is the inclination α of that of the XZ plane of the stability axis with the body axes. By putting $\psi = \phi = 0$ and $\theta = \alpha$ in the elements of matrix T in eqn. (3A. 10), one obtains the matrix T_{sb}:

$$T_{sb} = \begin{bmatrix} \cos\alpha & 0 & \sin\alpha \\ 0 & 1 & 0 \\ -\sin\alpha & 0 & \cos\alpha \end{bmatrix} \tag{3A.11}$$

The transpose of this matrix can be used to transform from body axis to stability axis.

3A.5 Transformation from Stability Axis to Wind Axis (w)

This transformation is given by the following matrix:

$$T_{ws} = \begin{bmatrix} \cos\beta & \sin\beta & 0 \\ -\sin\beta & \cos\beta & 0 \\ 0 & 0 & 1 \end{bmatrix} \tag{3A.12}$$

3A.6 Transformation from Body Axis to Wind Axis

The origin of wind and body axis is same and the two axes system are related through the flow angles α and β. Putting $\psi = \beta, \theta = \alpha$, and $\phi = 0$ in the elements of matrix T in eqn. (3A.10), one obtains the matrix T_{wb} to transform from body axis to wind axis.

$$T_{wb} = \begin{bmatrix} \cos\alpha\cos\beta & \sin\beta & \sin\alpha\cos\beta \\ -\cos\alpha\sin\beta & \cos\beta & -\sin\alpha\sin\beta \\ -\sin\alpha & 0 & \cos\alpha \end{bmatrix} \tag{3A.13}$$

Thus, using the transformation matrix T defined in eqn. (3A.10) and knowing angles ψ, θ, ϕ, one can obtain a transformation from one axis to another.

APPENDIX 3B HELICOPTER AERODYNAMICS

For a hovering rotorcraft, the action is the production of an upward rotor thrust and the reaction is in the downward velocity imparted to the air in the rotor wake. The strength of the airflow in the wake is measured by its dynamic pressure that turns out to be the disc loading; the rotor wake also produces downward-or-vertical drag force. In a hovering helicopter, the profile drag accounts for about one-quarter of the total power, and the remaining is induced power. The blade twist (6–12 deg.) is used to even out the induced flow across the rotor disc. Various important aspects of helicopter aerodynamics and control are capture here [14]:

i. The rotor is continuously pumping air into a big bubble under the rotor; this fills up and bursts every second or two, and causes large-scale disturbances in the surrounding flow field; this creates a variation in the rotor thrust, and the rotor flaps erratically in pitch and roll, this requires prompt action from the pilot.

ii. In the vortex-ring state, the power required is more, and also the collective pitch, apparently due to local blade stall during flow fluctuations.

iii. After the helicopter is descending fast enough to pass through the worst of the unsteadiness in the vortex-ring state, it will go into vertical autorotation, which is a stable condition and for one value of collective pitch the helicopter will settle on one rotor speed and one rate of descent; using the collective pitch, the pilot can control the rotor rpm; the safe range of which is 75%–110% of the normal power-on speed.

iv. The amount of flapping required to balance a given external moment depends upon the stiffness of the rotor. A teetering rotor whose only source is stiffness, is the tilt of the thrust vector that is 'soft', whereas a hingeless rotor is 'stiff'.

v. Once, the pilot finds the right trim condition, her/his ability to hold onto it, is helped if the vehicle has an effective horizontal stabilizer.

vi. The engine has to put out enough power to provide for the induced, profile, and the parasite components of the drag (and hence the proportional power) and 'miscellaneous' power components (for tail rotor, gearbox losses, hydraulic pumps, and electrical generators).

vii. The asymmetry of the velocity distribution (the variation in the local velocity that a blade sees as it goes round and round) is dominant factor in the forward flight of helicopters and this signifies most of the differences between helicopters and aircrafts.

viii. Any change in flight conditions that causes an imbalance in the lift force will result in the rotor flapping to a new equilibrium position.

ix. The use of the cyclic pitch makes it possible to eliminate the flapping hinges in a hingleless rotor system.

x. Eliminating the lead-lag hinges by using sufficient structural material will convert a fully articulated rotor to a hingeless one.

xi. The blade flapping contributes to the stability and control of a helicopter. If flapping occurs due to a change in the flight condition, with controls fixed, it is a stability condition; and if it occurs due to pilot's action, then it is control characteristics.

xii. A significant destabilizing effect is generated by an aerodynamic interference between the wake of the main rotor and the horizontal stabilizer; as the speed increases the downwash behind the main rotor decreases; this causes the AOA of the stabilizer to increase, and hence the lift, this produces the destabilizing effect.

xiii. The tail rotor is an effective device for the directional stability of a helicopter.

xiv. In most helicopters, the rotor is producing the positive dihedral, while the rest of the airframe has a negative effect.

xv. Like an aircraft, a helicopter also experiences the two modes of the longitudinal dynamic stability: (a) short period, and (b) Phugoid (porpoising). It also has DR and spiral modes.

APPENDIX 3C TYPES OF HELICOPTERS AND CONTROLS

A conventional helicopter has a large main rotor that rotates in a normally horizontal plane and a smaller tail rotor that rotates in a nominally vertical plane. There are three types of the main rotors: (i) rigid, (ii) semi-rigid (teetering), and (iii) fully articulated. The tail rotors are: (i) teetering, (ii) fully articulated, (iii) rigid, or (iv) fan-in-fan [15]. The tandem helicopters have two large main rotors, one in the front, and one in the rear; for which the synchronization between two rotors is important, and the rotor system is of the fully articulated type. In the coaxial type, the two main rotors are mounted on the same mast, but they rotate in the opposite directions. In a Synchropters type, there are two rotor heads at a slight angle to each other driven by a common transmission. The rotor heads consist of teetering systems; the blades intermesh, but interestingly avoid collision.

3C.1 Rotor Systems

The information is applicable to both the main and the tail rotor systems, the tail system does not have cyclic control, but its operation is similar to collective control of the main rotor, even though it provides the yaw reaction to the main rotor torque on the airframe; its operation can be linked to that of a variable pitch propeller.

3C.1.1 Fully Articulated Rotor

It allows each blade to feather, i.e. to rotate about the pitch axis to change lift, lead and lag, and flap, independent of the other blades; such systems are with more than two blades. The magnitude of the lift force is based on the collective input that changes pitch on all the blades in the same direction at the same time. The feathering angle of each blade (proportional to its own lifting force) changes as it rotates with the rotor, hence the 'cyclic control'. Also, the centrifugal force of the weight of the blade and flapping force, would cause the 'coning' of the blades. As the blade flaps, its C. G. changes, and also the local MOI (moment of inertia), and its speeding or slowing down is accommodated by the lead-lag hinge. Thus, as load increases from increased feathering, it will flap up and lead forward; and while it continues around, it will flap down and lag behind. Because the rotor is a large rotating mass, it will behave like a gyroscope; and the applied control input will be realized at 90 deg. behind the control input; however, the mechanism is so provided such that the forward input will result in the forward motion of the helicopter.

3C.1.2 Semi-Rigid/Teetering Rotor and Rigid Rotor

These rotors are found on the aircraft with two rotor blades; one blade flaps up and the other down; however, there is no lead-lag between them. The rotor will not 'cone', but tilt up with more lift and tilt down on the other; the flapping is self-balancing; and the issues of gyroscopic precession, and flap coupling, though present, can be easily handled.

The rigid rotor does not provide flapping or lead-lag hinges; the blade roots are rigidly fixed to the hub; and the motions are accommodated by bending; and high vibrations would results, since the secondary motions are not allowed to take place. Such systems are easier to design and operate.

Other systems are the swashplate/star assembly to provide for the transmission of control from the pilot's position to the rotating system. Also, attached to the swashplates are 'scissors' mechanisms; the upper ones force the swashplates to turn with the rotor head, and can hinge to accommodate control motions. The blades nowadays are fabricated almost completely of composites; and their aerodynamics are symmetrical airfoils, which offer a reasonable constant center of lift for optimization of control loads. Some blades employ twisting along their length to optimize lifting efficiency.

3C.2 Helicopter Controls

The helicopters are inherently somewhat unstable and hence small movement of the control inputs is required. All the helicopters have the same basic controls: (i) a collective stick to enable power changes and vertical motion, (ii) a cyclic stick to enable turns and forward speed, and (iii) anti-torque (known as rudder or tail rotor) pedals for anti-torque to the main rotor, for turn-coordination, and for turns about a vertical axis while in hover or performing a ground reference maneuver. The advanced helicopters would have hydraulic controls, automatic flight control system for stabilization, fly-by-wire (FBW) controls, and collective-yaw coupling.

Pedals/anti-torque control: In helicopter these are like ones in an aircraft, and they can be operated to change the pitch in the tail rotor, for anti-torque control, and turns about the vertical axis while in hover. The pedals are used to trim the aircraft. The right pedal is for yawing to the right and the left is for yawing to left. The turn coordination in a helicopter is almost automatic, requiring minimal adjustments of the pedals. If collective pitch is increased, the engine and the rotor torque will also increase, hence some pedal input will be required.

Collective and throttle: The collective lever collectively increases the pitch in the main rotor. One can raise the collective in the hover condition to gain the altitude. The throttle also rotates backwards. The collective control thus enables the power change and vertical motion.

Cyclic: This is used for changing the position in hover or forward flight. If the rotor disc is tilted forward with the cyclic control, the forward motion is the outcome. If the cyclic stick is pulled backwards, the rearwards motion is generated. Thus, the cyclic stick enables turn maneuvers and forward movements. It also provides for change in roll and pitch attitude.

To achieve forward or backward motion, the shaft is tiled in a desirable direction. To maneuver, the rotor plane is tilted relative to the horizontal plane to generate the necessary moments. During descent, the flow on the main rotor is reversed and lift is produced even in engine-off condition. This phenomenon of autorotation can be used to make landings with failed engine.

To fly a helicopter, four types of control are used. The vertical control is achieved by increasing or decreasing the pitch angle of the rotor blades. This increases or reduces the thrust, thus changing the altitude of the helicopter. In a single main rotor helicopter, by changing the thrust at the tail rotor, a yawing moment is created about the vertical axis that is used to provide the directional control. The longitudinal and lateral controls are applied by tilting the main rotor thrust vector.

The mixer: This is a collection of mechanical linkages that take lateral, longitudinal, and collective inputs from the cockpit and transform to the three inputs to the swashplate and the rotor system.

EPILOGUE

An extensive treatment of EOMs is found in one of the earliest books [1]. The extensive helicopter research work on aeromechanics is reported in Ref. [16]. It takes as such a system's approach and deals with three major aspects of the helicopter research which is also equally applicable and suitable to fixed wing aircraft research and development: (i) reality and conceptual model linking thru' wind tunnel simulation–this is analysis route, (ii) linking of conceptual model and computerized model via model verification–this is the programming route, and (iii) linking of computerized model and back to the reality via system identification/model validation procedure- this is the computer simulation route.

EXERCISES

3.1 If the total momentum is given as $H = \sum r\delta m \times V + \sum [r \times (\omega \times r)]\delta m$, where r is the distance from the center, m is mass of the body, and x defines the vector cross product. However, the first term would be zero. Explain why?

3.2 If $\omega = pi + qj + rk$ and $r = xi + yj + zk$, then obtain the cross product $(\omega \times r)$.

3.3 What are merits and demerits of small disturbance theory?

3.4 Is $(r \times \omega) = (\omega \times r)$?

3.5 By rotating the stability axis through positive AOA which axis system is obtained? About which axis is this rotation?

3.6 How many and what consecutive rotations are required to transform a vector from the Earth reference axis system to the body axis system?

3.7 What are the normal limits of the Euler angles?

3.8 Use eqn. (3.19) and see what rate is experienced by the pilot [17] for the following flight condition at a given Mach number and altitude? $\psi = 0\,\text{deg.}, \theta = 0\,\text{deg.}, \phi = 90\,\text{deg.}$ and $\dot{\psi} = 0\,\text{deg./s}, \dot{\theta} = 10\,\text{deg./s}, \dot{\phi} = 0\,\text{deg./s}$.

3.9 Use the equation $F = \text{mass} \times \text{acceleration}$ and mass flow concept for the momentum theory of the helicopter/rotor and compare sizes of the rotor (disc area) and the induced velocities. The mass is the mass flow of the air being pumped through the rotor every second.

3.10 Can you guess what would be (helicopter) disc loading, and its relation with the dynamic pressure? Explain by formulae.

3.11 There would be aircraft with variable wing area. Define wing loading so that the comparison between such aircraft would be easy.

3.12 Obtain the formula for the landing speed of an aircraft.

3.13 What is the significance of this landing speed?

3.14 What is the relation between the coefficients of normal force and Z force? Under what condition $C_L = C_N$?

3.15 What is the relation between the coefficients of axial force and X force?

3.16 What is the result of the cross product of the angular (rate) and linear velocity: $kR \otimes (iU + jV)$

3.17 The plane *X-Z* is regarded as a plane of symmetry. What is its significance from the moment of inertias in this plane wrt to *y*-axis?
3.18 From Figure 3.13, show that $\tan\beta = \tan\alpha_f \cos\alpha$.
3.19 What are the effects of the gravity forces on a missile under the following conditions?: (i) Only pitch attitude is zero; (ii) Only bank angle is zero; and (iii) Both the pitch and bank angle are zero.
3.20 Determine an approximate relation for the rate of change of sideslip angle $\dot{\beta}$ under the following conditions: (i) angle-of-attack is zero and sideslip is very small (ii) bank angle is also zero or small.
3.21 In the polar coordinate velocity form of the EOM, if α and β are both negligibly small, obtain an approximate expression for the total velocity.
3.22 If gravity and thrust effects are assumed small, how does the roll rate contribute to sideslip?

REFERENCES

1. Mcruer, D. T., Ashkenas, I., and Graham, D. *Aircraft Dynamics and Automatic Control.* Princeton University Press, NJ, 1973.
2. Nelson, R. C. *Flight Stability and Automatic Control.* WCB/Mc-Graw Hill Book Company, New York, 1998.
3. Klein, V., and Morelli, E. A. *Aircraft System Identification – Theory and Practice.* AIAA Education Series, Reston, VA, 2006.
4. Cook, M. V. *Flight Dynamics Principles.* John Wiley & Sons, Inc., New York, 1997.
5. Maine, R. E., and Iliff, K. W. Application of parameter estimation to aircraft stability and control – The output error approach. NASA RP 1168, 1986.
6. Chetty S., Madhuranath, P. (ed.) Aircraft flight control and simulation. NAL Special Publication 9717, FMCD, NAL, 1998.
7. Stevens, B. L., and Lewis, F. L. *Aircraft Control and Simulation.* John Wiley & Sons, Inc., New York, 2003.
8. Garnell, P. *Guided Weapon Control Systems*, 2nd Edn. Pergamon Press, Oxford, UK, 1980.
9. Seddon, J. *Basic Helicopter Dynamics*, 2nd Edn. AIAA Education Series, Reston, VA, 2001.
10. Venkatesan, C. (ed.). Short course on helicopter technology – Lecture notes. Department of Aerospace Engineering, IIT Kanpur, India, 15–18 October, 1997.
11. Hamel, P. G. (ed.). Rotorcraft system identification. AGARD LS-178, 1991.
12. Kaletka, J., Tischler, M. B., von Grünhagen, W., and Fletcher, J. Time and frequency identification and verification of BO 105 dynamic models. *Journal of the American Helicopter Society*, 36(4), 25–38, 1991.
13. Tischler, M. B., and Cauffman, M. G. Frequency response method for rotorcraft system identification: Flight applications to the BO 105 coupled rotor/fuselage Dynamics. *Journal of the American Helicopter Society*, 37(3), 3–17, 1992.
14. Prouty, R. W. *Practical Helicopter Aerodynamics.* PJS Publications, Peoria, IL, pages 86, 1982.
15. Anon. X-Plane POH-Supplements-Helicopter Tutorials. http://w3.nai.net/~psklenar/manual/supp/heli-01.htm. (-02 to -06), accessed December 2012.
16. Hamel, P., Gmelin, B., Kalekta, J., Pausder, H.-J., and Langer, H.-J. Helicopter aeromechanics research at DFVLR; Recent results and outlook. *Vertica*, 11(1/2), 93–108, 1987.
17. Yechout, T. R., Morris, S. L., Bossert, D. E., and Hallgren, W. F. *Introduction to Aircraft Flight Mechanics-Performace, Static Stability, Dynamic Stability, and Classical Feedback Control.* AIAA Education Series, Reston, VA, 2003.

4 Aerodynamic Derivatives

4.1 INTRODUCTION

Apart from the equations of motion (EOM) discussed in Chapter 3, the quantification and analysis of aerodynamic models is most important for flight mechanics analysis. The forces and moments that act on an aircraft can be distributed and described in terms of the so-called aerodynamic derivatives. The stability and control analysis is an integral part of the aircraft design cycles, design evaluations, and design of control systems. In this analysis, aerodynamic derivatives, often known as stability and control derivatives, form the basic inputs [1–3]. For the conventional stable aircraft configurations, these derivatives directly ensure the adequate levels of stability and controllability. For unstable aircraft configurations, these derivatives form the essential input parameters for the design of control laws, which, in turn, ensure the desired stability and controllability characteristics of the aircraft. The stability derivatives also play a very important role in the selection process of the aircraft configuration, as will be seen in Section 4.6 [4]. This aerodynamic model becomes an integral part of the EOM, which in totality represents the aerodynamics and dynamic behavior of the vehicle. The accuracy of the results of any flight mechanics analysis depends on the degree of completeness and approximation of aerodynamic models used for such analysis. Due to the assumption of linearity often made, the validity of these (the EOM and aerodynamic) models is limited to a small range of operating conditions, which are generally in terms of AOA and Mach number; often the excursion in AOSS is assumed to be very small. In general, aerodynamic derivatives are applicable to small perturbations of the motion about an equilibrium, and this leads to quasi-static aerodynamic derivatives. Large-amplitude motions/maneuvers require sophisticated and more complex aerodynamic modeling. In this chapter, we specifically deal with aerodynamic modeling that basically consists of the aerodynamic coefficients, which, in turn, expand to (or encompass) the aerodynamic derivatives. Certain stability and control aspects (of aircraft) are captured in certain aerodynamic derivatives. Primary knowledge of aerodynamic coefficients, especially of aerodynamic derivatives, can be obtained from some analytical means like DATCOM, CFD, and/or wind tunnel experiments. When flight tests are conducted on an airplane and if it is instrumented properly with sensors, then one can measure dynamic responses (p, q, r, AOA, AOSS, Euler angles, linear accelerations) of the aircraft when certain maneuvers (Chapter 7) are performed. These response data can be processed using parameter estimation methods to obtain these aerodynamic coefficients/derivatives (ACDs), as discussed in Chapter 9. The interplay of aerodynamic coefficients, derivatives, and EOM at the top level is shown in Figure 4.1.

DOI: 10.1201/9781003293514-5

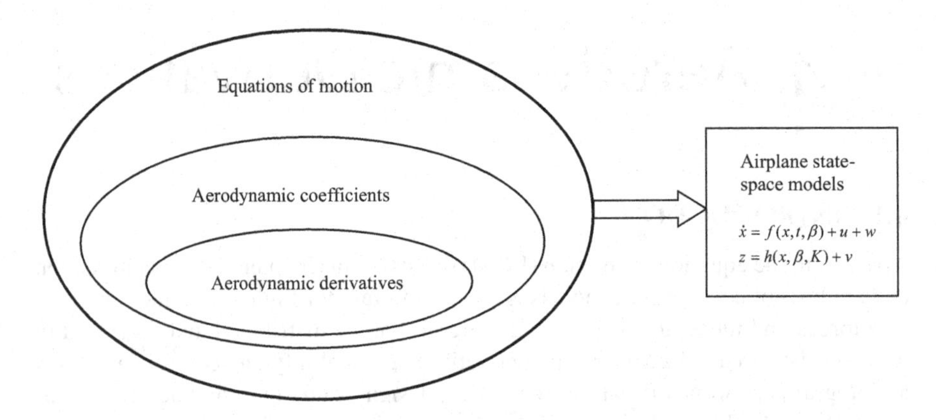

FIGURE 4.1 Relation between various levels of aerodynamic models.

4.2 BASIC AERODYNAMIC FORCES AND MOMENTS

It is apparent from Section 3.2 that if we specify certain geometric parameters of the aircraft like reference surface area of the wings as S, the mean aerodynamic chord $\overline{c}$, the wing span b, and the so-called dynamic pressure $\overline{q}$, then we have the following straightforward aerodynamic force and moment equations in terms of the component forces in x, y, and z directions:

$$\text{Force} = \text{Dynamic Pressure} \times \text{Area} \rightarrow F_x = C_x \overline{q}S; F_y = C_y \overline{q}S; F_z = C_z \overline{q}S \quad (4.1)$$

$$\text{Moment} = \text{Force} \times \text{lever arms}(\text{respective ones}) \rightarrow L$$

$$= C_l \overline{q}S\, b;\ M = C_m \overline{q}S\overline{c};\ N = C_n \overline{q}Sb \quad (4.2)$$

These expressions are in terms of the component moments: rolling moment L about the x-axis, pitching moments M about the y-axis, and the yawing moment N about the z-axis. Here, C_x, C_y, C_z, C_l, C_m and C_n are the non-dimensional body axis force and moment coefficients, also called the aerodynamic coefficients; these are the proportionality constants (that actually do vary with certain variables) in eqns. (4.1) and (4.2). The total velocity V (in the formula for the dynamic pressure) and the total angle can be resolved, as shown in Figure 4.2, in terms of component velocities: u, v, and w and angle of attack (AOA in the vertical x-z plane) and angle of side slip (AOSS in the horizontal x-y plane). From the geometry of the flow velocity, Figure 4.2, the following expressions of the so-called flow angles and component velocities emerge very naturally:

$$\text{AOA} \Rightarrow \alpha = \tan^{-1}\frac{w}{u};\ \text{AOSS} \Rightarrow \beta = \sin^{-1}\frac{v}{V} \quad (4.3)$$

By trigonometric transformation, we get equivalent expressions as follows:

$$u = V\cos\alpha\cos\beta;\ v = V\sin\beta;\ w = V\sin\alpha\cos\beta \quad (4.4)$$

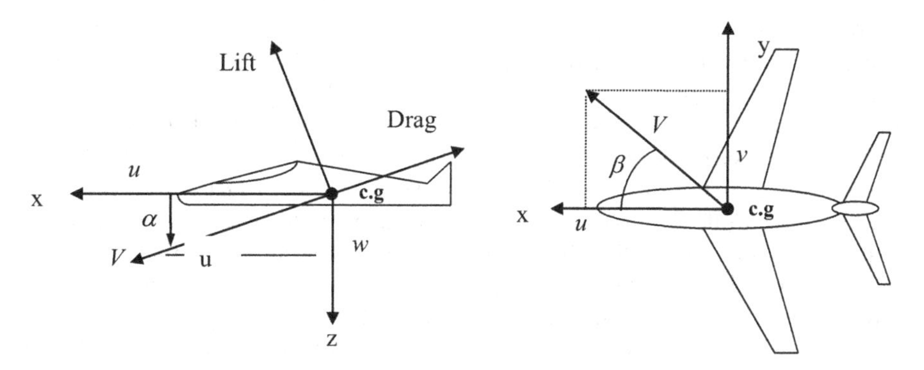

FIGURE 4.2 Velocity components and flow angles.

The total velocity V of the airplane is expressed as follows:

$$V = \sqrt{u^2 + v^2 + w^2} \tag{4.5}$$

From the basic definition of pressure and force, we can some more details of the interplay of flow angles, components of the total velocity, and aerodynamic coefficients. We must emphasize here that talking about the three components of the total velocity in the three directions is equivalent to saying that the airplane experiences motion along these three axes. Thus, we have put the airplane in the real motion in air! What actually keeps the airplane lifted in the air is the so-called lift force. Now, if we know the values of the aerodynamic coefficients, then we can easily determine the forces and moments acting on the aircraft since we know V and the geometrical parameters like area and lever arms. Alternatively, if we know the forces (like accelerations) by some measurements on the aircraft, then we can determine the aerodynamic coefficients. In fact, the latter is true of the wind tunnel experiments conducted on a scaled model of the actual aircraft. In a WT, the model is mounted and compressed air is released, thereby putting the model in the flow field. The experiments are conducted at various AOA and AOSS settings, and forces and moments are measured/calculated from these measurements. Since the flow velocity and the dynamic pressure are known, simple computations would lead to the determination of the aerodynamic coefficients. One can see that these coefficients can be obtained as a function of flow angles, Mach number, etc. Similar analysis can be carried out using CFD and other analytical methods. In order to determine the aerodynamic coefficients from experimental/flight data, we need to know V and measure the force, that is, force = mass × acceleration. Since mass, m, is known, we can get acceleration as $\frac{\bar{q}SC_x}{m}$. So if we measure the acceleration of an airplane in motion, we can get the aerodynamic coefficient C_x. Since the coefficients depend on aerodynamic derivatives and the dynamic variables, and if we measure these responses, then we can also get these derivatives. This calls for establishing formal relationship between these dynamic responses and the aerodynamic derivatives, thus leading to aerodynamic modeling.

What we have not said so far is the fact that the pilot will maneuver an aircraft by using pitch stick, roll stick, and rudder pedals besides using the throttle lever arm. The movements of these devices forward or backward or sideways are transmitted via the (servo motors and) control actuators to the moving/hinged surfaces of the aircraft. These surfaces are extensions of the main surfaces, like wing and rudder fin, and are called control surfaces. The movements of the control surfaces interact with the flow field, while the aircraft is in motion and alter the force and moment balance (equations), thereby imparting the "changed" motion to the aircraft until the new balance is achieved. During this changing motion (that can be called perturbation), the aircraft attains new altitude/orientation, etc. depending upon the energy exchanges and balance between kinetic, potential, and propulsive (thermal) energies. At the microscopic level (or scale), this behavior can be studied by exploring the effect of aerodynamic parameters discussed in the next section. At the macroscopic level (or scale), the behavior can be studied using the simplified EOM and resultant transfer function analysis (Chapter 5). Hence, we gradually get closer and deeper to flight motion and hence flight mechanics.

4.3 AERODYNAMIC PARAMETERS

The aerodynamic coefficients would, in general, be a function of not only flow angles but also of Mach number, angular rates, control surface deflections, and changes in thrust force due to the movement of the throttle lever arm. For example C_m from eqn. (4.2) can be expressed in terms of these independent response variables in a straight forward manner as follows:

$$C_m = \text{function of } (\alpha, q, \delta_e, M, F_T)$$

$$C_m = C_{m0} + C_{m\alpha}\alpha + C_{mq}\frac{q\bar{c}}{2V} + C_{m\delta_e}\delta_e + C_{mM}M + C_{mF_T}F_T \tag{4.6}$$

Here, the pitching moment aerodynamic coefficient is considered a function of several independent variables: AOA, pitch rate q, elevator control surface deflection, Mach number, and thrust. The effect on the pitching moment, due to small changes in these response variables, is captured in terms that are called aerodynamic derivatives (e.g., $C_{m\alpha}$), which can be considered as proportionality constants. However, these constants vary with certain variables. The quantity $\frac{q\bar{c}}{2V}$ is the non-dimensional angular rate (velocity), as can be seen from: (rad/s) (m/2) (s/m) => rad. The small perturbation theory when applied to the simplification of the EOM leads to the fact that the aerodynamic forces and moments depend on some constants that are called the stability derivatives. The parametric terms of eqn. (4.6) are called aerodynamic derivatives and often stability and control derivatives since many of these derivatives govern the stability and control of the aircraft (dynamics). For example, $C_{m\alpha}$ signifies $\Delta C_m/\Delta\alpha$ when other variables are set to zero.

4.3.1 Definition of Aerodynamic Derivatives

Essentially these derivatives are the sensitivities to changes in flight variables. If there is a change in AOA, there is a change in the pitching moment coefficient and appropriate change in the pitching moment M, see eqn. (4.2). How this change, in turn, alters the aircraft responses is the subject of the EOM, Chapters 3 and 5. There could be cross-coupling terms, that is, a change in an aerodynamic coefficient of one axis due to a change in the variable of the other axis. Several possibilities exist. The analysis is limited to straightforward effects. The force and moments coefficients of eqns. (4.1) and (4.2) are expressed in terms of aerodynamic derivatives, as seen in eqn. (4.6). These are often called stability and control derivatives. This is because they directly or implicitly govern the stability and control behavior of the vehicle. They are called derivatives since they specify the variation of aerodynamic forces and moments with respect to a small change in the perturbed variable. A particular derivative could vary with velocity or Mach number, altitude, and angle of attack. In turn, the coefficients that depend on these derivatives also vary with these and several other variables. However, the variation of these coefficients soon gets translated to variations with time due to the composition of the coefficients in terms of motion variables that vary with time. Table 4.1 gives the matrix of relationships in terms of the sensitivities between the forces/moments and the response variables u, v, w, p, q, r,

TABLE 4.1
Relationship Matrix for Forces/Moments and the Independent Response Variables

Response Variables→	With Respect to Component Velocities			With Respect to Angular Velocities (Rates)			With Respect to Control Surface Deflections		
Forces/ Moments↓	u (Axial)	v (Side)	w (Normal)	p (Roll)	q (Pitch)	r (Yaw)	δ_e (Elevon/ Elevator)	δ_a (Aileron)	δ_r (Rudder)
X (axial)	X_u C_{Du}		X_w $C_{D\alpha}$		X_q		$X_{\delta e}$ $C_{D\delta e}$		
Y (side)		Y_v $C_{Y\beta}$		Y_p C_{Yp}		Y_r C_{Yr}		Y_{δ_a} $C_{Y\delta_a}$	Y_{δ_r} $C_{Y\delta_r}$
Z (vertical/normal)	Z_u C_{Lu}		Z_w $C_{L\alpha}$		Z_q C_{Lq}		$Z_{\delta e}$ $C_{L\delta e}$		
L (rolling)		L_β $C_{l\beta}$		L_p C_{lp}		L_r C_{lr}		L_{δ_a} $C_{l\delta_a}$	L_{δ_r} $C_{l\delta_r}$
M (pitching)	M_u C_{mu}		M_w $C_{m\alpha}$		M_q C_{mq}		$M_{\delta e}$ $C_{m\delta e}$		
N (yawing)		N_β $C_{n\beta}$		N_p C_{np}		N_r C_{nr}		$N_{\delta a}$ $C_{n\delta a}$	$N_{\delta r}$ $C_{n\delta r}$

TABLE 4.2
Effective Consequential Relationships of the Forces and Moments and the Response Variables

	Linear Velocities	Angular Velocities (Rates)	Control Surface Deflections
Forces	Speed derivatives Static derivatives (with respect to AOA and AOSS)	Usually unimportant and often neglected	Control derivatives
Moments	Speed derivatives Static stability derivatives (with respect to AOA and AOSS)	Dynamic derivatives/ damping (related) derivatives	Control derivatives

and control surface deflections in terms of the aerodynamic derivatives. Table 4.2 depicts the major functional (effect) relationships between forces/moments and the aircraft response variables and classifies them broadly, as mentioned in Table 4.2. In other words, we have the following classification: (i) speed derivatives, (ii) static derivatives, (iii) dynamic derivatives, and (iv) control derivatives. The static derivatives are mainly with respect to AOA and AOSS. They govern the static stability of the vehicle. Dynamic derivatives are with respect to the rotational motion of the vehicle and mainly specify the damping (of certain modes of the vehicle) in the respective axis. The speed derivatives are with respect to linear velocities of the vehicle. The control derivatives specify the control effectiveness of the control surface movements in changing the forces and moments acting on the vehicle.

We see from Table 4.2 that the static and the dynamic derivatives as they are termed imply something definitely more in terms of the behavior of the aircraft from the control theory point of view. What is important to know is that these names themselves relate to certain stability and damping properties of an aircraft that is considered as a dynamic system.

The derivative effects are considered in isolation assuming that the perturbations occur in isolation. The aerodynamic derivatives are evaluated at a steady state equilibrium condition with small perturbation around it. Hence, these derivatives are called quasi-static aerodynamic derivatives. Interestingly enough, these derivatives are used in dynamically varying conditions also, mainly for small-amplitude maneuver analysis, thereby assuring the linear domain operations and analysis. For large-amplitude maneuvers where AOA is high, the unsteady and nonlinear effects need to be incorporated into the analysis for which additional aerodynamic derivatives must be included. For definitions of the dimensional aerodynamic derivative (DADs) and non-DAD (NDADs), we closely follow [2] since the definitions and procedures therein are consistent and straightforward. The present treatment is also enhanced by the treatment from [3,5–9]. Although there might be some differences in the formulae of certain derivatives between various sources [2,3,5–9], we feel that the presentation given in this book is more uniform and highly standardized, as in ref. [2]. The DADs are defined as follows:

$$M_w = (1/I_y)\frac{\partial M}{\partial w};\ I_y \text{ is moment of inertia (MOI)};\ Z_w = (1/m)\frac{\partial Z}{\partial w} \tag{4.7}$$

The derivatives specify the change in pitching moment and vertical force due to a small change in the vertical speed (w) of the aircraft. Next, we have the NDAD defined as follows:

$$C_{m_\alpha} = \frac{\partial C_m}{\partial \alpha} \tag{4.8}$$

Using $C_m = M/\bar{q}\,S\,\bar{c}$ and $\alpha = w/U$ (for small AOAs), we get

$$C_{m_\alpha} = \frac{\partial M}{\partial (w/U)}\frac{1}{\bar{q}\,S\,\bar{c}} = \frac{U}{\bar{q}\,S\,\bar{c}}\frac{\partial M}{\partial w} = \frac{UI_y}{\bar{q}\,S\,\bar{c}}M_w;$$

$$M_w = \bar{q}S\bar{c}/(I_yU)C_{m_\alpha};\ M_w = \rho US\bar{c}/(2I_y)C_{m_\alpha} \tag{4.9}$$

A similar procedure can be applied to all the DADs, and the equivalent NDAD derivatives can be easily obtained. The decoupling between longitudinal and lateral-directional derivatives is presumed. We also assume that during a small perturbation maneuver, the Mach number, Reynolds number, dynamic pressure, velocity, and engine parameters do not change much so that their effects can be neglected. The important longitudinal aerodynamic derivatives are presented in Tables 4.3a and 4.3b in a very comprehensive and compact manner along with brief explanations and indication of the influence of certain derivatives on the aircraft modes.

4.3.2 Longitudinal Derivatives

These derivatives are related to the sensitivities of the forces in axial and vertical (normal) directions and the pitching moment with respect to changes in forward speed, normal speed, pitch angle rate, and elevator/elevon control surface deflections.

4.3.2.1 Effect of Forward/Axial Speed *u* along the *X*-Axis: (X_u, Z_u, M_u)

When there is an increase in the axial velocity, the drag (D) and lift (L) forces will generally increase. Also, the pitching moment (M) will increase.

X_u is the change in forward force (often known as the axial force) due to a change in the forward (also known as the axial) speed. It is called the speed damping derivative. By definition, following eqn. (4.7) (for force derivatives, mass m is used for normalization, and u, as a perturbation, and U are interchangeably used for convenience, and appropriately U should be taken as V, the airspeed):

$X_u = \frac{1}{m}\frac{\partial X}{\partial u}$ or equivalently $X_u = -\frac{1}{m}\frac{\partial D}{\partial u}$; since the drag force acts in the negative direction, it contributes to a negative force X. Using the expression of the drag force D, we get

$$X_u = -\frac{1}{m}\frac{\partial\left(1/2\,\rho U^2 SC_D\right)}{\partial u};\ X_u = -\frac{\rho S}{2m}\frac{\partial\left(U^2 C_D\right)}{\partial u}$$

TABLE 4.3A
Longitudinal Aerodynamic Coefficients/Derivatives [2]

Non-Dimensional Form	Unit	Dim. Form	Unit	Remarks
$C_D = \text{drag force}/(\bar{q}S)$	–			Coefficient of drag force. Important for performance.
$C_{D_u} = \frac{U}{2}\frac{\partial C_D}{\partial u}$	–			Mach number/aeroelastic effect. Appropriate in the range 0.8<Mach<1.2
$-\frac{\rho SU}{m}(C_D + C_{D_u})$	1/s	$X_u = \frac{1}{m}\frac{\partial X}{\partial u}$	1/s	Negative in sign. Affects damping of Phugoid mode.
$C_{D_\alpha} = \frac{\partial C_D}{\partial \alpha}$	1/rad			Positive in sign. Often neglected/but can be important at low speeds (high AOA)
$C_L = \text{Lift force}/(\bar{q}S)$	–			Basic lift force coefficient.
$\frac{\rho SU}{2m}(C_L - C_{D_\alpha})$	1/s	$X_w = \frac{1}{m}\frac{\partial X}{\partial w}$	1/s	$X_\alpha = UX_w$
$C_{D_\delta} = \frac{\partial C_D}{\partial \delta}$	1/rad			Normally negligible. Could be positive or negative. For tailless a/c it is relatively large
$\frac{\rho SU^2}{2m}(-C_{D_\delta})$	$\frac{mt}{s^2}$	$X_\delta = \frac{1}{m}\frac{\partial X}{\partial \delta}$	$\frac{mt}{s^2}$	mt → meter
$C_{L_u} = \frac{U}{2}\frac{\partial C_L}{\partial u}$	–			Arises from Mach and aeroelastic effect. Sign can vary. Negligibly small for low Mach maneuvers
$-\frac{\rho SU}{m}(C_L + C_{L_u})$	1/s	$Z_u = \frac{1}{m}\frac{\partial Z}{\partial u}$ $Z_u = -\frac{1}{m}\frac{\partial L}{\partial u}$	1/s	Positive downward
$C_{L_\alpha} = \frac{\partial C_L}{\partial \alpha}$	1/rad			Lift curve slope. Mostly positive/could be very small or negative (implying "stall").
$-\frac{\rho SU}{2m}(C_{L_\alpha} + C_D)$	1/s	$Z_w = \frac{1}{m}\frac{\partial Z}{\partial w}$	1/s	
$C_{L_{\dot\alpha}} = \frac{\partial C_L}{\partial(\dot\alpha\bar{c}/2U)}$	1/rad			models the unsteady effects due to the lag in down wash on the horizontal tail.
$\frac{\rho Sc}{4m}(-C_{L_{\dot\alpha}})$	–	$Z_{\dot w} = \frac{1}{m}\frac{\partial Z}{\partial \dot w}$	–	Often neglected $X_{\dot\alpha} = U X_{\dot w}$

Note: (i) Thrust effects are neglected, (ii) "–" means rad/rad or no unit, and (iii) the symbol used for definition of derivatives is replaced by "=" sign for simplicity and uniformity.

TABLE 4.3B
Longitudinal Aerodynamic Coefficients/Derivatives [2]

Non-Dimensional Form	Unit	Dim. Form	Unit	Remarks
$C_{L_q} = \dfrac{\partial C_L}{\partial (qc/2U)}$	1/rad			Positive in sign for rigid body a/c & in low speed flights. In high speed flights sign could be ±ve
$\dfrac{\rho SUc}{4m}(-C_{L_q})$	mt/sec	$Z_q = \dfrac{1}{m}\dfrac{\partial Z}{\partial q}$	mt/s	Usually not significant.
$C_{L\delta} = \dfrac{\partial C_L}{\partial \delta}$	1/rad			Positive control deflection→positive lift for conventional a/c. Often unimportant. On tailless a/c with small effective elevator arm, it could be relatively large.
$\dfrac{\rho SU^2}{2m}(-C_{L\delta})$	$\dfrac{mt}{s^2}$	$Z_\delta = \dfrac{1}{m}\dfrac{\partial Z}{\partial \delta}$	$\dfrac{mt}{s^2}$	Also important for canard configurations.
$C_m = M/(\bar{q}Sc)$	-			Should have positive value at 0 deg. AOA for static stability & trim balance.
$C_{m_u} = U/2\dfrac{\partial C_m}{\partial u}$	-			Sign can change. Depends on Mach and elastic properties. Often ignored.
$\dfrac{\rho SU\bar{c}}{I_y}(C_m + C_{m_u})$	1/(mt.s)	$M_u = \dfrac{1}{I_y}\dfrac{\partial M}{\partial u}$	1/(mt.s)	
$C_{m\alpha} = \dfrac{\partial C_m}{\partial \alpha}$	1/rad			Preliminary design parameter. Defines the stick-fixed neutral point/static margin
$\dfrac{\rho SU\bar{c}}{2I_y}C_{m\alpha}$	1/(mt.s)	$M_w = \dfrac{1}{I_y}\dfrac{\partial M}{\partial w}$	1/(mt.s)	Defines the natural frequency of the SP mode. $M_\alpha = UM_w$
$C_{m\dot{\alpha}} = \dfrac{\partial C_m}{\partial (\dot{\alpha}\bar{c}/2U)}$	1/rad			Down wash lag on moments.
$\dfrac{\rho Sc^2}{4I_y}C_{m\dot{\alpha}}$	1/mt	$M_{\dot{w}} = \dfrac{1}{I_y}\dfrac{\partial M}{\partial \dot{w}}$	1/mt	$M_{\dot{\alpha}} = UM_{\dot{w}}$
$C_{m_q} = \dfrac{\partial C_m}{\partial (q\bar{c}/2U)}$	1/rad			Preliminary design parameter.
$\dfrac{\rho SU\bar{c}^2}{4I_y}C_{m_q}$	1/s	$M_q = \dfrac{1}{I_y}\dfrac{\partial M}{\partial q}$	1/s	Negative in sign, affects damping in pitch, main damping parameter.
$C_{m\delta} = \dfrac{\partial C_m}{\partial \delta}$	1/rad			Preliminary design parameter.
$\dfrac{\rho SU^2\bar{c}}{2I_y}C_{m\delta}$	$\dfrac{1}{s^2}$	$M_\delta = \dfrac{1}{I_y}\dfrac{\partial M}{\partial \delta}$	$\dfrac{1}{s^2}$	Elevator control effectiveness.

$$X_u = -\frac{\rho S}{2m}\left(U^2\frac{\partial C_D}{\partial u} + 2UC_D\right); X_u = -\frac{\rho SU}{m}\left(U/2\frac{\partial C_D}{\partial u} + C_D\right),$$

Also, by definition

$$C_{D_u} = U/2\frac{\partial C_D}{\partial u}; X_u = -\frac{\rho SU}{m}\left(C_{D_u} + C_D\right) + \frac{\partial T_x}{m\partial U};$$

$$T_x = \text{Thrust provided by the engine} \tag{4.10}$$

An increase in the thrust T contributes a positive X force. The drag is measured along the direction of the relative wind, and the drag coefficient is measured along the negative equilibrium x-axis (in the stability axis system). The drag coefficient is always positive in sign, and the derivative X_u is always negative in sign. The C_{D_u} derivative can arise from Mach number (Mn) and aeroelastic effects (usually small and negligible). The Mach effect will be significant in the range $0.8 < \text{Mach number} < 1.2$ (transonic region). The derivative C_{D_u} can be determined from the plot of the drag coefficient with respect to Mach number. We see that C_{D_u} in [3] is defined as $Mn\frac{\partial C_D}{\partial Mn}$, and using the definition of eqn. (4.10), we get C_{D_u} (present) = 1/2 C_{D_u} of ref. [3].

Z_u is the change in the normal force due to a change in the forward speed. Since the lift acts along the negative z-axis (the lift acts upward and that is the negative z-axis; Z_u is always measured along the positive z-axis and is positive downward), one gets

$$Z_u = -\frac{1}{m}\frac{\partial L}{\partial u}; \text{ by following the previous procedure one gets: } Z_u = -\frac{\rho SU}{m}\left(C_L + C_{L_u}\right) \tag{4.11}$$

The low values of C_L are associated with low AOA and high speeds, and high values are associated with high AOA and low speeds. C_{L_u} can arise from Mach number (Mn) and aeroelastic effects. For low Mach number, it is negligibly small, and it could be high at critical Mach number. The magnitude of C_{L_u} can vary considerably, and its sign can change, depending on airframe geometry and its elastic properties, but also on the Mach number and the dynamic pressure at which the aircraft is flying. We see that C_{L_u} in [3] is defined as $Mn\cdot\frac{\partial C_L}{\partial Mn}$, and using the definition, eqn. (4.10), C_{L_u} (present) = 1/2 C_{L_u} of ref. [3].

The derivative with respect to Mach number (here denoted as M) is derived next.

$$Z_{Mn} = -\frac{1}{m}\frac{\partial L}{\partial Mn}; Z_{Mn} = -\frac{1}{m}\frac{\partial\left(1/2\,\rho U^2 SC_L\right)}{\partial Mn}; Z_{Mn} = -\frac{1}{m}\frac{\partial\left(1/2\,\rho U^2 SC_L\right)}{\partial Mn}$$

$$Z_{Mn} = -\frac{\rho S}{2m}\left(U^2\frac{\partial C_L}{\partial Mn} + 2a^2 MnC_L\right); \text{ since } Mn = U/a = \text{airspeed/speed of sound,}$$

$$Z_{Mn} = -\frac{\rho S}{2m}\left(U^2\frac{\partial C_L}{\partial Mn} + 2U^2 C_L / Mn\right); Z_{Mn} = -\frac{\rho S}{m}\left(C_{L_{Mn}} + C_L / Mn\right)$$

Alternatively, it can be expressed by using eqn. (4.12) as follows:

$$Z_{Mn} = -\frac{\rho SUa}{m}\left(C_L + C_{L_u}\right), \text{Since } Z_{Mn} = Z_u a, \text{and } \frac{\partial Z}{\partial Mn} = \frac{\partial Z}{\partial u}\frac{u}{Mn}$$

M_u is the change in the pitching moment (M) due to a change in the forward speed. By definition, one gets from the basic equation of the pitching moment

$$M_u = 1/I_y\left[\frac{\rho SU^2\bar{c}}{2}\frac{\partial C_m}{\partial u} + \rho US\bar{c}C_m\right]$$

$$M_u = \frac{\rho SU\bar{c}}{I_y}\left(U/2\frac{\partial C_m}{\partial u} + C_m\right); C_{M_u} = U/2\left(\frac{\partial C_M}{\partial U}\right); C_M = \frac{M\,(\text{moment})}{qS\bar{c}}$$

Again by definition one gets

$$M_u = \frac{\rho SU\bar{c}}{I_y}\left(C_{m_u} + C_m\right) \tag{4.12}$$

C_m is the aerodynamic portion of the total trimmed pitching moment. In the presence of thrust asymmetry, C_m can change due to (change in) the airframe geometry and the elastic properties. Also, the sign can change due to Mach number and dynamic pressure variation. These derivatives originate from power, Mach number, and aeroelastic effects, and the latter two are more important. For subsonic flights, often M_u is considered negligible. However, at high speeds, bending due to aeroelastic effects can induce large changes in C_{m_u}. We see that C_{m_u} in [3] is defined as $M\frac{\partial C_m}{\partial M}$, and using the definition of eqn. (4.10), we get C_{m_u} (present) = 1/2 C_{m_u} of ref. [3].

4.3.2.2 Effect of Change in Vertical Speed *w* (Equivalently AOA) along the Vertical *Z*-Axis (X_w, Z_w, M_w)

According to the definition of stability axes, the relative wind during the steady flight condition is parallel to the *x*-axis. The only component of the linear velocity is U_0; when the aircraft is disturbed from the steady flight, so that it has a component of velocity w along the *z*-axis as well as a forward velocity *u*, and the relative wind shifts to a new position, resulting in an increase in AOA.

X_w is the change in the forward (axial) force due to the change in the vertical (normal) speed. By definition, we have

$$X_w = \frac{1}{m}\frac{\partial X}{\partial w} \text{ or } X_w = \frac{1}{mU}\frac{\partial X}{\partial \alpha}; X_w = \frac{1}{mU}\frac{\partial\left(1/2\,\rho U^2 SC_x\right)}{\partial \alpha}$$

We also have (see Exercise 4.4)

$$C_x = C_L \sin\alpha - C_D \cos\alpha; C_L = \frac{L}{\bar{q}S}; C_{D\alpha} \triangleq \frac{\partial C_D}{\partial \alpha}$$

Using the aforementioned expression, we obtain

$$X_w = \frac{\rho U^2 S}{2mU} \frac{\partial (C_L \sin\alpha - C_D \cos\alpha)}{\partial \alpha}$$

By evaluating the partials, we get

$$X_w = \frac{\rho S U^2}{2mU}\left(C_L \cos\alpha + C_{L\alpha} \sin\alpha + C_D \sin\alpha - \cos\alpha\ C_{D\alpha}\right)$$

Assuming that the AOA is very small, after simplification, we get

$$X_w = \frac{\rho U S}{2m}\left(C_L - C_{D\alpha}\right) \tag{4.13}$$

When the AOA (or vertical speed w) increases the total drag increases, $C_{D\alpha}$ is positive in sign; the main contribution is from the wing, and other from tail and fuselage and is a nonlinear function of the AOA.

Z_w is the change in the normal force due to the change in the vertical (normal) speed. By definition, we have

$$Z_w = \frac{1}{m}\frac{\partial Z}{\partial w} \text{ or } Z_w = \frac{1}{mU}\frac{\partial Z}{\partial \alpha}; Z_w = \frac{1}{mU}\frac{\partial\left(1/2\ \rho U^2 S C_z\right)}{\partial \alpha}$$

Also, we have (see Exercise 4.4),

$$C_z = -(C_L \cos\alpha + C_D \sin\alpha)$$

By using the previous equation, we obtain

$$Z_w = \frac{1}{mU}\frac{\partial\left(-1/2\ \rho U^2 S(C_L \cos\alpha + C_D \sin\alpha)\right)}{\partial \alpha}$$

By evaluating the derivatives and simplifying (for small AOA), we get

$$Z_w = -\frac{\rho S U}{2m}\left(-C_L \sin\alpha + C_{L\alpha} \cos\alpha + C_D \cos\alpha + \sin\alpha\ C_{D\alpha}\right); Z_w = -\frac{\rho S U}{2m}\left(C_{L\alpha} + C_D\right) \tag{4.14}$$

$C_{L\alpha}$ (known as the lift curve slope) is positive below "stall-α", and 80%–90% contribution comes from wing, fuselage, and tail.

M_w is the change in the pitching moment due to a change in the vertical (normal) speed. A change in the vertical speed causes a change in AOA, which, in turn, causes a change in lift and drag of the corresponding surfaces (like wing and horizontal tail). By definition, we have

$$M_w = \frac{1}{I_y}\frac{\partial M}{\partial w} \text{ also } M_w = \frac{1}{I_y U}\frac{\partial M}{\partial \alpha};\ M_w = \frac{\rho S U^2 \bar{c}}{2 I_y U}\frac{\partial C_m}{\partial \alpha};\ M_w = \frac{\rho S U \bar{c}}{2 I_y} C_{m\alpha} \quad (4.15)$$

4.3.2.3 Effect of Change in Pitch Rate q: (X_q, Z_q, M_q)

X_q is the change in axial force due to a change in the pitch (angle) rate. This is mainly due to the change of AOA of the horizontal tail (T) that contributes to the total drag. The contribution from the wing is usually negligible, and the resulting drag increase is neglected in the first approximation, and the derivative is also assumed to be zero.

$$X_q = 1/m\left[-\frac{\rho U^2 S_T}{2}\frac{\partial C_{D_T}}{\partial q}\right];\ \text{By definition; } C_{D_{Tq}} = \frac{\partial C_{D_T}}{\partial(q\bar{c}/2U)},$$

$$\text{we obtain } X_q = -\frac{\rho S_T U \bar{c}}{4m} C_{D_{Tq}} \quad (4.16)$$

Z_q is the change in normal force due to a change in the pitch rate. Pitching causes a change in the AOA of the horizontal tail, thereby resulting in the change of the lift at the tail. Following the previous development for X_q ($X=-sL$), we obtain

$$Z_q = -\frac{\rho S U \bar{c}}{4m} C_{L_q} \quad (4.17)$$

We can obtain C_{Z_q} (equivalently C_{L_q}) from the known variables and constants as follows realizing the change of tail lift:

$$\Delta L_t = \frac{\rho U_t^2 S_t}{2} C_{L\alpha}\Delta\alpha;\ \Delta Z = -\Delta L_t = -\frac{\rho U_t^2 S_t}{2} C_{L\alpha}\left(\frac{q l_t}{U_0}\right);$$

The last term is the change of the AOA of the tail due to the pitch rate-induced vertical speed of the tail. Next, we have from eqn. (4.1):

$$C_Z = \frac{2Z}{\rho U^2 S};\ \Delta C_Z = -\frac{2\rho S_t U_t^2}{2\rho U^2 S} C_{L\alpha}\cdot\left(\frac{q l_t}{U_0}\right)$$

Defining $C_{Z_q} = \dfrac{\partial C_Z}{\partial(q\bar{c}/2U_0)}$ and substituting from the previous equation, we get

$$C_{Z_q} = -2\left(\frac{2\rho U_t^2}{2\rho U^2}\right)C_{L\alpha}\cdot\left(\frac{l_t S_t}{S\bar{c}}\right);\ C_{Z_q} = -2\left(\frac{U_t^2}{U^2}\right)C_{L\alpha}\cdot\left(\frac{l_t S_t}{S\bar{c}}\right) \Rightarrow C_{Z_q} = -2C_{L\alpha}\eta V_H \quad (4.18)$$

Here, η is the ratio of the dynamic pressure at the tail and the main dynamic pressure and is often called the tail efficiency. V_H is horizontal tail volume ratio, the term in the second parenthesis. As is with $C_{L\dot{\alpha}}$, the effect of C_{L_q} on longitudinal stability is usually very small and is neglected in the dynamic analysis.

M_q is the change in pitching moment due to a change in the pitch rate. By definition, we get

$$M_q = \frac{1}{I_y}\frac{\partial M}{\partial q}; M_q = \frac{\rho S U^2 \bar{c}}{2I_y}\frac{\partial C_m}{\partial q}; M_q = \frac{\rho S U \bar{c}^2}{4I_y}\frac{\partial C_m}{\partial (q\bar{c}/2U)}; M_q = \frac{\rho S U \bar{c}^2}{4I_y} C_{m_q} \tag{4.19}$$

Following the steps as for C_{Z_q} we can obtain the formula for C_{m_q} realizing the change in pitching moment at the $C.G.$ due to a change in the lift force at the tail and the lever arm as follows:

$$\Delta M = -\Delta L_t l_t = -\frac{\rho S_t U_t^2}{2} C_{L\alpha} \Delta\alpha; C_{m_q} = \frac{\partial C_m}{\partial (q\bar{c}/2U)} = \frac{2U \partial C_m}{\bar{c}\,\partial q} = -2C_{L\alpha}\eta V_H \frac{q}{\Delta q}\frac{l_t}{\bar{c}}$$

Since we are dealing with only small perturbations, q and Δq are identical, hence cancel out, resulting in

$$2C_{m_q} = -2C_{L\alpha}\eta V_H \frac{l_t}{\bar{c}} \tag{4.20}$$

The formula (4.20) gives the damping derivative in terms of the known or measurable constants. It is very important because it contributes a large portion of the damping of the short period mode for a conventional aircraft. This damping effect comes from the AOA changes at the horizontal tail, which are proportional to the tail length; that is also a lever arm converting the tail length into the moment.

Next, we consider the effects of aircraft control surfaces, which include (i) elevators, (ii) stabilizers, (iii) flaps, (iv) slats, and (v) dive brakes which pertain to longitudinal control. The positive direction is taken as that giving positive lift for convenience.

4.3.2.4 Effect of Change in Elevator Control Surface Deflection: (X_{δ_e}, Z_{δ_e}, M_{δ_e})

These are control effectiveness derivatives. However, $C_{m\delta_e}$ derivative is called the primary control derivative or often control power. If the absolute value of the derivative is higher, then for a given deflection, more moment is generated. This can be regarded as higher control sensitivity for a given moment of inertia. The Z_{δ_e} and M_{δ_e} derivatives are closely related via the moment arm.

The X_{δ_e} derivative is defined/given as (since the axial force X is in the negative direction of the drag force D) follows:

$$X_{\delta_e} = \frac{1}{m}\frac{\partial X}{\partial \delta_e}; X_{\delta_e} = \frac{\rho S U^2}{2m}\left(-C_{D\delta_e}\right); C_{D\delta_e} = \frac{\partial C_D}{\partial \delta_e}$$

Similarly, since lift force is in the opposite direction to z-normal force, we get

$$X_{\delta_e} = \frac{1}{m}\frac{\partial X}{\partial \delta_e}; X_{\delta_e} = \frac{\rho S U^2}{2m}\left(-C_{D\delta_e}\right); C_{D\delta_e} = \frac{\partial C_D}{\partial \delta_e}$$

$$Z_{\delta_e} = \frac{1}{m}\frac{\partial Z}{\partial \delta_e} = \frac{\rho SU^2}{2m}(-C_{L\delta_e}); C_{L\delta_e} = \frac{\partial C_L}{\partial \delta_e} \tag{4.21}$$

M_{δ_e} is the change in pitching moment due to a small change in the elevator/elevon control surface deflection.

$$M_{\delta_e} = \frac{1}{I_y}\frac{\partial M}{\partial \delta_e} = \frac{\rho SU^2\bar{c}}{2I_y}\frac{\partial C_m}{\partial \delta_e} = \frac{\rho SU^2\bar{c}}{2I_y}C_{m\delta_e}; C_{m\delta_e} = \frac{\partial C_m}{\partial \delta_e} \tag{4.22}$$

The $C_{m\delta_e}$ derivative can be considered as a primary design parameter, and it specifies longitudinal axis control power. The important and routinely used longitudinal DADs and NDADs (stability and control derivatives) are given in Tables 4.3 in a most comprehensive format.

Some less important and often neglected longitudinal derivatives (due to change in the vertical speed) are briefly discussed here for the sake of completion:

$X_{\dot{w}}$ is a change in the axial force due to the rate of change of speed along the z-axis (w); the effective change of the AOA of the horizontal tail causes changes in the lift and drag acting on the tail. This change in drag is the main contributor to the change in the X force, but the drag on the tail is generally small in comparison to the total drag, and the increment in the tail drag due to change in speed (w) is even smaller; therefore, the derivative $X_{\dot{w}}$ is considered as zero in the first approximation.

$Z_{\dot{w}}$ is a change in lift on the tail due to speed (w) change and sometimes causes an important change in the total Z force, which in terms of the total lift coefficient can be written as follows:

$$Z = -1/2\ \rho U^2 SC_L (\text{because } Z = -L, \text{ the lift force}); Z_{\dot{w}} = \frac{1}{m}\frac{\partial Z}{\partial \dot{w}} = \frac{1}{mU}\cdot\frac{\partial Z}{\partial \dot{\alpha}}$$

$$Z_{\dot{w}} = \frac{1}{mU}\cdot\frac{\partial Z}{\partial \dot{\alpha}} = -\frac{\rho U^2 S}{2mU}\cdot\frac{\partial C_L}{\partial \dot{\alpha}};$$

Since $C_{L\dot{\alpha}} = \dfrac{\partial C_L}{\partial \dot{\alpha}(\bar{c}/2U)}$ (by definition), we have

$$Z_{\dot{w}} = -\frac{\rho S\bar{c}}{4m}\cdot C_{L\dot{\alpha}}$$

The derivative $C_{L\dot{\alpha}}$ arises from an aerodynamic time lag effect and various deadweight aeroelastic effects; and the effect on the longitudinal dynamics is essentially the same as if the airframe's mass or inertia were changed in the equation relating the forces in the Z direction; this effect is very small, and this derivative is often neglected in the analysis.

$M_{\dot{w}}$ is relatively important in the analysis because it does have a significant, if not powerful, effect in the damping of the short period mode of the aircraft:

$$M_{\dot{w}} = \frac{1}{I_y U}\cdot\frac{\partial M}{\partial \dot{\alpha}} = \frac{\rho U^2 S\bar{c}}{2I_y U}\cdot\frac{\partial C_M}{\partial \dot{\alpha}} = \frac{\rho S\bar{c}^2}{4I_y}C_{M\dot{\alpha}}; \text{Since } C_{M\dot{\alpha}} = \frac{\partial C_M}{\partial \dot{\alpha}(\bar{c}/2U)}$$

Some most important longitudinal derivatives are explained next.

a. C_{L_α} is defined as a change in the lift coefficient for a unit change in AOA. The lift force can be easily given by $L = C_L \bar{q}S$; C_{L_α} represents the lift curve slope with respect to AOA. This is a very important derivative because it almost directly determines the contribution to the lift force. It also signifies the fact that as AOA increases, the lift force increases proportionally, in the linear region up to a certain AOA. Beyond this AOA, the lift would remain somewhat constant and would even decrease further. This condition is called "wing stall"; this means that the aircraft loses some lift after it stalls.
b. C_{m_α} is the basic (longitudinal) static stability derivative and is referred to as pitch stiffness parameter [2]. A negative value of C_{m_α} indicates that the aircraft is statically stable, that is, if the AOA increases, then the pitching moment becomes more negative, thereby decreasing the AOA and hence restoring the stability. The positive change in AOA increases the lift on the tail inducing negative pitching moment about the C.G. and increases the lift of the wing inducing the positive moment if the lift acts fore of the C.G., and negative moment if the lift acts aft of the C.G. Thus, for a stable equilibrium, C_{m0} must be positive, and C_{m_α} negative; and for an airfoil with positive Camber, C_{m0} is negative; with zero Camber, it is 0; and with negative Camber it is positive. The C_{m_α} derivative is proportional to the distance between the aerodynamic center (better referred and used as neutral point, NP) of the aircraft and the C.G. This distance is related to the static margin as follows: static margin = (NP distance, $h_{NP} - h_{CG}$)/MAC; i.e., $C_{m_\alpha} = C_{L_\alpha}(h_{CG} - h_{NP})$. The distances (of NP and C.G.) are from some reference point in the front of the aircraft on the x-axis. If the static margin is positive, then the aircraft is said to have static stability; this means that the NP distance is larger than that of the C.G. distance from the reference point. If the C.G. is continuously moved (by some means) rearward, then at one point, the aircraft will become neutrally stable (neutrally unstable!, here $C_{m_\alpha} = 0$). This point is called the neutral point of the aircraft or this C.G. position or location is called the neutral point. The NP must be always behind the C.G. location for the guaranteed static stability (C_{m_α} is negative). A slightly more rearward movement of the C.G. will render the aircraft statically unstable (the C.G. is aft and C_{m_α} is positive). If the aircraft is dynamically, stable then it must have been statically stable. For dynamic stability, the static stability is a must, but the converse is not true, meaning that if the aircraft is statically stable, it could be dynamically unstable. This derivative is of primary importance for the longitudinal stability of the atmospheric vehicles. If the aircraft is inherently statically unstable (by design or so for gaining certain benefits in case of a relaxed static stability aircraft like FBW and many modern-day high-performance fighter aircraft), then artificial stabilization is required in order that the aircraft can fly. The level of instability is dictated by (i) available state-of-the-art technology (computers, actuators, sensors, and control law design tools), (ii) possible reduction in the size of

the aircraft and subsequent decrease in weight, (iii) increase in weight due to additional hardware (redundancy, computer, etc.), (iv) obtainable performance, (v) maneuverability, and (vi) agility. This is the subject of design, development, and analysis of flight control laws (Chapter 8).

c. C_{m_q} signifies a change in the pitching moment coefficient due to a small change in the pitch rate q. Being the (angular) rate-related derivative, it implies, from the control theory point of view, that it must have something to do with damping; as such, it contributes to the damping in pitch. Usually, more negative values of C_{m_q} signify increased damping. The sign is generally negative for both stable and unstable configurations. Since the statically unstable configurations will have stability augmentation control laws, the lower values of this derivative (less negative values) are acceptable. The aircraft flight control system (AFCS) also provides some artificial damping. This derivative often has higher prediction uncertainty levels (scatter in estimates); however, this is not a problem since the AFCS generally tolerates these uncertainties.
d. Longitudinal control effectiveness derivative, $C_{m\delta_e}$, is the elevator control effectiveness. Its function is through the applied pitching moment to control the AOA of the airframe in equilibrium and maneuvering flight, and this function is the most important. In conventional sense, more negative value means more control effectiveness. It also helps determine the sizing/design of the control surface. More elevator (elevons for FBW delta-wing aircraft configuration) control power means more effective control in generating control moment for the aircraft. This also applies to horizontal tails, canard, or some combination of these surface movements. The knowledge of the available elevator/elevon power at all the flight conditions (Chapter 7) in the flight envelope is very important. The design value of this derivative is determined by the anticipated fore and aft travel of C.G. The larger C.G. travel and higher lift coefficient demand larger value of this derivative. Supersonic flight may impose additional requirements because of the attendant aft movement of the a.c. The sign of this derivative depends on the location of the elevator, fore, or aft of the C.G.; for the aft locations and the elevator sign convention used here, the derivative has a negative value. The unstable configurations demand higher control power than the stable ones.

The $C_{L\delta_e}$ is always positive in accordance with the convention used here. For the conventional aircraft with the horizontal tail mounted at an appreciable distance aft of the C.G., this derivative is usually small and its effect is relatively unimportant.

The $C_{D\delta_e}$ is invariably smaller than $C_{L\delta_e}$ (because of the usual of drag with lift), and it is normally neglected. Its sign could be positive or negative, depending on the trim position of the elevator and the trim AOA.

Numerical values of certain longitudinal derivatives are given in Table 4.4.

The important lateral-directional aerodynamic derivatives are provided in Table 4.5 in a very comprehensive and compact manner along with brief explanations and indication of influence of certain derivatives on the aircraft modes.

TABLE 4.4
Typical Values of Some Derivatives for a Small Trainer Aircraft

Derivative	Analytical/Wind Tunnel Value	Flight Determined Derivatives (at 100 knots/1.87 km alt/4.21 deg. AOA trim)
$C_{L\alpha}$(/rad)	5.014	5.254
$C_{m\alpha}$(/rad)	−0.57	−0.495
C_{m_q}	−13	−14.49
$C_{m\delta_e}$(/rad)	−0.802	−0.911

TABLE 4.5
Lateral-Directional Aerodynamic Derivatives (Stability and Control Derivatives) [2]

Non-Dimensional Form	Unit	Dim. Form	Unit	Remarks
$C_{y\beta} = \frac{\partial C_y}{\partial \beta}$	1/rad			Usually –ve in sign, the side force opposes the sideward motion,→ total damping
$\frac{\rho US}{2m} C_{y\beta}$	1/s	$Y_v = \frac{1}{mU}\frac{\partial Y}{\partial \beta}$	1/s	Origin: major from vertical tail and small from wing/fuselage
$C_{y_{\dot{\beta}}} = \frac{\partial C_y}{\partial(\dot{\beta}b/2U)}$	1/rad			
$\frac{\rho Sb}{4m} C_{y_{\dot{\beta}}}$	-	$Y_{\dot{v}} = \frac{1}{mU}\frac{\partial Y}{\partial \dot{\beta}}$	-	
$C_{y_r} = \frac{\partial C_y}{\partial(rb/2U)}$	1/rad			Of little importance, usually neglected.
$\frac{\rho SbU}{4m} C_{y_r}$	mt/rad.s	$Y_r = \frac{1}{m}\frac{\partial Y}{\partial r}$	mt/rad.s	
$C_{y_p} = \frac{\partial C_y}{\partial(pb/2U)}$	1/rad			Positive/negative, of little importance, often neglected.
$\frac{\rho SbU}{4m} C_{y_p}$	mt/rad.s	$Y_p = \frac{1}{m}\frac{\partial Y}{\partial p}$	mt/rad.s	
$C_{y\delta} = \frac{\partial C_y}{\partial \delta}$	1/rad			$C_{y\delta_a}$, $C_{y\delta_r}$ are unimportant and often small or neglected.
$\frac{\rho SU^2}{2m} C_{y\delta}$	$\frac{\text{mt}}{\text{rad}\cdot\text{s}^2}$	$Y_\delta = \frac{1}{m}\frac{\partial Y}{\partial \delta}$	$\frac{\text{mt}}{\text{rad}\cdot\text{s}^2}$	
$C_{n\beta} = \frac{\partial C_n}{\partial \beta}$	1/rad			Static directional or weather cock stability. Preliminary design parameter. Origin from vertical tail.
$\frac{\rho SU^2 b}{2I_z} C_{n\beta}$	$\frac{1}{s^2}$	$N_\beta = \frac{1}{I_z}\frac{\partial N}{\partial \beta}$	$\frac{1}{s^2}$	Natural frequency of the DR oscillatory mode. Important for spiral stability. Mostly positive.

(Continued)

TABLE 4.5 (*Continued*)
Lateral-Directional Aerodynamic Derivatives (Stability and Control Derivatives) [2]

Non-Dimensional Form	Unit	Dim. Form	Unit	Remarks
$C_{n_{\dot\beta}} = \frac{\partial C_n}{\partial(\dot\beta b/2U)}$	1/rad			
$\frac{\rho SUb^2}{4I_z} C_{n_{\dot\beta}}$	1/s	$N_{\dot\beta} = \frac{1}{I_z}\frac{\partial N}{\partial\dot\beta}$	1/s	
$C_{n_r} = \frac{\partial C_n}{\partial(rb/2U)}$	1/rad			
$\frac{\rho SUb^2}{4I_z} C_{n_r}$	1/s	$N_r = \frac{1}{I_z}\frac{\partial N}{\partial r}$	1/s	From wing and vertical tail. Yaw damping derivative, negative in sign→ DR oscillatory mode, spiral mode
$C_{n_p} = \frac{\partial C_n}{\partial(pb/2U)}$	1/rad			
$\frac{\rho SUb^2}{4I_z} C_{n_p}$	1/s	$N_p = \frac{1}{I_z}\frac{\partial N}{\partial p}$	1/s	DR damping, usually –ve.
$C_{n\delta} = \frac{\partial C_n}{\partial\delta}$	1/rad			$C_{n\delta_r}$, negative, rudder effectiveness
$\frac{\rho SU^2 b}{2I_z} C_{n\delta}$	$\frac{1}{\text{rad}\cdot\text{s}^2}$	$N_\delta = \frac{1}{I_z}\frac{\partial N}{\partial\delta}$	$\frac{1}{\text{rad}\cdot\text{s}^2}$	$C_{n\delta_r}$ design parameter. $C_{n\delta_a}$ is an interesting derivative
$C_{l_\beta} = \frac{\partial C_l}{\partial\beta}$	1/rad			Preliminary design parameter. Origin is vertical tail/pylons/dihedral; "effective" dihedral.
$\frac{\rho SU^2 b}{2I_x} C_{l_\beta}$	$\frac{1}{s^2}$	$L_\beta = \frac{1}{I_x}\frac{\partial L}{\partial\beta}$	$\frac{1}{s^2}$	negative values→damping in DR & spiral modes. The lateral static stability of the a/c.
$C_{l_{\dot\beta}} = \frac{\partial C_l}{\partial(\dot\beta b/2U)}$	1/rad			
$\frac{\rho SUb^2}{4I_x} C_{l_{\dot\beta}}$	1/s	$L_{\dot\beta} = \frac{1}{I_x}\frac{\partial L}{\partial\dot\beta}$	1/s	
$C_{l_r} = \frac{\partial C_l}{\partial(rb/2U)}$	1/rad			Positive/negative
$\frac{\rho SUb^2}{4I_x} C_{l_r}$	1/s	$L_r = \frac{1}{I_x}\frac{\partial L}{\partial r}$	1/s	Affects the aircraft spiral mode
$C_{l_p} = \frac{\partial C_l}{\partial(pb/2U)}$	1/rad			Affects the design of ailerons
$\frac{\rho SUb^2}{4I_x} C_{l_p}$	1/s	$L_p = \frac{1}{I_x}\frac{\partial L}{\partial p}$	1/s	Roll damping derivative, negative. Origin: from wings, vertical & horizontal tail surfaces
$C_{l\delta} = \frac{\partial C_l}{\partial\delta}$	1/rad			$C_{l\delta_a}$ is positive, aileron control effectiveness, design parameter
$\frac{\rho SU^2 b}{2I_x} C_{l\delta}$	$\frac{1}{\text{rad}\cdot\text{s}^2}$	$L_\delta = \frac{1}{I_x}\frac{\partial L}{\partial\delta}$	$\frac{1}{\text{rad}\cdot\text{s}^2}$	$C_{l\delta_r}$ is neglected

4.3.3 Lateral-Directional Derivatives

These derivatives are defined as the sensitivities of the side force (Y), the rolling moment (L), and the yawing moment (N) with respect to small changes in side speed, roll rate, yaw rate, and aileron and rudder control surface deflections. These derivatives are, in general, difficult to estimate with a good degree of confidence.

4.3.3.1 Effect of Change in Side Speed v (Equivalently the AOSS): (Y_v, L_β, N_β)

Y_v is the change in the side force due to a small change in the side velocity v:

$$Y = \frac{\rho U^2 S}{2} C_y; \; \partial Y / \partial v = \frac{\rho U^2 S}{2} \frac{\partial C_y}{\partial v}; \text{ Using } v = \beta U, \text{ we obtain}$$

$$Y_v = (1/m)\partial Y / \partial v = \frac{\rho US}{2m} C_{Y\beta}; \; C_{y\beta} = \frac{\partial C_y}{\partial \beta} \tag{4.23}$$

This derivative is difficult to estimate. It is usually negative since the side slipping causes the side force that opposes the sideward motion. In general, several contributory forces are generated: (i) the one, say F_1, that arises from the change of the AOA of the vertical tail, (ii) side force acting on the fuselage, F_2, and (iii) the two forces (F_3, F_4) due to differential lift forces acting on each semi-span of the wing due to its "effective dihedral". Small or positive values of the derivative are undesirable because the resulting small (or reversal) side forces make the detection of SS difficult, and accordingly, coordination of banked turns becomes a piloting problem. Also, such values of the derivative contribute little to the damping of the Dutch roll (DR), where as normal (negative) values of $C_{y\beta}$ can contribute substantially to the total damping.

L_β is the change in the rolling moment due to a change in side slip. The rolling moment L about the X-axis is caused mainly by the components (F_3, F_4) that act normal to the wing, and by F_1 at the fin center of pressure, which can be either above or below the X-axis. We have

$$L_v = \frac{1}{I_x} \frac{\partial L}{\partial v} = \frac{\rho U^2 Sb}{2I_x} \frac{\partial C_l}{\partial v}; L_v = \frac{\rho USb}{2I_x} \left(\frac{\partial C_l}{\partial \beta} \right); L_\beta = \frac{\rho U^2 Sb}{2I_x} C_{l\beta} \tag{4.24}$$

$C_{l\beta}$ is the "effective" dihedral, and it is negative for the positive dihedral. It is very important for lateral stability and control and hence considered as a preliminary design parameter.

N_β is the change in the yawing moment N due to a change in the side velocity, v. It is mainly caused by the side force F_1 on the vertical tail. Following the previous development for L_v, we obtain

$$N_v = \frac{\rho USb}{2I_z} C_{n\beta}; \; N_\beta = \frac{\rho U^2 Sb}{2I_z} C_{n\beta}; \; C_{n\beta} = \frac{\partial C_{N\,(\text{or } n)}}{\partial \beta} \tag{4.25}$$

The vertical tail and related lever arm contribute to $C_{n\beta}$; the contribution is positive and signifies the fact that due to positive side slip, this force creates a positive yawing moment, which, in turn, tries to reduce the side slip, thereby maintaining the static stability.

4.3.3.2 Effect of Change in the Time Rate Change in the Side Velocity ($\dot{v}$): ($Y_{\dot{v}}$, $L_{\dot{v}}$, $N_{\dot{v}}$)

In fact nothing much is known of these derivatives; hence, generally these are neglected, but their expressions are given here for the sake of completion.

$$Y_{\dot{v}} = \frac{\rho Sb}{4m} C_{Y_{\dot{\beta}}};\ C_{y_{\dot{\beta}}} = \frac{\partial C_y}{\partial(\dot{\beta}b/2U)}$$

$$L_{\dot{v}} = \frac{\rho Sb^2}{4I_x} C_{l_{\dot{\beta}}};\ C_{l_{\dot{\beta}}} = \frac{\partial C_l}{\partial(\dot{\beta}b/2U)}$$

$$N_{\dot{v}} = \frac{\rho Sb^2}{4I_z} C_{n_{\dot{\beta}}};\ C_{n_{\dot{\beta}}} = \frac{\partial C_n}{\partial(\dot{\beta}b/2U)}$$

4.3.3.3 Effect of Change in Roll Rate p: (Y_p, L_p, N_p)

These derivatives arise due to the change in rolling angular velocity and roll rate p; the latter creates a linear velocity of the vertical, horizontal, and wing surfaces, causing a local change in the AOA of these surfaces and the lift distribution, which, in turn, causes the moment about the C.G.

Y_p is the change in side force due to a change in the roll rate. We have

$$Y_p = (1/m)\partial Y/\partial p = \frac{\rho USb}{4m}\frac{\partial C_y}{\partial(pb/2U)} = \frac{\rho USb}{4m} C_{Yp} \tag{4.26}$$

This derivative is not significant and is often neglected.

L_p is the change in the rolling moment due to a change in the roll rate p, known as a roll damping derivative. The induced rolling moment is such that it opposes the rolling moment, thereby providing the dynamic stability to the aircraft. We have

$$L_p = (1/I_x)\partial L/\partial p = \frac{\rho USb^2}{4I_x}\frac{\partial C_l}{\partial(pb/2U)};\ L_p = \frac{\rho USb^2}{4I_x} C_{l_p} \tag{4.27}$$

The derivative C_{l_p} (negative in sign) is composed of contributions from the wing and the horizontal/vertical tails; the major being from the wing, and the variations in the derivative with Mach number, and AOA closely follows the variations in the wing lift curve slope $C_{L\alpha}$.

N_p is a change in the yawing moment due to a change in the roll rate. The lift forces acting on the down-going and up-going semi-spans are rotated forward and backward, respectively; the change in the direction of these forces results in a negative yawing moment about the Z-axis. For the flight near the stall, the drag forces may become important and would result in a yawing moment of the opposite sign:

$$N_p = \frac{\rho USb^2}{4I_z} C_{n_p} \tag{4.28}$$

The derivative C_{n_p} arises mainly from the wing, and the vertical tail positive or negative), depending on its geometry.

4.3.3.4 Effect of Change in Yaw Rate *r*: (Y_r, L_r, N_r)

The yawing angular velocity and yaw rate cause these derivatives. The yaw (angle) rate causes the change in the side force F_1 that acts on the vertical tail surface, and the fact that effective AOA of the vertical tail increases.

Y_r is the change in side force due to a change in the yaw rate:

$$Y_r = (1/m)\partial Y/\partial r = \frac{\rho USb}{4m}.C_{y_r};\ C_{y_r} = \frac{\partial C_y}{\partial (rb/2U)} \tag{4.29}$$

This derivative is of little importance, and it is generally neglected.

L_r is the change in the rolling moment due to a change in the yaw rate *r*. The forward speed of a station/point of a distance d_1 toward the positive *Y*-axis from the *X*, *Z* plane on the semi-span of the wing is decreased by an amount d_1r, resulting in a decrease in lift at this point, and similarly the forward speed of a point distance d_2 normal to the *X*, *Z* plane on the semi-span of the wing is increased (because of the yawing moment and the rotation of the airframe clockwise about the Z-axis) by an amount d_2r, resulting in an increase in lift at this station; and this differential lift yields a rolling moment (usually positive). The corresponding derivative is as follows:

$$L_r = (1/I_x)\partial L/\partial r = \frac{\rho USb^2}{4I_x}C_{l_r};\ C_{l_r} = \frac{\partial C_l}{\partial (rb/2U)} \tag{4.30}$$

N_r is a change in the yawing moment due to a change in yaw rate. The side force F_1 also causes a moment about the *Z*-axis since the vertical tail is some distance aft of the C.G.; this moment is usually negative. The derivative is given as with the usual definition of the dimensional/non-dimensional derivatives:

$$N_r = \frac{\rho USb^2}{4I_z}C_{n_r} \tag{4.31}$$

4.3.3.5 Effect of Change in Aileron and Rudder Control Surface Deflection: (Y_{δ_a}, L_{δ_a}, N_{δ_a}, Y_{δ_r}, L_{δ_r}, N_{δ_r})

The L_{δ_a} and N_{δ_r} are the primary control derivatives or control power. If the absolute magnitude of these derivatives are higher, then for a given deflection, more rolling and yawing moments are generated. This is regarded as higher control sensitivity for a given moment of inertia.

$$Y_{\delta_a} = (1/m)\partial Y/\partial \delta_a = \frac{\rho U^2 S}{2m}C_{y\delta_a},\ Y\text{-axis force created by deflections ailerons}$$

$$Y_{\delta_r} = (1/m)\partial Y/\partial \delta_r = \frac{\rho U^2 S}{2m}C_{y\delta_r},\ Y\text{-axis force created by deflections of rudder pedals.} \tag{4.32}$$

The side force due to positive δ_r (rudder deflection) is positive $C_{y\delta r}$. Its effects are relatively unimportant in lateral stability and control, except when considering lateral acceleration feedback to an autopilot.

N_{δ_a} and L_{δ_r} are cross-derivatives (the change in yawing moment due to aileron deflection; and the change in the rolling moment due to the deflections of the rudder pedals, respectively) and will be useful if there is rudder–aileron interconnect for certain high performance fighter aircraft or rotorcrafts.

$$N_{\delta_a} = (1/I_z)\partial N/\partial\delta_a = \frac{\rho U^2 Sb}{2I_z} C_{n\delta_a};\ N_{\delta_r} = (1/I_z)\partial N/\partial\delta_r = \frac{\rho U^2 Sb}{2I_z} C_{n\delta_r}$$

$$L_{\delta_a} = (1/I_x)\partial N/\partial\delta_a = \frac{\rho U^2 Sb}{2I_x} C_{l\delta_a};\ L_{\delta_r} = (1/I_x)\partial N/\partial\delta_r = \frac{\rho U^2 Sb}{2I_x} C_{l\delta_r}$$

Some most important L-D derivatives are explained as follows:

a. $C_{n\beta}$ represents the directional static stability; if $C_{n\beta} > 0$, then the aircraft possesses static directional stability; the aircraft will always point to the relative wind direction, and hence, this stability is called the "weathercock" stability, synonymous with the concept of the weathercock at airports, etc. At high angles of attack (HAOA), the $C_{n\beta}$ will be affected significantly since the fin will get submerged in the wing-body wake because the major portion of this derivative comes from the vertical tail area and the lever arm that stabilize the airframe just as the tail feathers of an arrow stabilize the arrow shaft. The contribution to $C_{n\beta}$ from the tail is positive (stabilizing) and that from the body is negative (signifying instability); the one from the wing is usually positive but very small. Also, since the major portions depend on the dimensions of the body, the C.G. variation does not affect the derivative much since this variation would be a small fraction of the body length. This derivative is of importance in dynamic lateral stability and control. It governs the natural frequency of the DR oscillatory mode of the aircraft. It also contributes to the spiral stability of the aircraft. The high values $C_{n\beta}$ aids the pilot in affecting coordinated turns and prevent excessive sideslip and yawing motion in extreme flight maneuver.
b. $C_{l\beta}$ is the rolling moment derivative (due to a small change in AOSS). This provides the dihedral stability for $C_{l\beta} < 0$ for the positive dihedral. The contributions to this derivative come from: (i) wing (as a function of geometric dihedral, sweep, aspect ratio, and AOA; (ii) wing location on the fuselage; if it is high, then it is negative increment to the derivative, and if it lower, then there is a positive increment, and (iii) direct force on the vertical tail contribute decreasingly negative increments as the fin moves down with respect to the X-axis with increasing trim AOA. The dihedral instability will contribute to the spiral divergence. A stable dihedral will tend to decrease the AOSS by rolling into the direction of yaw. Both $C_{l\beta}$ and $C_{n\beta}$ affect the aircraft Dutch roll (DR) mode and spiral mode, as will be discussed in Chapter 5. It is also involved in maneuvering characteristics of an

aircraft, especially with regard to lateral control with the rudder alone near stall. To improve the DR damping of an aircraft, small negative values of the derivative are sought.

c. C_{l_p} is the damping-in-roll derivative and determines roll subsidence. The change in rolling velocity causes the change in rolling moment. The positive roll rate induces a restoring/opposing moment, and hence, the derivative has generally a negative magnitude. It can also be considered as a design criterion since it directly affects the design of ailerons. Since C_{l_p} in conjunction with the $C_{l\delta_a}$ establishes the airframe's maximum available rolling velocity, this is an important criterion of flying qualities. The positive values occur only when the wing or portions thereof are stalled, and such flight situations (including the separated flows) are avoided.
d. C_{n_p} is a cross-derivative, usually negative, that influences the frequency of DR damping, The larger its negative value, the greater is the reduction in the DR damping; also, this causes higher (un-coordinated) side slipping motions accompanying turn entry or exit; hence positive values of C_{n_p} are to be desired.
e. C_{n_r} is the "yaw damping derivative", and it is made up of contributions (all of negative sign), from the wing, the fuselage, and the vertical tail. It is the main contributor to the damping of the DR oscillatory mode, and it is important to the spiral mode; for each mode, large negative values of the derivative are sought.
f. C_{l_r} mainly affects the aircraft's spiral mode, and for the stability, it is desirable that its positive value be as small as possible. For a conventional airframe configuration, the changes in C_{l_r} of reasonable magnitude show only slight effects on the DR damping.
g. $C_{l\delta a}$, $C_{n\delta a}$, $C_{n\delta r}$ are important control derivatives.

$C_{l\delta_a}$ is the aileron control effectiveness, which in conjunction with damping in roll (C_{l_p}) establishes the maximum available roll rate of an airframe, which is a very important consideration in fighter tactics at high speeds. $C_{n\delta_r}$ is the "rudder effectiveness" derivative or the rudder power. $C_{n\delta_a}$ arises in part from the difference in drag due to the down aileron compared with the drag of the up aileron. If it is negative, then it is called the adverse yaw because it causes the airframe to yaw initially in the direction opposite to that desired by the pilot when she deflects the ailerons for turn; if it is positive, then it produces favorable or "proverse yaw" in the turning maneuver. In an overall sense, for the proverse or stable control of an aircraft (including a spacecraft), the following should be true:

$$C_{n\beta} > \frac{C_{n\delta_a} C_{l\beta}}{C_{l\delta_a}}$$

Numerical values of certain lateral-directional derivatives are given in Table 4.6.

TABLE 4.6
Typical Values of Some L-D Derivatives for a Small Trainer Aircraft

Derivative	Analytical/Wind Tunnel Value	Flight Determined Derivatives (at 100 Knots/1.64 km alt/4.21 deg. AOA Trim)
C_{l_p}(/rad)	−0.532	−0.4366
$C_{n\beta}$(/rad)	0.0688	0.0734
$C_{l\beta}$(/rad)	−0.0573	−0.037
$C_{l\delta_a}$(/rad)	−0.206	−0.2126

4.3.4 Compound Lateral-Directional Derivatives

The DR coupling is a dynamic lateral-directional stability of the stability axis; the body axis yaw and roll moments' coupling with the SS can produce LD instability or PIO (Chapter 10). The more pronounced DR coupling appeared due to the mass concentration in fuselage and decreased damping with increase speed and altitude. Nearly seven decades ago, the body axis directional stability $C_{n\beta}$ was calculated from the stability axis frequency of the DR oscillations, and the influence of inertia ratios, AOA, and effective dihedral were neglected in these calculations; and during that period, the values of $C_{n\beta}$ obtained from the flight test data were invariably less than the WT values. The correct computational formula is given next; two important compound derivatives are defined as follows [8]:

a. Later on, the researchers included the inertias, AOA, and body axis stability and derived the more accurate formula of $C_{n\beta}$, and the DR frequency:

$$C_{n\beta\,\text{dynamic}} = C_{n\beta}\cos\alpha - (I_{zz}/I_{xx})C_{l\beta}\sin\alpha;\ \omega^2_{\text{DR dynamic}} = \frac{N_\beta}{I_{zz}}\cos\alpha - \frac{L_\beta}{I_{xx}}\sin\alpha \quad (4.33)$$

This compound derivative contains $C_{n\beta}$ and $C_{l\beta}$ static stability derivatives. This derivative will be the same as the weathercock stability if the AOA is zero or negligibly very small. For reasonably small AOA, we have: $C_{n\beta\,\text{dynamic}} = C_{n\beta} - (I_{zz}/I_{xx})C_{l\beta}\alpha$

For DR static stability, $C_{n\beta\,\text{dynamic}}$ should be positive. At high AOA, it might become negative, as can be seen from the above expression, thereby affecting the stability adversely.

b. Lateral Control Divergence Parameter (LCDP)

$$\text{LCDP} = C_{n\beta} - (C_{n\delta_a}/C_{l\delta_a})C_{l\beta} \quad (4.34)$$

This parameter should be positive to avoid reverse responses due to adverse aileron yaw.

TABLE 4.7
Typical Values of the Compound Derivatives for a Supersonic Fighter Aircraft (Flight Condition: Mach = 0.55, Altitude = 6 km, and Trim AOA = 6 deg.)

Derivative	Analytical/Wind Tunnel Values	Flight Determined Derivatives
$C_{n\beta\,\text{dynamic}}$ (/rad)	0.164	0.154
LCDP (/rad)	0.138	0.125

The positive values of the compound derivatives are necessary up to a maximum AOA for ensuring departure free behavior of the aircraft. For non-zero AOA, both the derivatives should be studied, and the stability should be guaranteed. Typical numerical values of these compound derivatives are given in Table 4.7.

The DADs are directly related to the forces and moments as well as to the aircraft response variables. The DADs are useful since they directly appear in the simplified EOM/differential equations that describe the vehicle dynamic responses as will be seen in Chapter 5. That is to say that the DADs appear in the TF models that can be used to describe the dynamic responses/Bode diagrams for the aircraft in a particular axis in an approximate way. The NDADs are directly related to the aerodynamic coefficients and non-dimensional velocities (see Tables 4.3 and 4.4). The merits of the non-dimensional forms are: (i) easy correlation between the results from different test methods like wind tunnel, flight tests, theoretical predictions (CFD, DATCOM), (ii) correlation between the results from different flight test conditions, and (iii) possible comparison/correlation of the results of different configurations of the same aircraft or the different aircraft of similar kind (or even the other kinds!). The DADs and NDADs defined and studied here can be very conveniently used in the aerodynamic model of the atmospheric flight vehicles.

4.4 MISSILE AERODYNAMIC DERIVATIVES

Due to the axis symmetry of a missile, pitch derivatives are identical to yaw derivatives in magnitude with some sign changes due to asymmetry of an orthogonal right-handed axis system [6]. As in the case of aircraft, the force derivatives are divided by mass m and the moment derivatives by respective moment of inertia. A complete linear derivative model of a missile could have 30 derivatives [9], as shown in Table 4.8.

Interestingly, these derivatives of Table 4.8 are called acceleration derivatives. The reason will be clear from the following development:

$$\ddot{\theta} = \frac{\text{pitching moment}}{\text{pitch inertia}} = \frac{\rho U^2 S\bar{c}}{2I_y}\frac{\partial C_m}{\partial \alpha} = \frac{\delta\ddot{\theta}}{\delta\alpha} = \frac{C_m\bar{q}S\bar{c}}{I_y} = \frac{\rho U^2 S\bar{c}}{2I_y}C_{m\alpha} \quad (4.35)$$

TABLE 4.8
Missile Derivatives in a Linear Model [9]

Changes in Variables	Angle of Attack α (Pitch)	Angle of Attack β (yaw)	Fin Deflection		
			Pitch	Yaw	Roll
Pitch angular acceleration	$\frac{\Delta\ddot{\theta}}{\Delta\alpha}$	$\frac{\Delta\ddot{\theta}}{\Delta\beta}$	$\frac{\Delta\ddot{\theta}}{\Delta\delta_e}$	$\frac{\Delta\ddot{\theta}}{\Delta\delta_r}$	$\frac{\Delta\ddot{\theta}}{\Delta\delta_a}$
Roll angular acceleration	$\frac{\Delta\ddot{\varphi}}{\Delta\alpha}$	$\frac{\Delta\ddot{\varphi}}{\Delta\beta}$	$\frac{\Delta\ddot{\varphi}}{\Delta\delta_e}$	$\frac{\Delta\ddot{\varphi}}{\Delta\delta_r}$	$\frac{\Delta\ddot{\varphi}}{\Delta\delta_a}$
Yaw angular acceleration	$\frac{\Delta\ddot{\psi}}{\Delta\alpha}$	$\frac{\Delta\ddot{\psi}}{\Delta\beta}$	$\frac{\Delta\ddot{\psi}}{\Delta\delta_e}$	$\frac{\Delta\ddot{\psi}}{\Delta\delta_r}$	$\frac{\Delta\ddot{\phi}}{\Delta\delta_a}$
Normal (pitch) acceleration, a_z	$\frac{\Delta a_z}{\Delta\alpha}$	$\frac{\Delta a_z}{\Delta\beta}$	$\frac{\Delta a_z}{\Delta\delta_e}$	$\frac{\Delta a_z}{\Delta\delta_r}$	$\frac{\Delta a_z}{\Delta\delta_a}$
Axial acceleration, a_y	$\frac{\Delta a_x}{\Delta\alpha}$	$\frac{\Delta a_x}{\Delta\beta}$	$\frac{\Delta a_x}{\Delta\delta_e}$	$\frac{\Delta a_x}{\Delta\delta_r}$	$\frac{\Delta a_x}{\Delta\delta_a}$
Yaw acceleration a_y	$\frac{\Delta a_y}{\Delta\alpha}$	$\frac{\Delta a_y}{\Delta\beta}$	$\frac{\Delta a_y}{\Delta\delta_e}$	$\frac{\Delta a_y}{\Delta\delta_r}$	$\frac{\Delta a_y}{\Delta\delta_a}$

We know from eqn. (4.7) that

$$M_\alpha = \frac{1}{I_y}\frac{\partial M}{\partial \alpha} = \frac{\rho U^2 S \bar{c}}{2I_y}\frac{\partial C_m}{\partial \alpha} = \frac{\delta\ddot{\theta}}{\delta\alpha} \tag{4.36}$$

Hence, $\frac{\delta\ddot{\theta}}{\delta\alpha} = M_\alpha$, and we see that the corresponding derivative is also called a pitch angular acceleration derivative. Similarly, all other derivatives of Table 4.8 can be shown equivalent in their formats by their definitions to aircraft derivatives. Some additional derivatives seem to have been defined in the case of a missile, as can be seen from Table 4.8 as compared to the aircraft.

4.4.1 Longitudinal Derivatives

The normal force is given by $Z = 1/2\ \rho U^2 S C_z$.

Here, the area S is the maximum body cross-section reference area. C_z is the normal force coefficient and is a function of AOA (called incidence in pitch) for a given altitude and Mach number.

Z_α is defined as follows:

$$Z_\alpha = \frac{1}{2m}\rho U^2 S \frac{\partial C_z}{\partial\ \alpha}; Z_\alpha = \frac{1}{2m}\rho U^2 S\ C_{z\alpha} \text{Or } Z_w = \frac{1}{2m}\rho US\ C_{z\alpha} \tag{4.37}$$

The actual values of the normal force coefficient depend on the shape of the missile and the reference area S. The normal force coefficient usually for a missile is expressed in the literature as [5]:

$$C_N = C_{N_0} + C_{N_\alpha}\alpha + C_{N_q}\frac{qc}{2V} + C_{N\dot{\alpha}}\left(\frac{\dot{\alpha}c}{2V}\right) + \cdots \tag{4.38}$$

Here, V is the resultant velocity or the cruising speed of the missile. The parameter c is the total body length of the missile. In case of the missile, N denotes the normal force unlike for the aircraft, where Z denotes the vertical force. We will use only Z for the normal force uniformly. The meanings of the derivatives as such do not change.

M_w and M_q have the same significance as in the case of aircraft aerodynamics. M_{δ_e} will have an opposite sign for a missile with a canard control surface (in the front). M_w will also change sign in case of an unstable missile configuration. Z_{δ_e} is related to M_{δ_e} in usual way, and both are elevator control surface effectiveness derivatives. The derivative Z_q is of no practical significance and hence neglected in missile aerodynamics.

4.4.2 Lateral-Directional Derivatives

There are only a few of such important derivatives.

4.4.2.1 Roll Derivatives

There are only two important derivatives in the roll degree of freedom.

L_p is the usual damping derivative in the roll, it is a change in the rolling moment due to a change in the roll rate. As usual it is negative in sign. It is often considered to have a constant value for a given Mach number and altitude. Due to positive roll rate, the change in the rolling moment is such that it will oppose this change in the roll rate and hence provide (increased) damping in the roll axis. It is usually defined as follows:

$$L_p = \frac{1}{I_x}\frac{\partial L}{\partial p} \tag{4.39}$$

In an aircraft, the major contribution to this derivative is from wing; however, in the case of a missile, this is not the case.

L_{δ_a} is the aileron control surface effectiveness derivative. It is defined as follows:

$$L_{\delta_a} = \frac{1}{I_x}\frac{\partial L}{\partial \delta_a} \tag{4.40}$$

4.4.2.2 Yaw Derivatives

4.4.2.2.1 Effect of Change in Side Speed v/AOSS: (Y_v, N_β); Y_v is Given as follows:

$$Y_v = (1/m)\partial Y/\partial v = \frac{\rho US}{2m}C_{Y\beta};\, Y_\beta = \frac{\rho U^2 S}{2m}C_{Y\beta} \tag{4.41}$$

It must be noted here that for a symmetrical missile, the normal force coefficient for yaw has the same value as the C_z. This lateral force derivative should be calculated from the total force from the wings, body and the control surfaces. It is presumed that the control surfaces are in a central position. N_β is the force derivative Y_v times the distance of center of pressure from the C.G. As in the case of an aircraft, this distance is known as the static margin. It is a measure of the static stability of the missile. In usual manner, the derivative is defines as follows:

$$N_v = \frac{\rho USb}{2I_z} C_{n\beta}; N_\beta = \frac{\rho U^2 Sb}{2I_z} C_{n\beta} \tag{4.42}$$

It will change sign for an unstable missile configuration.

4.4.2.2.2 *Effect of Change in Yaw Rate r:(Y_r, N_r)*

Y_r is a difficult derivative to calculate and measure, and hence, it is usually neglected.

$$N_r \text{ given as: } N_r = \frac{\rho USb^2}{4I_z} C_{n_r} \tag{4.43}$$

It is a damping derivative in yaw. It has a small value.

4.4.2.2.3 *Effect of Change of Rudder Control Surface Deflection (Y_{δ_r}, N_{δ_r}):*

Y_{δ_r} is the side force derivative due to deflection of the rudder control surface.

N_{δ_r} is the yawing moment derivative due to deflection of the rudder control surface. It has an opposite sign for a missile with the canard control surface. It is given as Y_{δ_r} times the distance of the center of pressure of the rudder from the C.G. The formats of these derivatives are usually similar to the ones for the aircraft.

Table 4.9 gives the most important missile aerodynamic derivatives; a missile has very few derivatives that are most important to describe its dynamics, especially from the control point of view. It will be a good idea and exercise to establish the equivalence between the formulations of Tables 4.5 and 4.8. We further illustrate this as follows since acceleration = force/mass:

$$\frac{\partial a_y}{\partial \beta} = \frac{\partial F_y}{m\partial \beta} = \frac{\partial(\bar{q}SC_y)}{m\partial \beta} = \frac{\rho U^2 S \partial C_y}{2m\partial \beta} = \frac{\rho U^2 SC_{y\beta}}{2m} = Y_\beta \tag{4.44}$$

4.5 ROTORCRAFT AERODYNAMIC DERIVATIVES

In rotorcraft flight dynamics analysis, the same stability and control derivative expressions as used in the aircraft analysis are used. The asymmetric cross-coupling derivatives are usually neglected in aircraft flight mechanics analysis; however, in rotorcraft analysis, these are also used. For a rotorcraft, we have

$$M_w = \frac{1}{I_y}\frac{\partial M}{\partial w} \tag{a}$$

TABLE 4.9
Most Important Missile Aerodynamic Derivatives

	Response Variables								
	With Respect to Component Velocities			With Respect to Angular Velocities (Rates)			With Respect to Control Surface Deflections		
Forces/ Moments	***u*** **(Axial)**	***v*** **(Side)**	***w*** **(Normal)**	***p*** **(Roll)**	***q*** **(Pitch)**	***r*** **(Yaw)**	δ_e **(Elevon/ Elevator)**	δ_a **(Aileron)**	δ_r **(Rudder)**
X(axial)									
Y (side)		Y_v, Nm^{-1}s (negative) $y_v = \frac{Y_v}{m}$, s^{-1} $C_{Y\beta}$				Y_r, Nsec (usually −ve ve) $y_r = \frac{Y_r}{m}$, ms^{-1} C_{Yr}			Y_{δ_r}, N (positive) $y_{\delta_r} = \frac{Y_{\delta_r}}{m}$, ms_{-2} $C_{Y\delta r}$
Z (vertical/ normal)			Z_w, Nm^{-1}s (negative) $z_w = \frac{Z_w}{m}$, s^{-1} $C_{z\alpha}$		$Z_q{}^{a}$, Nsec (usually −ve) $z_q = \frac{Z_q}{m}$, ms^{-1} C_{zq}		Z_{δ_e}, N (negative) $z_{\delta_e} = \frac{Z_{\delta_e}}{m}$, ms^{-2} $C_{z\delta e}$		
L (rolling)				L_p, Nms (negative) $l_p = \frac{L_p}{I_x}$, s^{-1} C_{lp}				L_{δ_a}, Nm (negative) $l_\delta = \frac{L_\delta}{I_x}$, s^{-2} $C_{l\delta a}$	

(Continued)

TABLE 4.9 (*Continued*)
Most Important Missile Aerodynamic Derivatives

	Response Variables								
	With Respect to Component Velocities			With Respect to Angular Velocities (Rates)			With Respect to Control Surface Deflections		
Forces/ Moments	u (Axial)	v (Side)	w (Normal)	p (Roll)	q (Pitch)	r (Yaw)	δ_e (Elevon/ Elevator)	δ_a (Aileron)	δ_r (Rudder)
M (pitching)			M_w, Ns (negative) $m_w = \frac{M_w}{I_y}$, $m^{-1}s^{-1}$ $C_{m\alpha}$		M_q, Nms (negative) $m_q = \frac{M_q}{I_y}$, s^{-1} C_{ma}		M_{δ_e}, Nm (negative) $m_{\delta_e} = \frac{M_{\delta_e}}{I_y}$, s^{-2} $C_{m\delta e}$		
N (yawing)		N_v, Nsec (positive) $n_v = \frac{N_v}{I_z}$, $m^{-1}s^{-1}$ $C_{n\beta}$				N_r, Nmsec (negative) $n_r = \frac{N_r}{I_z}$, s^{-1} C_{nr}			N_{δ_r}, Nm (negative) $n_{\delta_r} = \frac{N_{\delta_r}}{I_z}$, s^{-2} $C_{n\delta r}$

In dimensional, say Z, and semi non-dimensional, say z, forms.

[a] Not important or negligible derivative; Newton (unit of weight = kg mass × g, ms^2); m-meter.

In this case, the controls are held at the trim values. The higher order rotor DOFs are allowed to reach a new equilibrium. The conventional 6 DOF derivative model is a poor approximation for rotorcrafts since the dynamics of the rotor response are not well separated from the airframe dynamic responses [7]. The use of the 6 DOF quasi-static derivatives is found to give an impression of greater aircraft stability. This is so because the regressing flapping mode (of the rotor) is neglected. In flight, the rotor is continuously being excited by control and turbulence inputs, and the estimates from the flights might differ considerably from the analytical perturbation derivatives (e.g., wind tunnel derivatives). The main reason is that the linear rotor plus body models are not easily derived by analytical means. These models should be determined from more comprehensive nonlinear models. Also, because of the lack of physical interpretation of the higher order rotor plus body models, the use of the quasi-static derivatives persists in rotorcraft analysis [7]. In general, forces/moment equations are given as follows, by representing the changes about the trim conditions as the linear functions of the rotorcraft's states:

$$\begin{bmatrix} \dot{u} \\ \dot{v} \\ \dot{w} \\ \dot{p} \\ \dot{q} \\ \dot{r} \end{bmatrix} = \begin{bmatrix} X_u & X_v & X_w & X_p & X_q & X_r \\ Y_u & Y_v & Y_w & Y_p & Y_q & Y_r \\ Z_u & Z_v & Z_w & Z_p & Z_q & Z_r \\ L_u & L_v & L_w & L_p & L_q & L_r \\ M_u & M_v & M_w & M_p & M_q & M_r \\ N_u & N_v & N_w & N_p & N_q & N_r \end{bmatrix} \cdot \begin{bmatrix} u \\ v \\ w \\ p \\ q \\ r \end{bmatrix} + \begin{bmatrix} X_{\delta_0} & X_{\delta_x} & X_{\delta_y} & X_{\delta_p} \\ Y_{\delta_0} & Y_{\delta_x} & Y_{\delta_y} & Y_{\delta_p} \\ Z_{\delta_0} & Z_{\delta_x} & Z_{\delta_y} & Z_{\delta_p} \\ L_{\delta_0} & L_{\delta_x} & L_{\delta_y} & L_{\delta_p} \\ M_{\delta_0} & M_{\delta_x} & M_{\delta_y} & M_{\delta_p} \\ N_{\delta_0} & N_{\delta_x} & N_{\delta_y} & N_{\delta_p} \end{bmatrix} \begin{bmatrix} \delta_{\text{lon}} \\ \delta_{\text{lat}} \\ \delta_{\text{ped}} \\ \delta_{\text{col}} \end{bmatrix} \tag{4.45}$$

$$+ \text{biases} + \text{gravity terms} + \text{centrifugal specific(forces)terms}$$

If the coupling between various axes is strong, then the rotorcraft has many more derivatives (say up to 36 force and moment derivatives and 24 control derivatives, (4.45)) than an aircraft. For example, if the coupling between roll and pitch axes is quite strong, then L_q and M_p will have significant values. Also, X_u is the drag damping derivative. The complete sets of EOM of a rotorcraft were discussed in Chapter 3, and the simplified equations and linear models that can be easily used for generating dynamic responses will be discussed in Chapter 5, where the meanings and utility of all these derivatives will become very clear. It is emphasized again here that the meanings of these aerodynamic derivatives across aircraft, missile, and rotorcraft should remain the same, with some specific interpretations depending on the special interactions between various axes of the airplane. This will depend on the extra degrees of freedom entering the EOM, for example, for a rotorcraft. As we have seen in Chapter 3, the longitudinal modes for helicopters with zero or low forward speeds are very different from the short period and phugoid modes (see also Chapter 5). This is because certain derivatives disappear for the hovering flight. The derivatives X_w, M_w, and $M_{\dot{w}}$ are negligible. Also, in the hovering condition $U_0 = N_p = L_r = N_v = Y_p = Y_r = 0$, under certain conditions that will be discussed in Chapter 5. In [10], the numerical values of the 60 (sixty) aerodynamic derivatives estimated from the flight test data by using parameter estimation method for three helicopters (AH-64, BO 105, and SA-330/PUMA) are given. However, a good number of derivatives has very small or negligible values.

4.6 ROLE OF DERIVATIVES IN AIRCRAFT DESIGN CYCLE AND FLIGHT CONTROL LAW DEVELOPMENT

The aerodynamic derivatives play a very important role in selection of the configuration to meet performance, agility, and maneuverability of the vehicle [4]. Some design criteria that a typical fighter aircraft is expected to comply with in order to realize the acceptable stability and control performance characteristics are discussed next. The stability and control derivatives provide very important input to the design process.

Some derivative-related criteria:

Common criterion for stable/unstable configurations: $C_{n_\beta}=0.13$/rad up to high AOA. Also, at C_L max, $C_{n\delta_r}$ should be equal/greater than $C_{n\delta_r}$ at zero AOA.

Criterion for stable configuration: damping ratio of the short period should be between 0.35 and 1.3. It must be recalled that the SP damping is very closely dependent on C_{m_q}.

Criterion for unstable configuration: for absolute AOSS up to 15 deg., there should not be pitch up, i.e. $C_{m\beta} \leq 0$.

The lateral-directional derivatives are not influenced by longitudinal instability. However, the change in pitching moment due to change in AOSS ($C_{m\beta}$) plays an important role in the design cycle of unstable configuration. If it has a positive value, then it signifies the pitch up at moderate AOSS and can affect the $C_{m\alpha}$ and $C_{m\delta}$ derivatives. Thus, $C_{m\beta}$ should be zero or negative up to ±15 degrees of AOSS. From Figure 4.3, we see that there are certain criteria termed as gross and detailed criteria that are required to be met to obtain the final configuration of the vehicle. The gross criteria are based on the expert designer's feel translated to some numerical values of the derivatives that the configuration should possess. Once these criteria are satisfied, one can go to meet the detailed criteria. Some compound derivatives, as discussed in Section 4.3.3 based on basic derivatives, are to be met in the gross criteria. The positive values of these compound derivatives are needed to ensure departure-free dynamic behavior of the aircraft up to maximum AOA. The detailed criteria need detailed mathematical analysis and/or simulation of dynamics to further ascertain the compliance of these criteria. The derivatives play an indirect role in this process as they appear in TFs. These studies are also very important for the assessment of the stability and control characteristics of the aircraft for which flight control laws need to be developed, especially for the inherently unstable configuration. The derivatives form important inputs to the control law development process and are also important for determining air loads on the vehicle. The study of the aerodynamic derivatives is also important to assess the effect of changes in aircraft configurations (air brakes, slates, leading edge vortex control devices, etc.) and assessment of stores (like missiles and drop tanks). Further role of the stability derivatives is in

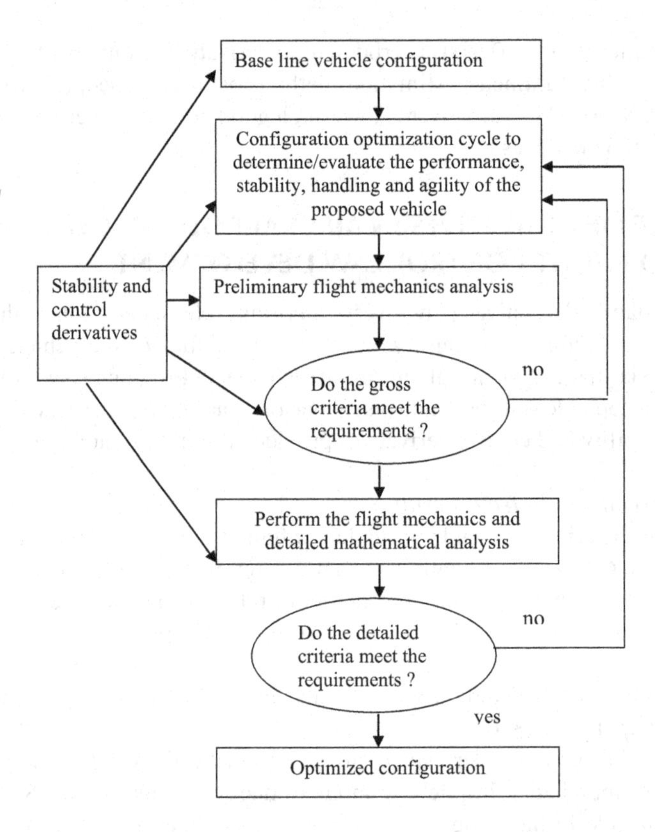

FIGURE 4.3 Role of aerodynamic derivatives in selection of an aircraft configuration.

realizing the combat effectiveness of any fighter aircraft. The combat effectiveness is a function of the following parameters:

a. Performance: sustained turn rate, specific excess power, and level speeds
b. Agility: ability to change the state vector in the shortest possible time
c. Maneuverability: roll performance, pull up, and roll pull out
d. Operational boundaries: AOA-AOSS envelope restricted by recovery margins, lateral-directional divergence angles, and pitch up

The primary dependence of these on the stability derivatives is qualitatively, as presented in Table 4.10. From the foregoing discussions, we see that the aerodynamic derivatives often called as stability and control derivatives or in short called stability derivatives (and very often only as derivatives!) play a very crucial role in aircraft design cycles, performance optimization, design and development of flight control laws, flight simulation, and many related analyses of aircraft (missiles) from flight mechanics point of view. We see that the information and knowledge from control and system theories play very important role in understanding and appreciating this design and development process apart from understanding the flight mechanics modeling and analysis.

TABLE 4.10
Relation of Derivatives to the Combat Effectiveness of a Typical Fighter Aircraft [4]

Significant Effects[a]	Performance	Agility	Maneuverability	Operational Boundaries
$C_{m\alpha}$	a	a		
$C_{m_q}, C_{n_r}, C_{l_p}$			a	a
$C_{n\beta}, C_{l\beta}$		a		a
$C_{m\delta_e}, C_{n\delta_r}, C_{l\delta_a}$		a	a	a

[a] Besides the usual role these derivatives play.

4.7 AIRCRAFT AERODYNAMIC MODELS

As we have seen earlier, aerodynamic forces and moments are expressed in Taylor's series expansion of the instantaneous (numerical) values of the aircraft variables. Specifically, the aerodynamic forces and moments (in fact the aerodynamic coefficients) can be expressed as functions of Mach number Mn, engine thrust F_T, and other aircraft motion and control variables α, β, $p, q, r, \varphi, \theta, \delta_e, \delta_a$, and δ_r. A complete set of the aerodynamic coefficients can be represented as follows [11]:

$$
\begin{aligned}
C_x &= C_{x0} + C_{x\alpha}\alpha + C_{x_q}\frac{q\bar{c}}{2V} + C_{x\delta_e}\delta_e + C_{xMn}Mn + C_{xF_T}F_T \\
C_y &= C_{y0} + C_{y\beta}\beta + C_{y_p}\frac{pb}{2V} + C_{y_r}\frac{rb}{2V} + C_{y\delta_a}\delta_a + C_{y\delta_r}\delta_r \\
C_z &= C_{z0} + C_{z\alpha}\alpha + C_{z_q}\frac{q\bar{c}}{2V} + C_{z\delta_e}\delta_e + C_{zMn}Mn + C_{zF_T}F_T \\
C_m &= C_{m0} + C_{m\alpha}\alpha + C_{m_q}\frac{q\bar{c}}{2V} + C_{m\delta_e}\delta_e + C_{mMn}Mn + C_{mF_T}F_T \\
C_l &= C_{l0} + C_{l\beta}\beta + C_{l_p}\frac{pb}{2V} + C_{l_r}\frac{rb}{2V} + C_{l\delta_a}\delta_a + C_{l\delta_r}\delta_r \\
C_n &= C_{n0} + C_{n\beta}\beta + C_{n_p}\frac{pb}{2V} + C_{n_r}\frac{rb}{2V} + C_{n\delta_a}\delta_a + C_{n\delta_r}\delta_r
\end{aligned}
\tag{4.46}
$$

The non-dimensional coefficients of lift and drag are denoted by C_L and C_D. Due to disposition of forces in x and z directions, we need to resolve the relevant coefficients as follows (see Exercise 4.4):

$$
\begin{aligned}
C_x &= C_L \sin\alpha - C_D \cos\alpha \\
C_z &= -(C_L \cos\alpha + C_D \sin\alpha)
\end{aligned}
\tag{4.47}
$$

Here, the effect of AOSS is neglected. If AOA is small, we can further simplify (4.47) as follows:

$$C_x = C_L\alpha - C_D, \qquad \text{If AOA is zero or very small } C_x = -C_D, \text{ and } C_z = -C_L$$

$$C_z = -(C_L + C_D\alpha),$$

This shows the straightforward relations between the axial force coefficients and the drag and lift coefficients. Furthermore, we get

$$C_L = -C_z\cos\alpha + C_X\sin\alpha, \quad C_D = -C_X\cos\alpha - C_z\sin\alpha \tag{4.48}$$

In terms of several independent variables we have

$$C_D = C_{D_0} + C_{D_\alpha}\alpha + C_{D_q}\frac{q\bar{c}}{2V} + C_{D_{\delta_e}}\delta_e + C_{D_M}M + C_{D_{FT}}F_T$$

$$C_L = C_{L_0} + C_{L_\alpha}\alpha + C_{L_q}\frac{q\bar{c}}{2V} + C_{L_{\delta_e}}\delta_e + C_{L_M}M + C_{L_{FT}}F_T \tag{4.49}$$

To account for nonlinear and unsteady aerodynamic effects, the higher order derivatives and down wash effects should be included. Also, some additional derivatives can be included to account for aerodynamic cross-coupling effects:

$$C_m = C_{m0} + C_{m\alpha}\alpha + C_{m_q}\frac{q\bar{c}}{2V} + C_{m\delta_e}\delta_e + C_{mM}M + C_{mFT}F_T + C_{m_{\beta^2}}\beta^2 + C_{m_{\beta^2\alpha}}\beta^2\alpha$$

$$+ C_{m_{\delta^2_r}}\beta^2\delta_r^{\,2} + C_{m_p}|pb/(2V)| + C_{m_r}|rb/(2V)| + \cdots + \tag{4.50}$$

However, the model of eqn. (4.50) is quite complex, and it will be difficult to determine the additional aerodynamic derivatives from flight test data of an aircraft. When the change in airspeed is not significant during the flight maneuver, the forces X, Y, and Z and the moments L, M, and N can be expanded in terms of the dimensional derivatives, rather than non-dimensional derivatives, for parameter estimation. Let us derive these expressions. Starting from eqn. (4.50), considering only three important terms and substituting the formula for the non-DADs (in terms of the dimensional derivatives) from Table 4.3, we obtain

$$\frac{M}{\bar{q}S\bar{c}} = \frac{2I_y}{\rho SU\bar{c}}M_w\alpha + \frac{4I_y}{\rho SU\bar{c}^2}M_q q\bar{c}/(2U) + \frac{2I_y}{\rho SU^2\bar{c}}M_{\delta e}\delta_e \tag{4.51}$$

Canceling out the common terms on both the sides, we get

$$M = I_y(M_w w + M_q q + M_{\delta e}\delta_e)$$

Substituting the defining terms for the dimensional derivatives from Table 4.3, we obtain

$$M = \frac{\partial M}{\partial w}w + \frac{\partial M}{\partial q}q + \frac{\partial M}{\partial \delta_e}\delta_e$$

We see that the pitching moment is expressible in terms of the dimensional derivatives in a very straight forward manner. Similarly, other moments and forces can be expressed in terms of the corresponding dimensional derivatives (with certain simple assumptions, see Exercise 4.11) as follows:

$$\begin{aligned} X &= m\left(X_u u + X_w w + X_q q + X_{\delta e}\delta_e\right) \\ Y &= m\left(Y_v v + Y_p p + Y_q q + Y_r r + Y_{\delta_a}\delta_a + Y_{\delta_r}\delta_r\right) \\ Z &= m\left(Z_u u + Z_w w + Z_q q + Z_{\delta e}\delta_e\right) \\ L &= I_x\left(L_v v + L_p p + L_q q + L_r r + L_{\delta_a}\delta_a + L_{\delta_r}\delta_r\right) \\ M &= I_y\left(M_u u + M_w w + M_q q + M_{\delta e}\delta_e\right) \\ N &= I_z\left(N_v v + N_p p + N_q q + N_r r + N_{\delta_a}\delta_a + N_{\delta_r}\delta_r\right) \end{aligned} \tag{4.52}$$

Starting from the natural phenomenon that an airplane experiences aerodynamic (air flow/interactions) forces and moments in flight, we have come to gradually appreciate the concept of the aerodynamic coefficients/derivatives that form the key elements in understanding the flight dynamic responses of an aircraft. The aerodynamic coefficients can (i) also be expressed in terms of derivatives due to change in forward speed, for example, $C_{L_u}, C_{D_u}, C_{z_u}$ and C_{m_u}; (ii) use higher order derivatives (e.g., $C_{x_{\alpha^2}}, C_{z_{\alpha^2}}$, and $C_{m_{\alpha^2}}$) to account for nonlinear effects; and (iii) $C_{L\dot{\alpha}}$ and $C_{m\dot{\alpha}}$ derivatives to account for unsteady aerodynamic effects. The choice is problem-specific. Some simple and moderately complicated forms are used in parameter estimation in Chapter 9.

APPENDIX 4A AIRCRAFT'S STATIC AND DYNAMIC STABILITY

If the sum of all the resultant forces and moments acting on the aircraft is zero about its C.G., then the aircraft is in a state of equilibrium. In this condition, the aircraft is said to be flying at a trim condition in a steady and uniform flight (with constant speeds); and the translational and angular accelerations are normally zero. An aircraft is said to be (statically) stable if on encountering a disturbance during steady uniform flight is able to come back to its initial trim state. There are two types of stability: static and dynamic. An aircraft possesses positive static stability if it *tends* to return to its equilibrium position after disturbance. An example to illustrate this point is shown in Figure 4.4. The ball in position 1 is stable as it will return to its equilibrium after disturbance. The ball in position 2 is unstable as any disturbance will cause it to roll off. In position 3, the ball possesses neutral stability because, being on a flat surface, any disturbance will cause it to seek a new position. So, the static stability does not say anything about whether the system ever reaches its equilibrium position or how it gets there; the latter aspect is dealt with in the realm of dynamic stability, and it deals with the *time history* of the aircraft motion (after ensuring that it has the static stability). After a disturbance, the aircraft may tend to

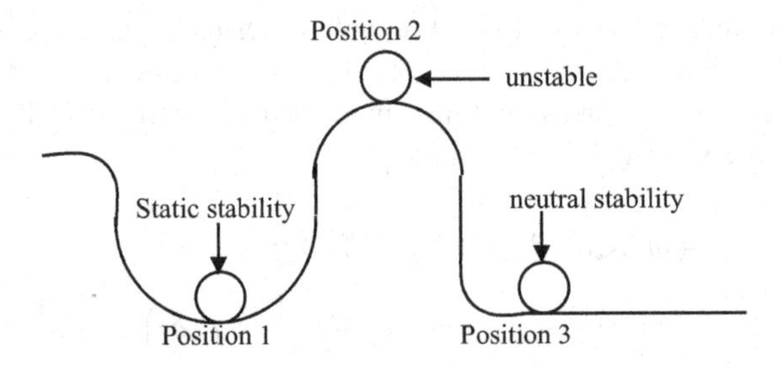

FIGURE 4.4 Types of static stability.

move back (because of its static stability) and settle down either monotonically or after a few oscillations. In either case, if it goes back to its equilibrium position on its own over a period of time, it is said to be dynamically stable. It should be noted that a dynamically stable aircraft must always be statically stable. On the other hand, static stability does not automatically ensure dynamic stability of the aircraft. Dynamic stability is very important for the flight control system design and, of course, for its smooth and safe flight.

From the foregoing explanation, it is clear that if the angle of attack increases, the aircraft develops a negative pitching moment so that it has tendency to return back to its equilibrium position. In other words, for an aircraft to possess static longitudinal stability, the slope of the pitching moment curve should be negative, that is, $\frac{dC_m}{d\alpha} < 0$ or $C_{m_\alpha} < 0$. In addition to having a negative pitching moment slope, the aircraft must also have a positive pitching moment at zero angle of attack to achieve trim at positive AOA, that is, C_{m0} should be positive (Figure 4.5). The value of C_{m_α} depends upon the location of the C.G. The point for which $C_{m_\alpha} = 0$ defines the aft limit of the C.G. At this point, the aircraft possesses neutral stability; this location of C.G. is called the neutral point. Estimation of the neutral point from flight test data is discussed in the next section. Controllability is the ability to control the aircraft to maneuver and

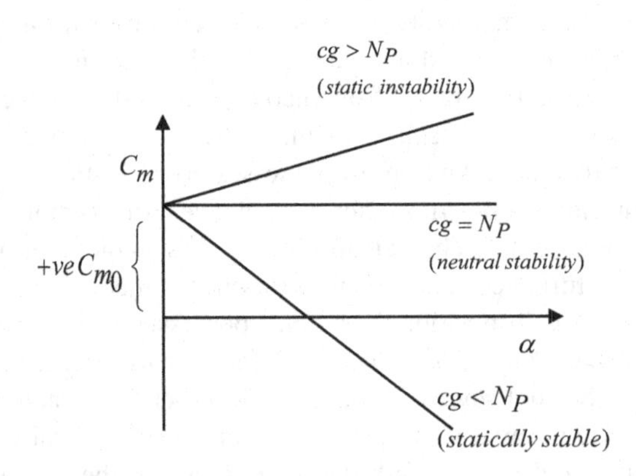

FIGURE 4.5 Longitudinal static stability as affected by C.G. location.

change the course of the flight path. Controllability depends on the pedal/stick force and the displacement needed to achieve the task. The faster the aircraft response to the force and displacement, the higher is the controllability.

4A.1 NEUTRAL AND MANEUVER POINTS

These points/parameters are a function of speed, AOA, external store configuration, control surface deployment (slats), etc.; extensive flight tests are conducted to accurately determine these critical stability parameters. Existing methods based on steady-state trim flights turn out to be time-consuming and are error-prone due to the results are dependent on air data and aircraft weight. An alternative approach, based on system theoretic concepts, is to estimate aircraft longitudinal static and dynamic stability from flight dynamic maneuvers in terms of neutral and maneuver points. In this procedure, the stability information is extracted from the short period dynamic response of the aircraft, which leads to substantial reduction in flight test time compared with the conventional steady-state flight tests. Since this method does not use air data information or mass/inertia data, the resulting estimates of the neutral and maneuver points are generally more accurate, as discussed in [9]. The neutral point is related to short-period static stability parameter M_α and, hence, the natural frequency. We estimate M_α values (using the parameter estimation method) from short-period maneuver flight data of an aircraft, flying it for different C.G. positions (minimum three) and then plotting with respect to C.G. and extend this line to the x-axis. The point on the x-axis when this line passes through "zero" on the y-axis is the neutral point (Figure 4.6). The distance between the neutral point N_P and the actual C.G. position is called the static margin, and when this margin is zero, the aircraft has neutral stability. As far as the maneuverability is concerned, the net normal force $L-W=(n-1)W$ is upward, the normal acceleration is $(n-1)g$, and the angular velocity is $q=(n-1)g/V$. The point where $\frac{\Delta\delta_e}{(n-1)}$, that is, the elevator angle/g is "zero" is called the control-fixed maneuver point (MP). The difference $(h_{MP}-h_{CG})$ is called the control-fixed maneuver margin.

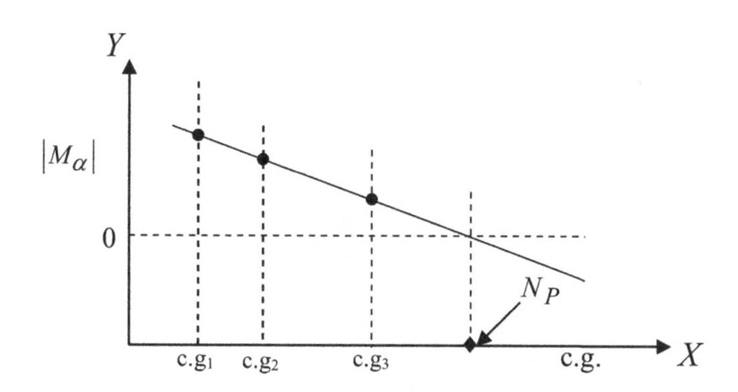

FIGURE 4.6 3-C.G. method for neutral point determination.

APPENDIX 4B TRANSFORMATIONS OF AERODYNAMIC DERIVATIVES

Derivatives from stability to the body axis are given as follows:

$$C_{N_\alpha} = C_{L_\alpha}\cos\alpha + C_{D_\alpha}\sin\alpha + C_c$$

$$C_{c_\alpha} = C_{D\alpha}\cos\alpha - C_{L_\alpha}\sin\alpha - C_N$$

$$C_{m_\alpha} = \left(C_{m\alpha}\right)_s$$

$$C_{l\beta} = \left(C_{l\beta}\right)_s\cos\alpha - \left(C_{n\beta}\right)_s\sin\alpha$$

$$C_{l_p} = \left(C_{l_p}\right)_s\cos^2\alpha + \left(C_{n_r}\right)_s\sin^2\alpha - \left(C_{n_p} + C_{l_r}\right)_s\sin\alpha\cos\alpha$$

$$C_{l_{\dot\beta}} = \left(C_{l_{\dot\beta}}\right)_s\cos\alpha - \left(C_{n_{\dot\beta}}\right)_s\sin\alpha$$

$$C_{l_r} = \left(C_{l_r}\right)_s\cos^2\alpha - \left(C_{n_p}\right)_s\sin^2\alpha - \left(C_{n_r} - C_{l_p}\right)_s\sin\alpha\cos\alpha$$

$$C_{l\delta} = \left(C_{l\delta}\right)_s\cos\alpha - \left(C_{n\delta}\right)_s\sin\alpha$$

$$C_{n\beta} = \left(C_{n\beta}\right)_s\cos\alpha + \left(C_{l\beta}\right)_s\sin\alpha$$

$$C_{n_r} = \left(C_{n_r}\right)_s\cos^2\alpha + \left(C_{l_p}\right)_s\sin^2\alpha + \left(C_{n_p} + C_{l_r}\right)_s\sin\alpha\cos\alpha$$

$$C_{n_{\dot\beta}} = \left(C_{n_{\dot\beta}}\right)_s\cos\alpha + \left(C_{l_{\dot\beta}}\right)_s\sin\alpha$$

$$C_{n_p} = \left(C_{n_p}\right)_s\cos^2\alpha - \left(C_{l_r}\right)_s\sin^2\alpha - \left(C_{n_r} - C_{l_p}\right)_s\sin\alpha\cos\alpha$$

$$C_{n\delta} = \left(C_{n\delta}\right)_s\cos\alpha + \left(C_{l\delta}\right)_s\sin\alpha$$

Derivatives from body to stability axes are given as follows:

$$C_{L\alpha} = C_{N_\alpha}\cos\alpha - C_{c_\alpha}\sin\alpha - C_D$$

$$C_{D\alpha} = C_{c\alpha}\cos\alpha + C_{N_\alpha}\sin\alpha + C_L$$

$$\left(C_{m\alpha}\right)_s = C_{m\alpha}$$

$$\left(C_{l\beta}\right)_s = C_{l\beta}\cos\alpha + C_{n\beta}\sin\alpha$$

$$\left(C_{l_r}\right)_s = C_{l_r}\cos^2\alpha - C_{n_p}\sin^2\alpha + \left(C_{n_r} - C_{l_p}\right)\sin\alpha\cos\alpha$$

$$\left(C_{l_{\dot\beta}}\right)_s = C_{l_{\dot\beta}} \cos\alpha + C_{n_{\dot\beta}} \sin\alpha$$

$$\left(C_{l_p}\right)_s = C_{l_p} \cos^2\alpha + C_{n_r} \sin^2\alpha + \left(C_{l_r} + C_{n_p}\right)\sin\alpha\cos\alpha$$

$$\left(C_{l_\delta}\right)_s = C_{l_\delta} \cos\alpha + C_{n_\delta} \sin\alpha$$

$$\left(C_{n_\beta}\right)_s = C_{n_\beta} \cos\alpha - C_{l_\beta} \sin\alpha$$

$$\left(C_{n_r}\right)_s = C_{n_r} \cos^2\alpha + C_{l_p} \sin^2\alpha - \left(C_{l_r} - C_{n_p}\right)\sin\alpha\cos\alpha$$

$$\left(C_{n_{\dot\beta}}\right)_s = C_{n_{\dot\beta}} \cos\alpha - C_{l_{\dot\beta}} \sin\alpha$$

$$\left(C_{n_p}\right)_s = C_{n_p} \cos^2\alpha - C_{l_r} \sin^2\alpha + \left(C_{n_r} - C_{l_p}\right)\sin\alpha\cos\alpha$$

$$\left(C_{n_\delta}\right)_s = C_{n_\delta} \cos\alpha - C_{l_\delta} \sin\alpha$$

APPENDIX 4C WIND TUNNEL EXPERIMENTAL METHOD FOR AERODYNAMIC COEFFICIENTS

Once the base line configuration of an aircraft is selected, the wind tunnel experiments are conducted on the (physically) scaled (down) models of the vehicle. The wind tunnel testing yields mainly the aerodynamic coefficients. Static rigs are used to obtain the steady-state aerodynamic characteristics. The tests are conducted in low-/high-speed wind tunnels. The effectiveness of various control surfaces is tested. To determine the dynamic derivatives, the forced oscillation test rigs are used. The aircraft physical model is oscillated, and force/moment measurements are made. Rotary rigs are used to determine the characteristics at high AOA and AOSS. To extend the force and moment coefficient measurements made in the wind tunnel for small-scale models to full-scale models, it is necessary to match the similarity parameters, for example, Reynolds number, Mach number, and Froude number. The Froude number is related to the ratio of inertia force to gravity force given as follows: $\frac{\text{mass} * \text{acceleration}}{\text{mass} * g} = \frac{V^2}{d\,g}$, which is the dimensionless quantity. The matching of Reynolds number is essential at low-speed tests, while the Mach numbers are matched for wind tunnel tests carried out at high speeds. Generally, it suffices to match these two similarity parameters for models held fixed within the wind tunnel. For a free flight model, however, it becomes necessary to match the Froude number. Also, the dynamic similarity (at the level of rotary motion, that is, frequency of oscillations of the model) is represented by the scaling called the helix angle: $\frac{\omega_{fs} \cdot l_{fs}}{V_{fs}} = \frac{\omega_m \cdot l_m}{V_m}$; here, l is the length, V is velocity, m stands for the model, and fs for the full-scale aircraft; if the velocities are the same, then the model will oscillate at a higher frequency than the actual aircraft.

EPILOGUE

The concept of aerodynamic derivatives as it is defined is very central to the flight mechanics modeling and analysis and addresses the micro-level behavior of the atmospheric vehicles. The concept of stability derivatives to represent the aerodynamic forces and moments was introduced by Bryan [1]. A very systematic way of defining the derivatives, detailed explanation of longitudinal and lateral-directional parameters/derivatives, and their origins are treated in [2]. Aerodynamic derivatives play a very important role in design and modification of vehicle's configuration and development of flight control laws. Also, determination of these derivatives from flight test data sheds considerable light on the actual behavior of the vehicle and helps validation of the aero data base (predictions based on analytical/wind tunnels sources) [5], [11]. In [12], airplane stability and control were stated to be viewed as an applied science and could benefit from several related fields.

EXERCISES

Exercise 4.1 Which is the most important merit of the non-dimensional form of the aerodynamic derivatives?

Exercise 4.2 Substitute eqn. (4.4) in eqn. (4.5) and verify eqn. (4.5).

Exercise 4.3 Verify that the unit for M_w is 1/mt-s by doing dimensional analysis.

Exercise 4.4 Based on the proper disposition of lift force L and drag force D, obtain the relationships

i. $X = L\sin\alpha - D\cos\alpha$ and $C_x = C_L\sin\alpha - C_D\cos\alpha$ and

ii. $Z = -(L\cos\alpha + D\sin\alpha)$ and $C_z = -(C_L\cos\alpha + C_D\sin\alpha)$.

Exercise 4.5 Derive the expression for X_w, eqn. (4.17), by using $X = L\sin\alpha - D\cos\alpha$ in the definition $X_w = \dfrac{1}{mU}\,\dfrac{\partial X}{\partial\,\alpha}$.

Exercise 4.6 Derive the expression for Z_w, eqn. 4.18, by using $Z = -(L\cos\alpha + D\sin\alpha)$ in the definition $Z_w = \dfrac{1}{mU}\dfrac{\partial Z}{\partial\alpha}$.

Exercise 4.7 Let the yawing moment derivative with respect to yaw rate be defined as follows:

$N_r = \dfrac{1}{I_z}\dfrac{\partial N}{\partial\,r}$, define, $C_{n_r} = \dfrac{\partial C_n}{\partial(\mathrm{br}/2\mathrm{U})}$, then obtain N_r in terms of C_{n_r}.

Exercise 4.8 Although in Tables 4.3, 4.4, and 4.6, the aerodynamic derivatives are defined in terms of forces and moments, why are they called the accelerations derivatives in Table 4.8?

Exercise 4.9 Establish relation between $\dfrac{\Delta\ddot{\psi}}{\Delta\beta}$ and N_β.

Exercise 4.10 Establish relation between $\dfrac{\Delta a_z}{\Delta\alpha}$ and $Z_{\dot{w}}$.

Exercise 4.11 Given $Z_M = -\dfrac{\rho S}{2m}\left(U^2\dfrac{\partial C_L}{\partial M} + 2U^2C_L/M\right)$, obtain the relation $Z_u a$.

Exercise 4.12 Let eqn. (4.6.) be written as $C_m = C_{m_0} + C_{m_\alpha}\alpha + C_{m_{\delta_e}}\delta_e$. An airplane is trimmed if forces and moments acting on the airplane are in equilibrium. From the above pitching moment coefficient equation, determine the elevator angle required to trim the airplane.

Exercise 4.13 From polar coordinate velocity form of EOM, obtain the expression for total velocity in terms of the aerodynamic derivatives in the wind axis when angle of attack and angle of sideslip are very small. Ignore thrust and gravity effects.

Exercise 4.14 Determine an approximate relation between the roll rate damping derivative C_{l_p} and the cross-derivative C_{n_p}. Assume cross-coupling inertia is negligible and roll rate and yaw rate are small, under steady-state roll and yaw motion.

REFERENCES

1. Bryan, G. H. *Stability in Aviation.* Macmillan, London, 1911.
2. McRuer, D. T., Ashkenas, I., and Graham, D., *Aircraft Dynamics and Automatic Control*, Princeton University Press, Princeton, NJ, 1973.
3. Nelson, R. C. *Flight Stability and Automatic Control*, 2nd Edn. McGraw Hill International Editions, New York, 1998.
4. Singh, K. P. and Sreenivasan, M. N. *Aerodynamic Stability Derivatives*, Personal communications and notes. Hindustan Aeronautics Limited, Bangalore, 1990.
5. Cook, M. V. *Flight Dynamics Principles.* Arnold, London, 1997.
6. Garnell, P. and East, D. J. *Guided Weapon Control Systems*, Pergamon Press Ltd, England, 1977.
7. Hansen, R. S. *Toward a Better Understanding of Helicopter Stability Derivatives.* Eighth European Rotorcraft Forum, Aix-En-Provence, France, Aug 31–Sept 3, 1982.
8. Yechout, T. R., Morris, S. L., Bossert, D. E., and Hallgren, W. F. *Introduction to Aircraft Flight Mechanics – Performance, Static Stability, Dynamic Stability, and Classical Feedback Control.* AIAA Education Series, AIAA, VA, 2003.
9. Driscoll, T. R., Stockdale, R. C., and Schelke, F. J. Determination of aerodynamic coupling derivatives through flight test. *AIAA Guidance and Control Conference,* Boston, MA, 1975.
10. Anon. *Rotorcraft System Identification.* AGARD Report No. AGARD-AR-280, AGARD, France, Sept. 1991.
11. Raol, J. R., Girija, G. and Singh, J. *Modeling and Parameter Estimation of Dynamic Systems.* IEE Control Book Series, Vol. 65, IEEE, London, Aug. 2004.
12. Roskam, J. Evolution of airplane stability and control: A designer's viewpoint. *J. Guidance*, 14(3), 481–491, May-June 1991.

5 Mathematical Modeling and Simplification of Equations of Motion

5.1 INTRODUCTION

An aircraft or any flying vehicle or a vehicle in motion can be looked upon as a dynamic system while in any type of dynamic/flight operation. This is because the flight-related control functions to be performed are necessarily: (i) the deflection of (aerodynamic) control surfaces, and (ii) the change in the engine thrust, which acts as an input to the dynamic system, and the responses of the aircraft are the desired outputs: an equilibrium state of the aircraft's motion, stabilized aircraft motion after a disturbance, and a change of one equilibrium state to another, if required. Aircraft is made of structures/material components, has to have an engine for providing thrust, avionics systems for guidance/navigation, control and monitoring of subsystems, and control mechanisms; however, it has to ultimately fly in the air, and these operations are called upon to be performed for any successful flight mission. Hence, the subject of flight mechanics can be and should be studied from control system point of view, a well-thought-out larger system consisting of several subsystems working together in unison. This study involves modeling, analysis, flight simulation, and flight control of an atmospheric vehicle.

A study of dynamic system can be performed via analytical or experimental means. Analytical approach yields a mathematical model, based on several assumptions (because the real-life dynamic systems are mostly/often non-linear and distributed in nature) and application of theory and concepts of modeling. In an experimental approach, one has to design an appropriate experiment and generate data for analysis. This analysis is mainly carried out by using signal processing, system identification, and parameter estimation techniques. Mathematical modeling of dynamic systems is very essential for analysis, design, and simulation of control systems and prediction of behavior of these systems. A mathematical model characterizes the static and dynamic behavior of a system. Some practical examples are use of mathematical models in flight simulation (Chapter 6) and design of control systems (Chapter 8) for aerospace vehicles (like aircraft, rotorcraft, missile, UAVs, micro-air vehicles (MAVs), and spacecraft), weather forecasting, control of robotic systems, use in optimization of chemical processes in industry, and analysis of communications and biomedical signals.

A system is a set of components and/or functional activities that interact with each other to achieve a specified task or a mission. Some examples of systems are chemical plants, robotic systems, atmospheric vehicles, computer networks, and economy (processes) of a country. Most of the dynamic systems are described by differential/difference equations. There is exchange or change of energy and

DOI: 10.1201/9781003293514-6

dissipation in (damped) systems. The real-life dynamic system can be modeled using a mathematical model that should be obtained, refined, and validated based on the acquired/actual data from the (system identification) experiments. Choice of a gross model is dictated by the previous experience and the physics of the problem at hand. A fair amount of knowledge base is generally accessible to the model developer/analyst. A system, in general, consists of a collection of variables needed to describe its dynamic behavior at some point in time. A system could be continuous-time or discrete-time (or hybrid, mixed) when the so-called state variables change either continuously or at discrete point in time (in sampled data systems), respectively. In aerospace applications, mixed-type models are used, continuous-time state model, and discrete-time measurement model (for aircraft parameter estimation, Chapter 9). A typical atmospheric vehicle (control) system is shown in Figure 5.1. The terms 'system', 'plant', and 'process' are used interchangeably. Aircraft can be called a system, whereas a chemical system is called a plant. This plant will involve various processes for which again some mathematical models will be required. A telephone/computer network can also be called a system. In many problems in science and engineering, one finds a crucial need to represent a dynamic system by its mathematical model. This model is used for a variety of reasons: (i) to study basic dynamic characteristics of the system, (ii) to use the model for the prediction of future behavior of the system, e.g., weather forecasting, (iii) to analyze the stability and control characteristics of the system so as to help monitor/modify/regulate/control the behavior of the dynamic process/plant/system, e.g., chemical reactors, (iv) to use the model for simulation of responses of the system for various kinds of inputs/environment, e.g., flight dynamics simulator (Chapter 6), and (v) to use the model for optimization of cost and reduction of losses, e.g., dynamics of economy of a country.

Thus, mathematical models are used in research for the interpretation of acquired knowledge/measurements that provide clues for further investigations [1], e.g., first a simplified model of the system is used and then, based on the analysis of experimental data, this model is revised. A mathematical model for the purpose of design of a new system or design of a controller for the same system is expressed in

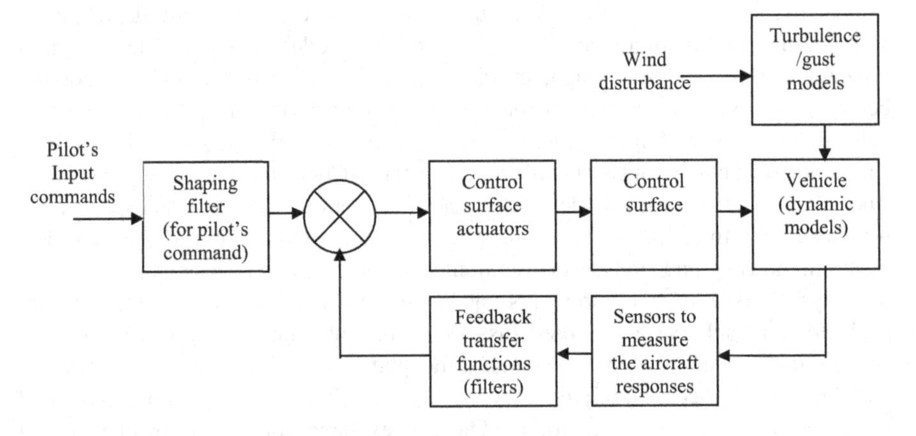

FIGURE 5.1 Aircraft dynamic system and other control system elements.

a manner that is compatible with the design criteria: guaranteeing the basic static/dynamic stability, overall performance, tracking error minimization, safety, and cost. For example, a time domain model will be more suitable for use in the control optimization methods based on time domain methods, and hence, the knowledge of components and subsystems should be in such usable forms. Finally, the control actions on a system depend on the knowledge about the system.

A mathematical model is a simplified representation of a system that helps to understand, predict, and possibly control the behavior of the system [1,2]. A model presents the knowledge of the system in a very usable form. In fact, the model simulates or mimics the behavior of the system. The models could be of several types: (i) hardware models (like scaled model of an aircraft, a missile, or a spacecraft that is used in wind tunnel testing), (ii) software expressions/computer algorithms, (iii) conceptual or phenomenological, (iv) physical, and (v) symbolic like maps, graphs, and pictures. The model represents a reality, the system/phenomenon/process under study, with a reduced complexity. Scientists and engineers, all over the world, spend major efforts to understand various natural phenomena as well as properties and characteristics of man-made systems. Mathematical modeling is one of the very fascinating and intriguing fields of research. It encompasses the study of deterministic and stochastic systems. The stochastic systems are, in general, difficult to model. Hence, a midway approach is usually preferred: deterministic systems affected by additive or multiplicative random noise.

There are three basic reasons why the usual deterministic systems/control theories do not provide a totally sufficient means of performing this analysis and design [3]: (i) no mathematical model of a given system is perfect, effects are modeled approximately by using some mathematical model, there are many sources of uncertainty in any mathematical model of a system, certain types of uncertainties can often be described by some random phenomena, some uncertainties can be described by bounded but unknown disturbances that are non-random in nature, between-the-events uncertainties can be axiomatically defined by probability theory that is fundamentally based on binary or crisp logic, within-the-events uncertainties can be modeled by using the concepts of fuzzy logic that is multi-valued logic; (ii) dynamic systems are driven not only by deterministic control inputs but also by disturbances that are neither controllable nor deterministic, such disturbances are random and are usually modeled by white Gaussian process; however, non-Gaussian/uniform/or other kind of noise processes are sometimes present, atmospheric turbulence and gust act on a flying vehicle; and (iii) measurements (sensors) do not provide perfect and complete data about a system; they introduce their own system dynamics and distortions, and are also corrupted with noise.

In light of these observations, the following questions become very relevant and interesting [3]: (i) what kind of models should be used to explain such uncertainties? e.g., Gaussian or uniform probability distributions, bounded disturbance models; (ii) with such noisy data, how do we use the system models to optimally estimate the quantities of interest to us? Here, system identification, state estimation, and specifically parameter estimation [4], and estimation of spectra of signals play an important role in analysis; (iii) how do we control the system's behavior in the face of these uncertainties? e.g., design of a suitable optimal controller; and (iv) how do we evaluate the performance of such estimators and control? e.g., quality or goodness-of-fit criteria, stability (gain and phase) margins, and error/robustness performance.

Some of the foregoing aspects are studied in the context of flight mechanics modeling and analysis in this book. A two-way approach to mathematical modeling is illustrated in Figure 5.2 [1]. The left side is the analytical route for model building. The right side is the empirical data-based model-building route. The block diagram illustrates a very comprehensive process of building a mathematical model. Most systems are non-linear and distributed in nature: there are two independent variables – time and spatial one. Such systems are called distributed systems and can be described by partial differential equations. When the linearization process is accomplished, linearizing errors will occur due to approximation. If the spatial effects are lumped, then we get linear models described by ordinary differential equations (ODE). Such systems/models are called lumped systems. At this stage, we obtain a model that can be postulated for further analysis and revised based on the systematic application of the right-side path. The measured data are used in system identification/parameter estimation procedures to arrive at an adequate model structure and parameters of the mathematical model [4]. This path involves measurement noise, data handling errors, quantization errors, and estimation errors.

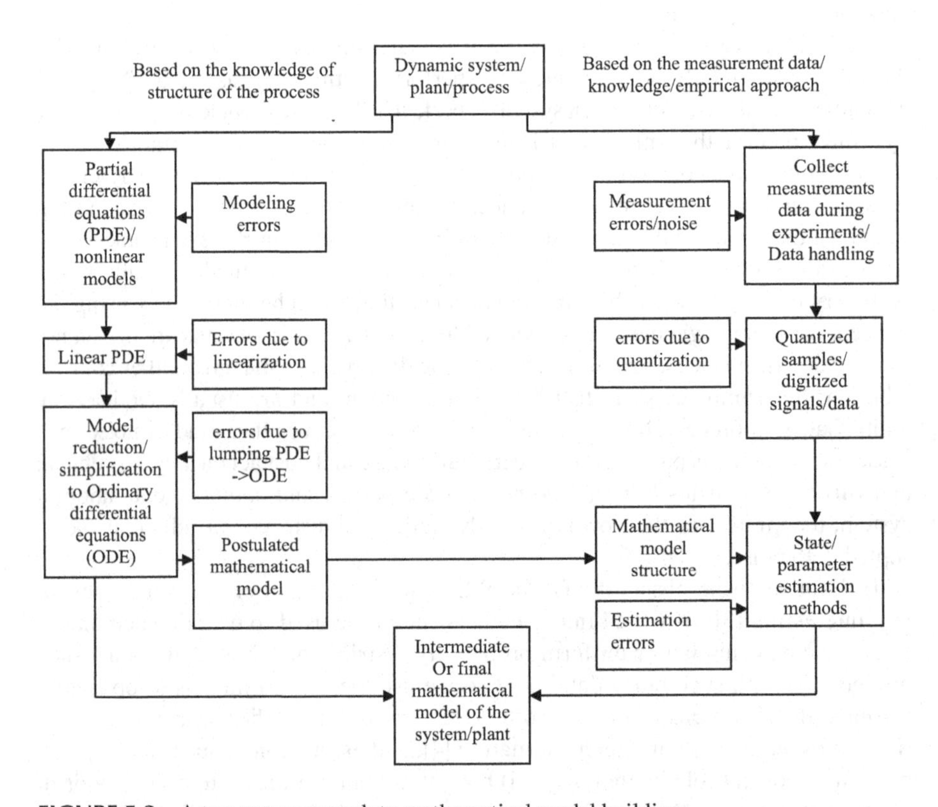

FIGURE 5.2 A two-way approach to mathematical model building.

5.2 MATHEMATICAL MODEL STRUCTURES

In one way, the model building is closely connected to experiments and observations. This is called empirical model building. Formulation of a theory can be called 'model building' because it gives an analytical basis to study the system. Theory is a proposed concept of the system being studied, e.g., the theory of Newton's laws of motion lends to the EOM of an aerospace vehicle. These EOM form the basis of mathematical models of such a vehicle and are backbone for flight control, flight simulation, parameter estimation from flight data and handling qualities analysis. This is the process of application of basic laws of physics: force/moment balance for mechanical systems, Kirchhoff's laws/Maxwell's electromagnetic laws for electrical/electronic systems, and energy balance for thermal/propulsion systems. Thus, the model building consists of: (i) selection of a mathematical structure based on knowledge of the physics of the problem, (ii) fitting of parameters to the experimental data (estimation), and (iii) verification/testing of the model (diagnostic checks), and finally its use for a given purpose.

In most of the aircraft identification and parameter estimation applications [4], the model can be represented by a set of rigid body equations in the body axis system [5]. The basic equations of motion, derived from the Newtonian mechanics, define the aircraft characteristic motion. They involve the fundamental assumption that the forces and moments acting on the aircraft can be synthesized. In many practical cases, the forces and moments, which include the aerodynamic, inertial, gravitational, and propulsive forces, are approximated by terms in the Taylor's series expansion. This invariably leads to a model that is linear in parameters. Aircraft mathematical models are treated in Chapters 3–5, and used in Chapters 6, 8–10. A more complete model can be justified for the correct description of the aircraft dynamics. The degree of relationship between the model complexity and measurement information needs to be ascertained. To arrive at a complex and complete model, one would need lot of repeated experimentation under varieties of environmental conditions and lot of data. This will ensure the statistical consistency of the modeling results in face of the uncertainties of measurement noise/errors. But, how much of this additional effort/data will be useful in building the complete model will depend on the methods of reduction of the data and engineering judgment.

There are various types of models: parametric and non-parametric. In a parametric model, the behavior of a system is captured by certain coefficients or parameters: state-space, time-series, and transfer function (TF) models. The model structure is assumed known or determined by processing the experimental data. A non-parametric model captures the behavior in terms of an impulse response or spectral density curve, and no model structure is assumed. There is near equivalence between these two types of models, especially for linear systems (or linearized systems). Given a set of data, an approach to construct a function, $y = f(x, t)$, is an integral part of the mathematical model building. This function should be parameterized in terms of a finite number of parameters. Thus, the function 'f' is a mapping from the set of x values to the set of values y. This parameterization could be a 'Tailor-made' [6], orthogonal function expansion, or based on artificial neural networks (ANNs) [7]. If a true system cannot be parameterized by a finite number of parameters, then one can use an expanded set of parameters as more and more data become available for analysis. This is a problem of parsimony: a set of minimum number of parameters is always preferable for good predictability by the identified model; a complex and very accurate model does not necessarily mean that it is an excellent (predictive) model of the system. Interestingly, one can say that the exact model of a system is the system itself! What is most important aspect of mathematical model building is to arrive at an adequate model that serves the purpose for which it is designed; one can look for a model that is probably approximately correct.

For the Tailor-made model structures (physical parameterization), an attempt is needed to bring physical modeling closer to system identification; this aspect is difficult to realize, since the physical parameters should be identifiable from the input/output data. There are models often in the form of state-space structures, and are based on the basic principles of the system's behavior and involve the physics of the process and are often termed as 'phenomenological models'. The simulation of these models is complex, but they have large validity range [6], e.g., complete simulation of flight dynamics of an aerospace vehicle. In the so-called black box models, like finite impulse response (FIR), autoregressive-exogenous inputs (ARX), and Box-Jenkin's models for linear systems [8], the a priori information about the system is very little or nil. The canonical state-space models and difference equation models have the parameters that have necessarily no physical meaning; these models, often called ready-made models, represent the approximate observed behavior of the process. For non-linear systems, one can use Volterra models and multilayer perceptrons (MLPT), i.e., neural networks [9]. Another black box non-linear structure is formed by fuzzy models that are based on fuzzy logic [7]. Since always some a priori information about a system will be available, these black box models can be considered as gray box models. Also, it is possible to preprocess the data to get some preliminary information on statistics of the noise affecting the system, thereby leading to gray box models.

5.2.1 Transfer Function Models

As many of us might be aware that the studies of electric/electronic circuits, basic linear control systems, mechanical systems, and related analyses lend to the so-called transfer function (TF) analysis [2]. A TF is defined, for single input single output (SISO) linear systems, as the ratio of the Laplace transform of the output to the Laplace transform of the input (in the continuous-time domain). It is often very important to quickly determine the frequencies and damping ratios of the dynamic systems. This is true also for the aircraft modes, and the aircraft-control system loop TF to assess the key dynamic characteristics (the stability/gain and phase margins of the feedback control system, Chapter 8) for longitudinal- and lateral-directional modes from small perturbation maneuvers (Chapters 6–10) [5]. For discrete-time systems, we use Z-Transform analysis; however, we will consider it as TF analysis in which case the coefficients of numerator and denominator polynomials (in 'z' or 's' – complex frequency) are estimated using some least-squares method from the real flight/test data. This estimation can be performed to obtain either discrete-time series models (to obtain pulse TF - in 'z' domain) or directly the continuous-time models (by using appropriate methods [10] in's' domain). Additionally, one can obtain Bode diagram, the amplitude ratio and phase of the TF as a function of frequency, and even the discrete Bode diagram. For linearized models from the non-linear systems, we need to use the so-called describing function analysis.

5.2.1.1 Continuous-Time Model

Let a system be described by linear constant coefficient ordinary n-order differential equation:

$$d^{n}y(t)/dt^{n}+a_{n-1}d^{n-1}y(t)/dt^{n-1}+\cdots+a_{0}y(t)=\tau_{p}d^{p}u(t)/dt^{p}+\cdots+\tau_{0}u(t) \quad (5.1)$$

Here, $u(t)$ is the system input and $y(t)$ its output. Taking the Laplace transform [2], and assuming that the initial conditions are zero, we obtain the continuous-time TF (CTTF):

$$y(s)=G(s)u(s)=\frac{\tau_{p}s^{p}+\tau_{p-1}s^{p-1}+\cdots+\tau_{1}s+\tau_{0}}{s^{n}+a_{n-1}s^{n-1}+\cdots+a_{1}s+a_{0}}u(s) \quad (5.2)$$

Here, s is a complex (frequency) variable. This is one type of the mathematical model of a SISO system. The linear theory of analysis can be applied to this model, and the system's properties can be analytically derived from this model itself. The natural dynamic behavior of the system, in fact now of the model, can be evaluated by computing the roots of the denominator of $G(s)$. These roots describe the various natural modes of the system, assuming the model is nearly a good representation of the system, and dictate the amplitude and the phase of the response. If the value of 'n' is equal to 2, then this TF is called 2nd-order T.F. and its denominator has two roots (real/complex). One can relate this to short-period (SP) modes of an aircraft wherein the SP frequency and damping are important parameters that are often talked about in dealing with simulation/control/parameter estimation of aircraft (Chapters 6, 8–10). These frequency and damping parameters in turn depend upon certain important aerodynamic (stability and control) derivatives of the aircraft, the description of which was given in Chapter 4.

Example 5.1

Obtain the Bode diagrams for the TF of an electrical circuit of Figure 5.3 for the time constant of 0.1 and 0.5 s. The resistor is R Ohm, and the capacitance is C microfarad. The time constant is given as RC = τ (s).

$$G(s)=\frac{V_o}{V_i}=\frac{1/\tau}{s+1/\tau} \quad (5.3)$$

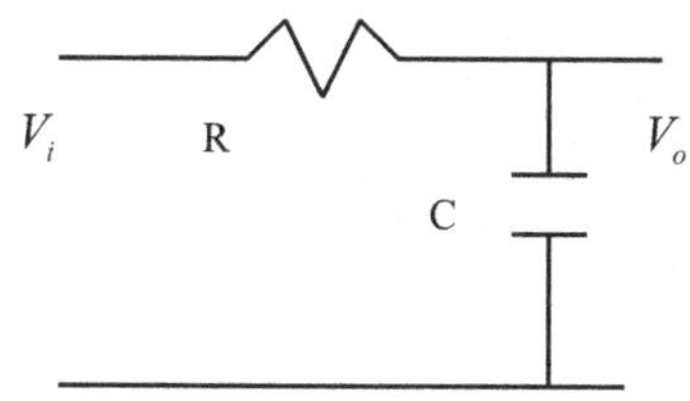

V_i R C V_o

FIGURE 5.3 RC electrical circuit.

Solution 5.1

Specify numerator and denominator of the TF as num = [10], den = [1 10] (in the command window of the MATLAB). Then, use sys = tf(num, den) and bode(sys) to obtain/plot the Bode diagram (Figure 5.4a and b). Repeat with 10 replaced by 2. At the cutoff frequencies (Cof, cutoff frequency = 1/time constant), the phase (lag) is −45 degs., the property of the first-order TF. The bandwidth of the system with Cof = 10 rad/s is broader than that of with Cof = 2 rad/s.

Example 5.2

The TF of a second-order mechanical system of Figure 5.5 is given as:

$$G(s) = \frac{X(s)}{f(s)} = \frac{K/M}{s^2 + (D/M)s + K/M} \tag{5.4}$$

Here, M is the mass, D is the damping constant, and K is the spring stiffness/constant. Use suitable values for M, D, and K and obtain the Bode diagrams. Study the improvement in the damping property by increasing the value of D by a factor of 2.

Solution 5.2

Let $M = 10$, $D = 10$, and $K = 100$, arbitrarily. Obtain numerator and denominator of the TF as num = [10], den = [1 10/10 100/10]. Then, use sys = tf(num, den) and bode(sys) to obtain/plot the Bode diagram. We see that the natural frequency is sqrt(K/M) = 3.1623 rad/s and the damping ratio = 0.1581 (of the second-order system) that is not the same as the damping constant. Due to low damping, the magnitude of the TF has a peaking value, see Figure 5.6a. Repeat the solution with $D = 20$, see Figure 5.6b for this case. The increased damping constant has increased the damping ratio (the magnitude of the peak is reduced). The step responses are improved, see Figure 5.6c. The step response is generated by step(sys). The first plot is held (by typing 'hold'), and then the step response with $D = 20$ gets automatically superimposed. We see that still the time response is not very well damped. With $D = 40$, the response (with arrow) is very well damped and acceptable. The time constant (Chapter 8) is approximately the same but the transients are suppressed very well when the value of D is increased. One can easily relate this TF to the SP mode of an aircraft. Thus, in (very) approximate sense, the aircraft's pitching motion dynamics are like a 2nd-order spring-mass system.

Example 5.3

Obtain the Bode diagrams and the step responses of the following T.Fs [11] and comment on the results.

$$y(s)/u(s) = \frac{0.787s + 1.401}{s^2 + 3.785s + 15.49}; \quad y(s)/u(s) = \frac{0.638s + 0.451}{s^2 + 4.67s + 10.30} \tag{5.5}$$

Solution 5.3

First, we obtain the bode diagrams by using the steps: num1 = [0.787 1.401]; den1=[1 3.785 15.49]; sys1 = tf(num1,den1); bode(sys1); hold; num2 = [0.638 0.451];

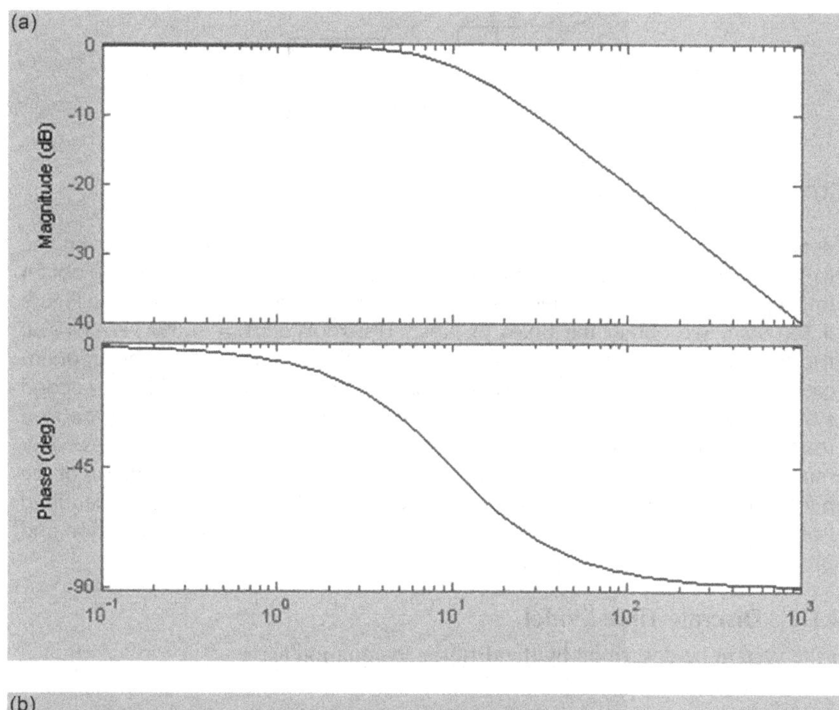

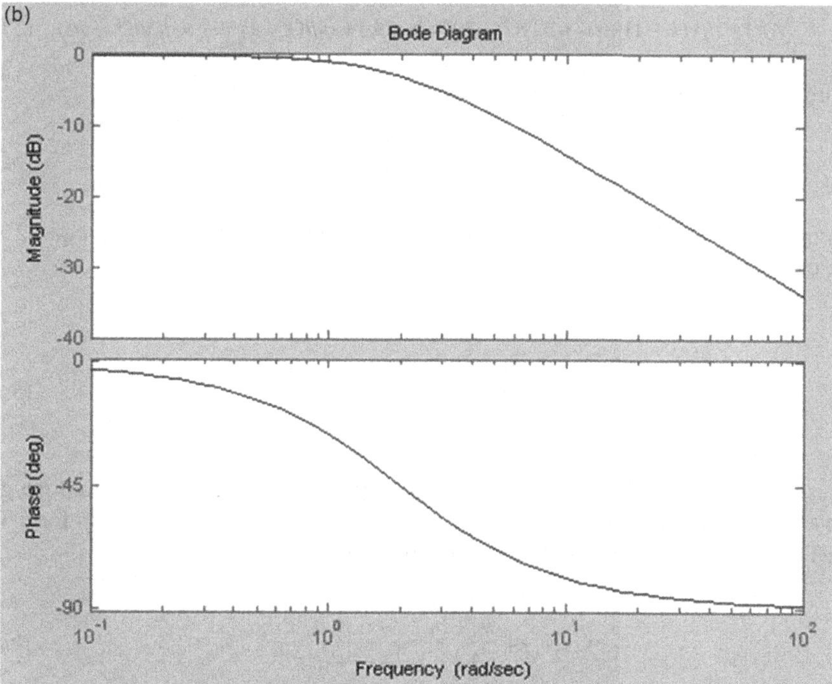

FIGURE 5.4 (a) Bode plot of the 1st-order continuous-time system with $\tau = 0.1$ s. (b) Bode plot of the 1st-order continuous-time system with $\tau = 0.5$ s.

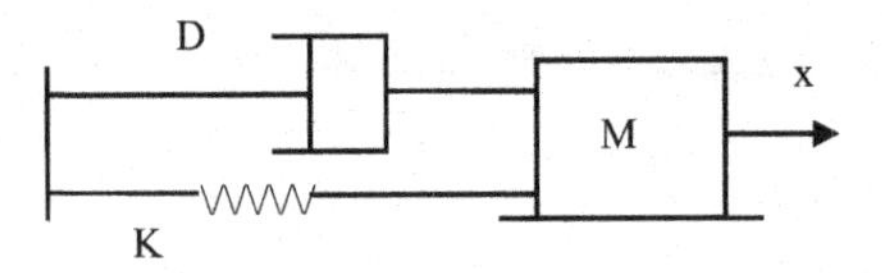

FIGURE 5.5 Mechanical spring mass systems.

den2 = [1 4.67 10.30]; sys2 = tf(num2,den2); bode(sys2) (in this sequence in the MATLAB command window). The plots are shown in Figure 5.7a. Then, we obtain the step responses by using step(sys1); hold; step(sys2). The plots are shown in Figure 5.7b. Next, we obtain the poles of these TFs: roots(den1) = –1.8925+/–3.4509; roots(den2) = –2.3350+/–2.2018. We also have the damping ratios and frequencies as: damp(sys1)=0.481, 3.94; damp(sys2) = 0.728, 3.21. We see that the second system has much better damping than the first one as is obvious from the step response of the second system, see Figure 5.7b (see arrow). Actually, these TFs/models are the lower-order equivalent system models (LOTF in Chapter 10) of the pitch rate to longitudinal input of F-18 HARV (HAOA research vehicle), identified from the measured data [11]. These LOTF models are used for the evaluation and prediction of the handling qualities of the aircraft (Chapter 10).

5.2.1.2 Discrete-Time Model

Let the system be described by the difference equation as:

$$y(k) + a_1 y(k-1) + \cdots + a_n y(k-n) = b_0 u(k) + b_1 u(k-1) + \cdots + b_m u(k-m) \quad (5.6)$$

Using the time shift operator [8], we obtain:

$$q^{-1} y(k) = y(k-1) \quad (5.7)$$

In eqn. (2.3), the corresponding polynomials, on the left and right sides of eqn. (2.3), are obtained as:

$$A(q^{-1}) = 1 + a_1 q^{-1} + \cdots + a_n q^{-n}$$

$$B(q^{-1}) = b_0 + b_1 q^{-1} + \cdots + b_m q^{-m} \quad (5.8)$$

Thus, we have $A(q^{-1})y(k) = B(q^{-1})u(k)$, and applying the Z-transformation, we get the so-called discrete transfer function (DTF/pulse transfer function or shift form TF (SFTF)) for the discrete-time linear dynamic system:

$$G(z^{-1}) = \frac{b_o + b_1 z^{-1} + \cdots + b_m z^{-m}}{1 + a_1 z^{-1} + \cdots + a_n z^{-n}} \quad (5.9)$$

Here, z is a complex variable, and m, n ($n > m$) are the orders of the respective polynomials. The discrete Bode diagram can also be obtained. The roots of the denominator

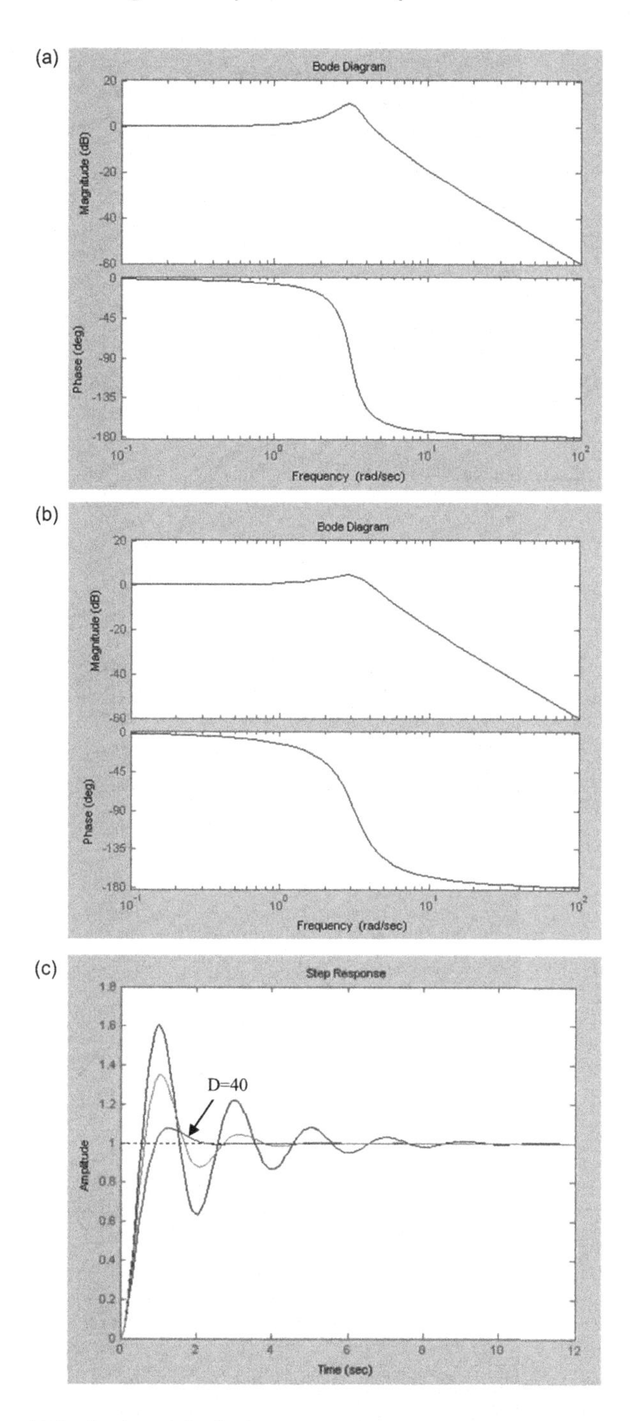

FIGURE 5.6 (a) Bode plot of the 2nd-order continuous-time system with $D = 10$. (b) Bode plot of the 2nd-order continuous-time system with $D = 20$. (c) Step responses of the 2nd-order system with $D = 10, 20, 40$.

(a)

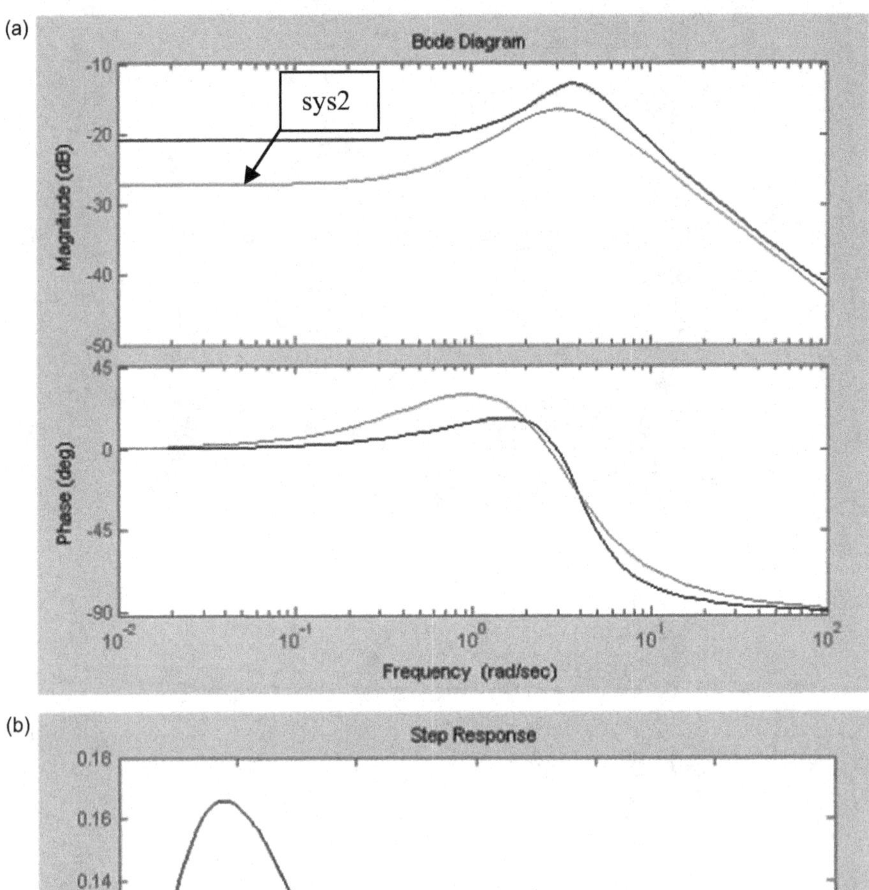

(b)

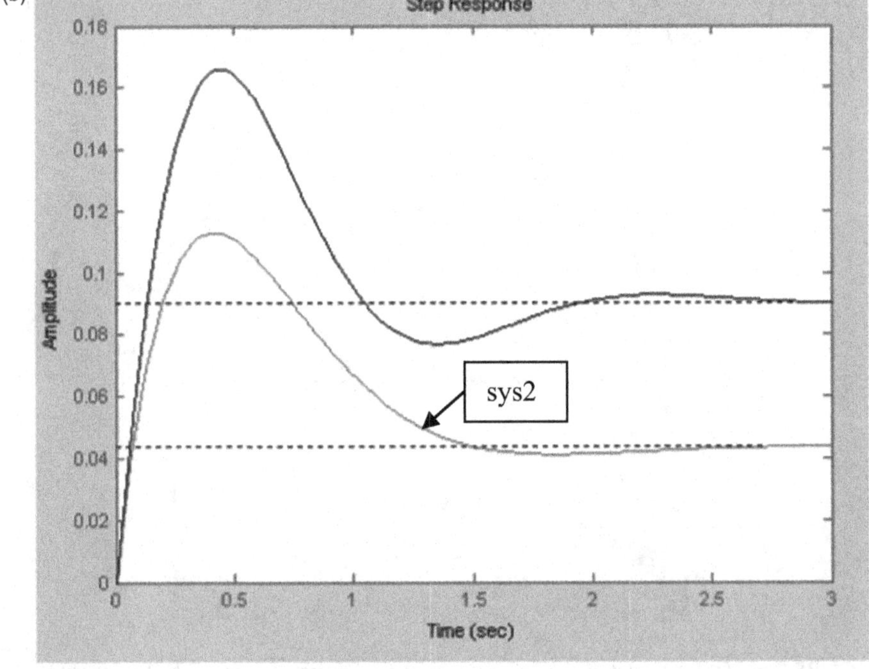

FIGURE 5.7 (a) Bode plots of the 2nd-order systems (sys1; sys2 with an arrow). (b) Step responses of the 2nd-order systems (sys1; sys2 with an arrow).

polynomial will give modes of the system in terms of frequency and damping ratio that are important parameters to be ascertained from real flight data as first-cut results (Chapter 9). However, these are in z domain and not in the continuous-time domain. For proper and formal interpretation, one should transform these into s-domain parameters. One major limitation of this approach is that there is no apparent correspondence between the numerical values of the coefficients of the DTF/SFTF and the corresponding CTTF of the real system. However, this correspondence is very closely obtained if the delta-operator form of the TF (DFTF) is employed for modeling and estimation as discussed in the sequel.

Example 5.4

Obtain the Bode diagram for the pulse TF [2] given as:

$$G(z) = \frac{3z^2 + 4z}{z^3 - 1.2z^2 + 0.45z - 0.05} \tag{5.10}$$

Solution 5.4

We use the function dbode([3 4 0], [1-1.2 0.45-0.05], 1) to obtain/plot the Bode diagram for the given discrete-time system. Here, sampling time (interval) is chosen as 1. Change the sampling time to 0.1, 0.01 and obtain the Bode diagrams, Figure 5.8. As the sampling time reduces, the frequency range increases.

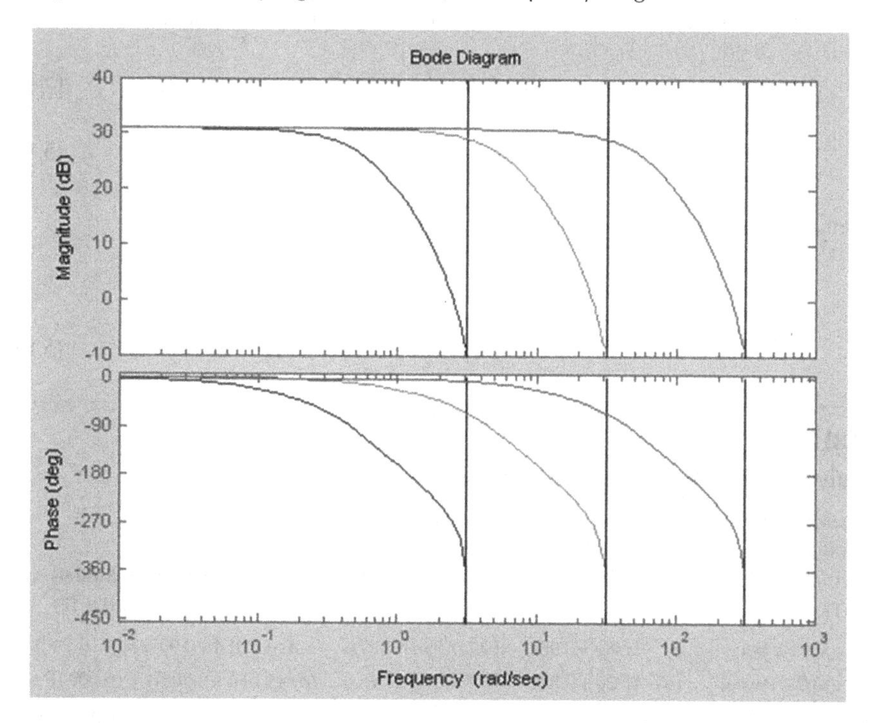

FIGURE 5.8 Bode plot of the 3rd-order discrete-time system with varying sampling times.

5.2.1.3 Delta Form TF

The 'q'-operator of eqn. (5.7) is as such not at all like the continuous-time operator $d(.)/dt$. If 'q' is replaced by a 'difference operator' that is like a 'derivative', the correspondence between discrete-time and continuous-time system will be definitely better. The delta operator with τ as the sampling interval is defined as [12]:

$$\delta = \frac{q-1}{\tau} \tag{5.11}$$

Table 5.1 gives equivalent TF models in various domains: $d/dt = D$, 'q' or shift form and delta form.

There is very good correspondence between the DFTF and the continuous-time system expressions (columns 1 and 3), whereas the correspondence between the continuous-time system and the SFTF (columns 1 and 2) is not obvious, even though these TFs are as such equivalent in their input/output behavior. The main point is that if we fit a DFTF to the discrete-time data, then from the DFTF, we can very easily discern the characteristics of the underlying continuous-time system. This cannot be done from the SFTF unless it is converted to the discrete bode diagram or time response is obtained. This is the main advantage of the DFTF compared to SFTF. Equivalently, the DFTF can be converted to the CTTF, but we will see that the difference between the DFTF and CTTF is inherently very small as such and the original DFTF can be interpreted in just similar manner as the CTTF, which will not be the case with the SFTF in general. We emphasize here that τ should be quite small. In shift form, we have:

$$q\,x(k) = F\,x(k) + G\,u(k) \tag{5.12}$$

$$y(k) = H\,x(k) \tag{5.13}$$

Then, in the delta form, we have:

$$q = 1 + \tau\delta;\ \delta\,x = F'\,x + G'\,u;\ y = H'\,x \tag{5.14}$$

$$F' = (F - I)/\tau;\ G' = G/\tau;\ H' = H \tag{5.15}$$

TABLE 5.1
Equivalent Transfer Function Models [12]

Systems in Term of $d/dt = D$ (Continuous Time/ CTTF)	Equivalent Discrete-Time/Shift Models (τ = 0.02 s) (SFTF)	Delta Form Models (τ = 0.02 s) (DFTF)
$(D^2+D+1)y = u$	$(q^2-1.98q+0.98)y = 10^{-4}(1.99q+1.97)u$	$(\delta^2+1.01\delta+0.99)y = (0.01\delta+0.99)u$
$(D^2+10D-1)y = u$	$(q^2-1.82q+0.82)y = 10^{-4}(1.87q+1.75)u$	$(\delta^2+9.05\delta-0.906)y = (0.0094\delta+0.906)u$
$(D^2-D+10)y = u$	$(q^2-2.02q+1.02)y = 10^{-4}(2.01q+2.03)u$	$(\delta^2-0.808\delta+10.1)y = (0.0101\delta+1.01)u$

Example 5.5

Obtain the roots of the numerator (zeros) and the denominator (poles) of the model forms [12] given in Table 5.1.

Solution 5.5

The zeros and the poles are obtained by using roots([1 1 1]), etc., as given in Table 5.2. Thus, we see striking similarity between the poles of CTTF (rows 1,2,3) and DFTF (rows 7,8,9), respectively. However, the shift TFs have additionally one zero each, and there is no correspondence between poles of CTTF and SFTF! Thus, one can easily see the merits of the delta form TFs.

5.2.2 State-Space Models

In the so-called modern control/system theory, the dynamic systems are described by the state-space representation. Besides having the input and output variables, the system's model will have internal states. These models are in time domain and can be used to represent linear, non-linear, continuous-time, and discrete-time systems with almost equal ease and are applicable to SISO as well as MIMO systems. The state-space models are very conveniently used in design and analysis of control systems. They are mathematically tractable for varieties of optimization, system identification, parameter estimation, state estimation, simulation, and control analysis/design problems. The TF models can be obtained easily from the state-space models and they will be unique, because TF is input/output behavior of a system. However, the state-space model from the TF model may not be unique, since inherently the states are not necessarily unique. This means that internal states could be defined in several ways, but the system's TF from those state-space models would be unique. By definition, the state of a system at any time t is a minimum set of values $x_1,\ldots,x_n$ that along with the input to the system for all time T, $T \geq t$, is sufficient to determine

TABLE 5.2
The Zeros and the Poles of the Equivalent Transfer Function Models

Systems in Terms of $d/dt = D$ (Continuous Time)	Zeros	Poles
$(D^2+D+1)y=u$	–	−0.5000+0.8660i; −0.5000 − 0.8660i
$(D^2+10D-1)y=u$	–	−10.0990; 0.0990
$(D^2-D+10)y=u$	–	0.5000+3.1225i; 0.5000 − 3.1225i
$(q^2-1.98q+0.98)y=10^{-4}(1.99q+1.97)u$	−0.9899	1.0000; 0.9800
$(q^2-1.82q+0.82)y=10^{-4}(1.87q+1.75)u$	−0.9358	1.0422; 0.7868
$(q^2-2.02q+1.02)y=10^{-4}(2.01q+2.03)u$	−1.0100	1.0200; 1.0000
$(\delta^2+1.01\delta+0.99)y=(0.01\delta+0.99)u$	–	−0.5050+0.8573i; −0.5050 − 0.8573i
$(\delta^2+9.05\delta-0.906)y=(0.0094\delta+0.906)u$	–	−9.1490; 0.0990
$(\delta^2-0.808\delta+10.1)y=(0.0101\delta+1.01)u$	–	0.4040+3.1523i; 0.4040 − 3.1523i

the behavior of the system for all (future) $T \geq t$. In order to specify the solution to an n-order differential equation completely, we must prescribe 'n' initial conditions at the initial time 't_0' and the forcing function for all (future times) $T \geq t_0$ of interest: there are 'n' quantities required to establish the system 'state' at t_0; but 't_0' can be any time of interest, so one can see that n variables are required to establish the state at any given time. Let a 2nd-order differential equation be given as:

$$\frac{d^2z(t)}{dt^2} + a_1\frac{dz(t)}{dt} + a_0z(t) = u(t) \tag{5.16}$$

Let $z(t) \to$ output; and $z(t) = x_1$ and $\dot{z}(t) = x_2$ be the two 'states' of the system. Then, we have from eqn. (5.16):

$$\begin{aligned} \dot{x}_1 &= \dot{z}(t) = x_2 \\ \dot{x}_2 &= \ddot{z}(t) = u(t) - a_1x_2 - a_0x_1 \end{aligned} \tag{5.17}$$

In vector/matrix form, this set can be written as:

$$\begin{bmatrix} \dot{x}_1 \\ \dot{x}_2 \end{bmatrix} = \begin{bmatrix} 0 & 1 \\ -a_0 & -a_1 \end{bmatrix}\begin{bmatrix} x_1 \\ x_2 \end{bmatrix} + \begin{bmatrix} 0 \\ 1 \end{bmatrix}u(t); \text{ and } z(t) = \begin{bmatrix} 1 & 0 \end{bmatrix}\begin{bmatrix} x_1 \\ x_2 \end{bmatrix} \tag{5.18}$$

In compact form, we have:

$$\dot{x} = Ax + Bu; \; z = Cx \tag{5.19}$$

with appropriate equivalence between eqns. (5.18) and (5.19). The part of eqn. (5.19) without the $Bu(t)$ term is called the homogeneous equation.

Example 5.6

Let the state-space model of a system have the following matrices [2]:

$$A = \begin{vmatrix} -2 & 0 & 1 \\ 1 & -2 & 0 \\ 1 & 1 & -1 \end{vmatrix}; B = \begin{vmatrix} 1 \\ 0 \\ 1 \end{vmatrix}; C = \begin{bmatrix} 2 & 1 & -1 \end{bmatrix} \tag{5.20}$$

Obtain the TF of this system, and the Bode diagram.

Solution 5.6

Use the function [num, den] = ss2tf([-2 0 1;1–2 0;1 1-1], [1;0;1], [2 1-1], [0], 1) to obtain num=[0 1.0 4.0 3.0] and den=[1.0 5.0 7.0 1.0], and hence, the following TF can be easily formed:

$$\frac{s^2+4s+3}{s^3+5s^2+7s+1} \tag{5.21}$$

The Bode diagram is obtained as in Example 2.1, see Figure 5.9. We see that the Bode diagram of this seemingly 3DOF system looks like that of a first-order system. To understand the reason, we obtain the roots of the denominator as: $-2.4196+0.6063i$; $-2.4196-0.6063i$; -0.1607 and the roots of the numerator: -3.0 and -1.0. Although there is no exact cancellation of the (numerator) zero at 3.0 and the (denominator) complex pole, effectively the frequency response degenerates to that of a TF with reduced order. The eigenvalues of the system obtained by eig(A) are exactly the roots of the denominator. This example illustrates a fundamental equivalence between SISO TF and the state-space representation of a dynamic system as well as effective (though not exact) zero/pole cancellation of the TF. This example also hints at the aspect that LOTFs can be obtained from some higher-order TFs by some approximation methods as long as the overall frequency responses (as represented by Bode diagram) and time responses are equivalent in the range of interest.

As we have said earlier the state-space representation is not unique, we consider now four major representations next [2]. These representations would be useful in certain optimization, control design, and parameter estimation methods.

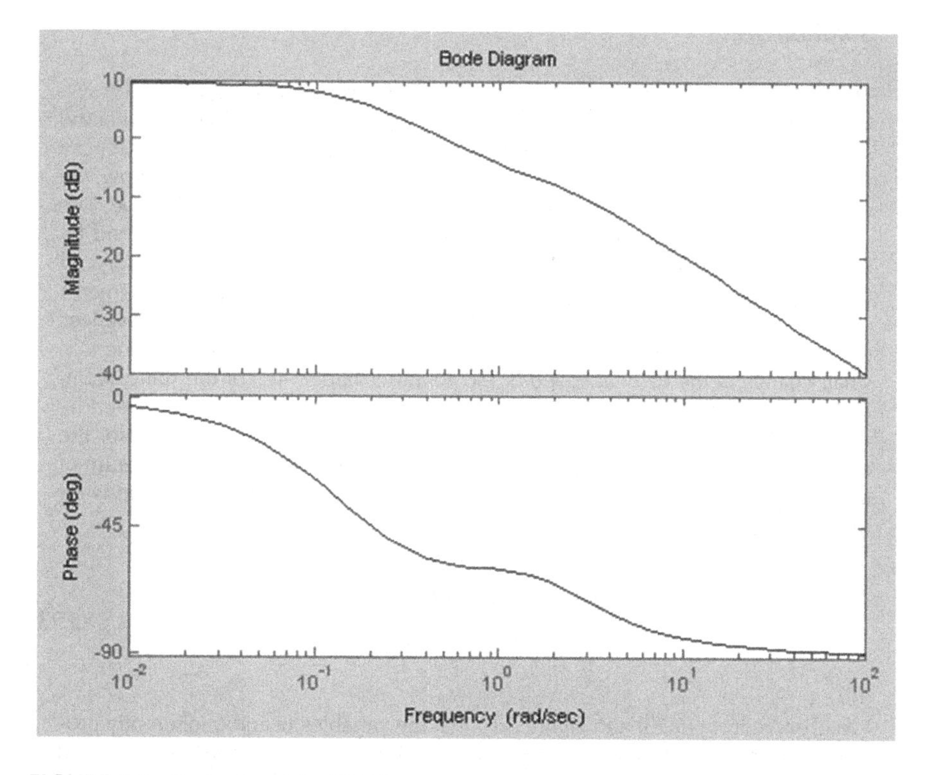

FIGURE 5.9 Bode plot of the 3DOF state-space system.

5.2.2.1 Physical Representation

A state-space model of an aircraft has the states as actual response variables like position, velocity, accelerations, Euler angles, flow angles, and so on (Chapter 3). In such a model, the states have physical meaning and they are measurable (or computable from measurable) quantities. State equations result from physical laws governing the process, e.g., aircraft EOM based on Newtonian mechanics, and satellite trajectory dynamics based on Kepler' laws. These variables are desirable for applications where feedback is needed, since they are directly measurable; however, the use of actual variables depends on specific system.

Example 5.7

Let the state-space model of a system be given as:

$$\dot{x} = \begin{bmatrix} -1.43 & (40-1.48) \\ 0.216 & -3.7 \end{bmatrix} x + \begin{bmatrix} -6.3 \\ -12.8 \end{bmatrix} u(t); \tag{5.22}$$

Let C = identity matrix with dimension of 2 and D = a null matrix. Study the characteristics of this system by evaluating the eigenvalues.

Solution 5.7

The eigenvalues are eig(A) = 2.725, −4.995. This means that one pole of the system is on the right half plane of the complex 's' plane, thereby indicating that the system is unstable. If we change the sign of the term (2,1) of the A matrix (i.e., use −0.216) and obtain the eigenvalues, we get eig(A) = −1.135+/−1.319. Now the system has become stable, because the real part of the complex roots is –ve. The system has damped oscillations, since the damping ratio is 0.652 (obtained by using damp(A), and the natural frequency of the dynamic system is 1.74 rad/s). In fact, this system describes the SP motion of a light transport aircraft. The elements of the A matrix are directly related to the dimensional aerodynamic derivatives (Chapter 4). In fact, the term (2,1) is the change in the pitching moment due to a small change in the vertical speed of the aircraft (Chapter 4). For the static stability of the aircraft, this term should be negative. The states are vertical speed (w m/s) and the pitch rate (q rad/s) and have obvious physical meaning. Also, the elements of the matrices A and B have the physical meaning. A block diagram of this state-space representation is given in Figure 5.10; the detailed equations for this example are:

$$\begin{aligned} \dot{w} &= -1.43w + 38.52q - 6.3\delta_e \\ \dot{q} &= 0.216w - 3.7q - 12.8\delta_e \end{aligned} \tag{5.23}$$

We observe from (5.23) and Figure 5.10 that the variables w and q inherently provide some feedback to state variables ($\dot{w}, \dot{q}$) and hence, signify internally connected control system characteristics of dynamic systems.

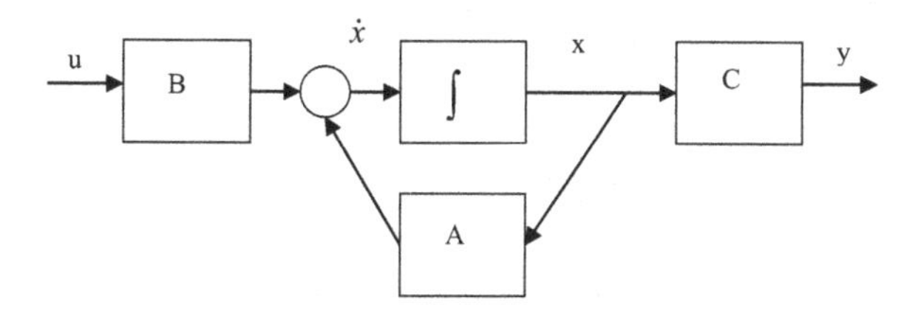

FIGURE 5.10 Block diagram for the state-space model.

5.2.2.2 Controllable Canonical Form

In this form, the state variables are the outputs of the integrators (in the classical sense of analog computers, which are now almost obsolete). The inputs to these integrators are differential states ($\dot{x}$ etc). It can be easily obtained from a given TF of the system by inspection:

$$\dot{x} = \begin{bmatrix} 0 & & \\ 0 & I & \\ \cdot & & \\ \cdot & & \\ \cdot & & \\ 0 & 0 \ldots & 0 \\ -a_0 & -a_1 \cdots & -a_{n-1} \end{bmatrix} x(t) + \begin{bmatrix} 0 \\ 0 \\ \cdot \\ \cdot \\ \cdot \\ 1 \end{bmatrix} u(t);\ x(t_0) = x_0 \tag{5.24}$$

$$y(t) = \begin{bmatrix} c_0 & c_1 & \ldots & c_m \end{bmatrix} x(t) \tag{5.25}$$

The corresponding TF is:

$$G(s) = \frac{c_m s^m + c_{m-1} s^{m-1} + \cdots + c_1 s + c_0}{s^n + a_{n-1} s^{n-1} + \cdots + a_1 s + a_0} \tag{5.26}$$

One can see that the last row of matrix A of eqn. (5.21) is the denominator coefficients in the reverse order. This form is so-called since this system (model) is controllable as can be seen from Example 5.8.

Example 5.8

Let the TF of a system be given as:

$$\frac{2s^2 + 4s + 7}{s^3 + 4s^2 + 6s + 4} \tag{5.27}$$

Obtain the controllable canonical state-space form from this TF.

Solution 5.8

Let

$$y(s)/u(s) = (2s^2 + 4s + 7)\frac{1}{s^3 + 4s^2 + 6s + 4} \quad (5.28)$$

$$y(s)/u(s) = (2s^2 + 4s + 7)v(s)/u(s) \quad (5.29)$$

Hence, $y(s) = (2s^2 + 4s + 7)\ v(s)$ and equivalently we have:

$$y(t) = 2\ddot{v} + 4\dot{v} + 7v \quad (5.30)$$

We define: $v = x_1$; $\dot{v} = x_2$ and $\ddot{v} = x_3$ and we get:

$$y(t) = 2x_3 + 4x_2 + 7x_1 \quad (5.31)$$

and

$$\dot{x}_1 = x_2 \text{ and } \dot{x}_2 = x_3$$

$$\text{Also, we have } v(s) = \frac{u(s)}{s^3 + 4s^2 + 6s + 4} \quad (5.32)$$

$$\dddot{v} = u(t) - 4\ddot{v} - 6\dot{v} - 4v \quad (5.33)$$

$$\dot{x}_3 = u(t) - 4x_3 - 6x_2 - 4x_1 \quad (5.34)$$

Finally, in the compact form, we have the following state space from:

$$\dot{x} = \begin{bmatrix} 0 & 1 & 0 \\ 0 & 0 & 1 \\ -4 & -6 & -4 \end{bmatrix} x + \begin{bmatrix} 0 \\ 0 \\ 1 \end{bmatrix} u(t); \text{ and } y = \begin{bmatrix} 7 & 4 & 2 \end{bmatrix} x \quad (5.35)$$

where $x = \begin{bmatrix} x_1\ x_2\ x_3\ x_4 \end{bmatrix}$. We see that the numerator coefficients of the TF appear in the *y*-equation, i.e., the observation model and the denominator coefficients appear in the last row of the matrix *A*. It must be recalled here that the denominator of the TF mainly governs the dynamics of the system, whereas the numerator mainly shapes the output response. Equivalently, the matrix *A* signifies the dynamics of the system and the vector *C* the output of the system. We can easily verify that the eig(*A*) and the roots of the denominator of TF have the same numerical values. Since there is no cancellation of any pole with zero of the TF, zeros=roots([2 4 7]).

$$\text{The controllable matrix } C_0 = \begin{bmatrix} 0 & 0 & 1 \\ 0 & 1 & -4 \\ 1 & -4 & 10 \end{bmatrix} \quad (5.36)$$

$$\text{is obtained from the above state space form as } C_0 = \begin{bmatrix} B\ AB\ A^2\ B\ldots \end{bmatrix} \quad (5.37)$$

For the system to be controllable, the rank of the matrix C_0 should be 'n' (the dimension of the state vector), i.e., rank $(C_0) = 3$. Hence, this state-space form/system is controllable, hence the name for this state-space form.

5.2.2.3 Observable Canonical Form

This state-space form is given as:

$$\dot{x}(t) = \begin{bmatrix} 0 & 0 & \dots & 0 & -a_0 \\ 1 & 0 & \dots & 0 & -a_1 \\ 0 & 1 & \dots & 0 & -a_2 \\ . & & & & \\ . & & & & \\ 0 & 0 & \dots & 1 & -a_n \end{bmatrix} x(t) + \begin{bmatrix} b_0 \\ b_1 \\ . \\ . \\ . \\ b_{n-1} \end{bmatrix} u(t); \quad x(t_0) = x_0 \tag{5.38}$$

$$z(t) = [0 \;\; 0, \dots, 1] x(t) \tag{5.39}$$

This form is so-called since this system (model) is observable as can be seen from Example 5.9.

Example 5.9

Obtain the observable canonical state-space form of the TF of Example 2.8.

Solution 5.9

We see that the TF, $G(s)$ is given as:

$$G(s) = C(sI - A)^{-1} B \tag{5.40}$$

Since the TF in this case is a scalar, we have [2]:

$$G(s) = \{C(sI - A)^{-1} B\}^T$$

$$G(s) = B^T (sI - A^T)^{-1} C^T$$

$$G(s) = C_o (sI - A^T)^{-1} B_o \tag{5.41}$$

Thus, if we use the A, B, C of solution 5.8 in their 'transpose' form, we obtain the observable canonical form:

$$\dot{x} = \begin{bmatrix} 0 & 0 & -4 \\ 1 & 0 & -6 \\ 0 & 1 & -4 \end{bmatrix} x + \begin{bmatrix} 7 \\ 4 \\ 2 \end{bmatrix} u(t); \text{ and } y = \begin{bmatrix} 0 & 0 & 1 \end{bmatrix} x \tag{5.42}$$

The observability matrix of this system is obtained as O_b = obsv(A, C) (in MATLAB):

$$O_b = \begin{bmatrix} 0 & 0 & 1 \\ 0 & 1 & 4 \\ 1 & 4 & 0 \end{bmatrix} \tag{5.43}$$

Since the rank $(O_b) = 3$, this state-space system is observable, and hence the name for this canonical form.

5.2.2.4 Diagonal Canonical Form

It provides decoupled system modes. The matrix A is a diagonal with entries as the eigenvalues of the system.

$$\dot{x}(t) = \begin{bmatrix} \lambda_1 & 0\ldots0 \\ 0 & \lambda_2\ldots \\ 0 & \\ . & \\ . & \\ . & \\ . & \ldots.\lambda_n \end{bmatrix} x(t) + \begin{bmatrix} 1 \\ 1 \\ . \\ . \\ 1 \end{bmatrix} u(t); \tag{5.44}$$

$$z(t) = \begin{bmatrix} c_1 & c_2\ldots c_n \end{bmatrix} x(t); \; x(t_0) = x_0 \tag{5.45}$$

If the poles are distinct, then the diagonal form is possible. We see from this state-space model form that every state is independently controlled by the input and it does not depend on the other states.

Example 5.10

Obtain the canonical variable (diagonal) form of the TF:

$$y(s) = \frac{3s^2 + 14s + 14}{s^3 + 7s^2 + 14s + 8} \tag{5.46}$$

Also, obtain the state-space form from the above TF by using [A, B, C, D]=tf2ss(num, den) and comment on the results.

Solution 5.10

The TF can be expanded into its partial fractions [2]:

$$y(s) = u(s)/(s+1) + u(s)/(s+2) + u(s)/(s+4) \tag{5.47}$$

$$y(t) = \begin{bmatrix} 1 & 1 & 1 \end{bmatrix} \begin{bmatrix} x_1 \\ x_2 \\ x_3 \end{bmatrix} \tag{5.48}$$

Since $x_1(s) = u(s)/(s+1)$, we have:

$$\begin{aligned} \dot{x}_1 &= -1x_1 + u \\ \dot{x}_2 &= -2x_2 + u \\ \dot{x}_3 &= -4x_3 + u \end{aligned} \tag{5.49}$$

Putting these together, we get:

$$\dot{x} = \begin{bmatrix} -1 & 0 & 0 \\ 0 & -2 & 0 \\ 0 & 0 & -4 \end{bmatrix} x + \begin{bmatrix} 1 \\ 1 \\ 1 \end{bmatrix} u(t) \tag{5.50}$$

We see from the above diagonal form/representation that each state is independently controlled by input. The eigenvalues are eig(*A*) as : −1, −2, −4 and are the diagonal elements of the matrix *A*. The state-space form directly from the given TF is obtained by [A, B, C, D]=tf2ss([3 14 14], [1 7 14 8]) as:

$$A = \begin{bmatrix} -7 & -14 & -8 \\ 1 & 0 & 0 \\ 0 & 0 & 1 \end{bmatrix}; B = \begin{bmatrix} 1 & 0 & 0 \end{bmatrix}; C = \begin{bmatrix} 3 & 14 & 14 \end{bmatrix}; D = [0] \tag{5.51}$$

We see from the above results that the matrix *A* thus obtained is not the same as the one in the diagonal form. This shows that for the same TF, we get different state-space forms indicating the non-uniqueness of the state-space representation. However, we can quickly verify that the eigenvalues of this new matrix *A* are identical by using eig(*A*) = −1, −2, −4. We can also quickly verify that the TFs obtained by using the function [num, den]=ss2tf(A, B, C, D, 1) for both the state-space representations are the same. The input/output representation is unique as should be the case.

5.2.2.5 A General Model

The following type of the state-space model can, in general, describe a continuous-time non-linear dynamic system:

$$\begin{aligned} \dot{x} &= f(x,t,\beta) + u + w \\ z &= h(x,\beta,K) + v \end{aligned} \tag{5.52}$$

Here, x is the $(n \times 1)$ state vector, u is the $(p \times 1)$ control input vector, z is the $(m \times 1)$ measurement vector, w is a process noise with zero mean and spectral density (matrix) Q, and v is the measurement noise with zero mean and covariance matrix R.

The unknown parameters are represented by vectors β and K. x_0 is a vector of initial conditions $x(t_0)$ at t_0. This model is very suitable for representing many real-life systems, since inherently they are non-linear. The non-linear functions f and h are vector-valued relationships and assumed known for the analysis.

5.2.3 Time-Series Models

Time series are a result of stochastic/random input to some system or some inaccessible random influence on some phenomenon, e.g., the pressure variation at some point at a certain time. A time series can be considered as a stochastic signal. Time-series models provide external description of systems and lead to parsimonious representation of a process or a phenomenon. For this class of model, accurate determination of the order of the model is a necessary step. Many statistical tests for model structure/order are available in the literature [4]. The time-series/ TF models are special cases of the general state-space models. The coefficients of time-series models are the parameters that can be estimated by the methods discussed in Chapter 9. One aim of time-series modeling is its use for prediction of the future behavior of the system/phenomenon. One application is to predict rainfall-runoff. The theory of discrete-time modeling is very easy to appreciate. Simple models represent the discrete-time noise processes. This facilitates easy implementation of identification/estimation algorithms on a digital computer. A general linear stochastic discrete-time system/model is described here with the usual meaning for the variables in state-space form [4] as:

$$\begin{aligned} x(k+1) &= \Phi_k x(k) + Bu(k) + w(k) \\ y(k) &= Hx(k) + Du(k) + v(k) \end{aligned} \tag{5.53}$$

For time-series modeling, a canonical form known as Astrom's model is given as:

$$A(q^{-1})y(k) = B(q^{-1})u(k) + C(q^{-1})e(k) \tag{5.54}$$

Here, A, B, and C are polynomials in q^{-1}, which is a shift operator defined as:

$$q^{-n}y(k) = y(k-n) \tag{5.55}$$

For SISO system, we have the expanded form as:

$$\begin{aligned} y(k) + a_1 y(k-1) + \cdots + a_n y(k-n) &= b_0 u(k) + b_1 u(k-1) + \cdots + b_m u(k-m) \\ &\quad + e(k) + c_1 e(k-1) + \cdots + c_p e(k-p) \end{aligned} \tag{5.56}$$

Here, y is the discrete-time measurement sequence, u is the input sequence, and e is the random noise/error sequence. We have the following equivalence:

$$\begin{aligned} A(q^{-1}) &= 1 + a_1 q^{-1} + \cdots + a_n q^{-n} \\ B(q^{-1}) &= b_0 + b_1 q^{-1} + \cdots + b_n q^{-m} \\ C(q^{-1}) &= 1 + c_1 q^{-1} + \cdots + c_n q^{-p} \end{aligned} \tag{5.57}$$

It is assumed that the noise sequences/processes w and v are uncorrelated and white with Gaussian distributions, and the time series are stationary in the sense that first- and second-order statistics are not dependent on time 't' explicitly. These models are called time-series models, because the observation process is considered as a time series of data that has some dynamic characteristics, affected usually by unknown random process. The inputs should be able to excite the modes of the system. The input, deterministic or random, should contain sufficient frequencies to excite the dynamic system to assure that in the output, there is sufficient effect of the mode characteristics. This will ensure the possibility of good identification. Such an input is called persistently exciting. It means that the input signal should contain sufficient power at the frequency of the mode to be excited. The bandwidth of the input signal should be greater (broader) than that of the system bandwidth.

The TF form is given by:

$$y = \frac{B(q^{-1})}{A(q^{-1})}u + \frac{C(q^{-1})}{A(q^{-1})}e \tag{5.58}$$

This model can be used to fit time-series data that arise out of some system/phenomenon with a controlled input u and a random excitation. Many special forms of eqn. (5.58) are possible [4,9].

Example 5.11

Expand the TF of example 2.4 into a time-series model. Also, obtain the poles and zeros of the TF.

$$G(z) = y(z)/u(z) = \frac{3z^2 + 4z}{z^3 - 1.2z^2 + 0.45z - 0.05} \tag{5.59}$$

Solution 5.11

We use the complex variable z for denoting the Z-transform (in discrete domain). It must be noted here that q and z are interchangeably used in the literature. Interestingly enough, the variable z is in fact a complex variable and can represent a complex frequency parameter in z-domain. Use the forward shift operator $z^n y(k) = y(k+n)$ to obtain:

$$y(k+3) - 1.2y(k+2) + 0.45y(k+1) - 0.05y(k) = 3u(k+2) + 4u(k+1)$$

$$y(k+3) = 1.2y(k+2) - 0.45y(k+1) + 0.05y(k) + 3u(k+2) + 4u(k+1) \tag{5.60}$$

Use roots([3 4 0]) to get zeros as : 0, -1.333 and roots([1−1.2 0.45−0.05]) to obtain the poles as : 0.20, 0.50+/−0.0i. We see that the poles lie within a circle of radius $z = 1$, indicating that the model is a stable system.

Example 5.12

Generate a time series y using the equation $y(k-2)=0.9y(k-1)+0.05y(k)+2u(k)$. Let *u(k)* be a random Gaussian (noise) sequence with zero mean and unit standard deviation. Compute SNR as 20log10(var(y)/var(u)) (in MATLAB).

Solution 5.12

Generate Gaussian random noise sequence by u = randn(500,1). Then, standardize the sequence by un = (u-mean(u))/std(u). The sequence *u* has mean = −0.0629 and std = 0.9467. The sequence *un* has mean = −1.8958e-017 and std = 1. The noise sequence *un* is plotted in Figure 5.11a. The sequence *y(k)* is generated using the following steps: *y(1) = 0.0+2*un(1);y(2) = 0.0+un(2);* for k = 3:500; *y(k) = 0.9*y(k–1)+0.05*y (k–2)+2*un(k);* end; *plot(un);plot(y)*. See Figure 5.11b. The SNR = 20log10(var(y)/ var(un)) = 31.1754. We see that the model is stable one and hence generates a stable time series, since roots([1−0.9−0.05]) are 0.9525 and −0.0525 and both are within unit circle in the z-domain/complex plane. If any root is greater than 1, then the time series will be unstable. Generate the unstable time series by *y(k) = −1.01*y (k–1)+0.05*y(k–2)+2*un(k),* see Figure 5.11c. We see that the new model is unstable and hence generates unstable time series, since roots([1 1.01−0.05]) are −1.0573 and 0.0473 and one root is out of the unit circle in the z-complex plane.

5.3 MODELS FOR NOISE AND ERROR PROCESSES

Usual assumption is that noise is a white and Gaussian process. White noise has theoretically infinite bandwidth, and it is unpredictable process, and hence it has no model structure. It is described by a mean value and spectral density/covariance matrix. The noise affecting the system can often be considered as unknown but with bounded amplitude/uncertainty. One should search the models that are consistent with this description of the noise/error processes [13]. We consider the continuous-time white noise process with the spectral density, S, passing through a sampling process to obtain the discrete-time white noise process as shown in Figure 5.12. The output is a sequence with variance Q (some different numerical value) and is a discrete process. If the sampling time/interval is increased, then more and more of the power at more and higher frequencies of the input white process will be attenuated and the so-called filtered process will have a finite power.

$$\text{Variance} = S/\text{sampling interval } (\Delta t) \tag{5.61}$$

We see that as the sampling time tends to zero, the white noise sequence tends to be a continuous white noise and the variance tends to infinity; in reality, such a process does not exist; hence, it is a fictitious noise. In practice, the statistical characteristic of (continuous-time) white noise is described in terms of its spectral density, i.e., power per unit bandwidth. If we pass uncorrelated noise signal (w) through a linear first-order feedback system, a correlated noise signal (v) is generated:

$$\dot{v} = -\beta v + w \tag{5.62}$$

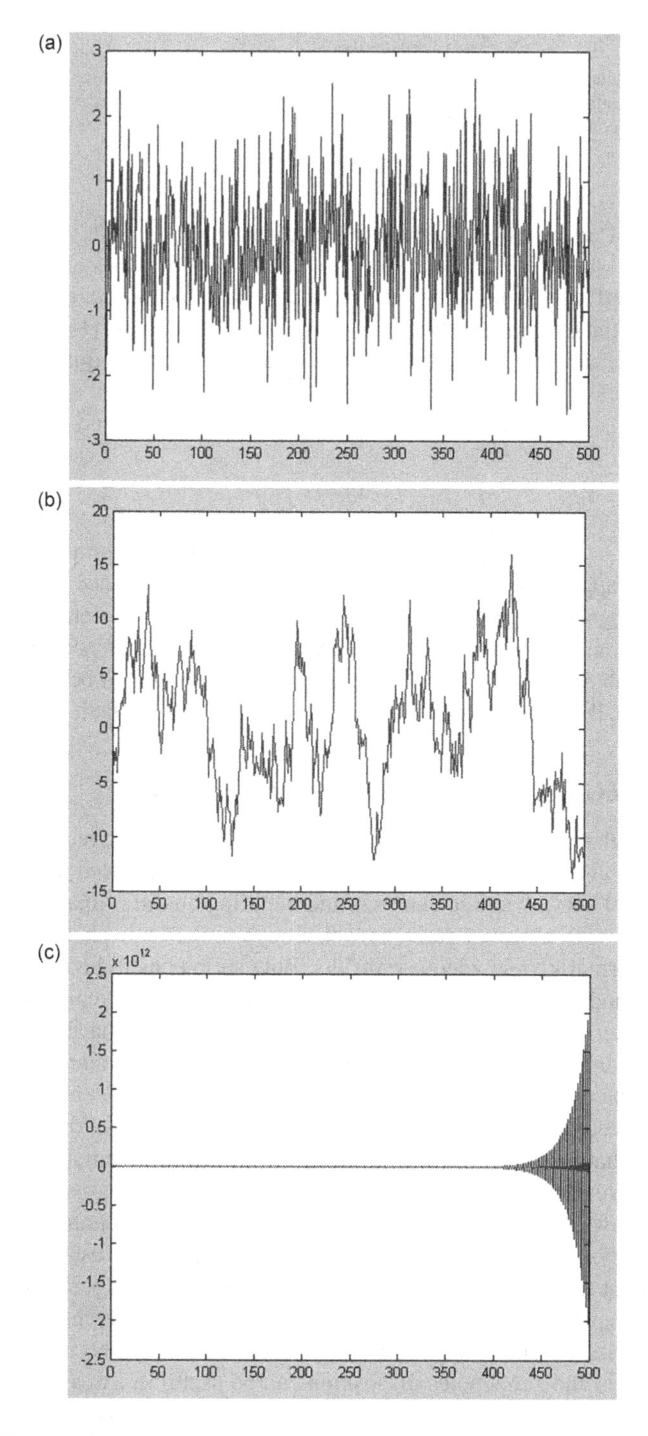

FIGURE 5.11 (a) The noise sequence 'un' with zero mean and std = 1.0. (b) The stable time series y. (c) The unstable time series y.

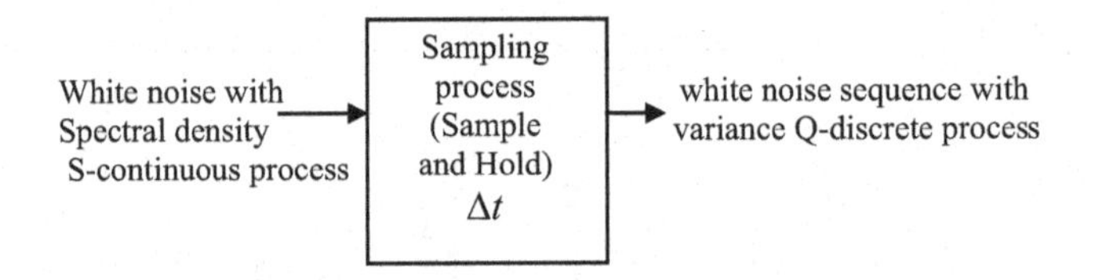

FIGURE 5.12 Continuous- to discrete-time white noise.

Such correlated signal/noise process would arise in practice if the process is affected by some dynamic system, e.g., turbulence on aircraft [14]. Assume that a signal has a component that drifts slowly with a constant rate equal to b. Then, the model is given as:

$$\begin{aligned} \dot{x}_1 &= x_2 \\ \dot{x}_2 &= 0 \end{aligned} \tag{5.63}$$

Assume that $x_2 = b$. The signal x_1 is also called the random ramp process. In sensor fault detection application [15], one can use the innovation sequence to detect the drift in the system/signal using this model. The point is that when a system/sensor is gradually deteriorating, in sensor fault detection/identification/isolation scheme, if the trend (drift) in the innovation (i.e., estimation residuals, Chapter 9) can be tracked by using the drift model, then the direction of occurrence or onset of the fault can be observed.

5.4 STRATEGIES FOR SIMPLIFICATION OF EOM

Although simulation of non-linear flight dynamics, in general, requires full 6DOF EOM, it is not always necessary to use very complex and coupled models in design of flight control laws, dynamic analysis, and handling qualities analysis procedures. Simple models with essential derivatives that capture the basic characteristics of the vehicle can normally yield adequate results. But caution has to be exercised in simplifying the model as too simple a model may not be able to represent the system dynamics properly. Some of the reasons why simplified models have found favor with flight analysts are: (i) full complex models would be difficult to interpret and analyze, (ii) it is possible to separate the model equations into independent subsets without much loss of accuracy, (iii) ease of linearization, (iv) no point in carrying small terms in the model that have little effect on system dynamics, (v) excursions during flight test maneuvers can be restricted to apply the assumption of linearity, and (vi) easy to implement software codes for handling qualities analysis and parameter estimation. The simplification of non-linear EOMs into workable and easily usable linear models can bring about the connections between various aircraft configurations via the TF analysis and frequency responses. The aircraft might be of different sizes, and wing-body configurations, but the dynamical characteristics might be of similar nature. The TF analysis and zeros/poles disposition can be useful in assessing the dynamic characteristics of these aircraft, rotorcraft, missiles, UAVs, MAVs, and airships. The simplified models are used for the design of autopilot control laws and evaluation of handling qualities of the vehicle. The detailed effect of particular derivatives can be observed. Also, it becomes easy to obtain the aerodynamic derivatives of the vehicle from the flight data using linear mathematical models in a parameter estimation algorithm. The TF analysis of the aircraft mathematical models gives a deeper insight into the effect of aircraft configurations on its dynamics. The aircraft configuration refers to the external shape of the aircraft and is related to the use of stores, C.G. movement (aft, mid, fore), flap settings, autopilot on or off condition, slat effect, airbrake deployment, etc. The aircraft dynamics for a particular configuration, at a particular flight condition, are characterized by the stability derivatives (Chapter 4). Understanding the flight mechanics models via TF helps to properly interpret the effects of the configuration on the flight responses as well as on the aerodynamic derivatives. The coupled non-linear 6DOF equations of motion discussed in Chapter 3 include non-linearities because of the gravitational- and rotation-related terms in the force equations and the appearance of products of angular rates in the moment equations. The dynamic pressure $\bar{q}$ also contributes to the non-linearity because it varies with the square of the velocity ($\bar{q} = \frac{1}{2}\rho V^2$). Apart from the non-linearities contributed by the kinematic terms, the aerodynamic coefficients $C_x, C_y, C_z, C_l, C_m, C_n$ may contain additional non-linearities, for example, the lift coefficient due to horizontal tail, when expressed in Taylor's series, may have the following form:

$$C_{LH} = C_{L\alpha H}\alpha_H + C_{L\delta_e}\delta_e + C_{L_{\delta_e^3}}\delta_e^3 + C_{L_{\delta_e\alpha_H^2}}\delta_e\alpha_H^2 \tag{5.64}$$

The above model form may be useful in explaining non-linearities arising from control surface deflections and flow effects at higher angles of attack. Analysis of a non-linear model would generally require the implementation of non-linear programs. However, it would be worthwhile to determine simplified models that can be used in flight regimes where non-linear effects are not important. In this chapter, we look at the various strategies adopted to simplify the aircraft EOM to obtain model forms that have practical utility and computation efficiency.

Some of the simplification techniques used to get workable models for flight data analysis are: (i) choice of coordinate systems, (ii) linearization of model equations, (iii) simplification using measured data, (iv) use of decoupled models, (v) neglecting small terms, and (vi) restricting excursions during maneuvers to permit the use of simplified models. Some of these simplification techniques are briefly discussed in the following section.

5.4.1 Choice of Coordinate Systems

The selection of appropriate coordinate system for equations of motion plays an important role in simplification of EOM. Some important points to be considered are [16]: (i) Body axes system is most suitable for defining rotational degrees of freedom. It is a normal practice to include the angular rates and the Euler angles in the aircraft state model and the equations in body axes, under certain assumptions, can be considerably simplified; (ii) either of the polar or rectangular coordinate system can be used for translational degrees of freedom; (iii) the polar axes system uses the (α, β, V) form, which involves wind axis coefficients C_L and C_D; the wind axis coefficients introduce non-linearities because of the way they are related to the body-axis force coefficients C_X, C_Y and C_Z. However, the α, β, V are commonly measured quantities, which encourage the use of polar coordinate system. The expressions for α, β, V in observation equations are simple if polar coordinates are used for data analysis; the liner accelerations are normally defined in terms of C_X, C_Y, C_Z, which suggests the use of u, v, w form of state equations. This form is particularly useful in case of the rotorcraft where the (α, β, V) form may lead to singularities. With the differential equations for u, v, w in the state model, the (α, β, V) expressions in the observation model will be non-linear.

5.4.2 Linearization of Model Equations

The non-linear coupled 6DOF equations of motion can be linearized using small disturbance theory [17]. In using this approach, we assume that the excursions about the reference flight condition are small. Thus, we assume small values of the variables $u, v, w, p, q, r, \phi, \theta$. The approach cannot be applied to situations with large change in motion variables. All the variables in the 6DOF model are expressed as sum of a reference value (steady state value) and a small perturbation, e.g.,

$$\begin{aligned} &U = u_0 + u;\; V = v_0 + v;\; W = w_0 + w \\ &P = p_0 + p;\; Q = q_0 + q;\; R = r_0 + r \end{aligned} \tag{5.65}$$

The non-linear translational accelerations from the full set of 6DOF equations (Chapter 3) are expressed as:

$$\begin{aligned} \dot{U} &= -qw + rv - g\sin\theta + a_x \\ \dot{V} &= -ru + pw + g\cos\theta\sin\phi + a_y \\ \dot{W} &= -pv + qu + g\cos\theta\cos\phi + a_z \end{aligned} \tag{5.66}$$

From eqns. (5.65) and (5.66), the $\dot{u}$ equation can be expressed in terms of perturbed variables as:

$$\dot{u} = -(q + q_0)(w + w_0) + (r + r_0)(v + v_0) - g\sin(\theta + \theta_0) + (a_x + a_{x0}) \tag{5.67}$$

Equation (5.67) can be simplified by neglecting products of the perturbations and using small angle approximation; also, assuming wings level, symmetric reference flight, we have:

$$w_0 = v_0 = p_0 = q_0 = r_0 = \phi_0 = \psi_0 = 0 \tag{5.68}$$

Setting all disturbances in the perturbed $\dot{U}, \dot{V}, \dot{W}$ equations to zero gives the reference flight conditions as:

$$a_{x0} = q_0 w_0 - r_0 v_0 - g\sin\theta_0$$
$$a_{y0} = r_0 u_0 - p_0 w_0 - g\cos\theta_0 \sin\phi_0 \quad (5.69)$$
$$a_{z0} = p_0 v_0 - q_0 u_0 - g\cos\theta_0 \cos\phi_0$$

Thus, the set of simplified linearized perturbation equations of motion can be expressed as:

$$\dot{u} = -g\theta\cos\theta_0 + a_x$$
$$\dot{v} = -ru_0 + g\phi\cos\theta_0 + a_y$$
$$\dot{w} = +qu_0 - g\theta\sin\theta_0 + a_z$$
$$\dot{p} = \frac{1}{I_x I_z - I_{xz}^2}(I_z L + I_{xz} N)$$
$$\dot{q} = \frac{1}{I_y} M \quad (5.70)$$
$$\dot{r} = \frac{1}{I_x I_z - I_{xz}^2}(I_x N + I_{xz} L)$$
$$\dot{\phi} = p$$
$$\dot{\theta} = q$$

The linearized implementation of eqn. (5.70) is frequently used for: (i) routine batch analysis of large amount of data in cruise flight regimes, (ii) for the analysis of small-amplitude maneuvers, and (iii) where computational efficiency is important and non-linear effects are minimal.

5.4.3 Simplification Using Measured Data

In this approach, we do away with some of the differential equations from the state model and work with a reduced set of equations. However, since the eliminated variables will mostly appear in other equations, measured values of these variables from flight data is used. A very good example is that of the model equations used to analyze data generated from longitudinal SP maneuver. It is well known that the during a SP pitch-stick-maneuver, there is only marginal change in velocity, implying that the longitudinal SP mode is more or less independent of velocity. Therefore, in the α, β, V form of equations, the $\dot{V}$ equation can be omitted from SP data analysis; instead, the measured value of V can be used in the model equations without any loss of accuracy. The measured data for the variables used in this manner are also sometimes referred to as 'pseudo-control inputs'. It is clear from this discussion that the use of this approach would require measurements of the motion variables omitted from analysis; the disadvantage of this approach is that it is sensitive to the noise in the measurements/data. Further simplification is possible, since most aircraft can be assumed to be symmetric about the XZ plane and fly at small AOSS, by separating the 6DOF coupled equations into nearly independent sets: longitudinal- and lateral-directional. Each set has nearly half the number of differential equations in the state model compared to the full coupled non-linear 6DOF state equations. The simplified non-dimensional and dimensional models normally used for longitudinal- and lateral-directional flight data analysis are discussed next.

5.5 LONGITUDINAL MODELS AND MODES

Assuming that the aircraft is not performing maneuvers with large excursion, it is possible to characterize the aircraft response to pitch-stick inputs by considering only the lift force, drag force, and pitching moment equations. Assuming the aircraft to be symmetric about the XZ plane, i.e., I_{xy} and I_{yz} are zero, the general free longitudinal motion can be described by eliminating the $\dot{\beta}, \dot{p}, \dot{r}$ and $\dot{\phi}$ equations from the 6DOF state model (Chapter 3); thus, we have:

$$\dot{V} = g(\cos\phi\cos\theta\sin\alpha\cos\beta + \sin\phi\cos\theta\sin\beta - \sin\theta\cos\alpha\cos\beta)$$
$$-\frac{\bar{q}S}{m}C_{D\text{wind}} + \frac{T}{m}\cos(\alpha+\sigma_T)\cos\beta$$

$$\dot{\alpha} = \frac{g}{V\cos\beta}(\cos\phi\cos\theta\cos\alpha + \sin\theta\sin\alpha)$$
$$+q - \tan\beta(p\cos\alpha + r\sin\alpha) - \frac{\bar{q}S}{mV\cos\beta}C_L - \frac{T}{mV\cos\beta}\sin(\alpha+\sigma_T) \quad (5.71)$$

$$\dot{q} = \frac{1}{I_y}\left\{\bar{q}ScC_m - \left(p^2 - r^2\right)I_{xz} + pr\left(I_z - I_x\right) + T(l_{tx}\sin\sigma_T + l_{tz}\cos\sigma_T)\right\}$$

$$\dot{\theta} = q\cos\phi - r\sin\phi$$

In (5.71), although the differential equations for $\dot{\beta}, \dot{p}, \dot{r}$ and $\dot{\phi}$ are omitted from analysis, these terms still appear in the right-hand side of the equations; measured values of β, p, r, ϕ can be used to solve (5.71). Equation (5.71) can be further simplified by assuming these quantities to be small during longitudinal maneuvers and neglecting them altogether, and we obtain:

$$\dot{V} = g(\cos\theta\sin\alpha + \cos\theta - \sin\theta\cos\alpha) - \frac{\bar{q}S}{m}C_{D\text{wind}} + \frac{T}{m}\cos(\alpha+\sigma_T)\cos\beta$$

$$\dot{\alpha} = \frac{g}{V}(\cos\theta\cos\alpha + \sin\theta\sin\alpha) + q - \frac{\bar{q}S}{mV}C_L - \frac{T}{mV}\sin(\alpha + \sigma_T) \tag{5.72}$$

$$\dot{q} = \frac{1}{I_y}\{\bar{q}ScC_m + T(l_{tx}\sin\sigma_T + l_{tz}\cos\sigma_T)\}$$

$$\dot{\theta} = q$$

We also know from eqn. (3.29) of Chapter 3, that:

$$C_{D\text{wind}} = C_D\cos\beta - C_Y\sin\beta;\ \text{For small values of } \beta, C_{D\text{wind}} = C_D \tag{5.73}$$

Substituting eqn. (5.73) back in eqn. (5.72) and assuming the thrust to be acting at C. G., i.e., $l_{tx} = l_{tz} = 0$, the differential equations for longitudinal motion now take the following form:

$$\begin{aligned}
\dot{V} &= g\sin(\theta - \alpha) - \frac{\bar{q}S}{m}C_D + \frac{T}{m}\cos(\alpha + \sigma_T) \\
\dot{\alpha} &= \frac{g}{V}\cos(\theta - \alpha) + q - \frac{\bar{q}S}{mV}C_L - \frac{T}{mV}\sin(\alpha + \sigma_T) \\
\dot{q} &= \frac{1}{I_y}\bar{q}ScC_m \\
\dot{\theta} &= q
\end{aligned} \tag{5.74}$$

In the u, v, w form, the longitudinal motion of the aircraft in flight can be described by the following 4th-order model:

$$\begin{aligned}
\dot{u} &= \frac{\bar{q}S}{m}C_X - qw - g\sin\theta \\
\dot{w} &= \frac{\bar{q}S}{m}C_Z + qu + g\cos\theta \\
\dot{q} &= \frac{\bar{q}S\bar{c}}{I_y}C_m \\
\dot{\theta} &= q
\end{aligned} \tag{5.75}$$

Here, C_X, C_Z and C_m are the non-dimensional aerodynamic coefficients. The set (5.75) is commonly used to analyze aircraft longitudinal motion. Certain aerodynamic derivatives rarely appear in the technical literature concerning with the aircraft dynamics, since these are of secondary importance because either they are very small (or their effects are negligible) or they are very difficult to determine:

$$X_{\dot{w}} = X_q = Z_{\dot{w}} = Z_q = 0$$

Also, in the steady-state flight conditions, the flight path of the airplane is assumed to be horizontal, $\gamma_0 = 0$. Using the foregoing two conditions, the longitudinal equations of motion (in the stability axes) become as follows, in the form of complex frequency 's', useful for the TF forms:

$$
\begin{aligned}
&(s - X_u)u - X_w w + g\theta = X_{\delta_e}\delta_e - X_u u_g - X_w w_g \\
&-Z_u u + (s - Z_w)w - U_o s\theta = Z_{\delta_e}\delta_e - Z_u u_g - Z_w w_g \\
&-M_u u - (M_{\dot{w}} s + M_w)w + s(s - M_q)\theta = M_{\delta_e}\delta_e - M_u u_g - [M_{\dot{w}} - M_q / U_0)s + M_w]w_g \\
&s\theta = q
\end{aligned}
\tag{5.76}
$$

This set can be further simplified, under certain assumptions, to describe the two basic oscillatory longitudinal modes: (i) the heavily damped SP mode and (ii) the lightly damped long period or phugoid (fleeing away) mode.

Example 5.13

The state-space equations in matrix form for LTV A-7A Corsair aircraft [18] are given as:

$$
\begin{bmatrix} \dot{u} \\ \dot{w} \\ \dot{q} \\ \dot{\theta} \end{bmatrix} = \begin{bmatrix} 0.005 & 0.00464 & -73 & -31.34 \\ -0.086 & -0.545 & 309 & -7.4 \\ 0.00185 & -0.00767 & -0.395 & 0.00132 \\ 0 & 0 & 1 & 0 \end{bmatrix} \begin{bmatrix} u \\ w \\ q \\ \theta \end{bmatrix} + \begin{bmatrix} 5.63 \\ -23.8 \\ -4.52 \\ 0 \end{bmatrix} \delta
$$

At the flight condition: 4.57 km (15 kft), $M_n = 0.3$, obtain the frequency responses for all the variables and unit step responses for pitch rate and pitch attitude, and see the discerning effects.

Solution 5.13

The TFs are obtained by: [num, den]=ss2tf(a, b, c, d, 1); sysu=tf(num(1,:), den); sysw=tf(num(2,:), den); sysq=tf(num(3,:), den); and systheta=tf(num(4,:), den). The Bode diagrams are shown in Figure 5.13. We notice from these frequency responses that the aircraft dynamics has two distinct modes: one at lower frequency and another at relatively higher frequency. This shows the feasibility of modeling these two modes separately as they are relatively well separated. One can also discern from these plots that the low-frequency mode is lightly damped compared to the higher-frequency mode, which is confirmed from Table 5.3.

We see from Figure 5.14 that the pitch rate response is well damped and the pitch attitude response takes more time to settle. This leads to the two distinct modes discussed in the next section.

Some observations about the possible TFs arising from eqn. (5.76) are (importance is indicated by '*'): (i) $|u/\delta_e|$ → the amplitude ratio is much smaller at the natural

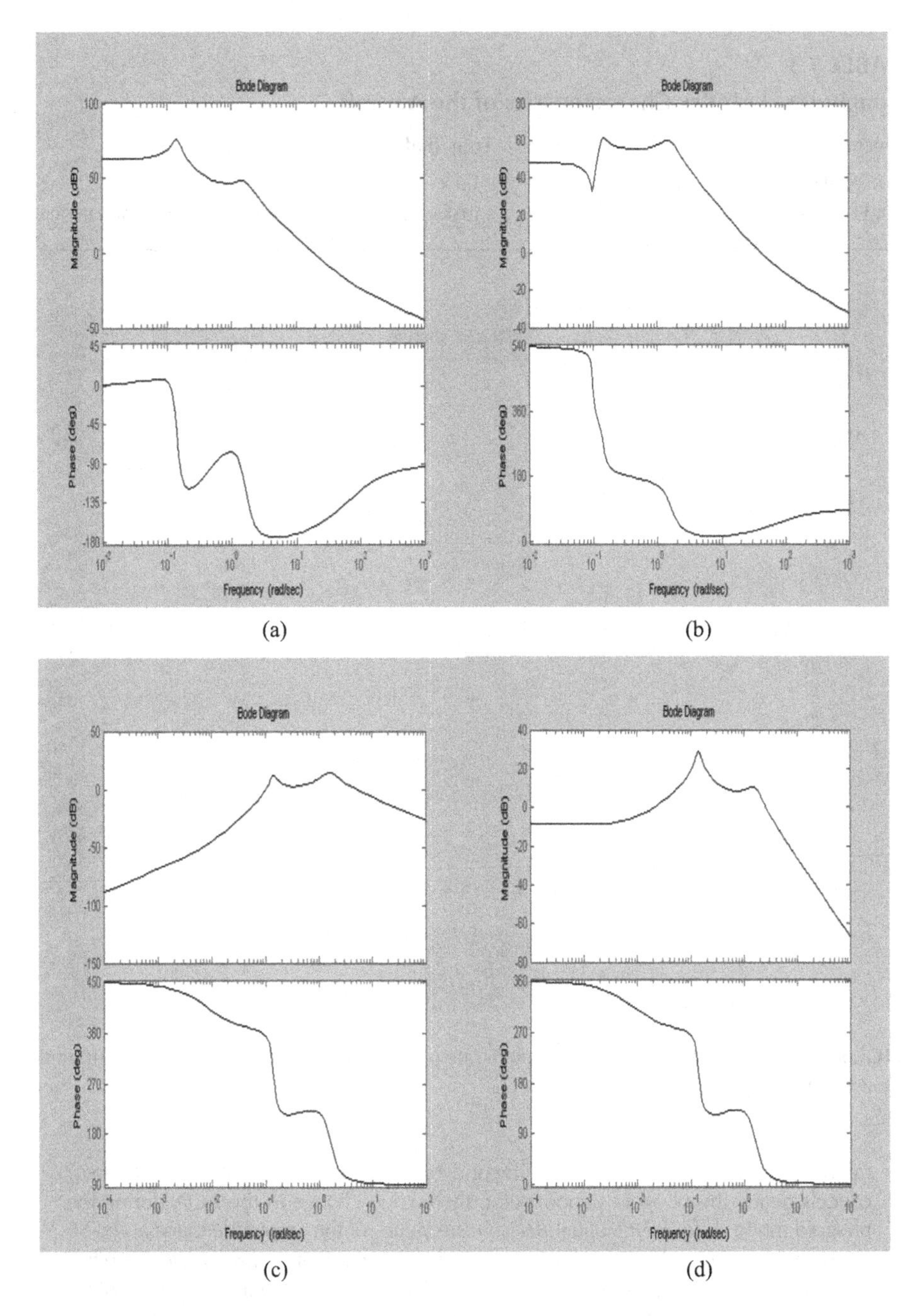

FIGURE 5.13 Frequency responses/Bode diagrams of the combined longitudinal modes of the LTV A-7A Corsair aircraft. (a) Forward velocity, (b) vertical velocity, (c) pitch rate, (d) pitch attitude.

TABLE 5.3
Longitudinal Modes Characteristics of the Aircraft

Eigenvalues	Damping	Freq. (rad/s)	Mode
−1.66e−002 +/− 1.39e−001i	1.18e−001	1.40e−001	Lightly damped (Phugoid)
−4.51e−001 +/− 1.57e+000i	2.76e−001	1.64e+000	Relatively highly damped (short period)

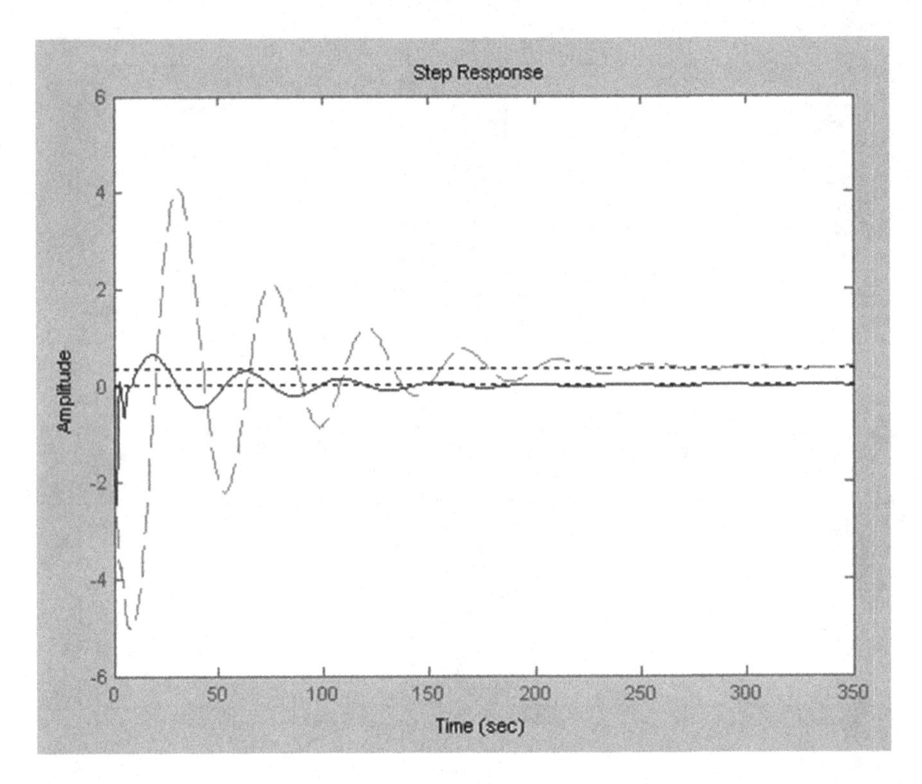

FIGURE 5.14 Unit step responses for pitch rate (-) and pitch attitude (--) of the longitudinal modes of the LTV A-7A Corsair aircraft.

frequency of the SP (*) than at that of the phugoid (**), (ii) $|\alpha/\delta_e| \rightarrow$ since the numerator cancels nearly the phugoid denominator, there is no change in the AOA during the phugoid mode (*; for SP **/2), (iii) $|\theta/\delta_e| \rightarrow$ the values of the amplitude ratios at the SP (**/2) and the Phugoid (**) modes are nearly the same as that of $|u/\delta_e|$ at the same frequencies, and (iv) $|a_z/\delta_e| \rightarrow$ the vertical (normal) acceleration amplitude at Phugoid is only somewhat higher than that at the SP; however, considering the large differences in the frequencies, a given oscillatory acceleration will involve much higher excursions in h and (h-dot) at Phugoid frequency (**) than at SP frequency (*). Successive integration of the normal acceleration to obtain h-dot and then h involve successive clockwise rotations of the $|\alpha/\delta_e|$ TF-Bode diagram each of 20 dB/decade, and such rotation progressively suppresses the SP hump relative of the phugoid peak.

5.5.1 Short-Period Mode

The SP is a relatively well-damped, high-frequency oscillation mode. Chapter 7 shows some typical time history plots of longitudinal SP motion. The amplitude ratio V/δ_e is generally much smaller at the natural frequency of the SP mode. In other words, the velocity is weakly related to the SP mode and only small changes in speed and altitude occur during SP mode. As such, the $\dot{V}$ equation can be eliminated from the set of differential equations for SP mode, and the measured velocity can be used wherever it appears in the $\dot{\alpha}$ and $\dot{q}$ equations. The simplified set for SP mode is written as:

$$\begin{aligned}
\dot{\alpha} &= \frac{g}{V}\cos(\theta-\alpha)+q-\frac{\bar{q}S}{mV}C_L-\frac{T}{mV}\sin(\alpha+\sigma_T)\\
\dot{q} &= \frac{1}{I_y}\bar{q}ScC_m \\
\dot{\theta} &= q
\end{aligned} \tag{5.77}$$

Although pitch angle θ changes during a SP maneuver, the effect of the change on SP dynamics is small. Approximating the gravity term in $\dot{\alpha}$ equation by setting θ equal to α and neglecting $\dot{\theta}$ equation and thrust, we get:

$$\begin{aligned}
\dot{\alpha} &= \frac{g}{V}+q-\frac{\bar{q}S}{mV}C_L\\
\dot{q} &= \frac{1}{I_y}\bar{q}ScC_m
\end{aligned} \tag{5.78}$$

Equation (5.78) is the simplest useable form for characterizing SP mode. In the u, v, w form of state equations and in terms of dimensional stability and control derivatives, the following simplified model is generally used for longitudinal SP motion (including in TF form):

$$\begin{aligned}
\dot{w} &= Z_w w+(u_0+Z_q)q+Z_{\delta_e}\delta_e \\
\dot{q} &= M_w w+M_q q+M_{\delta_e}\delta_e
\end{aligned} \quad ; \quad
\begin{aligned}
(s-Z_w)w-u_0 s\theta &= Z_{\delta_e}\delta_e \\
-(sM_{\dot{w}}+M_w)w+(s-M_q)s\theta &= M_{\delta_e}\delta_e
\end{aligned} \tag{5.79}$$

Here, w and q are the vertical velocity and pitch rate, respectively, and u_0 is the forward speed under steady-state condition. The stability and control derivatives are defined as [19]:

$$Z_w=\frac{1}{m}\frac{\partial Z}{\partial w};\ Z_q=\frac{1}{m}\frac{\partial Z}{\partial q};\ Z_{\delta_e}=\frac{1}{m}\frac{\partial Z}{\partial \delta_e};\ M_w=\frac{1}{I_y}\frac{\partial M}{\partial w};\ M_q=\frac{1}{I_y}\frac{\partial M}{\partial q};\ M_{\delta_e}=\frac{1}{I_y}\frac{\partial M}{\partial \delta_e} \tag{5.80}$$

Since $\alpha \approx \frac{w}{u_0}$, eqn. (5.80) can also be written in terms of α instead of w:

$$\dot{\alpha}=\frac{Z_\alpha}{u_0}\alpha+\left(1+\frac{Z_q}{u_0}\right)q+\frac{Z_{\delta_e}}{u_0}\delta_e;\ \dot{q}=M_\alpha\alpha+M_q q+M_{\delta_e}\delta_e \tag{5.81}$$

$$Z_w=\frac{Z_\alpha}{u_0};\ M_w=\frac{M_\alpha}{u_0} \tag{5.82}$$

From (5.81), we have the following TFs:

$$\frac{\alpha(s)}{\delta_e(s)}=\frac{1}{U_0}\frac{Z_{\delta_e}s+(U_0M_{\delta_e}-Z_{\delta_e}M_q)}{(s^2-(U_0M_{\dot{w}}+Z_w+M_q)s+(M_qZ_w-U_0M_w)}$$

$$\frac{\theta(s)}{\delta_e(s)}=\frac{(Z_{\delta_e}M_{\dot{w}}+M_{\delta_e})s+(Z_{\delta_e}M_w-M_{\delta_e}Z_w)}{s(s^2-(U_0M_{\dot{w}}+Z_w+M_q)s+(M_qZ_w-U_0M_w)}$$

Putting the SP 2DOF model, from (5.81) in the state-space form $\dot{x}=Ax+Bu$, and neglecting Z_q, we get:

$$\begin{bmatrix}\dot{\alpha}\\ \dot{q}\end{bmatrix}=\begin{bmatrix}\frac{Z_\alpha}{u_0} & 1\\ M_\alpha & M_q\end{bmatrix}\begin{bmatrix}\alpha\\ q\end{bmatrix}+\begin{bmatrix}\frac{Z_{\delta_e}}{u_0}\\ M_{\delta_e}\end{bmatrix}\delta_e \tag{5.83}$$

The characteristic equation of the form $(\lambda I-A)=(sI-A)$ for (5.83) will be:

$$\lambda^2-\left(M_q+\frac{Z_\alpha}{u_0}\right)\lambda+\left(M_q\frac{Z_\alpha}{u_0}-M_\alpha\right)=0 \tag{5.84}$$

Solving for the eigenvalues of the characteristic equation yields the following frequency and damping ratio for the SP mode:

$$\text{Frequency}\quad \omega_{n_{sp}}=\sqrt{\frac{Z_\alpha M_q}{u_0}-M_\alpha} \tag{5.85}$$

$$\text{Damping ratio}\quad \zeta_{sp}=-\frac{M_q+\frac{Z_\alpha}{u_0}}{2\omega_{n_{sp}}} \tag{5.86}$$

The SP occurs before there is any appreciable change in the forward speed.

Example 5.14

The aerodynamic derivatives determined from flight test conducted with 3-2-1-1 pilot-stick command input for a transport aircraft are given as:

$Z_0=-0.0085$, $Z_\alpha=-0.480$, $Z_q=0.102$, $Z_{\delta_e}=0.652$, $M_0=0.472$, $M_\alpha=-4.916$,

$M_q=-1.946$, $M_{\delta_e}=-7.011$

The *Z*-force derivatives are assumed to be normalized with the longitudinal velocity component u_0.

Write an appropriate mathematical model, and obtain TF and frequency response of the SP model.

Solution 5.14

Based on the derivatives given, we can build the following longitudinal SP model:

$$\begin{aligned} \dot{\alpha} &= Z_0 + Z_\alpha \alpha + (1 + Z_q) q + Z_{\delta_e} \delta_e \\ \dot{q} &= M_0 + M_\alpha \alpha + M_q q + M_{\delta_e} \delta_e \end{aligned} \tag{5.87}$$

The state-space form is given as:

$$\begin{bmatrix} \dot{\alpha} \\ \dot{q} \end{bmatrix} = \begin{bmatrix} -0.482 & (1+0.102) \\ -4.916 & -1.946 \end{bmatrix} \begin{bmatrix} \alpha \\ q \end{bmatrix} + \begin{bmatrix} 0.652 \\ -7.011 \end{bmatrix} \delta_e + \begin{bmatrix} -0.0085 \\ 0.472 \end{bmatrix} \tag{5.88}$$

The TFs are obtained as: [numsp, densp]=ss2tf(a, b, c, d, 1) and sysalpha=tf(numsp(1,:), densp) and sysq=tf(numsp(2,:), densp). The bias/'nought' derivatives are ignored. The two TFs are:

$$\frac{\alpha(s)}{\delta_e(s)} = \frac{0.652s - 6.457}{s^2 + 2.428s + 6.355} \text{ and } \frac{q(s)}{\delta_e(s)} = \frac{-7.011s - 6.585}{s^2 + 2.428s + 6.355} \tag{5.89}$$

The SP frequency and damping ratio are obtained as: [spfreq, spzeta]=damp (sysalpha)= [2.521 rad/s, 0.4816]. The frequency/Bode diagrams are shown in Figure 5.15. Thus, we see the distinct SP mode, though for a different aircraft.

Example 5.15

The aerodynamic derivatives for an air force medium transport aircraft are given as:

$$Z_\alpha / u_0 = -0.66, \ Z_{\delta_e} / u_0 = 0.01, \ M_\alpha = -1.74, \ M_q = -0.67, \ M_{\delta_e} = -5.33$$

Write an appropriate mathematical model, and obtain TFs and frequency responses of the SP model. Also, obtain the unit step responses.

Solution 5.15

Based on the derivatives given, we can build the following longitudinal SP model:

$$\begin{aligned} \dot{\alpha} &= (Z_\alpha / u_0) \alpha + q + (Z_{\delta_e} / u_0) \delta_e \\ \dot{q} &= M_\alpha \alpha + M_q q + M_{\delta_e} \delta_e \end{aligned} \tag{5.90}$$

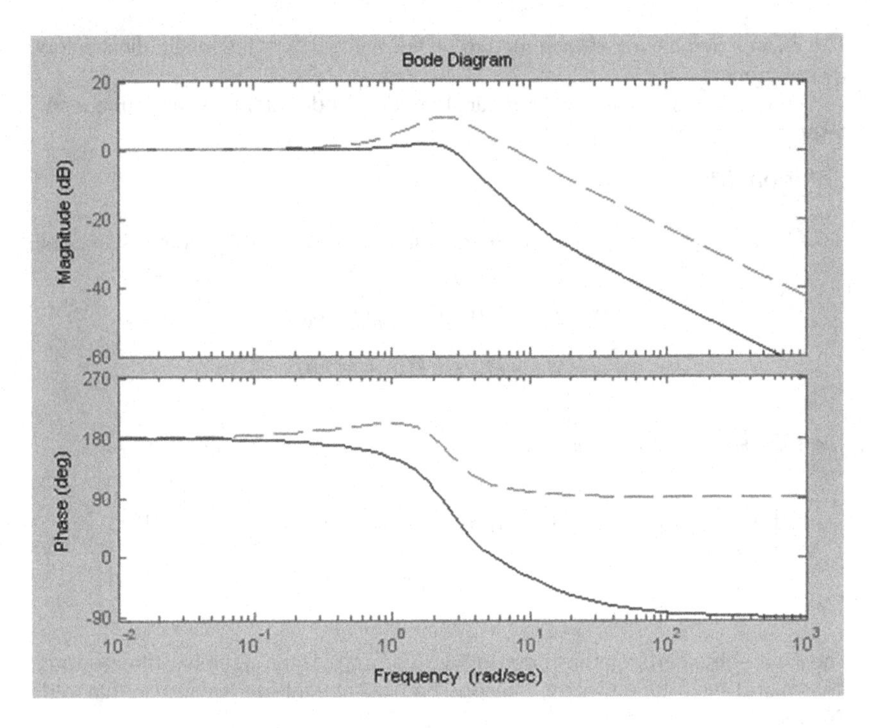

FIGURE 5.15 Bode diagrams-frequency responses for short-period mode of a transport aircraft alpha (-) and q (--) TFs.

The state-space form is given as:

$$\begin{bmatrix} \dot{\alpha} \\ \dot{q} \end{bmatrix} = \begin{bmatrix} -0.66 & 1 \\ -1.74 & -0.67 \end{bmatrix} \begin{bmatrix} \alpha \\ q \end{bmatrix} + \begin{bmatrix} 0.01 \\ -5.33 \end{bmatrix} \delta_e \tag{5.91}$$

The TFs are obtained as: [num, den]=ss2tf(a, b, c, d, 1) and sysalpha=tf(num(1,:), den) and sysq=tf(num(2,:), den). The two TFs are:

$$\frac{\alpha(s)}{\delta_e(s)} = \frac{0.01s - 5.323}{s^2 + 1.33s + 2.182} \text{ and } \frac{q(s)}{\delta_e(s)} = \frac{-5.33s - 3.535}{s^2 + 1.22s + 2.182} \tag{5.92}$$

The SP frequency and damping ratio are obtained as: [spfreq, spzeta]=damp (sysalpha)= [1.48 rad/s, 0.45]. The frequency/Bode diagrams are shown in Figure 5.16. The unit step responses are shown in Figure 5.17.

5.5.2 Phugoid

The phugoid mode is a lightly damped low-frequency oscillation. The aircraft responses in phugoid are very slow compared to the changes in the motion parameters in SP mode. Chapter 7 shows a time history plot for the phugoid mode. This mode describes the long-term translatory motions of the vehicle center of mass with practically no change in the angle of attack. This mode involves fairly large

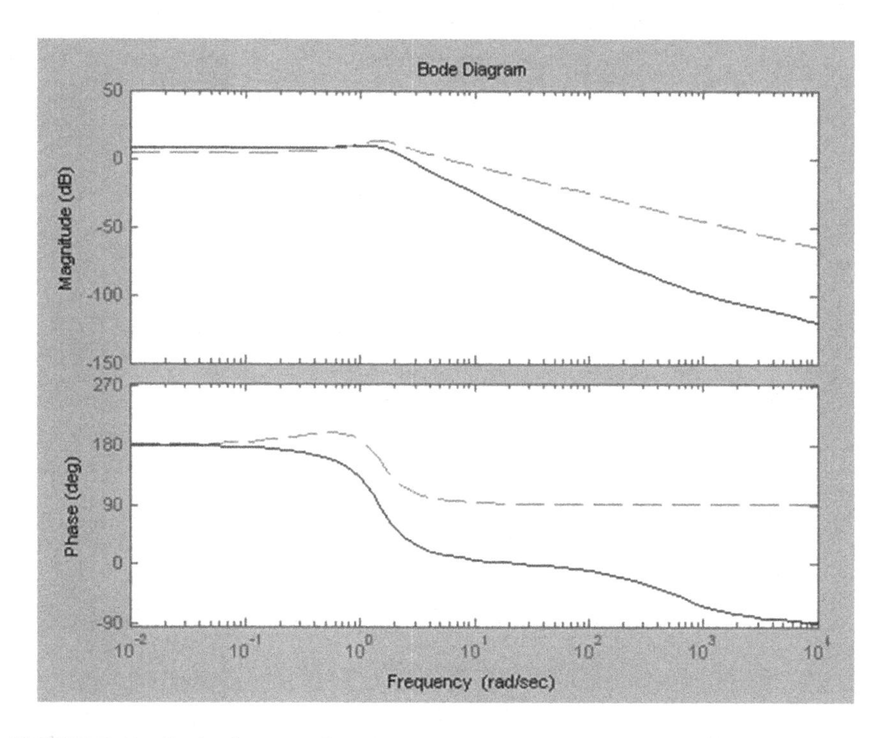

FIGURE 5.16 Bode diagrams-frequency responses for short-period mode of the medium transport aircraft: alpha (-) and q (--) TFs.

oscillatory changes in forward speed, pitch angle θ, and altitude h at constant AOA. An approximation to phugoid mode can be made by omitting the pitching moment equation and having only the 2DOF Phugoid mode state equations, from eqn. (5.76):

$$\begin{aligned}(s - X_u)u &= -g\theta + X_{\delta_e}\delta_e \\ -U_o\, s\theta &= Z_u\, u + Z_{\delta_e}\,\delta_e\end{aligned} \quad \text{(The TF Form);} \rightarrow \begin{bmatrix} \dot{u} \\ \dot{\theta} \end{bmatrix}$$

$$= \begin{bmatrix} X_u & -g \\ -\dfrac{Z_u}{U_0} & 0 \end{bmatrix} \begin{bmatrix} u \\ \theta \end{bmatrix} + \begin{bmatrix} (X_{\delta_e} =)0 \\ -Z_{\delta_e} / U_0 \end{bmatrix} \delta_e \tag{5.93}$$

Forming the characteristic equation and solving for the eigenvalues yield the following expressions for the phugoid natural frequency and damping ratio:

$$\text{Frequency} \quad \omega_{n_{\text{ph}}} = \sqrt{-\frac{Z_u g}{u_0}}; \text{ with } M_u = 0.0 \tag{5.94}$$

$$\text{Damping ratio} \quad \zeta_{\text{ph}} = -\frac{X_u}{2\,\omega_{n_{\text{ph}}}} \tag{5.95}$$

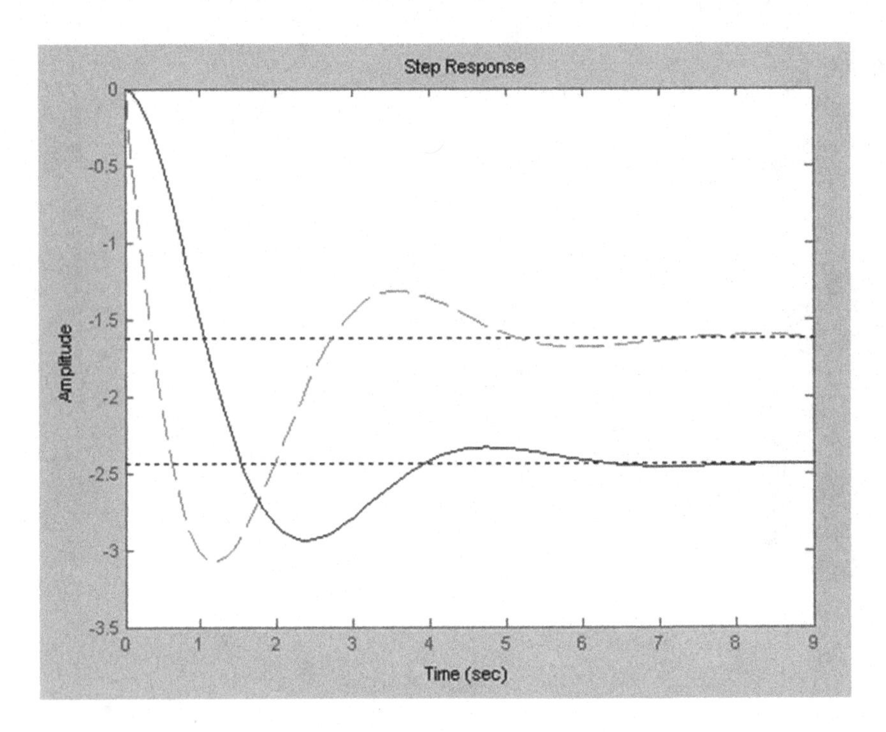

FIGURE 5.17 Unit step responses of short-period mode of the medium transport aircraft: alpha (-) and q (--).

Since $Z_u = -\dfrac{\rho SU}{m}(C_L + C_{L_u})$; for $C_{L_u} = 0$; we have $Z_u = -2L/(mU_0) = -2g/U_0$; (since $L = mg$). Thus, $\omega_{n_{ph}} \simeq \sqrt{2}(g/U_0)$.

Example 5.16

The aerodynamic derivatives of a transport aircraft in phugoid mode are given as:

$$X_u = -0.015,\ Z_u = -0.1,\ u_0 = 60 \text{ m/s} \tag{5.96}$$

Use the phugoid model of eqn. (5.93), and obtain the frequency responses and other characteristics.

Solution 5.16

The state-space model is written as:

$$\begin{bmatrix} \dot{u} \\ \dot{\theta} \end{bmatrix} = \begin{bmatrix} -0.015 & -g \\ \dfrac{0.1}{60} & 0 \end{bmatrix} \begin{bmatrix} u \\ \theta \end{bmatrix} + \begin{bmatrix} 0.01 \\ 0 \end{bmatrix} \delta_e \tag{5.97}$$

Based on the procedure of the previous examples, the TFs and the frequency responses are obtained for this phugoid model. The natural frequency of this mode is 0.128 rad/s, and the damping ratio is 0.0587. The frequency and unit step responses are shown in Figure 5.18. Since the damping ratio of this mode is very

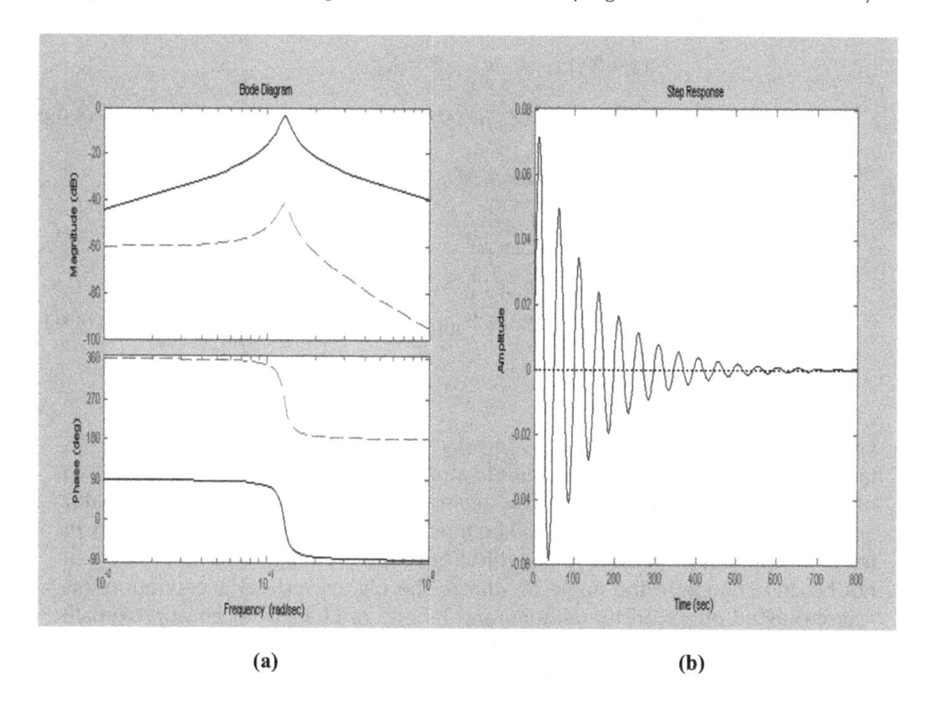

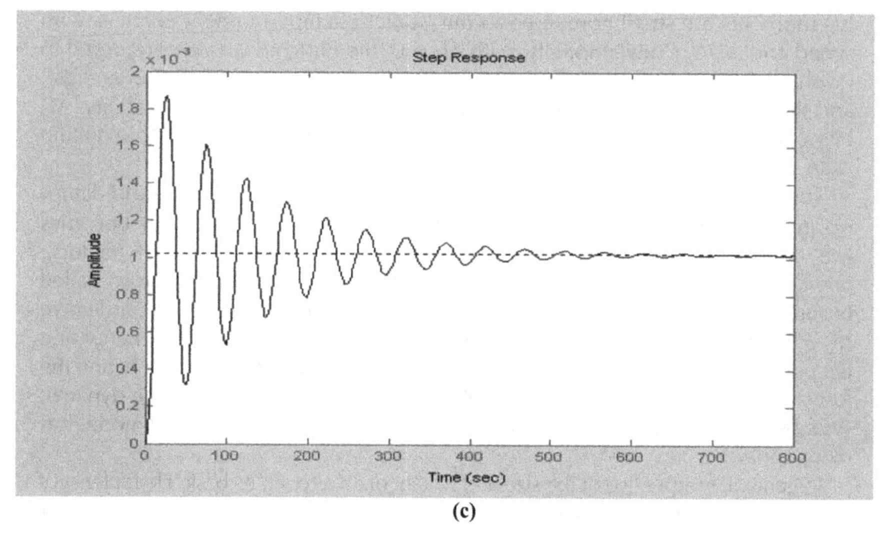

FIGURE 5.18 Phugoid mode characteristics for a transport aircraft. (a) Frequency responses for forward velocity/elevator (-) theta/elevator (--), (b) unit step response for u, (c) unit step response for theta.

small, the oscillations last for several seconds. Since the amplitudes are small and oscillations are of low frequency, the pilot would be able to manage these oscillations very well.

There is possibility of 3 DOF phugoid approximation model. This can be done by retaining the derivatives related to the vertical speed:

$$
\begin{aligned}
&(s - X_u)u - X_w w + g\theta = X_{\delta_e}\delta_e \\
&-Z_u u + (s - Z_w)w - U_o s\theta = Z_{\delta_e}\delta_e \\
&-M_u u - M_w w = M_{\delta_e}\delta_e
\end{aligned}
\tag{5.98}
$$

$$
\text{With } \omega_{n_{\text{ph}}} = \sqrt{-\frac{g\left(Z_u - \dfrac{M_u}{M_w}Z_w\right)}{u_0}} \text{ and } \zeta_{\text{ph}} = \frac{\left\{-X_u + \dfrac{M_u(X_\alpha - g)}{M_\alpha}\right\}}{2\,\omega_{n_{\text{ph}}}}
\tag{5.99}
$$

Generally, the phugoid frequency and damping increase in proportion to M_u. In summary, we see that free longitudinal motions of the aircraft comprise two oscillatory-mode characteristics. This means that the forward velocity to elevator amplitude/magnitude (of the TF) will be much smaller at the natural frequency of the SP compared to that at the phugoid frequency. In phugoid mode, there will not be any change in the angle of attack. The higher vertical acceleration has a pronounced effect on the altitude and its rate of change. In an approximate sense, the Phugoid mode describes the long-term motion of the vehicle center of mass (C.G.), whereas the SP describes rotation about the C.G.; for practical purposes, in the Phugoid motion, the 'dynamic' pitching, inertial, and damping moments are small compared to the 'static' pitching moment changes with speed and AOA. Conventionally, with $M_u = 0$, the Phugoid was/is considered to involve fairly large oscillatory changes in forward speed u, pitch attitude/angle, and the altitude, with approximately constant AOA; and the static stability, M_α being sufficiently large to decouple the Phugoid and SP modes, and to maintain AOA perturbations small.

For the normally positive values of (Z_w/M_w), the Phugoid frequency and damping increase proportionally to M_u; for sufficiently –ve values of M_u, ω_p^2 becomes –ve, and the Phugoid mode is then characterized by two first-order TF (factors): one convergent and the other a divergent mode, i.e., 'tuck' mode, the so-called because, as speed increases, the aircraft's nose has a tendency to 'tuck' under –ve M_u. Phugoid modal response is such that roughly, the sum of kinetic energy and the potential energy remains constant, mainly the change in the altitude, h, and the forward speed, u; and similar energy play would be happening in all the dynamic systems, including, say, SP of an aircraft; however, here in the Phugoid mode, it is more apparent.

In general, irrespective of the size and weight of an aircraft, its basic characteristics can be described in very simple manner in terms of only few parameters, say, two or four aerodynamic derivatives. This is made possible due to the fact that linear system theory is applicable to simplified dynamics and the related TF/control system concepts.

The qualitative importance of normal airframe stability parameters (derivatives) to longitudinal elevator-input TF qualities is shown in Table 5.4.

TABLE 5.4
Contributions of the Derivatives to Frequency and Damping of Modes

	M_u	M_α	$M_{\dot{\alpha}}$	M_q	Z_u	Z_α	X_u
ω_{SP}		***		*		*	
ζ_{SP}			**	**		**	
Classical Phugoid, $\omega_{Ph} \doteq -X_u$							***
Normal Phugoid, ω_{Ph}	***	*		*	**	*	
Normal or 'tuck' damping	**	*		*		*	

Note: Blank→little or no effect; *→moderate; **→important; ***→predominant effect.

5.6 LATERAL AND LATERAL-DIRECTIONAL MODELS AND MODES

From the non-linear coupled differential equations defined in (3.22), (3.23), and (3.27), Chapter 3, the general free lateral-directional (LD) motion of an aircraft can be described by eliminating (more of the longitudinal state variable) $\dot{V},\dot{\alpha},\dot{q}$ and $\dot{\theta}$ equations from the 6DOF state model:

$$\dot{\beta} = \frac{g}{V}(\cos\beta\sin\phi\cos\theta + \sin\beta\cos\alpha\sin\theta - \sin\alpha\cos\phi\cos\theta\sin\beta) + p\sin\alpha - r\cos\alpha$$
$$+\frac{\bar{q}S}{mV}C_{Y\text{wind}} + \frac{T}{mV}\cos(\alpha+\sigma_T)\sin\beta$$
$$\dot{p} = \frac{1}{I_xI_z - I_{xz}^2}\left\{\bar{q}Sb\left(I_zC_l + I_{xz}C_n\right) - qr\left(I_{xz}^2 + I_z^2 - I_yI_z\right) + pqI_{xz}\left(I_x - I_y + I_z\right)\right\}$$
$$\dot{r} = \frac{1}{I_xI_z - I_{xz}^2}\left\{\bar{q}Sb\left(I_xC_n + I_{xz}C_l\right) - qrI_{xz}\left(I_x - I_y + I_z\right) + pq\left(I_{xz}^2 + I_x^2 - I_xI_y\right)\right\}$$
$$\dot{\phi} = p + q\tan\theta\sin\phi + r\tan\theta\cos\phi$$
$$\dot{\psi} = r\cos\phi\sec\theta + q\sin\phi\sec\theta \tag{5.100}$$

If the excursions (of the major longitudinal-axis state variable) are small, one can neglect the longitudinal motion altogether. Also, from equation (3.29) of Chapter 3, we have:

$$C_{Y\text{wind}} = C_Y\cos\beta + C_D\sin\beta \tag{5.101}$$

$$\text{For small values of } \beta\text{, we obtain } C_{Y\text{wind}} = C_Y \tag{5.102}$$

Since the variable ψ does not appear in any of the equations for $\dot{\beta},\dot{p},\dot{r}$ and $\dot{\phi}$, the $\dot{\psi}$ equation is generally omitted from the analysis; using these assumptions in $\dot{\beta}$ equation, a simplified set to describe the LD motion can be written as:

$$\dot{\beta} = \frac{g}{V}(\sin\phi\cos\theta) + p\sin\alpha - r\cos\alpha + \frac{\bar{q}S}{mV}C_Y$$

$$\dot{p} = \frac{1}{I_xI_z - I_{xz}^2}\left\{\bar{q}Sb\left(I_zC_l + I_{xz}C_n\right) - qr\left(I_{xz}^2 + I_z^2 - I_yI_z\right) + pqI_{xz}\left(I_x - I_y + I_z\right)\right\} \quad (5.103)$$

$$\dot{r} = \frac{1}{I_xI_z - I_{xz}^2}\left\{\bar{q}Sb\left(I_xC_n + I_{xz}C_l\right) - qrI_{xz}\left(I_x - I_y + I_z\right) + pq\left(I_{xz}^2 + I_x^2 - I_xI_y\right)\right\}$$

$$\dot{\phi} = p + q\tan\theta\sin\phi + r\tan\theta\cos\phi$$

One can find several approximate forms of the equations of motion in the literature [1–4]. The coefficients C_Y, C_l and C_n can be expressed in terms of stability and control derivatives using Taylor series expansion as discussed in Chapter 4. Expressing the side force, rolling, and yawing moments in terms of dimensional derivatives, the following state-space model can be used to describe aircraft LD motion for most applications:

$$\begin{bmatrix} \dot{\beta} \\ \dot{p} \\ \dot{r} \\ \dot{\phi} \end{bmatrix} = \begin{bmatrix} \frac{Y_\beta}{u_0} & \frac{Y_p}{u_0} & \frac{Y_r}{u_0} - 1 & \frac{g\cos\theta_0}{u_0} \\ L_\beta & L_p & L_r & 0 \\ N_\beta & N_p & N_r & 0 \\ 0 & 1 & 0 & 0 \end{bmatrix} \begin{bmatrix} \beta \\ p \\ r \\ \phi \end{bmatrix} + \begin{bmatrix} \frac{Y_{\delta_a}}{u_0} & \frac{Y_{\delta_r}}{u_0} \\ L_{\delta_a} & L_{\delta_r} \\ N_{\delta_a} & N_{\delta_r} \\ 0 & 0 \end{bmatrix} \begin{bmatrix} \delta_a \\ \delta_r \end{bmatrix} \quad (5.104)$$

Here, u_0 and θ_0 are the forward speed and pitch angle, respectively, under steady-state condition. This state-space model can be used to obtain the lateral-directional TF characteristic equation, solving which for the eigenvalues will yield two real roots corresponding to the spiral mode and roll subsidence, while a pair of complex roots defines the Dutch roll (DR) mode. These three/four distinct lateral-directional modes are explained using the following example:

Example 5.17

The Douglas DC-8 aircraft state-space lateral-directional model [4] is given as:

$$\begin{bmatrix} \dot{v} \\ \dot{p} \\ \dot{r} \\ \dot{\phi} \end{bmatrix} = \begin{bmatrix} -0.1 & 0 & -468 & 32 \\ -0.0058 & -1.232 & 0.397 & 0 \\ 0.0028 & -0.0346 & -0.257 & 0 \\ 0 & 1 & 0 & 0 \end{bmatrix} \begin{bmatrix} v \\ p \\ r \\ \phi \end{bmatrix} + \begin{bmatrix} 0 & 13.48 \\ -1.62 & 0.392 \\ -0.0188 & -0.864 \\ 0 & 0 \end{bmatrix} \begin{bmatrix} \delta_a \\ \delta_r \end{bmatrix} \quad (5.105)$$

Obtain all the eight TFs using MATLAB. Obtain the characteristics modes of the aircraft dynamics.

Solution 5.17

The TFs are obtained by: [numail, denail]=ss2tf(a, b, c, d, 1); [numrud, denrud]=ss2tf (a, b, c, d, 2); for aileron input: sysva=tf(numail(1,:), denail); syspa=tf(numail(2,:), denail); sysr=tf(numail(3,:), denail); and dampsystpha=tf(numail(4,:), denail). For rudder input: sysvrud=tf(numrud(1,:), denail);sysprud=tf(numrud(2,:), denrud); sysrrud=tf(numrud(3,:), denrud); and systphrud=tf(numrud(4,:), denrud). These TFs are given in Table 5.5.

In fact, Tables 5.5 and 5.6 show the feasibility of separately representing these modes in different three characteristic modes, since these modes seem to be reasonably well separated.

Next, we discuss the lateral equations of motion referred to stability axes and written in terms of AOSS, with assumptions $Y_{\dot{v}} = Y_p = Y_r = L_{\dot{v}} = N_{\dot{v}} = 0$:

$$(s - Y_v)\beta - \frac{g}{U_0}\frac{p}{s} + r = Y_\delta^* \delta; \quad Y_\delta^* = Y_\delta / U_0$$

$$-L_\beta \beta + (s - L_p)p - \left(\frac{I_{xz}}{I_x} s + L_r\right) r = L_\delta \delta$$

$$-N_\beta \beta - \left(\frac{I_{xz}}{I_x} s + N_p\right) p + (s - N_r) r = N_\delta \delta$$

TABLE 5.5
Lateral Modes Transfer Functions for the Chosen Aircraft

	Aileron Control Input	Rudder Control Input
Side velocity	–5.107e–015 s^3 + 8.798 s^2 – 67.23 s – 13.56	13.48 s^3 + 424.4 s^2 + 521.5 s – 7.752
	--	--
	s^4 + 1.589 s^3 + 1.78 s^2 + 1.915 s + 0.01238	s^4 + 1.589 s^3 + 1.78 s^2 + 1.915 s + 0.01238
Roll rate	–1.62 s^3 – 0.5858 s^2 – 2.201 s – 3.123e–017	0.392 s^3 – 0.2813 s^2 – 1.865 s – 1.301e–016
	--	--
	s^4 + 1.589 s^3 + 1.78 s^2 + 1.915 s + 0.01238	s^4 + 1.589 s^3 + 1.78 s^2 + 1.915 s + 0.01238
Yaw rate	–0.0188 s^3 + 0.03101 s^2 + 0.003289 s – 0.1476	–0.864 s^3 – 1.127 s^2 – 0.05891 s – 0.1255
	--	--
	s^4 + 1.589 s^3 + 1.78 s^2 + 1.915 s + 0.01238	s^4 + 1.589 s^3 + 1.78 s^2 + 1.915 s + 0.01238
Roll angle	4.441e–016 s^3 – 1.62 s^2 – 0.5858 s – 2.201	8.882e-016 s^3 + 0.392 s^2 – 0.2813 s – 1.865
	--	--
	s^4 + 1.589 s^3 + 1.78 s^2 + 1.915 s + 0.01238	s^4 + 1.589 s^3 + 1.78 s^2 + 1.915 s + 0.01238

TABLE 5.6
Lateral-Directional Characteristics of the DC-8 Aircraft

Eigenvalues	Damping Ratio	Freq. (rad/s)	Mode
–1.27e-001 +/– 1.19e+000i	0.106	1.2	2nd-order lightly damped (DR)
–0.0065	1	-	A very large time constant mode (spiral)
–1.33	1	-	Roll subsidence with very small time constants

Also, $p = s\phi$, $r = s\varphi$; $a_{y_{cg}} = U_0\dot{\beta} - g(p/s) + U_0 r = \dot{v} - g\phi + U_0 r$, and the possible TFs are: $\frac{\beta(s)}{\delta(s)}, \frac{\phi(s)}{\delta(s)}, \frac{r(s)}{\delta(s)}$, and $\frac{a_y(s)}{\delta(s)}$ for aileron and rudder inputs : δ_a, δ_r

5.6.1 Dutch Roll Mode

2DOF DR: It is a relatively lightly damped oscillatory mode that consists of primarily the sideslip and yawing motions. Neglecting equations for $\dot{p}$ and $\dot{\phi}$ in eqn. (5.104) and with no control inputs, the simplified form of state-space model for the DR oscillatory mode can be expressed as [2]:

$$\begin{bmatrix} \dot{\beta} \\ \dot{r} \end{bmatrix} = \begin{bmatrix} \frac{Y_\beta}{u_0} & \frac{Y_r}{u_0} - 1 \\ N_\beta & N_r \end{bmatrix} \begin{bmatrix} \beta \\ r \end{bmatrix}; \text{Sometimes } Y_r \text{ is neglected} \quad (5.106)$$

The characteristic equation for the above model will be:

$$\lambda^2 - \left(\frac{Y_\beta + u_0 N_r}{u_0} \right)\lambda + \left(\frac{Y_\beta N_r - N_\beta Y_r + u_0 N_\beta}{u_0} \right) = 0 \quad (5.107)$$

Solving eqn. (5.106) for the eigenvalues of the characteristic equation yields the following expressions for the natural frequency and damping ratio for this oscillatory mode:

$$\text{Frequency} \quad \omega_{n\text{DR}} = \sqrt{\frac{Y_\beta N_r - N_\beta Y_r + N_\beta u_0}{u_0}} \quad (5.108)$$

$$\text{Damping ratio} \quad \zeta_{\text{DR}} = -\left(\frac{Y_\beta + N_r u_0}{u_0} \right) \frac{1}{2\omega_{n\text{DR}}} \quad (5.109)$$

From eqns. (5.108), and (5.109), we can make the following comments: Even though $N_\beta + Y_v N_r \to 0$, it is still possible to have +ve stiffness and a finite DR frequency if I_{xz} and L_β are of the same sign (i.e., for normally –ve L_β and nose-up inclination of the principal axis of inertia to the direction of the flight). Such situations are not uncommon for HAOA conditions, where a negative N_β, e.g., due to fin immersion in the wing/body wake, can be stabilized by overriding negative values of I_{xz} and L_β. The DR damping is almost always +ve because usually both the derivatives are negative, see the numerator of the bracketed term of (5.109). Thus, the possible motion of 2DOF simplified equations must therefore be either a damped (or overdamped) oscillation or a subsidence/divergence combination.

Example 5.18

The state-space model of the DR mode of a supersonic fighter aircraft is given as:

$$\begin{bmatrix} \dot{\beta} \\ \dot{r} \end{bmatrix} = \begin{bmatrix} \frac{Y_\beta}{u_0} & \frac{Y_r}{u_0} - 1 \\ N_\beta & N_r \end{bmatrix} \begin{bmatrix} \beta \\ r \end{bmatrix} + \begin{bmatrix} Y_{\delta_a}/u_0 & Y_{\delta_r}/u_0 \\ N_{\delta_a} & N_{\delta_r} \end{bmatrix} \begin{bmatrix} \delta_a \\ \delta_r \end{bmatrix} \quad (5.110)$$

With the derivatives incorporated into this model, we obtain:

$$\begin{bmatrix} \dot{\beta} \\ \dot{r} \end{bmatrix} = \begin{bmatrix} -0.139 & 2.218/75-1 \\ 1.125 & -0.571 \end{bmatrix} \begin{bmatrix} \beta \\ r \end{bmatrix} + \begin{bmatrix} 0.056/75 & 1.134/75 \\ 0.083 & -0.343 \end{bmatrix} \begin{bmatrix} \delta_a \\ \delta_r \end{bmatrix} \quad (5.111)$$

Obtain the corresponding TFs.

Solution 5.18

Following the procedure outlined in previous examples, four TFs are obtained, see Table 5.7. The frequency responses are shown in Figure 5.19. The unit step responses for side slip and roll rate are shown in Figure 5.20; the lightness of the damping is evident from these responses.

3DOF DR: Although the rudder TFs are well matched, the 2DOF approximation may depart reasonably from the complete 3DOF case, which is given with its characteristic equation as:

$$\begin{aligned} &(s-Y_v)\beta + r = Y_\delta^*\delta; \quad Y_\delta^* = Y_\delta / U_0 \\ &-L_\beta\beta + s(s-L_p)\phi = L_\delta\delta \qquad ; \; s(s-L_p)[s^2+(-Y_v-N_r)s+(N_\beta+Y_vN_r)] \\ &-N_\beta\beta + (s-N_r)r = N_\delta\delta \end{aligned}$$

We see the following modes from the characteristics equation: (i) $s \rightarrow$ the spiral mode, (ii) $(s - L_p) \rightarrow$ roll subsidence mode, and (iii) quadratic DR mode that is the same as in 2DOF. The TFs obtained are: $\frac{\beta(s)}{\delta(s)}, \frac{r(s)}{\delta(s)}$, and $\frac{\phi(s)}{\delta(s)}$.

5.6.2 3DOF Spiral and Roll Subsidence Modes

For both the modes, the side slip motions are relatively small, and the term $(s - Y_v)\beta$ for the spiral mode is relatively negligible; hence, we obtain along with its characteristics equation:

TABLE 5.7
Dutch roll TFs for the Supersonic Fighter Aircraft

	Aileron Control Input	Rudder Control Input	DR Frequency and Damping Ratio
Beta TF	0.0007467 s – 0.08012	0.01512 s + 0.3415	1.08 rad/s
	---------------------------	-----------------------	
	s^2 + 0.71 s + 1.171	s^2 + 0.71 s + 1.171	
Roll rate TF	0.083 s + 0.01238	–0.343 s – 0.03067	0.328
	-----------------------	-----------------------	
	s^2 + 0.71 s + 1.171	s^2 + 0.71 s + 1.171	

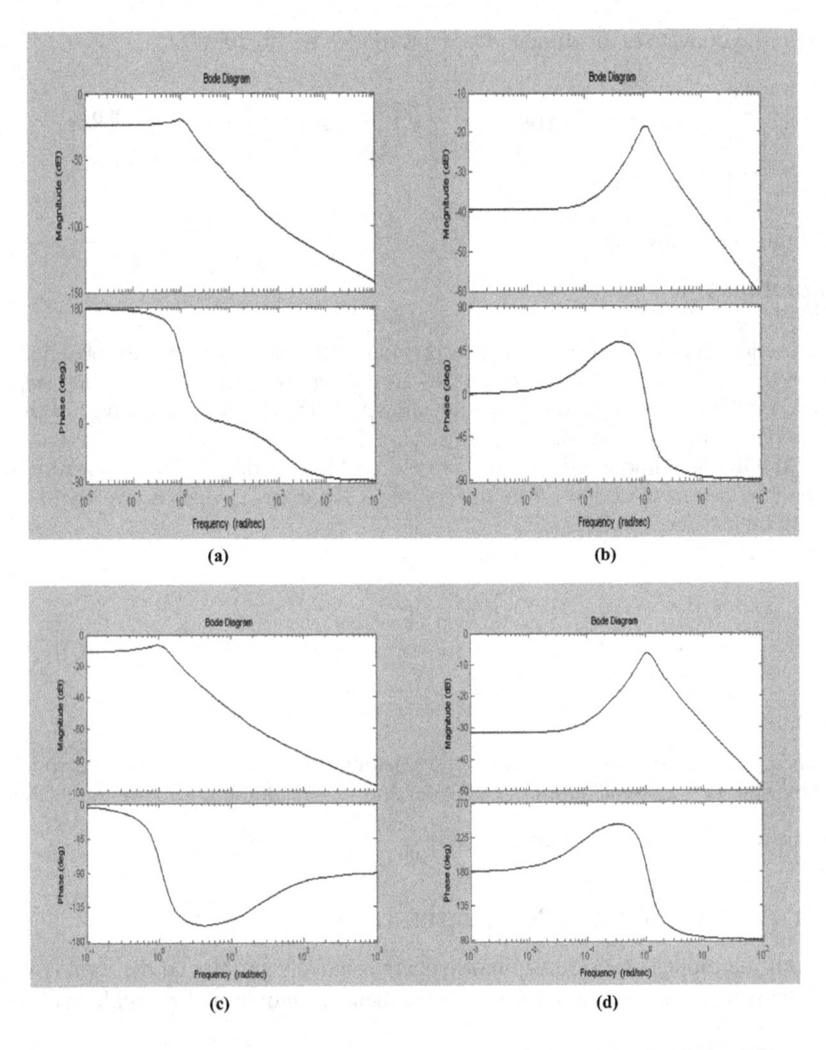

FIGURE 5.19 Dutch roll mode frequency responses of a fighter aircraft. (a) Beta/aileron, (b) roll rate/aileron, (c) beta/rudder, (d) roll rate/rudder.

$$-\frac{g}{U_0}\phi + r = Y_\delta^* \delta; \quad Y_\delta^* = Y_\delta / U_0$$

$$-L_\beta \beta + s(s - L_p)\phi - \left(\frac{I_{xz}}{I_x} s + L_r\right) r = L_\delta \delta;$$

$$s^2 + \left[-L_p + \frac{L_\beta}{N_\beta}\left\{\left(\frac{I_{xz}}{I_x} s + N_p\right) - g/U_0\right\}\right] s + g/U_0 \left(\frac{L_\beta}{N_\beta} N_r - L_r\right)$$

$$-N_\beta \beta - \left(\frac{I_{xz}}{I_x} s + N_p\right) s\phi + (s - N_r) r = N_\delta \delta$$

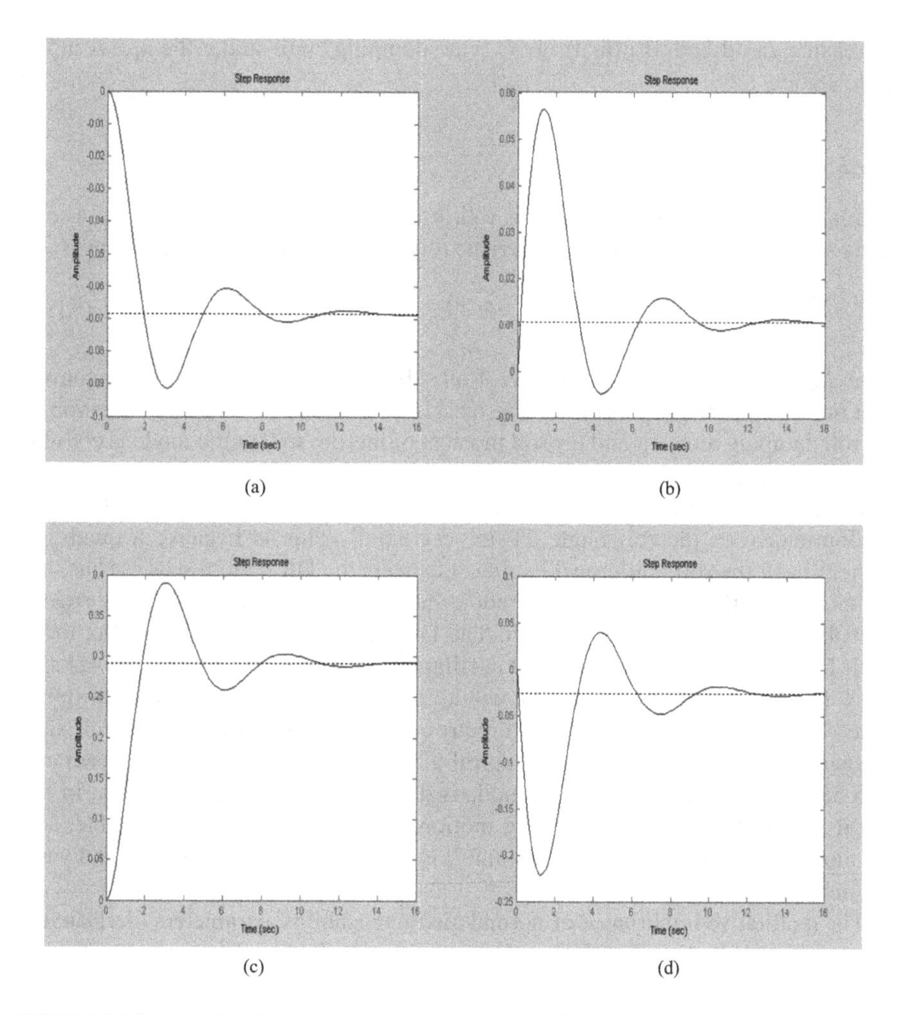

FIGURE 5.20 Dutch roll mode unit step responses of a fighter aircraft. (a) Beta/aileron, (b) roll rate/aileron, (c) beta/rudder, (d) roll rate/rudder.

The possible TFs are: $\frac{\beta(s)}{\delta(s)}, \frac{r(s)}{\delta(s)}$, and $\frac{\phi(s)}{\delta(s)}$ for aileron and rudder pedal inputs.

5.6.3 Spiral Mode

One of the real roots, having a small value (relatively long time period), indicates the spiral mode. The root can have a negative or positive value, making the mode convergent or divergent. This mode is dominated by rolling and yawing motions; sideslip is almost non-existent; and this mode could be unstable, yet it is nearly coordinated. The characteristic root λ for the spiral mode is given by:

$$\lambda = \frac{L_\beta N_r - L_r N_\beta}{L_\beta} \tag{5.112}$$

Increasing L_β (dihedral effect) or N_r (yaw damping) will make the spiral mode more stable.

5.6.4 Roll Mode

In this mode, the dominant motion is roll. It is a highly damped mode with a relatively short time period. The characteristic root λ for the roll mode is given by:

$$\lambda = L_p \tag{5.113}$$

Here, L_p is the roll damping derivative. Roll subsidence mode is a first-order convergent mode of relatively short time constant. In roll mode, there is a combination of the roll damping moment and the roll moment of inertia, so that the mode is conventionally considered to be essentially of single DOF.

In summary, the DR is the dominant characteristics of any aircraft. However, its dominance in the roll angle TF is very small. This is because a quadratic numerator in the roll – aileron TF, almost cancels the DR denominator. Thus, the DR excited by the aileron is more predominant in roll rate and AOSS. The rudder contribution to the roll rate is small. The DR damping is usually +ve. So, the 2 DOF DR mode would be a damped oscillation, or worst-case divergence/subsidence. The 3DOF DR approximation would incorporate roll angle as an additional state variable. This will then have three roots: the spiral mode ('s'), the roll subsidence mode, and DR. The 3DOF spiral and roll subsidence approximation can also be made. The roll subsidence mode is dominant in the rolling motion. In the spiral mode, the rolling and yawing motions predominate, mostly unstable, and the mode has a very large time constant. It is a fairly coordinated rolling and yawing motion mode.

The qualitative importance of normal airframe stability parameters (derivatives) to longitudinal elevator-input TF qualities is shown in Table 5.8.

TABLE 5.8
Contributions of the Derivatives to Frequency and Damping of Modes

	L_β	L_p	L_r	N_β, Y_v	N_p	N_r
Spiral mode; $1/T_s$	***		**	**		**
Roll subsidence mode; $1/T_R$	*	***		*	*	
DR ω_{DR}	**			***		
DR $2\zeta_{DR}\omega_{DR}$	**			**	*	***
Lateral Phugoid (LP) ω_{Ph}	***		*	***		**
LP, $2\zeta_{Ph}\omega_{Ph}$	**	***		**	*	

Blank→ little or no effect; *→ moderate; **→ important; ***→ predominant effect.

5.7 MISSILE AERODYNAMIC TRANSFER FUNCTIONS

It has already been emphasized that linearization of the equations of motion can be carried out only if the small perturbation assumption is not violated. The same is true in the case of a missile: linearity constraint can be invoked if the changes in the incidence angles and angular body rates are small. To design a control system for a missile, the procedure is to define an operating point (in terms of Mach and altitude) and consider the aerodynamic derivatives pertaining to the operating point to remain unchanged within a small region of the selected flight condition. The control system is then designed to meet the specific requirements of system phase lag, damping, and bandwidth. In this manner, several test conditions are investigated. Detailed analysis is carried out by also considering the configuration changes due to change in mass and shift in C. G. as the fuel gets consumed. The force equation along the x-axis is generally omitted because it neither affects the roll, pitch, or yaw motion. From eqns. (3.33) and (3.34) of Chapter 3, neglecting the gravity terms and considering only the Y-force, Z-force, and the equations for angular accelerations, the control equations for the missile can be expressed as [20]:

$$
\begin{aligned}
&\tilde{Y} = \dot{v} + rU = \tilde{Y}_v v + \tilde{Y}_r r + \tilde{Y}_{\delta_r}\delta_r && \Rightarrow \{ma_y = m(\dot{v} + ru)\} \\
&\tilde{Z} = \dot{w} - qU = \tilde{Z}_w w + \tilde{Z}_q q + \tilde{Z}_{\delta_e}\delta_e && \Rightarrow \{ma_z = m(\dot{w} - qw)\} \\
&\dot{p} = \tilde{L}_p p + \tilde{L}_{\delta_a}\delta_a && \Rightarrow \{I_x\dot{p} = L = (L_p p + L_{\delta_a}\delta_a)\} \Rightarrow \{\dot{p} = l_p p + l_{\delta_a}\delta_a\} \\
&\dot{q} = \tilde{M}_w w + \tilde{M}_q q + \tilde{M}_{\delta_e}\delta_e && \Rightarrow \{\dot{q} = m_w w + m_q q + m_{\delta_e}\delta_e\} \\
&\dot{r} = \tilde{N}_v v + \tilde{N}_r r + \tilde{N}_{\delta_r}\delta_r && \Rightarrow \{\dot{r} = n_v v + n_r r + n_{\delta_r}\delta_r\}
\end{aligned} \tag{5.114}
$$

Here, $\tilde{Y}$ and $\tilde{Z}$ denote the specific forces (accelerations) given by:

$$
\begin{aligned}
\tilde{Y} &= Y/m \\
\tilde{Z} &= Z/m
\end{aligned} \tag{5.115}
$$

and $\tilde{Y}_v, \tilde{Y}_r, \tilde{Y}_{\delta_r}, \tilde{Z}_w, \ldots, \tilde{N}_{\delta_r}$ are the specific derivatives (also, $I_y = I_z$). Full values of these derivatives can be computed given the information on mass and inertia characteristics of the missile. The aileron deflection in a missile is given by one of the following formulae:

$$
\frac{1}{4}(\delta_1 + \delta_2 + \delta_3 + \delta_4); \frac{1}{2}(\delta_1 + \delta_3); \frac{1}{2}(\delta_2 + \delta_4) \tag{5.116}
$$

If only two surfaces act in differential mode, then we have: elevator deflection: $\frac{1}{2}(\delta_1 - \delta_2)$; rudder deflection: $\frac{1}{2}(\delta_2 - \delta_4)$. Some of the aerodynamic TFs obtained by taking Laplace of eqn. (5.114) are given below [20]:

a. **Roll rate/aileron** $\left(\dfrac{p}{\delta_a}\right)$

Consider the $\dot{p}$ equation; taking Laplace, we have:

$$sp - \tilde{L}_p p = \tilde{L}_{\delta_a}\delta_a \tag{5.117}$$

or

$$\frac{p(s)}{\delta_a(s)} = \frac{\tilde{L}_{\delta_a}}{s - \tilde{L}_p} = \frac{-\tilde{L}_{\delta_a}/\tilde{L}_p}{1+\tau s} = \frac{(-l_{\delta_a})/l_p}{1+s(-1/l_p)} \tag{5.118}$$

Equation (5.118) gives steady-state gain equal to $-\tilde{L}_{\delta_a}/\tilde{L}_p$ and time constant τ equal to $-1/\tilde{L}_p$.

b. **Lateral acceleration/rudder** $(\tilde{Y}/\delta_r)$

Simplifying $\tilde{Y}$ and $\dot{r}$ equations in eqn. (5.114) by eliminating r and v and neglecting the $\tilde{Y}_r$ derivative, which is generally very small, we obtain the following expression for TF for $\tilde{Y}/\delta_r$

$$\frac{\tilde{Y}(s)}{\delta_r(s)} = \frac{s^2\tilde{Y}_{\delta_r} - s\tilde{N}_r\tilde{Y}_{\delta_r} + U\left(\tilde{N}_v\tilde{Y}_{\delta_r} - \tilde{N}_{\delta_r}\tilde{Y}_v\right)}{s^2 - s\left(\tilde{N}_r + \tilde{Y}_v\right) + \left(U\tilde{N}_v + \tilde{N}_r\tilde{Y}_v\right)} \Rightarrow \left\{\frac{a_y(s)}{\delta_r(s)} = \frac{s^2 y_{\delta_r} - s n_r y_{\delta_r} + U\left(n_v y_{\delta_r} - n_{\delta_r} y_v\right)}{s^2 - s\left(n_r + y_v\right) + \left(U n_v + n_r y_v\right)}\right\} \tag{5.119}$$

If ω_n denotes the undamped natural frequency and ς denotes the damping ratio, then we have:

$$\omega_n^2 = U\tilde{N}_v + \tilde{N}_r\tilde{Y}_v \equiv \left(U n_v + n_r y_v\right);\ 2\varsigma\omega_n = \tilde{N}_r + \tilde{Y}_v \tag{5.120}$$

In the expression for ω_n, the term $U\tilde{N}_v$ is generally much larger than $\tilde{N}_r\tilde{Y}_v$; hence, we have the weathercock frequency:

$$\omega_n^2 = \frac{\text{the restoring moment / unit angualr deflection} (= N_\beta)}{\text{moment of intertia about C.G.} (= I_z)} = \frac{UN_v}{I_z} = Un_v$$

In eqn. (5.119), the steady-state gain is given by $\dfrac{U\left(\tilde{N}_v\tilde{Y}_{\delta_r} - \tilde{N}_{\delta_r}\tilde{Y}_v\right)}{\omega_n^2}$.

c. **Yaw rate/rudder** (r/δ_r)

The TF (also called the body rate TF) r/δ_r can be obtained by eliminating v from $\tilde{Y}$ and $\dot{r}$ equations in eqn. (5.114):

$$\frac{r(s)}{\delta_r(s)} = \frac{s\tilde{N}_{\delta_r} + \left(\tilde{N}_v\tilde{Y}_{\delta_r} - \tilde{N}_{\delta_r}\tilde{Y}_v\right)}{s^2 - s\left(\tilde{N}_r + \tilde{Y}_v\right) + \left(U\tilde{N}_v + \tilde{N}_r\tilde{Y}_v\right)} \Rightarrow \frac{s n_{\delta_r} + \left(n_v y_{\delta_r} - n_{\delta_r} y_v\right)}{s^2 - s\left(n_r + n_v\right) + \left(U n_v + n_r y_v\right)}$$

$$\frac{r}{\delta_r} = \frac{-(n_{\delta_r} y_v - n_v y_{\delta_r})(T_i s + 1)}{\text{denominator}};\ T_i = \frac{n_{\delta_r}}{-n_{\delta_r} y_v + n_v y_{\delta_r}} \approx -\frac{1}{y_v} \tag{5.121}$$

Comparing eqns. (5.119) and (5.121), one observes that ω_n and ς are the same for both the TFs. We see that the body rate is the phase-advanced flight path rate, and the body rate (with 'U') is the phase-advanced lateral acceleration (latex); and hence, a feedback from a rate gyro in an autopilot tends to damp the weathercock mode.

d. **Sideslip/rudder** (β/δ_r)
Eliminating r from $\tilde{Y}$ and $\dot{r}$ equations in eqn. (5.114) and, for small incidence angles, assuming $\beta = v/U$, the TF for β/δ_r can be expressed as:

$$\frac{\beta}{\delta_r} = \frac{s\frac{\tilde{Y}_{\delta_r}}{U} - \left(\frac{\tilde{N}_r \tilde{Y}_{\delta_r}}{U} + \tilde{N}_{\delta_r}\right)}{s^2 - s\left(\tilde{N}_r + \tilde{Y}_v\right) + \left(U\tilde{N}_v + \tilde{N}_r \tilde{Y}_v\right)} \tag{5.122}$$

Once again, the ω_n and ς for the above TF are same as those for the TFs expressed in eqns. (5.119) and (5.121).

Example 5.19

For a surface-to-air missile (SAM) with rear controls, the yaw aerodynamic derivatives for a certain flight condition (Mach = 1.4, H = 1.5 km, and U = 467 m/s) [5] are given as:

$$Y_v = -2.74;\ N_v = 0.309;\ Y_\delta = 197;\ N_\delta = -534;\ N_r = -2.89$$

The other details are: length = 2 m, mass = 53 kg, and moment of inertia, I_z = 13.8 kg m^2.

The missile latex (lateral acceleration) rudder control TF is given as:

$$\frac{a_y(s)}{\delta(s)} = \frac{Y_\delta s^2 - Y_\delta N_r s - U(N_\delta Y_v - N_v Y_\delta)}{s^2 - (Y_v + N_r)s + Y_v N_r + U N_v} \tag{5.123}$$

With the values given, obtain the missile TF, its characteristic modes, frequency responses, and step input responses.

Solution 5.19

The following TF is obtained:

$$\frac{a_y(s)}{\delta(s)} = \frac{197s^2 + 569.33s - 467(1463.16 - 60.87)}{s^2 + 5.63s + (7.92 + 144.3)} \tag{5.124}$$

The static margin $= \frac{N_v}{Y_v} = \frac{n_v I_z}{y_v m} = 0.029$ Or 1.5% of the length of the missile; a large static margin would result in: (i) a small steady-state gain, (ii) a high weathercock frequency, and (iii) a very low damping ratio. The latex dynamic mode is 2nd-order oscillatory:

−2.815+/−12.0123i; natural frequency = 12.4 rad/s; the damping ratio = 0.228 (this will decrease at higher altitudes); the steady-state gain is = −4300; this means that if the aerodynamics were assumed to be linear, then 0.1 rad of rudder deflection would produce a lateral acceleration in the −*y* direction of 430 m/s² Or 4.3 *g*. The numerator factors are: −655,000(1+*s*/59.1)(1−*s*/56.2); one can deduce from these factors that the phase contribution of the first will be cancelled by the second one; however, these zeros will be recognizable in the range of frequencies of 60 rad/s and would be troublesome in the design of an autopilot, since these would prevent the continuous attenuation with frequency (as is usually observed in any 2nd-order oscillatory dynamic system); all the time the rudder is moving, there will be a lateral acceleration whatever the frequency beyond 60 rad/s. The frequency response and the unit step rudder response for the latex are shown in Figure 5.21. The $T_i = 0.365$ s is the time constant of the zero of the yaw rate/rudder TF.

5.8 ROTORCRAFT LINEAR MODELING

A general state-space form of model to describe rotorcraft dynamics may be written as:

$$\dot{x} = f[x(t),u(t),\Theta] + g_1 w(t); \qquad y = h[x(k),u(k),\Theta]$$
$$z(k) = h[x(k),u(k),\Theta] + g_2 \upsilon(k); \quad x(0) = x_0 \tag{5.125}$$

Equation (5.125) is the mixed form of continuous-discrete state and observation equations, wherein Θ is the vector of unknown stability and control derivatives, g_1 is the state noise vector/matrix, and g_2 is the measurement noise vector/matrix. The linear form of (5.125) state-space model can be expressed as:

$$\dot{x} = Ax(t) + Bu(t) + G_1 w(t)$$
$$z(k) = Hx(k) + Du(k) + G_2 \upsilon(k); \; x(0) = x_0 \tag{5.126}$$

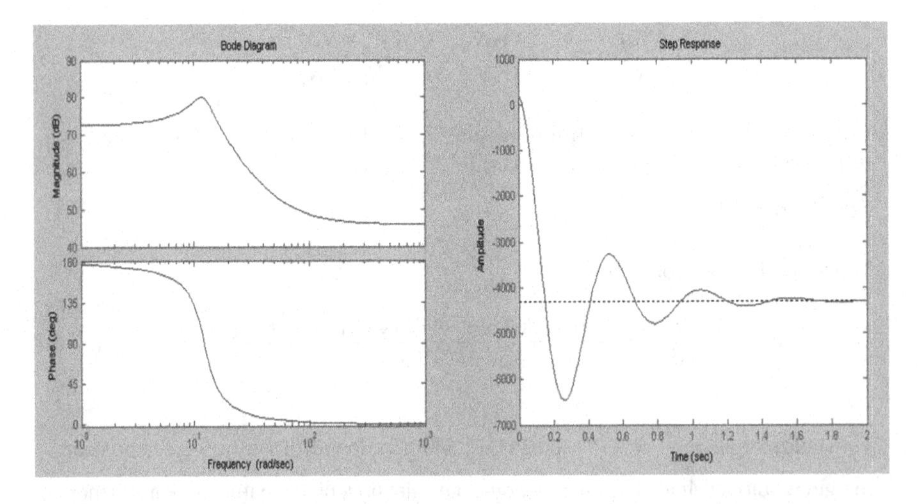

FIGURE 5.21 Latex frequency response and unit step input response for a surface to air missile.

In (5.126), A and B are the matrices containing the stability and control derivatives, and C and D are the matrices that relate the measured quantities to the rotorcraft states and control variables. The 6DOF equations of motion for modeling rotorcraft dynamics were discussed in Chapter 3; from which and from the state-space model described in eqn. (5.126), it is evident that the state, measurement, and control vectors will include the following motion variables:

$$\begin{aligned} x &= [u,v,w,p,q,r,\phi,\theta,h] \\ y &= [u_m,v_m,w_m,p_m,q_m,r_m,\phi_m,\theta_m,h_m,a_{xm},a_{ym},a_{zm},\dot{p}_m,\dot{q}_m,\dot{r}_m,] \\ u &= [\delta_{\text{lon}},\delta_{\text{lat}},\delta_{\text{col}},\delta_{\text{ped}}] \end{aligned} \tag{5.127}$$

Assuming small variations in u,v,w,ϕ and θ, the linearized form of eqn. (3.43), in the state-space formulation, can be expressed as [21]:

$$\begin{bmatrix} \dot{u} \\ \dot{v} \\ \dot{w} \end{bmatrix} = \begin{bmatrix} \tilde{X} \\ \tilde{Y} \\ \tilde{Z} \end{bmatrix} + g\begin{bmatrix} -\sin\theta_0 - \Delta\theta\cos\theta_0 \\ \Delta\phi\cos\theta_0 \\ \cos\theta_0 - \Delta\theta\sin\theta_0 \end{bmatrix} + \begin{bmatrix} -w_0q + v_0r \\ -u_0r + w_0p \\ -v_0p + u_0q \end{bmatrix} \tag{5.128}$$

In (5.128), the subscript '0' represents the quantity at the reference condition. The simplified form of the moment eqn. (3.44) is given by eqn. (3.49), assuming small values of angular speeds p,q and r. We have already studied that, for a fixed-wing aircraft, decoupled longitudinal and lateral model equations can be used to model the aircraft dynamics without much loss of accuracy. In the case of helicopters, the degree of coupling between the longitudinal- and lateral-directional motion is generally stronger, and therefore, a 6DOF model is preferred [6]. The dynamics of the main rotor in helicopter, however, introduces additional complexities, and it becomes necessary to go for the higher-order models.

5.8.1 Rotor Plus Body Models

As mentioned above, rotorcraft applications require at least a 6 DOF model to represent the rigid body motion. Such models are generally adequate to describe helicopter dynamics associated with low frequency (such as phugoid). But the prediction of flight dynamic behavior from such models in the high-frequency range is poor. This is due to the absence of the rotor degrees of freedom in the model, which affect the helicopter motion. Some improvements can be obtained by approximating rotor dynamics via equivalent time delay effects [22]. However, this approach cannot adequately represent the rotor influences [23]. The other alternative is to augment the 6 DOF models with additional degrees of freedom that explicitly model the rotor dynamic effects, e.g., longitudinal and lateral flapping and coning. Figure 3.16 (Chapter 3) shows the sub-matrices that define body, rotor, and body-rotor coupling. An extended model of 8 DOF can be obtained from eqn. (5.127) by including longitudinal and lateral flapping in the state and observation equations [24]:

$$x^T = (u, v, w, p, q, r, \phi, \theta, \psi, h, a_{1s}, b_{1s})$$

$$y^T = (a_{xm}, a_{ym}, a_{zm}, p_m, q_m, r_m, \phi_m, \theta_m, u_m, v_m, w_m, \dot{p}_m, \dot{q}_m, \dot{r}_m, \psi_m, h_m, a_{1s}, b_{1s}) \quad (5.129)$$

$$u^T = (\delta_{\text{lon}}, \delta_{\text{lat}}, \delta_{\text{col}}, \delta_{\text{ped}})$$

In comparison with the 6DOF rigid body model, this extended model structure can provide an in-depth insight into the helicopter dynamics. However, the augmented model will have larger number of unknowns to be estimated, which can lead to serious convergence problems and make identification difficult. The rotorcraft mathematical model should be such that it gives a realistic representation of the helicopter behavior, and at the same time, it is simple and mathematically tractable. To this end, it is essential to determine the lowest-order model that would best fit the flight test data. Linear models of different order, which may include body as well as rotor degrees of freedom, can be tried out in identification. Any one of the following model structures, consistent with the frequency range of interest, can be selected for characterizing rotorcraft dynamics [25,26]:

i. Body longitudinal dynamics alone-4th-order model (3 DOF)
ii. Body lateral dynamics alone-4th-order model (3 DOF)
iii. Body-coupled dynamics-8th-order model (6 DOF)
iv. Body dynamics with first-order flapping dynamics (longitudinal and lateral tip path plane tilts-11th-order model (9 DOF))
v. Body dynamics with rotor flapping dynamics-15th-order model (10 DOF)
vi. Body dynamics with rotor flapping and lead-lag dynamics-21st-order model (13 DOF)
vii. Body dynamics with rotor flapping and lead-lag dynamics and with inflow dynamics-24th-order model (16 DOF)

5.8.2 Stability-Derivative Models

The linearized equations for the rigid body and rotor dynamics, in the state-space form, can be expressed as:

$$\begin{aligned} \dot{x}_b &= A_b x_b + A_{rb} x_r + D_b u \\ \dot{x}_r &= A_r x_r + A_{rb} x_b + D_r u \end{aligned} \quad (5.130)$$

In (5.130), x_b represents the basic 6DOF rigid body state vector, x_r represents the higher-order dynamics that may include flapping and lead-lag dynamics, and u represents the control input vector, with the following descriptions:

$$\begin{aligned} x_b &= (u, v, w, p, q, r, \phi, \theta, \psi, h) \\ x_r &= (u, v, w, p, q, r, \phi, \theta, \psi, h, a_{1s}, b_{1s}, \ldots) \\ u &= (\delta_{\text{lon}}, \delta_{\text{lat}}, \delta_{\text{col}}, \delta_{\text{ped}}) \end{aligned} \quad (5.131)$$

The matrices A_b and D_b denote the rigid body dynamics in the absence of rotor dynamics, while the matrices A_r and D_r represent the rotor flapping dynamics. The matrices A_{rb} and A_{br} represent the rotor-body and body-rotor coupling. Taking Laplace transform of eqn. (5.130) and solving for x_r, we get:

$$sx_r = A_{rb}x_b + A_r x_r + D_r u$$

$$(sI - A_r)x_r = A_{rb}x_b + D_r u$$

$$x_r = (sI - A_r)^{-1} A_{rb}x_b + (sI - A_r)^{-1} D_r u \tag{5.132}$$

Expanding $(sI - A_r)^{-1}$ in a series form, we get:

$$(sI - A_r)^{-1} = -A_r^{-1}[I + A_r^{-1}s + A_r^{-2}s^2 + A_r^{-3}s^3 + \cdots] \tag{5.133}$$

Substituting the series expansion for $(sI - A_r)^{-1}$ in eqn. (5.132) and taking inverse Laplace yield the following solution for x_r:

$$x_r = -A_r^{-1}A_{rb}x_b - A_r^{-2}A_{rb}\dot{x}_b - A_r^{-3}A_{rb}\ddot{x}_b - \cdots - A_r^{-1}D_r u - A_r^{-2}D_r\dot{u} - A_r^{-3}D_r\ddot{u} - \cdots \tag{5.134}$$

Equation (5.134) contains the control input and body states along with their higher-order derivatives. At equilibrium, the higher-order derivatives vanish and the rotor state x_r can be expressed as:

$$x_r = -A_r^{-1}A_{rb}x_b - A_r^{-1}D_r u \tag{5.135}$$

Substituting x_r from eqn. (5.135) into eqn. (5.130) yields the quasi-static model for x_b:

$$\dot{x}_b = A_b x_b + A_{br}[-A_r^{-1}A_{rb}x_b - A_r^{-1}D_r u] + D_b u \tag{5.136}$$

Rearranging eqn. (5.136), we get:

$$\dot{x}_b = [A_b - A_{br}A_r^{-1}A_{rb}]x_b + [D_b - A_{br}A_r^{-1}D_r]u \tag{5.137}$$

$$\dot{x}_b = A_{QS}x_b + B_{QS}\,u \tag{5.138}$$

In (5.138) A_{QS} and B_{QS} represent the quasi-static state and control matrices that comprise the rotor and rigid body stability and control derivatives.

5.8.3 Rotor-Response Decomposition Models

These models are formulated by decomposing the response of higher-order states into two components: (i) self-induced response and (ii) response from rigid body motion. For example, the rotor state can be decomposed as:

$$x_r = x_{rMR} + x_{rB} \tag{5.139}$$

Here, $x_{r_{MR}}$ indicates the response due to main rotor dynamics and x_{r_B} is the response due to body dynamics; further, we get:

$$\dot{x}_{r_{MR}} + \dot{x}_{r_B} = A_r(x_{r_{MR}} + x_{r_B}) + A_{rb}x_b + D_r u \tag{5.140}$$

In the decomposed form, we have:

$$\begin{aligned} \dot{x}_{r_{MR}} &= A_r x_{r_{MR}} + D_r u \\ \dot{x}_{r_B} &= A_r x_{r_B} + A_{rb} x_b \end{aligned} \tag{5.141}$$

The approach used for obtaining the expression for x_r in eqn. (5.134) can be applied to find a solution to eqn. (5.141); the contribution to the rotor state from body motion is then given by:

$$x_{r_B} = -A_r^{-1} A_{rb} x_b - A_r^{-2} A_{rb} \dot{x}_b - A_r^{-3} A_{rb} \ddot{x}_b - \cdots \tag{5.142}$$

Assuming that the body dynamics are low-frequency dynamics compared to the rotor dynamics, the higher-order derivatives in eqn. (5.142) can be neglected to obtain the instantaneous rotor response to body motion. Substituting x_{r_B} in eqn. (5.139), by retaining only the first term of eqn. (5.142), we get:

$$x_r = x_{r_{MR}} - A_r^{-1} A_{rb} x_b \tag{5.143}$$

Substituting x_r from eqn. (5.143) into the expression for $\dot{x}_b$ in eqn. (5.130), we obtain:

$$\dot{x}_b = A_b x_b + A_{br}[x_{r_{MR}} - A_r^{-1} A_{rb} x_b] + D_b u \text{ or } \dot{x}_b = [A_b - A_r^{-1} A_{rb}]x_b + A_{br} x_{r_{MR}} + D_b u \tag{5.144}$$

The rotor-response decomposition model can now be expressed as:

$$\begin{aligned} \dot{x}_b &= A_{QS} x_b + A_{br} x_{r_{MR}} + D_b u \\ \dot{x}_{r_{MR}} &= A_r x_{r_{MR}} + D_r u \end{aligned} \tag{5.145}$$

The elements of the matrices in eqn. (5.145) can be identified either from wind tunnel testing or from flight testing.

5.8.4 Evaluation/Validation of Linear Flight Dynamics Models

Evaluation of a flight dynamic model implies checking out the accuracy and adequacy of the identified model against the real system, and to ensure that, to the extent possible, it is a true representation of the system. Evaluation methods can give valuable clues to the possible sources of error that might lead to discrepancies between the model output and flight-measured response. Having done the model evaluation, one can use it for to evaluate the aircraft dynamics, performance, and handling qualities. Such models are also useful in evaluating and/or updating flight control systems or investigating aircraft design modifications. Evaluation of the dynamic models can be done at two levels – (i) adequacy of the model structure and (ii) adequacy and accuracy of the identified derivatives. Since the dynamic models for fixed-wing aircraft and rotorcraft are based on phenomenological considerations, there is limited scope of making changes as far as the model structure is concerned. However, the set of derivatives to be identified from flight data is not always obvious, particularly if the system under consideration is operating in non-linear regions.

The presence of too many and secondary derivatives can lead to correlations among parameters that can adversely affect the identification results. On the other hand, too few parameters can cause the model to produce an inadequate response match. There is no unique method to arrive at a reduced parameter model. One approach to recognize and retain significant parameters in the estimation model is to use stepwise linear regression [27]. Another approach is based on the Cramer Rao Bounds (CRB) of the derivatives. The CRB of the identified parameters are obtained from the information matrix, which is anyway computed by the output error algorithm during the estimation process. Derivatives that are secondary in nature and show exceptionally high standard deviations are dropped (fixed to zero) from estimation each time the model is re-converged. Though time-consuming, this approach helps to provide a reduced parameter set for identification.

Compared to fixed-wing aircrafts, helicopters pose greater modeling problems due to their highly cross-coupled and non-linear dynamic behavior. Non-linearities in helicopters could arise from compressibility effects, tip vortices, blade elasticity, inflow dynamics, lead-lag effects, torsion, etc., and these significantly alter the flow field over the blades. The aerodynamic forces and moments generated at the main rotor strongly influence the helicopter dynamics. At the basic level, classical 6 DOF models are used assuming the blades to be rigid. The 6 DOF model treats the rotor dynamics in quasi-steady form. This may, at times, lead to degradation of the estimated results. Including additional degrees of freedom due to blade flapping and inflow dynamics allows for better representation of rotorcraft dynamics at higher frequency leading to improved estimates. Adequacy and accuracy of the identified models can be assessed by any one of the following methods:

- Comparing the estimated values of the identified derivatives with values obtained from other sources, like analytical values, CFD results, or wind tunnel data.
- Comparing the estimated derivative values from different sets of flight data gathered from similar maneuvers at the same flight condition.
- Comparing the model responses with the data not used in identification of the derivatives. Normally, it is advised to keep first-half of the data for model development and the second-half for model validation.
- Based on the physical plausibility of the identified derivatives. Physical knowledge of the system being modeled is essential to correctly interpret the results and validate the model.

It needs to be emphasized here that no mathematical model is going to be perfect. Therefore, model evaluation, to some extent, is subjective. Comparative plots of the estimated and measured time histories alone are not sufficient to evaluate a model. Thus, a combination of several criteria should be used. Fully evaluated and validated dynamic models are essential for flying quality evaluation, for piloted simulations, and for the design and up-gradation of helicopter flight control system.

5.9 UAV DYNAMICS

The UAVs (unmanned/uninhibited aerial vehicles) could perform the tasks that are more difficult to do by the manned aircraft, e.g., chemical and/or biological warfare missions. UAVs can also be used to test and verify technologies for UAV and for general aviation aircraft. Utilization and operation of UAV entails less risk to humans in combat. UAVs can also be freed from the limitation of human operators. This would improve the performance of UAVs. UAVs utilize various advanced technologies: (i) data/signal processing, (ii) off-board on-board sensors, (iii) communications links, and (iv) integrated avionics and flight control. UAVs could be remotely piloted or autonomous with an autopilot. They can operate at altitudes above 70,000 feet, have higher maneuverability and longer endurance, and can carry optical sensors and radars. They, however, need large communications bandwidths. The UAVs could also be equipped with aircraft like controls: elevators, ailerons, rudders, flaps, extensions of wings, and undercarriage.

In an autonomous UAV, human has no direct control over the vehicle. One form of autonomy is the radio-controlled model airplane. The most complicated UAV would be that it utilizes a rule-based fuzzy logic to detect, identify, and attack a moving target. Flight dynamics modeling of UAVs presents some unique challenges. UAVs are lighter compared to manned aircraft, and hence, UAVs have higher natural frequencies. To simulate the dynamic motion, the non-linear fully coupled equations of motion of an aircraft, discussed in Chapter 3, can be used. Simplified EOM discussed in this chapter can also be utilized to carry out preliminary studies relating to UAV performance evaluation. Any one of the following model forms can be selected for UAV dynamic analysis:

a. non-linear fully coupled equations as described in eqns. (3.22), (3.23), and (3.27)
b. linear coupled, eqn. (5.70)
c. non-linear decoupled, eqns.(5.71) and (5.100)
d. linear decoupled, eqns. (5.83), (5.93), (5.104)

Working with the fully coupled equations of motion in option (a) can be time-consuming and complex. Therefore, for control system design, the designers usually make use of linearized and/or decoupled models given in options (b), (c), and (d). Of course, one has to keep in mind the limitations of these models. The linearized equations of motion are valid in the small range around the trim point when excursions from the reference flight condition are small. Decoupling is also valid when there is negligible interaction between the longitudinal- and lateral-directional motion. In using the said set of EOM for UAVs, the following assumptions are made:

(i) It is a rigid body and flexibility effects are not considered; (ii) it has a conventional configuration (i.e., aft tail); (iii) symmetry about the *XZ* plane is assumed; and (iv) the effect of thrust on lateral-directional motion is neglected. Reference [28] gives more details about the dynamic modeling of UAVs.

5.10 MAV DYNAMICS

Biologically inspired MAVs are small autonomous flying vehicles. Their dimension could be smaller than 15 cms, or sometimes smaller than 300 cms, latter towards mini-air vehicles (MiAVs). The MAVs are used for reconnaissance over lands, in/over buildings, for collecting evidence of industrial faults, and survey over hazardous places or nuclear sites. For design and development of these vehicles, one needs to understand aerodynamics, structural aspects, and propulsion physics at a very small scale. Due to their small size, there would be limited or no space to carry avionics that would otherwise carry out various navigational tasks [29].

The operational challenges of MAVs are related to the low Reynolds number (~10e+4), presence of gust and obstacles at low altitudes, microsize and weight, need of compact power pack, microsensors and actuators, and aero elastic coupling. MAVs (and MiAVs) could be fixed wing, rotary wing, or flapping wing. Flapping (-wing) flight is exhibited by millions and millions of species (of bats, birds, and insects). A flapping wing MAV can be viewed as a single non-linear dynamical system with integrated fluid, structure, and control systems components.

The research in the areas of MAVs can offer new and computationally efficient solutions to problems of design, propulsion/energy efficiency, structural aspects, machine vision, and navigation. Currently, the research is focused in three main areas: (i) to study the aerodynamic characteristics of low aspect ratio wings at low Reynolds number; (ii) wind tunnel experiments for flow visualization, aerodynamic coefficients in the form of look-up tables of stability, dynamic, and control derivatives are also obtained from such experiments; and (iii) numerical simulation based on CFD and wind tunnel experiments.

Today, autonomous aerodynamic control of MAVs is of much interest. The issues of flight mechanics and flight control for MAVs are of particular concern. In general, MAV dynamics can be modeled using the well-established set of rigid body equations of motion for aircraft. Since MAVs are extremely light and susceptible to gusts, a possible way to avoid the degradation in their performance is to build MAVs with higher flexibility [30]. The flexible MAVs will have the ability to absorb the wind energy and reduce the effect of gust, thereby improving the MAV's overall performance. Models for simulation of MAV dynamics would therefore also require appropriate representation of the flexibility effects in the aerodynamic models. Figure 5.22 shows a simplistic layout of the simulation model for MAV.

5.11 LIGHTER-THAN-AIR VEHICLE/BLIMP DYNAMICS

Lighter-than-air vehicles (LAVs) have several advantages and find increasing applications in various tasks such as surveillance and advertising. Modern aerostats use helium for lift and consist of a power supply system. A two-way fiber-optic datalink

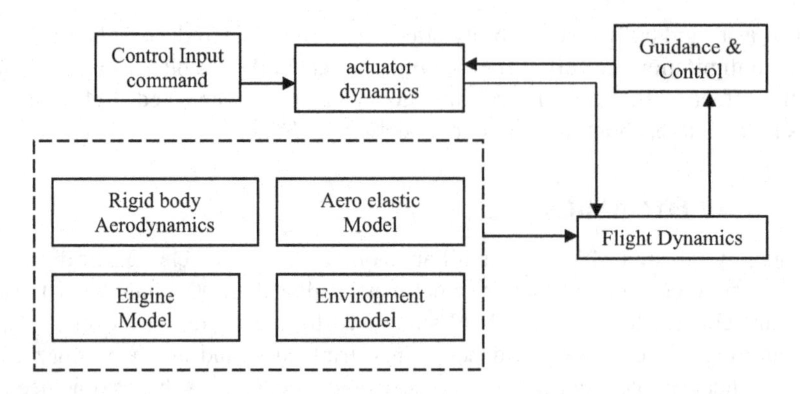

FIGURE 5.22 Schematic for MAV simulation.

is provided for sensor data and control. The low-pressure differential allows them to sustain for a long time. These are also difficult to detect with radar. Due to these features, the demand for the use of aerostats for surveillance is continually growing and, being less expensive, they offer a viable and inexpensive alternative to satellites.

In this section, we briefly discuss a mathematical model of an airship. As for the fixed-wing aircraft, the basic derivation of the equations of motion of an airship is based on rigid body dynamics [31]. One noticeable difference in developing the model equations for an aircraft and an airship is the consideration of buoyancy in the latter case. To account for the apparent mass effect due to the large volume of air mass displaced by an airship, the equations of motion are usually written about the center of buoyancy. In Ref. [32], effects of the change in the position of the C. G. as a result of venting or intake of the ballast air, are also modeled. Reference [33] suggested simplification in the model equations by determining the change in the position of the C. G. and translating all the forces, moments, and inertias to the C. G. to compute the airship dynamics. This results in EOM that are similar to those discussed in Chapter 3 for the fixed-wing aircraft. In state-space form, these can be expressed as:

$$\dot{x}(t) = f[x(t), u(t), \Theta] \tag{5.146}$$

Here, for a 12th-order system, the state vector x consists of:

$$x = [u, v, w, p, q, r, \phi, \theta, \psi, X, Y, H]$$

The vector Θ is a collection of the stability and control derivatives, and u is the control input vector. Computing the center of mass location and translating the moment of inertias to this center allow to decouple the accelerations and write the equations of motion in the conventional form. Since the variables X, Y and ψ do not couple with the rest of the equations, mostly a 9th-order linear model comprising the longitudinal states u, w, q, θ, H and the lateral states comprising v, p, r, ϕ can be used to analyze the modal characteristics of an airship.

Due to the buoyancy and apparent mass of the airship, its climb performance is different from that of the conventional fixed-wing aircraft [33]. For an aircraft, we have:

$$T - D - W\sin\gamma = 0$$
$$L - W\cos\gamma = 0 \tag{5.147}$$

With the buoyancy force B in the case of an airship, the force balance equations in climb are:

$$T - D + (B - W)\sin\gamma = 0$$
$$L - W\cos\gamma = 0 \tag{5.148}$$

The first of the above equation describes the force balance in the direction of flight path, while the second describes the force balance perpendicular to the flight path. The net buoyant force $(B - W)$ for an airship in climb is positive. The velocity of the airship will undergo a change if the first equation is not balanced, and the airship flight path angle will undergo a change if the second equation is imbalanced.

Linearized equations of motion provide the modal characteristics of the airship. The longitudinal modes of an airship generally consist of an oscillatory mode with low damping and a heave mode. Unlike a conventional aircraft, the airship has no mode equivalent to the phugoid mode as the coupling between the forward speed and heave mode is missing in the case of an airship. The lateral-directional modes comprise the DR mode, which is generally an oscillatory stable mode with high damping, and a roll mode, which is an oscillatory stable mode with very low damping. Airship can also show sideslip-subsidence and yaw-subsidence modes. Typical aerodynamic derivatives for an aerostat/blimp configuration are given in Table 5.9 [33].

Reference [34] discusses the airship modal characteristics in greater details.

TABLE 5.9
Some Aerostat Derivatives

Derivative/deg.	Numerical Value
$C_{L\alpha}$	0.02
$C_{m\alpha}$	−0.027
C_{m_q}	−0.00021
$C_{l\beta}$	−0.0000009
C_{l_p}	−0.00003
$C_{n\beta}$	0.00087

APPENDIX 5A EQUILIBRIUM, STABILITY, AND DAMPING

Equilibrium in essence means everything is in balance (like in the olden days' weighing balance), since there are no unbalanced forces. The stability has to do with the response of an object or a system, when it is disturbed by a small unbalanced force applied to it [35]: (i) positive stability is when the object generates a restoring force and comes back to its equilibrium, (ii) neutral stability is when the object is not affected by the disturbance, no force is generated, and (iii) in negative stability, the object generates a force that pushes it far away from its equilibrium. The damping is an effect such that an initial motion itself generates a force that opposes the motion itself. Negative damping tends to increase the motion, meaning somehow from somewhere some energy is pumped into the system. The stability refers to the force that arises depending on the position of the system, and the damping signifies the force arising due to velocity of the object.

To control the pitch angle/attitude, the pilot pushes or pulls the stick, and then uses the trim to trim off the stick; the aircraft is trimmed for a definite AOA. Since the airspeed indicator is closest to the AOA indicator, the trimming for airspeed is as sensible as that for AOA. An aircraft has a considerable AOA stability, it is trimmed for a definite AOA. Pitch stability is accurate than the AOA stability, and the larger contribution for the latter is decalage: the thing in back flies at the lower AOA, than that in the front; and the AOA stability reduces as the C. G. moves aft. The trimmed AOA corresponds to a definite airspeed.

Vertical damping: In normal situations, an aircraft is in equilibrium (and it is trimmed), since all the four forces are balanced. However, in case if the weight increases, then the unbalanced downward force makes the aircraft to accelerate downward, and this in turn leads to the downward velocity, thereby increasing the angle of attack (the air packets strike at a larger AOA); this increases the lift coefficient and hence the lift, and the aircraft restores the temporarily lost balance; this is the vertical damping. Ironically, this damping weakens at near the stall, since the increase in AOA does not increase the lift.

Roll damping: In an another situation, if a large number of some heavy weight passengers moved to, say on the left side suddenly, then the left wing will do down, and the right wing up, this will cause a large AOA of the left wing, and the reduced AOA of the right wing, and the lift of the left wing will increase and the roll balance will be restored; this is roll damping. It also suffers at or near the stall, because lift is not now proportional to AOA at stall. The roll damping can be increased by decreasing the incidence angle along the span from the root to the tip of the wing; hence, the tips will be un-stalled, and hence, there will be positive amount of roll damping. The (extending of) flaps have the following effects: (i) lowering of the stalling speed, (ii) increases the wing-section's AOA, effectively increases the washout, (iii) increases the drag that is useful during landing (works a 'brake'), and (iv) perturbs the trim speed.

Dutch Roll: The aircraft has a small amount of stability around the roll axis. The DR is a SP oscillation, and the spiral divergence is a long-period oscillation. The DR is a tricky combination of yawing and rolling motion, and side slipping; this combined motion is less damped than that of an individual motion; if it is only the former two, then there is no problem since there is a good amount of positive damping. However, the side-roll coupling (SRC) contributes a negative amount of damping to DR, though the SRC increases roll-axis stability. It is better to avoid any un-coordinated usage of ailerons and rudder.

APPENDIX 5B STALLS AND SPINS

Stall: If an aircraft is not stalled, then it will not spin [35]. The stall occurs at a critical AOA after which an increase in the AOA does not increase the lift coefficient (C_L). If one goes far beyond the critical AOA, the lift coefficient is greatly reduced, and the drag coefficient is increased. This is caused by flow separation on the wing surface. The slowest speed at which the aircraft can fly straight and level is called the stalling speed, $V_{\text{stall}} = \sqrt{\dfrac{2W}{\rho S C_{L_{\max}}}}$. The aim always is to make V_{stall} as small as possible. This is achieved by increasing $C_{L_{\max}}$ with the use of high lift devices, e.g., trailing edge flaps.

The aircraft will descent rapidly; of course, still the weight of the aircraft is supported by the wing; if the sections near the root(s) of the wing have their lift coefficient reaching maximum earlier, and the lift coefficient near the tips still produces the lift, then as a whole, the wing will not be stalled.

Spin: If an aircraft is in a steady side slip to the left, then the slip-roll coupling will cause it to spin to the right. If the aircraft is not in much of a slip, but if suddenly it is yawed to the right, then the left wingtip will be temporarily moving faster, and the right wingtip would be moving slower, then this difference in airspeeds will cause difference in the lift, and this in turn will cause slip to the right. The spin family could be: (i) onset of undamped rolling, the departure, (ii) one that has just started, the incipient spin, and (iii) well-developed spin – a steep spin or a flat spin. The recovery from spin can be done as follows: (i) retard the throttle to idle, (ii) retract the flaps, (iii) neutralize the ailerons, (iv) apply full rudder in the direction opposing the spin, and (v) briskly move the stick to select zero AOA.

EPILOGUE

Various models for representing dynamic systems have been considered here. The treatment has been kept as simple as possible. State-space and TF models are very appropriate for parameter estimation and system identification and also useful for evaluating the handling qualities of aircraft.

Since the direct use of the EOMs in their full form is usually avoided, except in the full dynamics flight simulation, the simplified models are useful in various applications: linear models for control law design, quick analysis to understand the dynamic characteristics of the vehicle, mathematical modeling from flight data, and related handling qualities analysis. This evokes the need of the system's approach in understating and use of the flight mechanics aspects in various aerospace engineering applications. In [36], the authors give the so-called literal approximate factors to obtain approximate relations for TF poles and zeros literally in terms of the stability and control derivatives of the atmospheric vehicle.

EXERCISES

5.1 Obtain a differential equation from the T.F. of example 5.1.

5.2 Give an expression for the damping ratio of the T.F. of example 5.2.

5.3 Obtain the roots of the numerator and denominator of the discrete-time T.F. of example 5.4, and comment on the stability of the system.

5.4 Given the state-space form A, B, C, D, obtain the formula for getting the T.F. of the system.

5.5 Convert the continuous-time model of the correlated process in the discrete-recursive form (see Section 5.3).

5.6 The sensitivity of a TF $G(s)$ with respect to parameter (variation) is defined as: $S_{G_a} = \frac{a}{G}\frac{dG}{da}$. Obtain a similar expression for a closed-loop TF. Interpret the result. Take $H(s)$ as the feedback TF.

5.7 $G_1(s)$ and $G_2(s)$ are in (a) cascade, (b) in parallel. Give the equivalent TFs.

5.8 Given $\dot{x} = \begin{bmatrix} 0 & 1 \\ 0 & -2 \end{bmatrix}\begin{bmatrix} x_1 \\ x_2 \end{bmatrix} + \begin{bmatrix} o \\ K \end{bmatrix} v$, obtain TF between v and x_1 by simplifying the state-space form.

5.9 Obtain the TF formula for the discrete state-space form: $x(k+1) = \phi x(k) + Bu(k)$; $y(k) = Cx(k)$. Hint: use $x(k+1) = z\,x(k)$.

5.10 Why the equation $\dot{x} = Ax$ is called the homogeneous equation?

5.11 Obtain the recursive form of eqn. (5.11).

5.12 The equation $\dot{v} = -\beta v + w$ can be used to generate the correlated noise process from the white noise w. Draw the block diagram from this equation.

5.13 Obtain the zeros and poles of the TF of example 5.8. Obtain the eigenvalues of the matrix A of the controllable canonical form of the TF. Verify the claims made in solution 5.8 of example 5.8.

5.14 Establish $\phi^{-1}(t,\tau) = \phi(\tau,t)$. Recall the properties of the transition matrix.

5.15 Obtain a compact formula for unit step response of a first-order system: $y(k) = \beta u(k) + \alpha y(k-1)$; Let $u(k) = 0$ for k–1, –2, –3 and $u(k) = 1$ for $k = 0,1,2$; and $y(-1) = 0$.

5.16 We obtain from the solution of exercise 5.16 the (easy to verify) following equation, $y(k) - \alpha y(k) = (1-\alpha^{k+1})\beta$; $y(k) = \frac{\beta(1-\alpha^{k+1})}{1-\alpha}$. Comment on the stability of the solution $y(k)$ wrt the values of α.

5.17 Establish that the state-space representation is not unique. Hint: use $x = Pz$. What assumption is required?

5.18 Is the following system controllable and stable?

$$\dot{x}_1 = -2x_1 + 4x_2$$

$$\dot{x}_2 = 2x_1 - x_2 + u$$

5.19 In the T.F. of example 5.1, put τ =0.1, and obtain discrete-form TF. (Hint: use bilinear transformation: $z^{-1} = e^{-Ts} \approx \frac{2-Ts}{2+Ts}$, where T is the sampling interval). Comment on the stability of the DFTF.

5.20 What does the recursion $y(k) = 0.5(y(k-1) + x(k)/(y(k-1))$ achieve? Hint: use $y(0) = 1$ and some numerical value for $x(k)$. Perform a few hand calculations.

5.21 Obtain discrete-time random numbers' time series by passing a random numbers' (continuous!) time series (this is also a sequence of numbers because these are generated using digital computer program/MATLAB) through sample and hold block (see Section 5.3 and Figure 5.12). Vary the sampling interval to study its effect on the output sequences.

5.22 The aerodynamic derivatives for a transport aircraft are given as:

$$L_0 = 0.0095,\ L_p = -2.0374,\ L_r = 0.876,\ L_{\delta_a} = -5.933,\ L_{\delta_r} = 1.034,$$

$$N_0 = 0.03,\ N_p = -0.177,\ N_r = -0.57,\ N_{\delta_a} = -0.292,\ N_{\delta_r} = -1.749$$

Write an appropriate mathematical model and obtain various TFs and frequency responses of the lateral-directional model using MATLAB functions. Determine the frequency and damping ratio of the mode.

5.23 What is the significance of the natural frequency, say of the short-period mode, from pitching moment point of view?

5.24 Recall eqn. (3.23), say for rolling moment and yawing moment. Assume the neutral static condition and neglect cross-coupling inertia terms, and obtain the relation between the lateral-directional static stability and the dihedral (static stability) derivatives in terms of the control effectiveness derivatives (see Chapter 4).

5.25 The aerodynamic derivatives of a transport aircraft in phugoid mode are given as:

$$X_u = -0.011,\ Z_u = -0.1583,\ u_0 = 111\ \text{m/s}$$

Use the phugoid model of eqn. (5.93), and obtain the frequency responses and other characteristics.

5.26 For an airplane with the following given characteristics, compute

i. the lateral directional dimensional derivatives.

$$C_{y\beta} = -1.125;\ C_{y_r} = 0.8;\ C_{n\beta} = 0.27;\ C_{n_r} = -0.5;$$

$$C_{y\delta_r} = 0.28;\ C_{n\delta_r} = -0.16;$$

$$C_{y_p} = 0.17;\ C_{l\delta_r} = 0.0697\ C_{l\beta} = -0.133;$$

$$C_{l_p} = -0.96;\ C_{l_r} = 0.42;$$

$$C_{n_p} = -0.1;\ C_{l\delta_r} = -0.2496$$

Other data are:

$$I_x = 105{,}500\ \text{kg}\,\text{m}^2;\ I_y = 250{,}000\ \text{kg}\,\text{m}^2;\ I_z = 340{,}000\ \text{kg}\,\text{m}^2;$$

$$I_{xz} = 11{,}500\ \text{kg}\,\text{m}^2$$

$$V = 72.114;\ \rho = 0.72851\ \text{kg/m}^3;\ b = 21.5\ \text{m};\ S = 65\ \text{m}^2;$$

$$\text{mass} = 5000\ \text{kg};$$

$$u_0 = 140\ \text{Knots};\ H = 5100\ \text{m}$$

ii. Using 4 DOF lateral-directional model, find the eigenvalues for spiral, roll, and Dutch roll mode, and iii) determine the roots using Dutch roll approximation. Compare the results with those obtained in ii) above and comment.

5.27 How would you get the flight path rate from the acceleration and subsequently the body rate?
5.28 Will large or small changes occur in the airspeed in the short-period transient mode?
5.29 What is the main distinction between the short-period and phugoid-mode characteristic from the motion of the aircraft and the interplay of the forces and moments?
5.30 If you want to add another state variable in the phugoid model, which one would you add?
5.31 If the moment-speed derivative in eqn. (5.99) becomes sufficiently negative, then what happens to the Phugoid-mode characteristic?
5.32 In short-period mode/model, which aerodynamic derivative would have the largest influence on the natural frequency?
5.33 In SP mode, which aerodynamic derivatives would have influence on the damping ratio?
5.34 What is inertial and aerodynamic 'damping' in the short-period mode?
5.35 In Phugoid mode, which is the most influential aerodynamic derivative and in what way?
5.36 Which aerodynamic derivatives have most significant influence on the DR frequency?
5.37 Which are most effective aerodynamic derivatives in spiral and roll subsidence modes?
5.38 Which is most effective DR damping aerodynamic derivative?
5.39 Why the name spiral divergence?
5.40 Derive the condition for the spiral stability.
5.41 Using small disturbance theory and assuming wings-level symmetric flight condition, show that the expression $\dot{W} = -PV + QU + g\cos\Theta\cos\Phi + A_z$ reduces to $\dot{w} = qu_0 - g\theta\sin\theta_0 + a_z$.

REFERENCES

1. Eykoff, P. *System Identification: Parameter and State Estimation*. John Wiley, London, 1972.
2. Sinha, N. K. *Control Systems*. Holt, Rinehart and Winston, Inc., New York, 1988.
3. Maybeck, P. S. *Stochastic Models, Estimation, and Control*, Vol. 2. Academic Press, Inc. Ltd. London, 1982.
4. Raol, J. R., Girija, G., and Singh, J. *Modeling and Parameter of Dynamic Systems*. IEE Control Series, Vol. 65, IEE, London, UK, 2004.
5. McRuer, D. T., Ashkenas, I., and Graham, D. *Aircraft Dynamics and Automatic Control*. Princeton University Press, Princeton, NJ, 1973.
6. Ljung, L. Issues in system identification. *IEEE Control Systems Magazine*, 12(1), 25–29, 1991.
7. Kosko, B. *Neural Networks and Fuzzy Systems–A Dynamical Systems Approach to Machine Intelligence*. Prentice Hall, Englewood Cliffs, NJ, 1992.
8. Box, G. E. P., and Jenkins, G. M. *Time Series: Analysis, Forecasting and Controls*. Holden Day, San Francisco, CA, 1970.
9. Irwin, G. W., Warwick, K., and Hunt, K. J. (Eds.). *Neural Network Applications in Control*, IEE Control Engineering Series 53, The IEE, London, U.K., 1995.
10. Raol, J. R., and Sinha, N. K. *Advances in Modeling, System Identification, and Parameter Estimation*. Special issue of SADHANA, Academy Proceedings in Engineering Sciences, (Indian Academy of Sciences), Bangalore, 25, April 2000.
11. Morelli, E. A., and Klein, V. Application of system identification to aircraft at NASA Langley Research Center. *Journal of Aircraft*, 42(1), 12–25, 2005.
12. Middleton, R.H., and Goodwin G.C. *Digital Estimation and Control: A Unified Approach*. Prentice Hall, New Jersey, 1990.
13. Gelb, A. (Ed.) *Applied Optimal Estimation*. The M.I.T. Press, Massachusetts Institute of Technology, Cambridge, MA, 1974.

14. Singh, J. Application of time varying filter to aircraft data in turbulence. *Journal of Institution of Engineers (Ind.), Aerospace, AS/I*, 80, 7–17, 1990.
15. Raol, J. R. Aerodynamic modeling and sensor failure detection using feed forward neural networks. *Journal of Aeronautical Society of India*, 47(4), 193–197, 1995.
16. Maine, R. E. and Iliff, K. W. Application of parameter estimation to aircraft stability and control – the output error approach, NASA RP-1168, 1986.
17. Nelson, R. C. *Flight Stability and Automatic Control*, 2nd Edn. McGraw-Hill International Editions, New York, 1998.
18. Cook, M. V. *Flight Dynamics Principles*. John Wiley & Sons, Inc., New York, 1997.
19. McRuer, D. T., Ashkenas, I., and Graham, D. *Aircraft Dynamics and Automatic Control*. Princeton University Press, Princeton, NJ, 1973.
20. Garnell, P. *Guided Weapon Control Systems*, 2nd Edn. Pergamon Press, Oxford, UK, 1980.
21. De Leeuw, J. H., and Hui, K. The application of linear maximum-likelihood estimation of aerodynamic derivatives for the Bell-205 and Bell-206, NAE-AN-48, NRC No. 28442, Ottawa, Canada, 1987.
22. Hamel, P. G. (Ed.) *Rotorcraft System Identification*, AGARD AR-280, 1991.
23. Kaletka, J., and von Grünhagen, W. Identification of mathematical models for the design of a model following control system. *Vertica*, 13(3), 213–228, 1989.
24. Anon. Rotorcraft system identification, AGARD LS-178, 1991.
25. Molusis, J. A. Rotorcraft derivative identification from analytical models and flight test data, AGARD-CP-172, 1974, pp. 24-1 to 24-31.
26. Molusis, J. A. Helicopter stability derivative extraction and data processing using Kalman filtering techniques. *28th AHS National Forum*, Washington, DC, 1972.
27. Klein, V. Estimation of aircraft aerodynamic parameters from flight data. *Progress in Aerospace Science*, 26(1), 1–77, 1990.
28. Sadraey, M., and Colgren, R. UAV flight simulation: Credibility of linear decoupled vs. nonlinear coupled equations of motion. *AIAA Modeling and Simulation Technologies Conference and Exhibit*, 15–18 August 2005, San Francisco, California, AIAA 2005-6425.
29. Anon. Indo-US Workshop on Micro Air Vehicles, National Aerospace Laboratories, August 1–2, 2005.
30. Stewart, K., Abate, G., and Evers, J. Flight mechanics and control issues for micro air vehicles. *AIAA Atmospheric Flight Mechanics Conference and Exhibit*, 21–24 August 2006, Keystone, Colorado, AIAA 2006-6638.
31. Khoury, G. A., and Gillet, J. D (Eds.) *Airship Technology*, Cambridge Aerospace Series: 10, ISBN 0 521 430 747. Cambridge University Press, Cambridge, UK, 1999.
32. An, J., et al. Aircraft dynamic response to variable wing sweep geometry. *Journal of Aircraft*, 25(1), 216–221, 1988.
33. Pashilkar, A. A. Modeling, analysis and autopilot design for an airship. *Journal of Aerospace Sciences and Technologies, Aeronautical Society of India*, 55(3), 186–192, 2003.
34. Cook, M. V., Lipscombe, J. M., and Goineau, F. Analysis of the stability modes of an airship. *The Aeronautical Journal*, 104, 279–290, 2000.
35. Anon. How Airplanes Fly. http://www.allstar.fiu.edu/aero/, accessed 2007.
36. McRuer, D. T., Myers, T. T., and Thompson, P. M., Literal singular-value-based flight control system design techniques. *Journal of Guidance*, 12(6), 913–919, 1989.

6 Flight Simulation

6.1 INTRODUCTION

Simulation of flight dynamics is a very important and crucial component of any large aircraft development program. Simulation is meant to replicate, mainly on the ground, the real-life environment/behavior of an aircraft that is present in flights [1–5]. When similar replication is done in the air, it is called an in-flight simulation (IFS). The wind tunnel itself is a type of simulation. Simulation can be used when the system as such does not exist, e.g., the development of a new aircraft. It is also used when experiments with real systems are very expensive and not very safe, especially for training new pilots/flight test engineers/crew. A simulator is thus a very useful system/device for the study of the dynamic behavior of aircraft (including rotorcraft, missiles, UAVs, and micro aerial vehicles); to try out difficult and sometimes risky experiments, conduct failure mode trials to test the flight control laws, and get the feel of various systems and robustness and reliability of other critical subsystems. It really calls for a system approach from theoretical and analytical modeling to practical evaluation/validation of flight mechanics/flight control/handling quality behavior of the simulated vehicle. One guesstimate is that in the world there would be more than 600 simulators that are used for training and maintaining the skills of military and civilian aircrew. Research flight simulators are used for evaluating the design options without primarily building the prototypes of aircraft, thereby saving considerable time, effort, and cost. Various design implications can be studied and evaluated in research simulators. There are several benefits of using flight/research simulators: (i) increased efficiency since there is no limitation of airspace or adverse weather conditions, (ii) enhanced safety aspect, (iii) training cost is drastically reduced, and (iv) several situations which cannot be practiced in real flights can be rehearsed. The design, development, and use of a flight simulator involve considerable modeling and analytic exercises. One goes from reality to a conceptual model via analysis, from this to a computer model via programming, which then simulates reality. Figure 6.1 depicts this aspect very clearly. Using procedures of verification, computer SW/program models are confirmed as an adequate representation of the conceptual model. The computer models are then validated for possessing satisfactory accuracy in comparison with reality by extensive simulation for the intended flight-like maneuvers and exercises. Thus, the process of conceptualization, development, performance evaluation, and utilization of a flight simulation is an iterative one; and it is an intensive task. Simulation can be at various levels: software, hardware, or both [1]. Mainly for the simulation of aircraft dynamics, one needs to mathematically model the dynamics of every subsystem that is deemed important and contributes to generating the time responses of the aircraft [2,3]. In order that the simulation is accurate, high-fidelity math models are needed. Simulation is thus pretending the reality and, for visual impact (like environment, runway conditions),

DOI: 10.1201/9781003293514-7

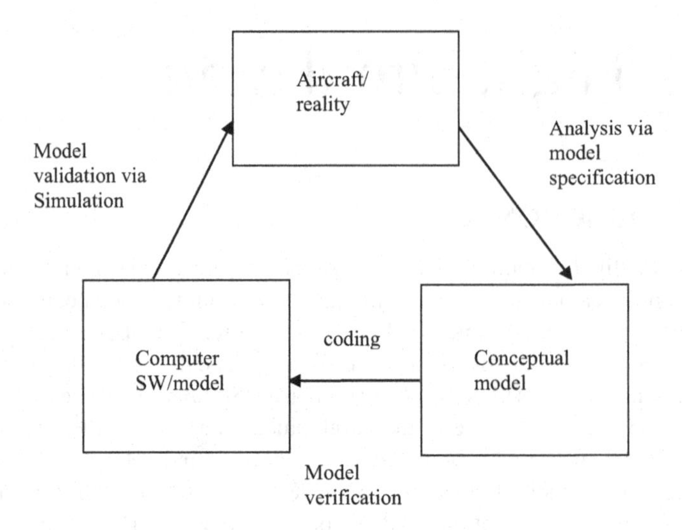

FIGURE 6.1 Modeling and analysis are triad for flight simulation.

one uses the concepts and techniques of visual software and virtual reality. Often many instruments on the cockpit instrument panel are 'soft instruments'. There are multifunction displays, and the actual instruments required in the simulator are just a minimum set. This reduces the maintenance efforts, volume, and cost. So, simulation is meant to provide almost a real-life environment, control, and feel of the actual aircraft/system that is being simulated. If one is in the ground-based simulator (say without motion) but with strong visual cues, one feels that she is in the actual aircraft as far as the dynamic behavior and visuals are concerned. If the simulator is used for aiding the design, development, and validation of flight control laws, then a test pilot, engaged for flight testing of the aircraft being designed, or a control engineer should 'fly' the simulator. This simulator is then called a pilot/engineer-in-the-loop simulator [4]. It consists of a cockpit, controls, throttle, switches, data acquisition system, and visualization/executive SW. This simulator can be run in a batch or real-time mode. A batch simulation works in a pre-programmed and multiuser mode, and computer-generated inputs can be used; as such, it does not need any hardware except a PC. The pre-programmed command could be given as input to the computer in which the flight dynamics and other math model computations are carried out. The computer monitor itself would serve as the display system for the batch simulation results.

In a hardware-in-loop simulator (HILS), the actual flight computer with control law software is also incorporated for testing. HILS is used for testing safety-critical flight control software/computers in an easily controlled laboratory environment at a relatively low cost. In an iron bird simulator (wherein 'iron angles/beams' are used), subsystems like actuators, hydraulic systems, computers, and cabling are incorporated and interconnected to represent the aircraft more completely for ground testing. A full-fledged simulator, with or without motion cues, is called a real-time simulator (RTS) that test pilots would fly and try out all flight test maneuvers on it, before flying

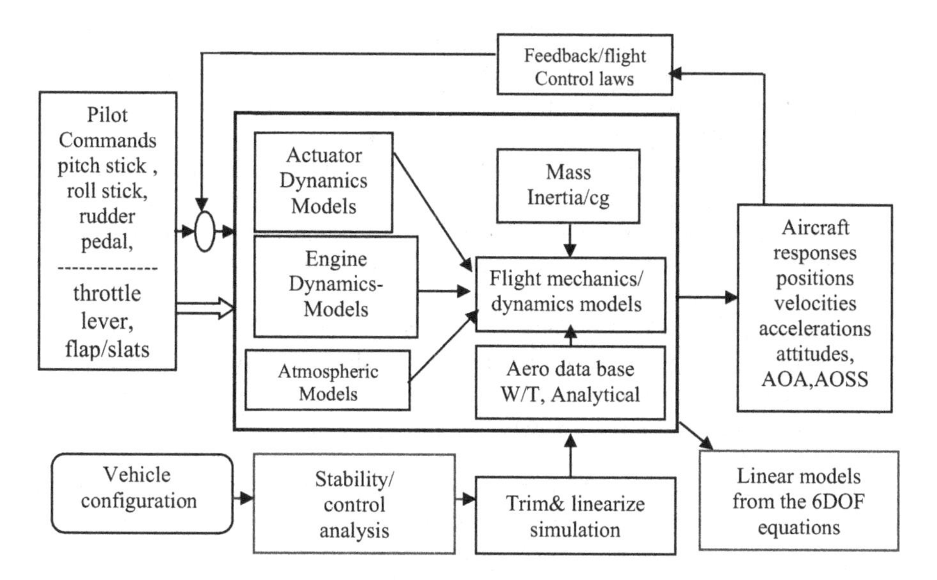

FIGURE 6.2 Components of a typical Soft-Flight Simulation scheme.

the actual aircraft (for which RTS is meant). It includes a cockpit, control stick, pedals, and throttle. Normally, it would be a single-user mode. This RTS can be used for certification of flight control laws, at least for some important flight conditions. Figures I.2 and 6.2 show a schematic of a 'soft' flight simulator with an associated control law development/validation process integrated with many subsystems that play an important role in the development of a simulator facility [4].

Both fixed and motion-based flight simulators play a very vital role in the evaluation of flight control laws and handling qualities of the aircraft right at the development stage and are used to gain enhanced confidence in the performance of the aircraft. The idea in the variable stability research aircraft/in-flight simulation (VSRA/IFS) [5] is that the static and dynamic stability characteristics of this simulator vehicle can be varied over a wide range by using feedback control laws and that this host simulator (aircraft) would be used to fly like the guest aircraft which is being simulated, hence the name in-flight simulator. A variable feel system should be included to simulate various control system characteristics. The other uses of this facility are pilot training for advanced vehicles and the solution of specific stability and control design problems. These simulators provide motion and visual cues that are more realistic than some ground-based simulators. In addition to this, the airborne simulators provide stresses and motivation for the pilot which can be attained in the actual flight environment. Also, the IFS can be used to evaluate ground-based simulators. The development of VSRA/IFS could proceed in three phases:

i. Literature survey on the process of the development and related methodologies, study of existing aircraft (fighters, trainers, and transport), systematic evaluation of these with a view to adopting one of these for the intended purpose, and other hardware/software (HW/SW) requirements and related cost and manpower skill requirement and whether all that is available and how to tap it. Maintenance and related aspects should be addressed at the outset itself. Finalization of the strategy of development and planning.

ii. Systematic development of the facility in a modular fashion, upgradation of the selected aircraft with HW/SW (avionics/instruments/flight control, etc.) and various integration studies, exercise, etc. Preliminary evaluation of the performance of the host aircraft (IFS) and improvements there from. Detailed analysis of the ongoing development process and feedback from the users of the facility.

iii. Refinement based on the trial runs, HW/SW configuration improvement if needed. Training of the pilots and research evaluations, extensive flight data analysis and control law evaluation and refinement if needed, performance evaluation thru' several mission-related tasks, etc.

The foregoing is also applicable to the development of other simulators, except that an actual aircraft is not required, instead its mock-up would be needed. Since for the IFS a different aircraft (host) is used to simulate the dynamics and control laws of the aircraft that is being developed and simulated, the flight dynamics/control bandwidth of the host aircraft has to be greater than that of the simulated aircraft. The IFS is an aircraft whose stability, feel, and flying characteristics can be changed to match those of the 'to-be-simulated' aircraft. The IFS is very useful in realistically and safely evaluating a new or modified aircraft before its first flight and before even committing the aircraft to production. A schematic of the IFS' response feedback system is shown in Figure 6.3. What would physically fly is the host aircraft, not the model aircraft that is simulated; however, the host aircraft is made to/will behave like the guest aircraft. The augmented dynamics of the simulated aircraft are matched by the host aircraft by feeding back appropriate motion variables to the control surfaces. One can easily imagine that the process from ground-based to final IFS, and for many other types of simulators, involves increasing complexity, sophistication and cost, and of course enhanced confidence in the performance/testing of the aircraft.

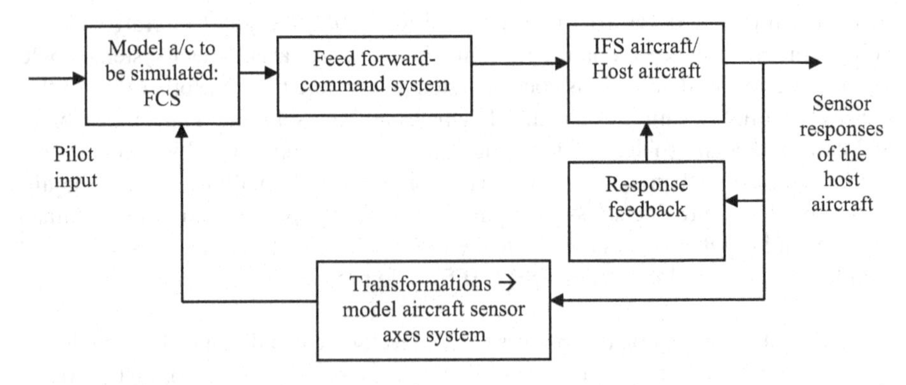

FIGURE 6.3 Augmented simulation: IFS command and control schematic for an aircraft to be simulated.

The main purpose of any simulator is to obtain the dynamic responses of the aircraft for specific inputs and in the presence of a simulated environment, e.g., turbulence and gust. Often the list of these specific inputs is very large. Many such practicable pilot command inputs (signals) are discussed in Chapter 7. All such maneuvers can be tried out in a RTS wherein the pilot sits in the cockpit and uses the command sticks/wheel and rudder pedals to excite the modes of the aircraft. Thus, flight simulation/simulator provides a reasonably accurate, sometimes high fidelity representation of the sensation of sound, touch, visual scenario, often called simply the visuals, and acceleration feels (for motion-based simulators) as encountered in a flight. Thus, flight simulation is very essential in aircraft design and development programs. It could be a totally and fully virtual reality simulator with of course powerful visuals. Training simulators are also used for pilot training before they can fly transport or fighter aircraft. For phases like landing, takeoff, and tracking tasks, a longer period of the simulation run (e.g., 10–15 minutes) is required and needs a designer or/and engineer/pilot in the loop.

Thus, the flight simulation process centers around (i) the study of flight dynamics (including the coordinate systems/transformations) and formulating these in a mathematical language that translates into mathematical models, in fact into equations of motion (EOMs) (Chapters 3 and 4) and (ii) the coding of these math models and implementing the codes/algorithms via suitable programming language (preferably C) in a computer. The overall functioning of the flight simulation (process) involves: (i) developing performance requirements, (ii) integrating subsystems/subcomponents, (iii) validating the design of a vehicle and the flight control laws, (iv) supporting the flight testing of the aircraft and reducing the cost of actual flight tests, (v) investigating inaccessible environments and practice, otherwise dangerous procedures and exercises, (vi) pilot/crew training, and (vii) supporting R&D in all the foregoing aspects via the flight simulation process and exercises.

6.2 AIRCRAFT SUBSYSTEM DATA AND MODELS

Aircraft consists of several systems that require mathematical models if a full dynamic simulation is needed: aircraft flight mechanics, actuation system, sensors, engine, navigation, landing gears, and control system. In addition, the environment and display systems are important. The mathematical models of these subsystems are supposed to be known and the fidelity of the simulation depends on how close are these mathematical models to the actual subsystems (dynamics) [1]. For a flight dynamic simulation, the model structure, EOM, and aerodynamic derivatives (often the knowledge of aerodynamic coefficients is enough) are assumed known. The HW/SW configuration should be strictly maintained for faithful reproduction of flight trials/conditions and repeatability of various simulated tasks being performed. Since several HW/SW modules need to be interconnected and integrated to realize the flight simulator's functioning, it is important not only to test individual subsystems and modules for their accurate functioning but also to see that the tasks performed by these modules are properly synchronized in time and the sequence of operation. In addition to these subsystems, the logic circuits, radio aids, and navigation-related components/SW are needed to realize the high fidelity and full-fledged-ness of flight simulators.

6.2.1 Aero Database

In general, the aero database obtained from the wind tunnels is supplied in tabular formats along with the so-called application formulae [1,6] that are used for the computation of force and moment coefficients; the latter forms part of the EOMs. The EOMs are then numerically integrated to obtain the dynamic responses of the aircraft which are called time histories. An application formula for the axial force coefficient of a typical aircraft is given as:

$$\begin{aligned} C_x &= C_{xa0ds}(mach,\delta_{slat}) + C_{xads}(mach,\alpha,\delta_{slat}) + C_{xuc}(\alpha,mach,UC) + C_{xAB}(mach,\alpha,\delta_{AB}) \\ &+ C_{xdlds}(mach,\alpha,\delta_{lob},\delta_{slat}) \times KC_{xdlob}(mach,\alpha,\delta_{lob}) + C_{xdlds}(mach,\alpha,\delta_{lib},\delta_{slat}) \\ &\times KC_{xdlib}(mach,\alpha,\delta_{lib}) + C_{xdlds}(mach,\alpha,\delta_{rib},\delta_{slat}) \times KC_{xdrib}(mach,\alpha,\delta_{rib}) \\ &+ C_{xdlds}(mach,\alpha,\delta_{rob},\delta_{slat}) \times KC_{xdrob}(mach,\alpha,\delta_{rob}) \end{aligned} \tag{6.1}$$

In the parentheses, the terms are independent variables that signify the dependencies of the coefficients on these terms. This application formula is translated into the following expression:

Coefficient for axial force $= C_x$ for $\alpha = 0 + C_x$ due to α with slat deflection + effect of undercarriage (up or down) + effect of airbrake deflection + C_x due to left outboard elevon (combined elevator/aileron) deflection * a factor on C_x due to left outboard elevon deflection + C_x due to left inboard elevon deflection * a factor on C_x due to left inboard elevon deflection + C_x due to right inboard elevon deflection * a factor on C_x due to right inboard elevon deflection + C_x due to right outboard elevon deflection * a factor on C_x due to right outboard elevon deflection. Similar application formulae are available for the remaining other five coefficients. The intermediate values of a given coefficient, for the specific independent parameters, can be obtained by interpolation (Exercise 6.1).

6.2.2 Mass, Inertia, and Center of Gravity Characteristics

Besides other requirements, adequate knowledge about the mass and inertia characteristics is also important for accurate flight simulation and identification of the aircraft stability and control derivatives of the vehicle. The nondimensional moment derivatives are directly influenced by the inertia calculations, while the force derivatives will be straightway affected by the errors in aircraft mass calculations [7]. The moment of inertia information is more difficult to obtain as it not only depends upon the placement of the test equipment but also on CG variation during flight. The way the fuel is consumed during the flight, e.g., the forward tank, followed by wing tanks, and so on, can affect the moments of inertia. Determination of inertia from special oscillation test rigs is not practical [8]. Manufacturer's data is mostly for the moment of inertia calculations. For a four-ton class of a trainer aircraft, typical values of the moment of inertia I_x, I_y, I_z, and I_{zx} will be like 4500, 18,000, 20,000, and 1500 kgm^2. The cross products of inertias I_{xy} and I_{yz} are small and hence are usually neglected.

Information on fuel consumption is useful to compute a CG travel and the actual mass of the aircraft at any time during the flight. To correct the flight data for the offset in sensor locations from C G, we need to compute the actual distance of C G from the reference vertical datum rather than expressing it in terms of the percentage of the mean aerodynamic chord (MAC). The actual distance of CG in meters from the vertical datum can be computed as follows:

$$x_{\text{C.G}} = x_{\text{L.E.}} + \frac{\text{C.G (in \% mac)}}{100}\bar{c} \tag{6.2}$$

where $\bar{c}$ is the mean chord length and $x_{\text{L.E.}}$ is the extreme L.E. location of the MAC.

6.2.3 Instrumentation System

In modern simulators, most of the traditional instrumentations/instrument panels would be replaced by multifunctional displays and soft instruments or the so-called virtual instruments. The point is that if an angle of attack (AOA) is needed to be displayed, it does not matter what is the mechanism that drives the AOA 'pointer', which could be a 'soft pointer' and not a physical hard pointer of the classical instrument. Thus, the computed values of flight variables are displayed by the soft instruments. This shifts a lot of burden from the hardware mechanisms of the driving motor, etc., to a software task. This reduces/eliminates the wear and tear of the instruments/driving motors as well as the maintenance and heating of the hardware components. However, for safety, there should be some redundancy provided via analytical means or some most important instruments should be still provided in the traditional hardware.

6.2.4 Inertial Navigation System - INS

The INS consists of a stable platform with gyros and accelerometers. The gyros maintain the platform's accelerometers in a level position in relation to the Earth's surface so that accurate measurements of accelerations are provided [1]. These acceleration data are integrated to produce aircraft velocities. The computer platform heading resolves the velocities through the heading angle to produce North/South and East/West components of the velocity which in turn are integrated to obtain the distance. This distance with the datum latitude and longitude gives an accurate present position. The INS platform and computer form the INU (inertial navigation unit). Software simulation of INU is often used. Software simulation of the INU has several merits: (i) ease of availability of the required data, (ii) ease of update of the SW, and (iii) simulation computer core and time requirements are reasonable.

6.2.5 Flight Management System

The flight management system (FMS) accepts information on INS and data from radio navigation, engine, and fuel sensors and contains preflight navigation and aircraft performance data. The crew uses these data to perform (i) navigation,

(ii) guidance, (iii) electronic flight instrument system support, and (iv) data displays. The outputs are command data to flight control. The usual practice is to replicate the (aircraft's) HW FMS for the simulation also and not to go for SW simulation of the FMS in flight simulators [1].

6.2.6 Actuator Models

A typical actuator comprises electric, hydraulic, and mechanical elements. A comprehensive mathematical model of an actuator is required for using in the design of flight control laws and nonlinear flight simulation. For control law design, a small amplitude linear model is sufficient. The large amplitude behavior is also very important for nonlinear simulation. This behavior is characterized by first-order system dynamics with limits of position, rate, and accelerations of the actuator (and the control surfaces), for which the typical actuator model constants for a transport aircraft are given in Table 6.1 [9].

The actuators used in aircraft are of two types: direct drive valve (DDV) and electrohydraulic servo valve (EHSV) [10]. DDV is designed to produce large forces to overcome forces due to flow, stiction, and acceleration or vibration, and the force motor is directly attached to and positions a spool valve (the force motor is directly coupled to the main control valve). Position feedback from the spool position to the input voltage is present. In the EHSV, the first stage of a hydraulic preamplifier is to multiply the force output of a torque motor to a sufficient level for overcoming the opposing forces. Direct position feedback through a flapper and nozzle arrangement is obtained in EHSV. The actuator consists of an electric force motor driving the main control valve, which in turn drives the main ram piston against the load. This is achieved by the flow of hydraulic fluid across it. The actuators themselves have two position feedback loops. The DDV and EHSV can be modeled as first-order transfer function (TF) as:

$$\frac{x}{e} = \frac{K/T}{s + 1/T} \tag{6.3}$$

Here, x is the valve position and e is the input voltage to the valve. The time constants in Table 6.1 maximally amount to the cut-off frequency of 20 rad/s, which is standard for the actuator model. If necessary, a second-order model can be used. The model of

TABLE 6.1
Control Surface Actuator Model Parameters

	Elevator	Aileron	Rudder
DC gain	0.89	0.73	0.795
Time constant (s)	0.06	0.07	0.07
Position limit (deg.)	−19 to 16	−21 to 21	−30 to 30
Rate limit (deg./s)	−30 to 30	−79 to 79	−50 to 50

the valve flow (Q) can be obtained as described next. The load flow can be expressed as a function of the valve position (x) and the load pressure (P), neglecting the return pressure and for a constant supply pressure:

$$Q = f(x,P) \tag{6.4}$$

The expression in the terms of small changes can be obtained as:

$$\Delta Q = \frac{\partial f}{\partial x}\Delta x + \frac{\partial f}{\partial P}\Delta P, \tag{6.5}$$

since 'f' is essentially a function for Q variable, we have

$$\Delta Q = \frac{\partial Q}{\partial x}\Delta x + \frac{\partial Q}{\partial P}\Delta P \tag{6.6}$$

Next, we define the sensitivities as $K_x = \frac{\partial Q}{\partial x}$ as the flow gain and $K_p = \frac{\partial Q}{\partial P}$ as the flow pressure coefficient to obtain

$$\Delta Q = K_x \Delta x + K_p \Delta P \tag{6.7}$$

We also can define the pressure sensitivity as $K_{px} = \frac{\partial P}{\partial Q}\frac{\partial Q}{\partial x} = \frac{1}{K_p}K_x = \frac{K_x}{K_p}$, and inserting this into linearized expression, we obtain,

$$\Delta Q = K_x \Delta x + \frac{K_x}{K_{px}}\Delta P \tag{6.8}$$

$$\Delta Q = K_x \left(\Delta x + \frac{\Delta P}{K_{px}} \right) \tag{6.9}$$

Finally, in the form of perturbation variables (using the same symbols) for a small amplitude variation, we get:

$$Q = K_x \left(x + \frac{P}{K_{px}} \right) \tag{6.10}$$

The main rotor ram assembly model can be found in Refs. [10,11]. The hydraulic subsystems are hydraulic pumps, accumulators, and hydraulic 'lines'. The pump can be modeled as first-order lag dynamics. The 'lines' dynamics can be approximated as lumped parameter coefficients.

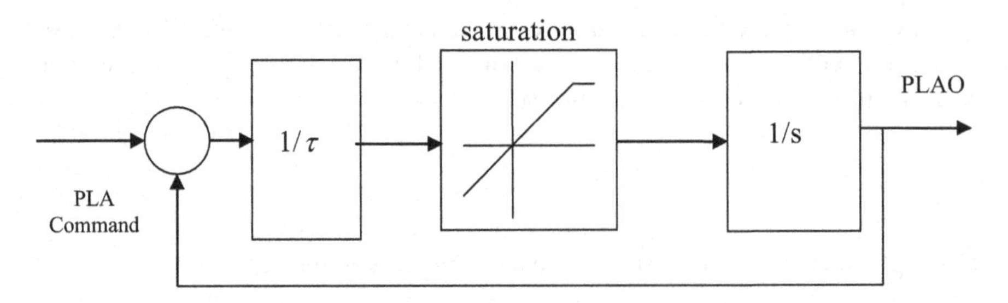

FIGURE 6.4 Shaping model for Power lever angle PLAO – PLA output to access static thrust database.

6.2.7 Engine Model

This is a crucial and main subsystem for any aircraft. Engine simulation helps in predicting the flight characteristics of an aircraft during climbing, cruising, gliding, engine failure, and maneuver accompanying throttle lever variation. The power plant is a jet engine or low-bypass turbofan. The thrust attained at the equilibrium depends on the power lever angle, altitude, and Mach number. It is very important to model the engine transient response during takeoff and landing. A jet engine has three major stages (i) the compressor, (ii) the combustion chamber, and (iii) the turbine stage. The intake and diffuser are the other important stages. For the purpose of a flight simulator, it is not necessary to use the complete nonlinear model that models all the stages of the engine. In a linear engine model, pilot throttle setting and flight and atmospheric conditions are inputted to obtain thrust power that is used in the complete flight simulation. In flight simulation, it is enough to excite the engine model at sufficient points in the flight envelope and determine the time constants. The predetermined time constants and interpolations can be used during the simulation; a first-order model can be used as shown in Figure 6.4 [10,12]. For the use in a research simulator, some basic data for the engine can be obtained by studying the software manual of a training simulator (of the same aircraft for the pilots); these modeling/data relate to the computation of thrust at four major elements: (i) propeller thrust: in a propeller aircraft, the driving force is generated by the engine shaft that rotates the propeller, this, in turn, moves the air-mass/-parcels/-packets through the propeller disc and accelerates the aircraft. The propeller force is a function of the flight condition, engine RPM, and throttle setting. The shaft horse power is a function of the fuel flow, engine RPM, and air inlet temperature and pressure. The fuel flow largely determines the turbine gas temperature; (ii) the gross jet thrust is approximately 15% of the total; (iii) Ram drag is a function of the intake airflow (that is dependent on the fuel flow) and varies linearly with the speed of the aircraft; and (iv) feather drag.

6.2.8 Landing Gear

The study of the landing gear subsystem is important in the simulation (and also in control analysis) of touchdown and ground roll. The landing gear dynamics play an important role during and after the touchdown and before takeoff. Also, the suspension geometry (cantilever or telescopic, lever or trailing link, semi-articulated), braking system, and nose wheel steering and other related aspects should be kept in mind. Some important typical landing gear elements to be modeled are described next [10].

Oleo-pneumatic strut: It contains two chambers telescoping into each other. It is partially filled with compressed air and oil. When the strut is extended or compressed, the oil flows between the chambers via a metering orifice. The damping is due to the viscosity of the fluid and the compressed air acts as a spring. The spring action is nonlinear and follows the Boyle's law: $PV^{\gamma} = \text{constant}$. Here, P is the chamber pressure of the gas, V is the chamber volume, and the exponent is the gas constant. The damping effect is based on the orifice flow as expressed by $Q_o = C_{\text{di}} A_o \sqrt{\frac{2\Delta P}{\rho}}$. Here, Q_o is the flow rate through the orifice, C_{di} is the discharge coefficient, A_o is the area of the orifice, ΔP is the differential pressure across the orifice, and ρ is the density of the oil. The composite force across the strut is the sum of the spring and damping forces, and this model is equivalent to spring and damper system in parallel.

Tire Dynamics: The pneumatic tire acts as a spring in the vertical axis. As a first approximation, the characteristics of an automobile tire can be used. The forward direction frictional force helps the aircraft to decelerate.

6.2.9 Control Loading and Sound Simulation

Normally when a pilot flies the aircraft and applies the control input, the control surface has to move in the flow field and it would experience resistance to its movement. This resistance is 'felt' by the pilot on the stick if it is a reversible system. Thus, the pilot gets the feel of the forces acting on the control surfaces when these are moved. This should be simulated in the flight simulator to enhance its fidelity. This aspect is achieved by the control loading system that produces the feel forces on the simulator flying controls [1]; this system is also known as the artificial feel system. To achieve the correct control feel, it would be necessary to utilize a good math model of the system. This math model can be formulated from the aircraft manufacturer's control data. The information on wiring diagrams, mechanical gearing, graphical representation of forces, and control surface deflection would be used. The models are derived from the analysis of these data with associated features. The numerical values of position limits, inertias, control surface gearings, and autopilot (AP) drive rates are included in the model. In most cases, the second-order models are used to represent both ends of the control arms. The aircraft surface actuator is modeled using a first-order-lag TF. The pilot's applied force is obtained from the load cell data and an AP drive is synthesized in the host computer. The minimum interaction rate required for the computer is 500 kHz so that the simulated feel is properly integrated with the pilot's senses.

If required, sound simulation should be incorporated into the flight simulator. The current trend is the use of a digital signal processor (DSP), and totally, a digital sound system can be integrated with other systems. Often, the actual sound recording is done from the aircraft (engine/propulsion systems), and the sound is synthesized using filtering algorithms and DSP.

6.2.10 Motion Cues

The pilot in the actual flight of the aircraft experiences the natural motion cues; this is more so in case of a fighter aircraft. Some motion (cues) aid the pilot in stabilizing and maneuvering the aircraft by confirming that the pilot's internal model response and aircraft responses match very well. Some other motions of the aircraft alert the pilot to ensuing system failures [1]. In a fixed base flight simulator, the motion cues are absent and the only cue is via the visual cues from the instruments and the simulated view of the outside world. These visual cues of motion are very important low-frequency cues. The vestibular system of humans is a sensor of motion and position. The semicircular canals in the inner ears are the rotational and motion sensors. They act as damped angular accelerometers. Three such canals form an almost orthogonal axes system in each ear and sense angular accelerations as low as 0.1 deg./s^2. The linear motion is sensed by the otoliths of the inner ears. They sense specific forces, the external or non-gravitational force, with the threshold as low as 0.02 m/s^2 and can detect a tilt of 2 deg. A system approach to human perception of orientation and motion has been developed. The vestibular sensors are modeled as mechanical systems. The parameters of these models have been validated by several experiments of human perception in flight simulators. The semicircular canal model is given as [1]:

$$\frac{\text{afferent firing rate (deg./s}^2)}{\text{physical stimuli (deg./s}^2)} = \frac{0.07s^3(s+50)}{(s+0.05)(s+0.03)} \tag{6.11}$$

The velocity threshold for this model is 2.5 deg./s. The otolith TF is given as:

$$\frac{y_{oto}}{f} = \frac{2.02(s+0.1)}{(s+0.2)} \tag{6.12}$$

The velocity threshold for this model is 0.2 m/s. There are other physiology-based cues like visual, proprioceptive, and tactile cues. In general, the proprioceptive sensors are associated with the vestibular system, joints, muscles, and internal organs of the human body. Very interestingly, the processing of the sensor signals by the central nervous system can be effectively modeled by a steady-state KF (Kalman filter). The spectrum of sensing spans a lower frequency (visual) to a higher band due to the proprioceptive and other cues; thus, the earliest recognition of the motion is possible, and hence, the usefulness and importance of the motion cues/sensing in the flight-training simulators. The motion system's bandwidth should be at least slightly greater than the simulated aircraft's rigid body dynamics. Also, various tasks which need to excite the motion cues (sensors) should be such that they fall within the bandwidth of 0.1–1.5 Hz, the body sensors' active sensitive region. In general, elaborate visual cues are effective for many piloting tasks in flight simulators; however, human eyes are slow in perceiving motions. Before any visual cue sensing takes place, the human brain receives the acceleration cues. Many body sensors detect the acceleration signals that are communicated to the brain in milliseconds. If motion cues are needed, then the motion system should be systematically designed and integrated. This will enhance the quality and utility of motion cues, with an increase in cost and maintenance effort.

6.2.11 Turbulence and Gust Models

The aircraft flies in an atmosphere where its responses are affected by atmospheric disturbances and these create additional forces on the aircraft. These disturbances are (i) gust, (ii) turbulence, (iii) wind shear, and (iv) crosswind. Turbulence is a random phenomenon (Example 6.3). The other effects are represented by suitable deterministic models that are usually obtained from experimental data.

6.2.12 Sensor Modeling

Sensors are used to derive measurements of aircraft's positions, angular rates, accelerations, and several other information of importance for monitoring the flight's performance as well as for the use of sensed variables in the computation of control laws. The rate and acceleration sensors can be modeled as first-order lag. The vanes used for deriving the measurement of flow angles can be modeled as second-order TF. The flow angles can also be determined from the pressure probes which measure the local and impact pressures. The computation of the flow angles from these data involves computational delay. A first-order lag model can be used for the flow angle simulation.

6.2.13 Flight Dynamics

The flight dynamics of a vehicle to be simulated should be fairly accurately known. It is important to know the aerodynamic coefficients at all the flight conditions in the aircraft flight envelope and these are obtained from the wind tunnel aero database and by application formulae. Figure 6.5 depicts a flight dynamics simulation block diagram for a typical missile. The fin axis aerodynamic derivatives are first specified.

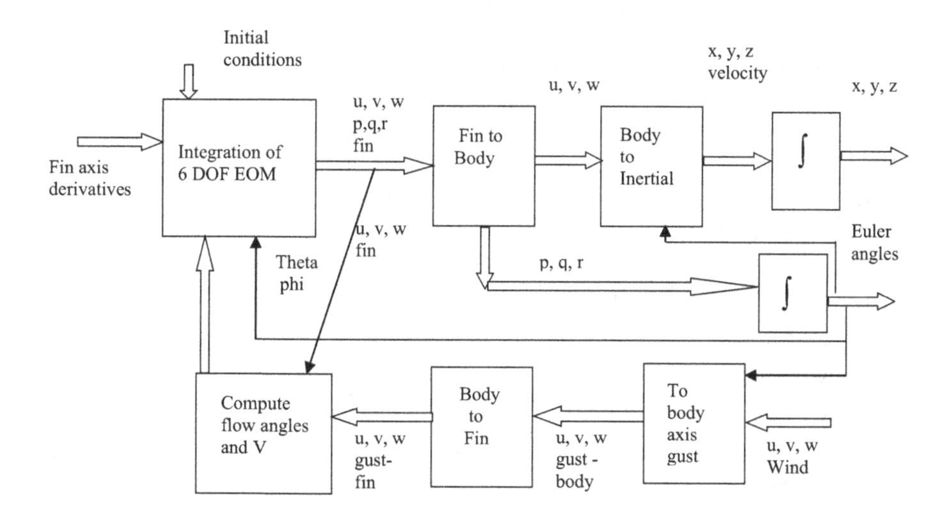

FIGURE 6.5 Typical missile flight dynamics transformations.

The fin-to-body (noted as '$b \leftarrow f$') transformation is done using the following matrix (angle ϕ is generally 45 deg.):

$$T_{bf} = \begin{bmatrix} 1 & 0 & 0 \\ 0 & \cos\phi & -\sin\phi \\ 0 & \sin\phi & \cos\phi \end{bmatrix}$$

The body velocities are converted into inertia by using a direction cosine matrix. The inverse transformation is obtained by transposing the respective matrices. The deterministic gust effects are transformed into a fin axis for the flight simulation. For numerical integration of nonlinear flight dynamics, generally, R-K (Runge-Kutta) method is used. It works well for a good number of such simulation problems, even with moderate discontinuities. In fact, it approximates the Taylor series method. Certain aircraft flight simulation aspects are illustrated by Examples 6.1 and 6.2, which demonstrate the complexity of the problem.

6.3 STEADY-STATE FLIGHT AND TRIM CONDITIONS

It is important to obtain steady-state flight conditions for obtaining the linearized mathematical models of the aircraft at various flight conditions, since these models are required for the design of control laws [2,9,13]. An aircraft is considered to be in an equilibrium state when all its 'state derivatives' become zero simultaneously, mass being assumed constant. This is a stringent requirement and may not be possible to attain in all types of flight conditions. Conditions like wings level and level turn, wherein the rotational and translational accelerations are zero and with Euler angle rates as constants, are practically feasible situations and hence useful to study. Such conditions are called quasi-steady states. To obtain a steady-state equilibrium of aircraft (dynamics), the trim state should be obtained by estimating the control settings such that the accelerations have zero values. The equilibrium point is $f(\dot{x},x,u) = 0$, with $\dot{x} = 0$, $u = 0$, or constant; due to this condition, the system is at rest. The aircraft then flies at steady wings level and can have a steady-turning flight. Also, a wings-level climb and climbing turn are possible. The steady flight is defined in terms of the following variables: $\dot{p},\dot{q},\dot{r},\dot{V},\dot{\alpha},\dot{\beta}=0$; u = constant.

The following flight-specific constraints are:

Steady wings-level:	$\phi,\dot{\phi},\dot{\theta},\dot{\psi} = 0$	(p, q, r =0)
Steady turning:	$\dot{\theta},\dot{\phi} = 0$	(turn rate $\dot{\psi}$)
Steady pull-up:	$\phi,\dot{\phi},\dot{\psi} = 0$	(pull-up rate $\dot{\theta}$)
Steady roll:	$\dot{\theta},\dot{\psi} = 0$	(roll rate $\dot{\phi}$)

The condition $\dot{p},\dot{q},\dot{r} = 0$ dictates that the angular rates and aerodynamic and thrust moments should be zero or constant. Also, $\dot{V},\dot{\alpha},\dot{\beta} = 0$ dictates that the aerodynamic forces should be zero or constant. In order to arrive at steady-state conditions, a set of nonlinear simultaneous equations need to be solved. Due to the complexity of the EOMs, a numerical routine is necessary to adjust the independent variables to meet the required criteria/constraints. The knowledge of aircraft behavior would be useful to specify steady-state conditions and achieve the required trim. A trim routine flow diagram is given in Figure 6.6. For steady translational flight, the constraint is given as $p,q,r,\phi = 0$. The trim surface positions and other variables are determined by numerically solving the nonlinear equations for the translational and

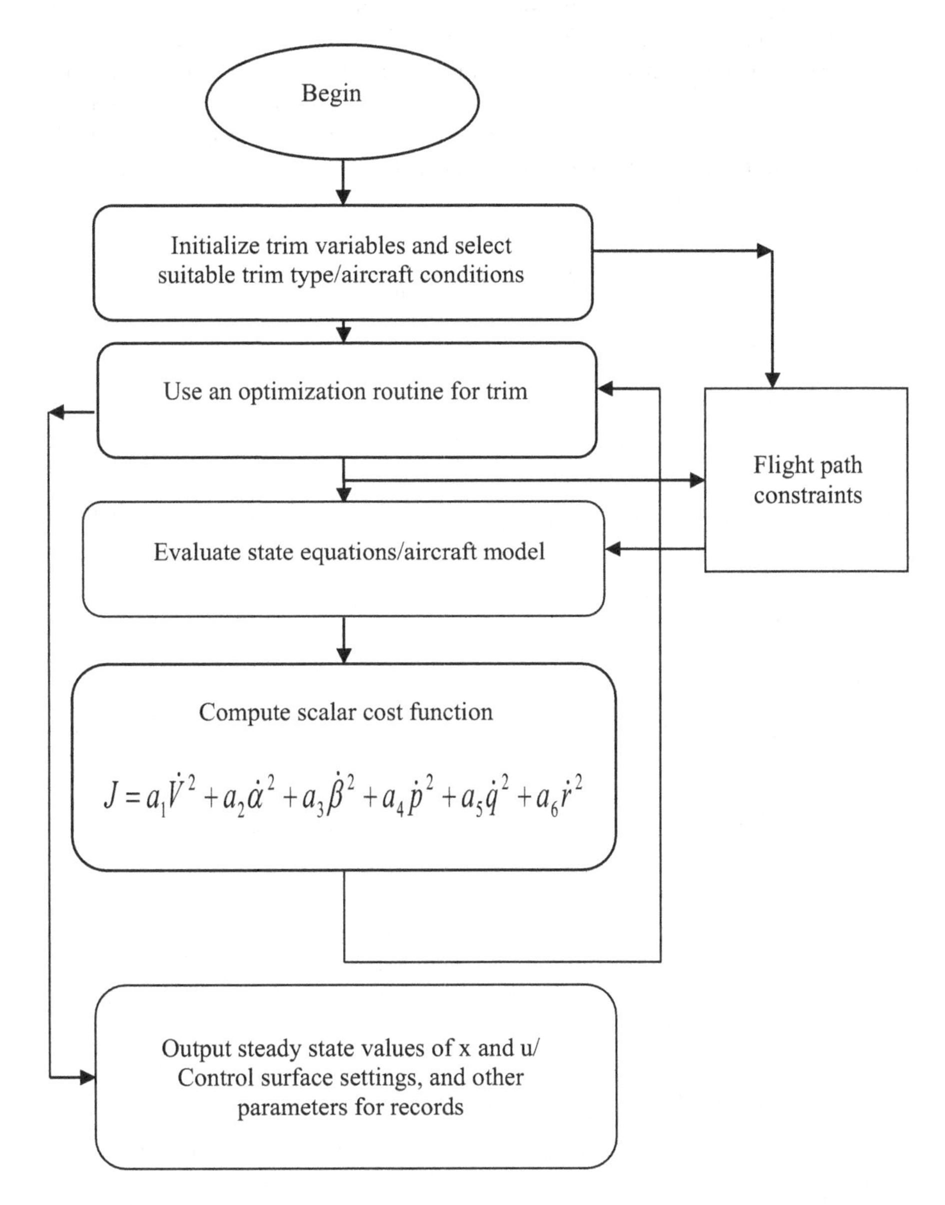

FIGURE 6.6 Flow diagram for trim routine.

rotational accelerations. The pitch attitude can be determined from the attitude rate. For pull-up/push-over, the trim is computed with the following conditions: $p, r, \phi = 0$. The pitch rate is computed as:

$$q = \frac{1}{V m \cos\beta}\{mg(n + \cos(\alpha - \theta)) - Z_T \cos\alpha + X_T \sin\alpha\} \quad (6.13)$$

There are other trim conditions possible: (i) thrust-stabilized turn trim, (ii) specific power trim, and (iii) β trim. The thrust-stabilized condition yields a constant non-wings-level turn with a nonzero altitude rate. The constraints are the same as the level turn. The altitude and Mach number are specified for the thrust-stabilized trim condition. The specific power trim situation yields a level turn at a specified altitude, thrust trim parameter, specific power, and Mach number. The specific (excess) power is given by (see Exercise 7.9)

$$P_s = dh/dt + \frac{V}{g}\frac{dV}{dt} \quad (6.14)$$

Then, the velocity rate is obtained, with the altitude being kept constant as:

$$\dot{V} = \frac{P_s g}{V} \quad (6.15)$$

The other states, $\dot{p}, \dot{q}, \dot{r}, \dot{\alpha}, \dot{\beta}$, would be zero if the required trim point is obtained. Other constraints need to be satisfied for this specific power trim [14].

Example 6.1

Develop a trim routine based on MATLAB following the flow diagram given in Figure 6.6. Aircraft geometry/configuration data need to be provided for the chosen aircraft. Also, aerodynamic derivatives (or the aero database/aerodynamic coefficients) are needed [2]. This builds up the math model of the aircraft. The cost function needs to be tailored to specific aircraft/sets of equations. MATLAB's 'fminsearch' can be used for the minimization of the chosen cost function.

Solution 6.1

The programs [for a transport aircraft/ 3 degrees of freedom (DOF) model] written in MATLAB are given in the directory **'ExamplesSolSW/Example 6.1Trim'**. The file 'main.m' needs to be run in MATLAB. You can specify speed and altitude. Some results are given in Table 6.2.

6.3.1 The Rate of Climb and Turn Coordination Flights

In some of these, the steady-state flight conditions do not apply, since with the change in the altitude the atmospheric density varies. In a steady-turning flight, the ϕ, p, q, r will not be zero. Once the rate of the climb is determined, linearized models can be

TABLE 6.2
Results of Aircraft Trim Routine

Altitude (ft)	Airspeed (ft/s)	Cost (approximate)	Throttle value	Elevator in deg.	Alpha in deg.
0	150	2.0e-25	0.254	−18.2	22.8
20k	300	2.02e-24	0.176	4.29	8.88

used for nonzero flight path angles. The turn is specified by $\dot{\psi} = \frac{V}{r_t}$, with r_t as the turn radius. The angular rates can be computed from kinematic equations. A coordinated turn is such that the aircraft is banked/rolled at a required angle in order that there is no Y-axis force. This will necessitate solving the constraints for roll angle and pitch attitude simultaneously. Following [2,15], we have the following equation for the flight path angle (assuming the wind velocity as zero):

$$\sin\gamma = a_1 \sin\theta - a_2 \cos\theta \tag{6.16}$$

Here, $a_1 = \cos\alpha\cos\beta$, and $a_2 = \sin\phi\sin\beta + \cos\phi\sin\alpha\cos\beta$. The pitch attitude is given as:

$$\tan\theta = \frac{a_1 a_2 + \sin\gamma\sqrt{\{a_1^2 - \sin^2\gamma + a_2^2\}}}{a_1^2 - \sin^2\gamma}, \text{ with } \theta \neq \pm\frac{\pi}{2} \tag{6.17}$$

In wings-level, for $\beta = 0$, we have $\theta = \alpha + \gamma$. In a coordinated turn, the force F_y is assumed zero, and with the steady-state condition $\dot{V} = 0$ (the *Y*-axis velocity), we obtain from Eq. (3.23), assuming full nonlinear EOM:

$$-RU + PW = g\sin\phi\cos\theta \tag{6.18}$$

We have the conditions $\dot{\phi} = \dot{\theta} = 0$, and we use the following equations

$$\begin{bmatrix} P \\ Q \\ R \end{bmatrix} = \begin{bmatrix} 1 & 0 & -\sin\theta \\ 0 & \cos\phi & \sin\phi\cos\theta \\ 0 & -\sin\phi & \cos\phi\cos\theta \end{bmatrix} \begin{bmatrix} \dot{\phi} \\ \dot{\theta} \\ \dot{\psi} \end{bmatrix} \tag{6.19}$$

to obtain the expressions for *P* and *R* as follows:

$$\begin{aligned} P &= -\dot{\psi}\sin\theta \\ Q &= \dot{\psi}\sin\phi\cos\theta \\ R &= \dot{\psi}\cos\phi\cos\theta \end{aligned} \tag{6.20}$$

We also have the following expressions for aircraft velocities:

$$\begin{bmatrix} U \\ V \\ W \end{bmatrix} = \begin{bmatrix} V_t \cos\alpha\cos\beta \\ V_t \sin\beta \\ V_t \sin\alpha\cos\beta \end{bmatrix} \tag{6.21}$$

Then, substituting for *R, U, P,* and *W* in Eq. (6.18) and just simplifying, we obtain the following equation:

$$g\sin\phi\cos\theta = \dot{\psi}\, V_t \cos\beta[\cos\phi\cos\alpha\cos\theta + \sin\theta\sin\alpha] \tag{6.22}$$

Finally, we obtain the required constraint condition for the coordinated turn:

$$\sin\phi = \frac{1}{g}\dot{\psi}\, V_t \cos\beta[\cos\phi\cos\alpha + \tan\theta\sin\alpha] \tag{6.23}$$

Next, the solution of Eqns. (6.17) and (6.23) together yields [2,15]:

$$\tan\phi = g_c\cos\beta\sec\alpha\frac{(a-b^2) + b\tan\alpha\sqrt{c(1-b^2) + g_c^2\sin^2\beta}}{a^2 - b^2(1 + c\tan^2\alpha)} \tag{6.24}$$

Here, $g_c = \dot{\psi}\frac{V_t}{g}$, the centripetal acceleration, $a = 1 - g_c\tan\alpha\sin\beta$, $c = 1 + g_c^2\cos^2\beta$, and $b = \sin\gamma\sec\beta$. It must be emphasized here that Eqn. (6.24) would be solved by the trim routine when other required information is supplied to it. Then, the value of pitch attitude can be obtained from Eqn. (6.17). When the flight path angle is zero and since b is zero, we obtain from Eqn. (6.24):

$$\tan\phi = \frac{g_c\cos\beta}{\cos\alpha - g_c\sin\alpha\sin\beta} \tag{6.25}$$

When the AOSS is very small, and since the flight path angle is zero, we have from Eqn. (6.26),

$$\tan\phi = \frac{g_c}{\cos\alpha} = \frac{\dot{\psi}\, V_t}{g\cos\theta} \tag{6.26}$$

6.3.2 Computation of Linear Models for Control Law Design

It is wiser and simpler to use a linear approach for the design of control laws. For this, one needs the linear dimensional models of aircraft dynamics at several flight conditions in the flight envelope. These linear models are computed by a sophisticated program within the structure of nonlinear flight simulation that uses the aero database.

The linear models are obtained by specifying the Mach number, trim AOA, altitude, slat deployment, if any, and undercarriage up or down position. The linear simulation software would give linearized models along with the modal characteristics like eigenvalues and damping ratios of the longitudinal and lateral directional modes. Several thousand linear models would be required due to a variety of flight conditions and configurations in the entire flight envelope as can be seen from the application formula, Eqn. (6.1). These linear models are then studied to arrive at important flight mechanic parameters (Chapter 4). It is important to evaluate these parameters of the aircraft from the aero database to assess the need for control law design. These parameters would help assess the aircraft's capability as well as deficiencies in the dynamics. These deficiencies then could be compensated by designing proper flight control algorithms/laws/systems. Also, it is important to evaluate the dynamic characteristics and responses of the aircraft and study its handling qualities by flight simulation exercises during the process of development of the aircraft (program). Flight mechanics parameters and linear models of the aircraft can be obtained using the equilibrium analysis of the aircraft dynamics.

In fact, an aircraft is required to be trimmed for a given flight condition and then linearization routine is used to obtain linear models at these flight conditions. Figures 6.6 and 6.7 give flow diagrams of these routines. Derivation and development of linear models for an aircraft and related programming aspects are detailed in Refs. [16,17]. Such a program would consist of several modules [14] for (i) input data, (ii) aero data, (iii) aerodynamic models, (iv) engine model, (v) trim routine that would satisfy the required trim constraints as per the trim conditions set; this routine would also convert the trim parameters into control surface deflections, and (vi) output result modules that will give linear model stability and control derivatives, all the state space matrices, and modal parameters of the aircraft dynamics at the chosen flight conditions.

Linearization can be done using the finite difference method. Let the nonlinear equation for the aircraft dynamics be given as:

$$\dot{x}(t) = f(x(t), u(t)) \tag{6.27}$$

Define a certain operating point as $\dot{x}_0 = f(x_0, u_0)$, and use a Taylor's series for linearization, to obtain

$$\dot{x}(t) \cong \dot{x}_0 + \frac{\partial f}{\partial x}(x - x_0) + \frac{\partial f}{\partial u}(u - u_0) + (\text{higher order terms neglected}) \tag{6.28}$$

$$\dot{x}(t) - \dot{x}_0 = \frac{\partial f}{\partial x}\tilde{x} + \frac{\partial f}{\partial u}\tilde{u} \tag{6.29}$$

The linearized equations would be represented as:

$$\dot{\tilde{x}}(t) = A'\tilde{x} + B'\tilde{u} \tag{6.30}$$

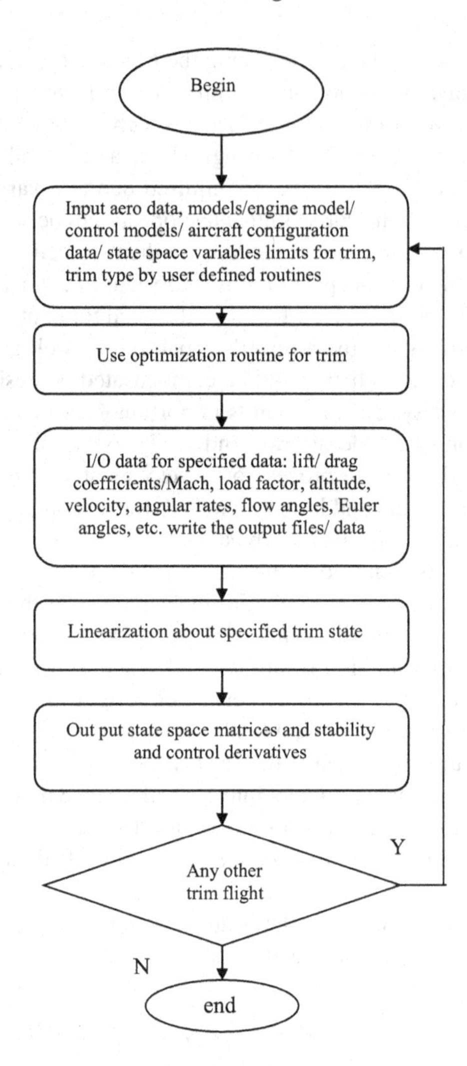

FIGURE 6.7 Flow diagram for a linearization routine.

The numerical linearization routine would obtain the elements of matrices A', B', (and C', D' of measurement equation) using a finite difference method, as shown for matrix A:

$$A' = \begin{bmatrix} \dfrac{f_1(x_0+\Delta x_1, u_0) - f_1(x_0, u_0)}{\Delta x_1} & \cdots & \dfrac{f_1(x_0+\Delta x_n, u_0) - f_1(x_0, u_0)}{\Delta x_n} \\ \cdot & & \\ \cdot & & \\ \cdot & & \\ \dfrac{f_n(x_0+\Delta x_1, u_0) - f_1(x_0, u_0)}{\Delta x_1} & \cdots & \dfrac{f_n(x_0+\Delta x_n, u_0) - f_n(x_0, u_0)}{\Delta x_n} \end{bmatrix} \tag{6.31}$$

Similarly, for matrix B, the point (x_0) is kept fixed and the point (u_0) is perturbed by a term (Δu_j) for each element appropriately. A similar procedure can be followed to obtain the elements of linearized measurement matrices C' and D'. The step size for evaluating the numerical difference is typically given as:

$$\Delta x_j \rightarrow 10^{-7} \times x_j$$

Forward differencing has been found to be quite accurate for numerical computation. Alternatively, a fixed value of 10e-7 can also be used for the step size.

6.4 SIX DOF SIMULATION AND VALIDATION

A six DOF flight simulator is required for a fighter aircraft development program to carry out full nonlinear flight dynamics simulation and for aiding the design of flight control laws, via iterative validation of these laws. Also, it is required for the evaluation of control laws during the design process and even during and after the flight tests of the prototype aircraft. A simulator would consist of (i) visual system hardware – projector, screen, instrument panel/virtual instruments, (ii) visual software – terrain, runway near airfield, cloud/rain visuals, (iii) cockpit with throttle handle, pitch-roll control stick, rudder pedals, and pilot's seat, (iv) computers mainly PCs, to solve flight dynamics equations, atmospheric models, and execution of software to link various subsystems/models, and (v) data acquisition system.

Generation of visual imagery is very crucial to obtaining good fidelity of the simulator [4]. Video films/CCTV (closed-circuit television) camera can be used for visual scene creation. The CCTV camera captures the terrain view (from its physical situation/model) and the view is projected on the screen of the simulator. CGI (computer-generated imagery) is used to generate the required imagery in real time. This is the most widely used system for visual generation/simulation. The steps for generating the visual images are (i) definition of objects: runway, roads, trees, buildings, clouds, terrain, and lights – this defines a viewing volume, (ii) create objects with appropriate texture and level of detailing required and storage of these data/information, (iii) create visuals by appropriate definitions of all the required/desired objects with respect to some reference coordinate system, and (iv) linking of the virtual world to the dynamics of the flight by appropriate timing and coordinating transformation and projecting the visuals on to a 2D screen form the pilot's eye point. This is called rendering-projecting-timing cycle for visual simulation with delays of less than 100 ms and screen refresh rate of about 60 Hz. If the monitor screen has to display the graphic picture of 1280×1024 pixels with a depth complexity of 4 units, then at a rate of 60 Hz, the pixel transfer rate needed is $4 \times 60 \times 1024 \times 1280 = 312$ Mpixels/s [4]. This can be provided by 4 raster managers with 80 Mpixels/s. The raster-type computer display is generally used. It draws images by dots or pixels and each pixel is represented by a sequence of bits. The map of one screen represented in the form of bits is termed as bitmap, and the location of the corresponding bit in the computer memory is called frame buffer. A display controller passes the contents of the frame buffer to the screen at a rate of 30 Hz to avoid flicker. Interestingly, many

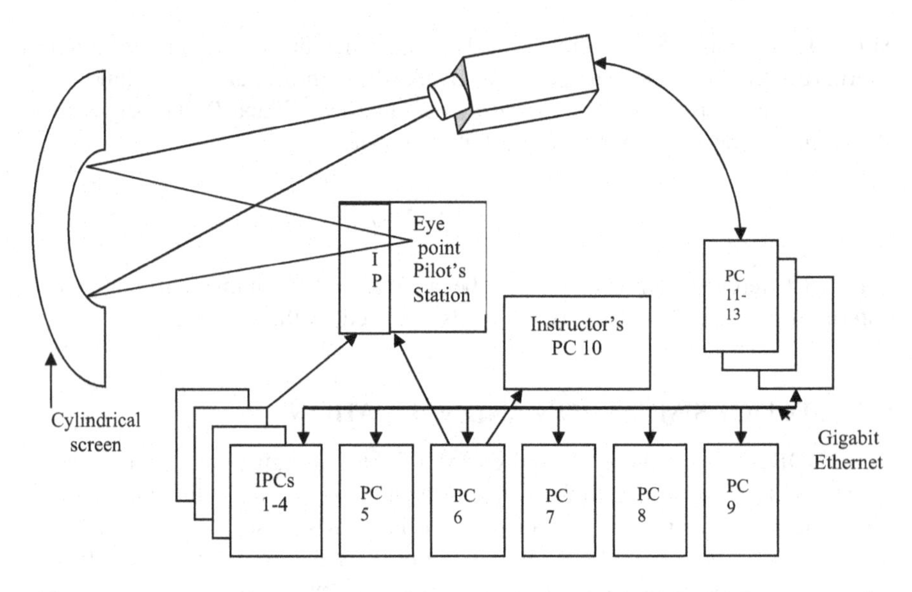

FIGURE 6.8 Schematic outline of a PC-based flight simulator. (IPCs 1-4: Instrumentation data/simulation; PC 5: Flight model, autopilot; PC 6: Data acquisition system; PC 7: Avionics simulation; PC 8: Data storage/feel system simulation; PC 9: Aural simulation; PC 10: Instructor's station; PC 11-13: Visual system/SW; IP: Instrument panel/displays.)

powerful visual software modules are available nowadays but could be very expensive. Development of such SW could be very time-consuming and involves a lot of effort. A good picture of interconnectivity of various HW/SW modules of a PC-based flight simulator is given in Figure 6.8. The pilot's eye point is rigidly linked to the aircraft's attitudes and positions in relation to the Earth. The variables that change during the flight progression update the visuals (visual scenery) as observed by the pilot.

6.4.1 Flight Simulation Model Validation for a Rotorcraft

The flight control system should be evaluated in a nonlinear simulation mode with a larger portion of the flight envelope and larger amplitude maneuvers. For this purpose, the nonlinear control system elements, like rate limiting, etc., should be modeled and included in the flight simulator evaluation experiments.

A systematic approach to flight simulation model validation for a rotorcraft would have the specific aims [18]: (i) identify a consistent set of vehicle states and inputs, the data processing methods should be employed to check the consistency and these data must be consistent to justify model changes based on the flight test data; (ii) the aerodynamic loads should be estimated from the flight test data, this is a major uncertainty in rotorcraft simulation, and these estimates must be compared with simulation results for identical flight conditions; and (iii) once the discrepancies are identified, the math model of the vehicle (rotorcraft) should be upgraded using the model structure determination and parameter estimation methods by matching the flight test results. The idea is to drive both simulation flight controls and simulation states with

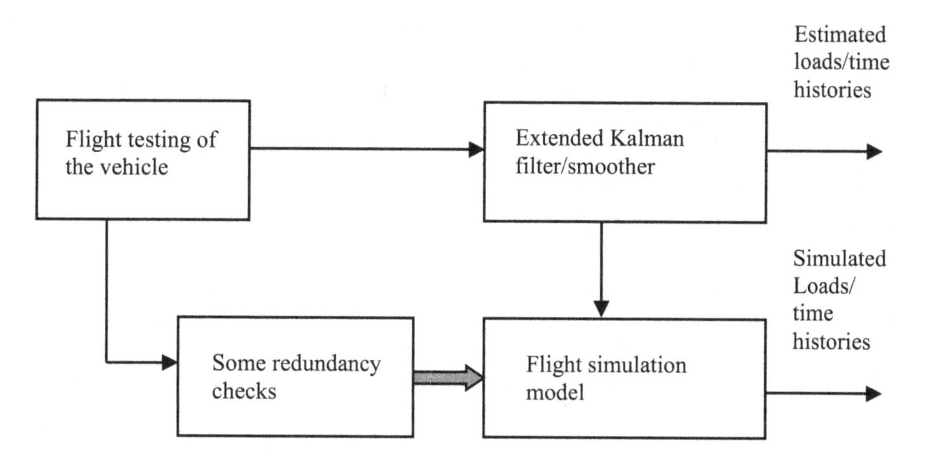

FIGURE 6.9 Flight simulation model validation scheme.

flight test time histories, Figure 6.9. If the aerodynamic loads compare very well, then the trajectories should match well when the fuselage DOF equations are integrated. If these loads do not match, then the simulation is not generating some of the aerodynamic loads. The KF/smoother estimates biases and misalignment errors and reconstructs the net aerodynamic forces and moments. If there is a mismatch, then a comparison between simulated and measured flapping should be done to isolate the problem. This could identify a problem in the main rotor. Regression method was used to identify the correlation between the simulation error and vehicle states, and the structure of the math model was upgraded for the US Army/McDonnell Douglas AH-64 Apache attack helicopter [18]. This is a practical example of the use of system identification and state estimation methods for the validation of nonlinear helicopter simulations from flight test data.

6.4.2 Flight Simulation Model Validation Using the Concept of Coefficient Matching

Although the flight control laws would have been designed using the linear math models of the aircraft, it is necessary to validate the entire design using nonlinear flight simulation procedures. If required, the models used in the design procedures should be modified based on the feedback from the flight tests conducted on the aircraft and its versions [19]. Most of the techniques for the validation of the simulation model depend on estimating the linear models of the aircraft from flight test data at several flight conditions. However, as seen in Section 6.2.1, the aerodynamic database is given in the coefficient form, and the direct comparison is difficult and this approach of validation does not give a scope of updating the database that is in 'lookup table' form. An alternative approach is to estimate the aerodynamic coefficients from the flight data and determine the differences between the estimated coefficients and the ones obtained from the tables by applying the application formulae. Then, an appropriate method can be used to model any discrepancies. The procedure requires a thorough validation of these models based on several sets of flight data and many flight conditions for getting consistent results. The discrepancy can be modeled by using either feed-forward neural network (FFNN) or polynomial models.

The nominal aerodynamic coefficients can be computed from the application formula which utilizes the aero database. The aerodynamic coefficients from flight (or flight-like) responses are estimated using the suitable parameter estimation/Kalman filtering method. Direct use of the flight responses to compute the coefficient would give inaccurate results due to noise in the measured data. The estimation technique would be more accurate due to filtering out the noise processes. It is obvious from Section 6.2.1 that all the signals/variables used for computing the coefficients from the application formula and parameter estimation would not be the same. However, within the structure of the rule used for these computations, consistent sets of flight variables/data should be used. Primarily, the comparison should be carried out for the same flight conditions. A procedure for using this method is outlined below:

i. compute the (nominal) force and moment coefficients using the application rule and related data
ii. compute the coefficients from the flight responses using an estimation technique
iii. obtain the differences between these coefficients
iv. fit FFNN or a polynomial model to these differences as functions of selected/required flight parameters
v. Incorporate these incremental models into the nominal database models; this is likely to bring the upgraded coefficients closer to the actual ones
vi. The incremental models are compact ways to incorporate any future data-dependent variations into the database/coefficient base, i.e., new flight data can be used to get updated database
vii. Model structure selection methods would be required to be used for getting adequate fit to different time histories of the coefficients

Figure 6.10 depicts the process of this method of validation and upgrading of the aero database. A typical result of the application of this process with simulated data is shown in Figure 6.11. Other important aspects to be kept in mind are: the flight data should be in proper time-stamped form, accurate flow angle data should be used, and variation of mass and CG should be taken into account. The task of determination of which terms in the aerodynamic model buildup should be revised is not easy. It is an iterative process and requires considerable engineering judgment. The team should have extensive knowledge of the aerodynamic effects on stability, control, and performance of the aircraft, for which this exercise is being done.

6.4.3 Flight Simulation Model Validation Using Direct Update

Another approach based on the parameter estimation technique that allows direct updates to the aerodynamic coefficient is briefly discussed [20]. It has the merits of capturing the nonlinearities in coefficients and joint dependence of a particular coefficient on one or more independent variables, like α, β, δ, and Mach number and can

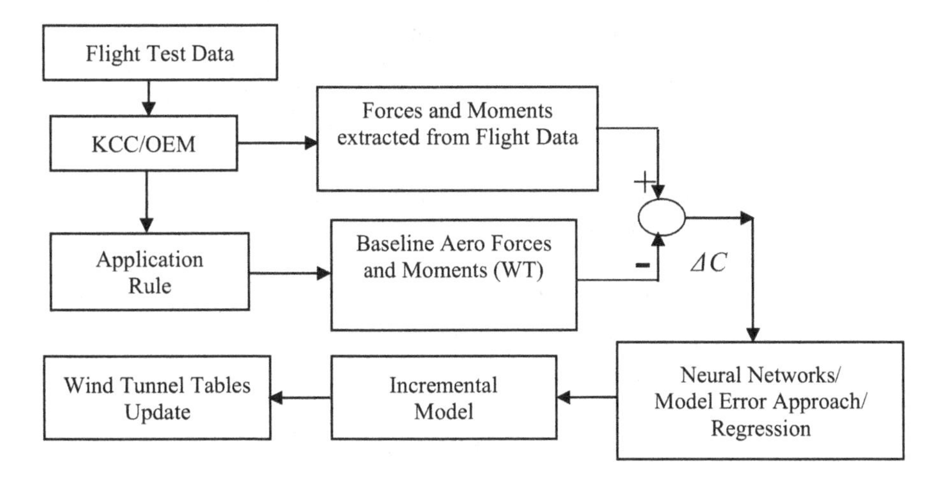

FIGURE 6.10 Process of validation/upgrading of aero database.

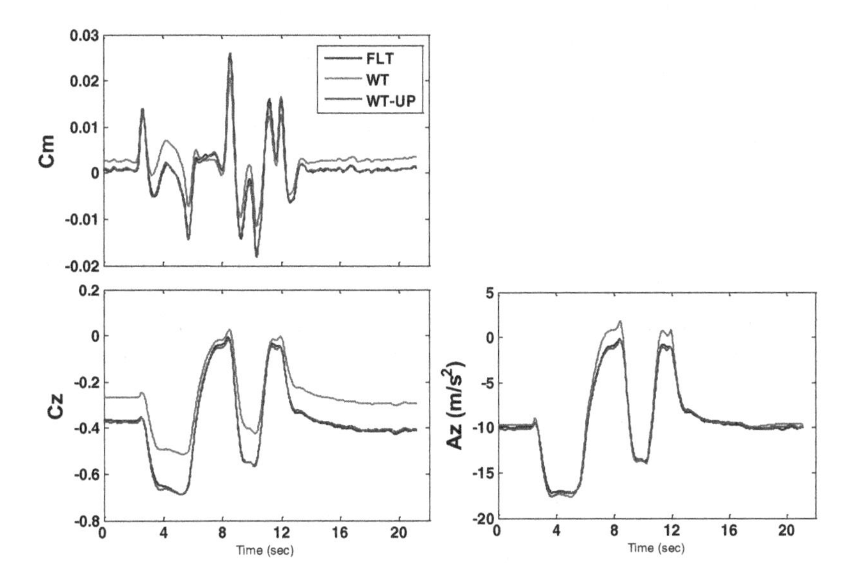

FIGURE 6.11 Time histories for coefficient matching method.

encompass a process of model structure determination using splines and modified regression methods. WT (wind tunnel) predictions using the method of incremental coefficients determined from the flight data have been studied in [21,22]. In the present approach, the nonlinear aerodynamic functions are obtained in the table look-up form. First, a nonlinear short-period model to account for nonlinearities is postulated:

$$\begin{bmatrix} \dot{\alpha} \\ \dot{q} \end{bmatrix} = \begin{bmatrix} Z(\alpha) & Z_q \\ M(\alpha) & M_q \end{bmatrix} \begin{bmatrix} 1 \\ q \end{bmatrix} + \begin{bmatrix} Z_{\delta_e} \\ M_{\delta_e} \end{bmatrix} \delta_e \tag{6.32}$$

Functional dependency on more than one variable can also be considered:

$$\begin{bmatrix} \dot{\alpha} \\ \dot{q} \end{bmatrix} = \begin{bmatrix} Z(\alpha,\delta_e) \\ M(\alpha,\delta_e) \end{bmatrix} + \begin{bmatrix} Z_q q \\ M_q q \end{bmatrix}, \text{ or } \begin{bmatrix} \dot{\alpha} \\ \dot{q} \end{bmatrix} = \begin{bmatrix} Z(\alpha,\delta_e,q) \\ M(\alpha,\delta_e,q) \end{bmatrix} \tag{6.33}$$

To represent a nonlinear term (say, between two breakpoints), the interpolation formula can be used for the pitching moment as:

$$M(\alpha) = M_1 + \frac{(M_2 - M_1)}{(\alpha_2 - \alpha_1)}(\alpha - \alpha_1) \tag{6.34}$$

$$M(\alpha) = \frac{(\alpha_2 - \alpha)}{(\alpha_2 - \alpha_1)} M_1 + \frac{(\alpha - \alpha_1)}{(\alpha_2 - \alpha_1)} M_2 \tag{6.35}$$

$$M(\alpha) = W_l(\alpha) M_1 + W_r(\alpha) M_2 \tag{6.36}$$

Here, the nonlinear function is expressed in the form of near-combination of weights and components of the moment variable at the breakpoints. Similarly, the expressions for 2-D formulae to represent the dependent variable as a function of two variables can be obtained:

$$M(\alpha,\beta) = W_{l1}(\alpha) W_{l2}(\beta) M_{i,j} + W_{l1}(\alpha) W_{r2}(\beta) M_{i,j+1}$$
$$+ W_{r1}(\alpha) W_{l2}(\beta) M_{i+1,j} + W_{r1}(\alpha) W_{r2}(\beta) M_{i+1,j+1}$$

With $\alpha \in [\alpha_i, \alpha_{i+1}]$ and $\beta \in [\beta_i, \beta_{i+1}]$. The weights can be computed separately for each dimension as done for the 1-D equation. The moment-related dependent variables are unknown. The nonlinear functions $M(\alpha), Z(\alpha)$ are estimated by treating them as function values at a selected breakpoint expressed in the form of the following equation:

$$\begin{bmatrix} \dot{\alpha}_1 & \dot{q}_1 \\ \dot{\alpha}_2 & \dot{q}_2 \\ \cdot & \cdot \\ \cdot & \cdot \\ \cdot & \cdot \\ \dot{\alpha}_m & \dot{q}_m \end{bmatrix} = \begin{bmatrix} 0 & \cdot & 0 & W_{l1} & W_{r1} & \cdot & \cdot & 0 & q_1 & \delta_{e_1} \\ W_{l2} & W_{r2} & 0 & \cdot & \cdot & 0 & 0 & 0 & q_2 & \delta_{e_2} \\ 0 & W_{l3} & W_{r3} & \cdot & \cdot & 0 & 0 & 0 & q_3 & \delta_{e_3} \\ \cdot & \cdot & \cdot & \cdot & \cdot & \cdot & \cdot & \cdot & \cdot & \cdot \\ 0 & 0 & \cdot & \cdot & W_{lm} & W_{rm} & \cdot & \cdot & q_m & \delta_{e_m} \end{bmatrix}$$

$$\begin{bmatrix} Z_1 & M_1 \\ Z_2 & M_2 \\ \cdot & \cdot \\ \cdot & \cdot \\ \cdot & \cdot \\ Z_n & M_n \\ Z_q & M_q \\ Z_{\delta_e} & M_{\delta_e} \end{bmatrix} \tag{6.37}$$

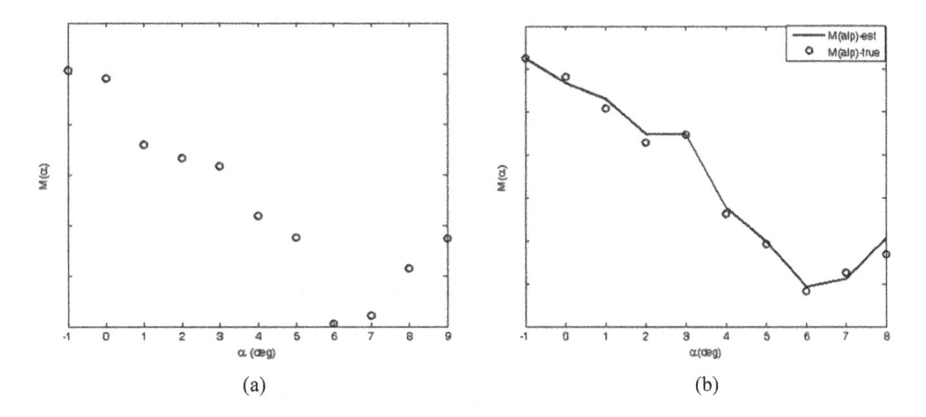

FIGURE 6.12 Aero database validation: comparisons using the direct method. (a): estimate from flight data, (b): estimate compared with true value.

Equation (6.37) can be written in a shorthand form as $Y = X\,b$, with b as the parameter matrix which is estimated using the LS method: $b = (X^T X)^{-1} X^T Y$. Figure 6.12a shows a typical result obtained from the flight data of a supersonic fighter aircraft, and Figure 6.12b shows the validation of the concept. The procedure is versatile and it should be validated using various flight database.

6.5 PC MATLAB- AND SIMULINK-BASED SIMULATION

The workstation/PC-based simulation of a light transport aircraft was carried out to assess the short period, Phugoid and related HQs to see if these met the HQ specifications/requirements before even the aircraft was flight tested.

Example 6.2

The simplified block diagram of a flight control system of a high-performance fighter aircraft is given in Figure 6.13 [23]. Use MATLAB/SIMULINK tool to realize the interconnected blocks and obtain doublet input responses of the closed loop system. The state and measurement models are:

$$\begin{bmatrix} \dot{\alpha} \\ \dot{q} \\ \dot{\theta} \\ \dot{v}/v_0 \end{bmatrix} = \begin{bmatrix} Z_{\alpha/v_0} & 1 & 0 & Z_{v/v_0} \\ M_\alpha & M_q & 0 & M_{v/v_0} \\ 0 & 1 & 0 & 0 \\ X_\alpha & 0 & X_\theta & X_{v/v_0} \end{bmatrix} \begin{bmatrix} \alpha \\ q \\ \theta \\ v/v_0 \end{bmatrix} + \begin{bmatrix} Z_{\delta_e} \\ M_{\delta_e} \\ 0 \\ X_{\delta_e} \end{bmatrix} \delta_e \tag{6.38}$$

$$\begin{bmatrix} \alpha \\ q \\ a_x \\ a_z \end{bmatrix} = \begin{bmatrix} 1 & 0 & 0 & 0 \\ 0 & 1 & 0 & 0 \\ C_{31} & 0 & 0 & C_{34} \\ C_{41} & 0 & 0 & C_{44} \end{bmatrix} \begin{bmatrix} \alpha \\ q \\ \theta \\ v/v_0 \end{bmatrix} + \begin{bmatrix} 0 \\ 0 \\ D_{31} \\ D_{41} \end{bmatrix} \delta_e \tag{6.39}$$

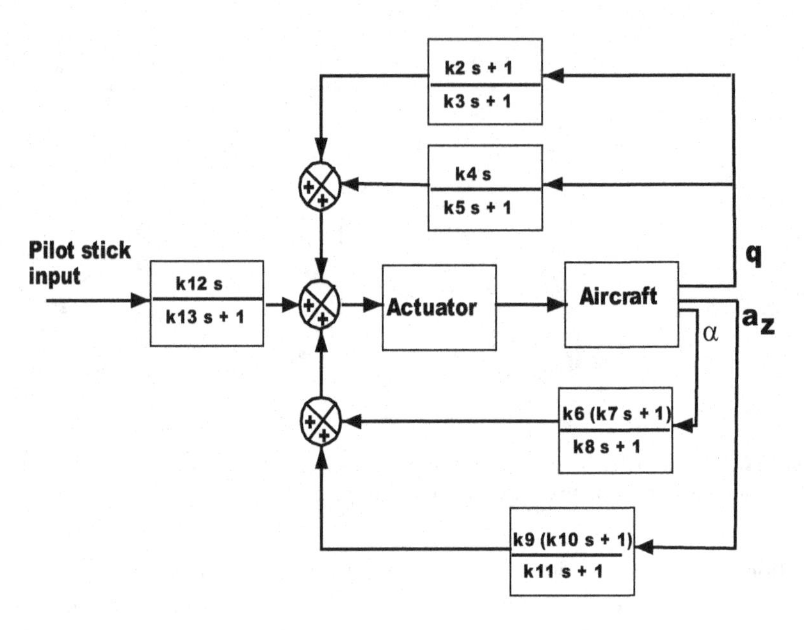

FIGURE 6.13 An unstable aircraft operating in a closed loop [23].

Here, the $Z_{(.)}$, $X_{(.)}$, $M_{(.)}$, $C_{(.)}$, and $D_{(.)}$ are the aerodynamic parameters. The equations with the parameters/constants used in the simulation are given as:

$$\begin{bmatrix} \dot{\alpha} \\ \dot{q} \\ \dot{\theta} \\ \dot{v/v_0} \end{bmatrix} = \begin{bmatrix} -0.771 & 1 & 0 & -0.1905 \\ 0.3794 & -1.259 & 0 & 0.116 \\ 0 & 1 & 0 & 0 \\ -0.937 & 0 & -0.096 & -0.0296 \end{bmatrix} \begin{bmatrix} \alpha \\ q \\ \theta \\ v/v_0 \end{bmatrix} + \begin{bmatrix} -0.299 \\ -9.695 \\ 0 \\ -0.0422 \end{bmatrix} \delta_e \tag{6.40}$$

$$\begin{bmatrix} \alpha \\ q \\ a_x \\ a_z \end{bmatrix} = \begin{bmatrix} 1 & 0 & 0 & 0 \\ 0 & 1 & 0 & 0 \\ 2.7546 & 0 & 0 & -0.0176 \\ -79.9 & 0 & 0 & -19.68 \end{bmatrix} \begin{bmatrix} \alpha \\ q \\ \theta \\ v/v_0 \end{bmatrix} + \begin{bmatrix} 0 \\ 0 \\ 0 \\ 0 \end{bmatrix} \delta_e \tag{6.41}$$

The actuator TF is considered as $\frac{kA}{kA\,s+1}$. The gain values (in the block diagram) used are: $k2=0.125$, $k3=0.212$, $k4=0.519$, $k5=2.0$, $k6=0.519$, $k7=0.13$, $k8=0.3$, $k9=0.0057$, $k10=0.01$, $k11=0.289$, $k12=1.0$, $k13=0.25$, and $kA=20$.

Solution 6.2

The control blocks and plants given in Figure 6.13 are realized in SIMULINK. The relevant files are in **'ExamplesSolSW/Example 6.2FS'.** The simulation responses are obtained by using a doublet input signal with a sampling interval of 1 s. Figure 6.14 shows the time histories of output dynamics of the aircraft with a stabilized controller. It can be observed from the figure that the output responses settle after some transients; hence, the controller stabilizes the aircraft. Moreover, the response of the aircraft can be further modified by adjusting gain values judiciously.

Example 6.3

A block diagram representation to generate the aircraft time histories in the presence of turbulence is given in Figure 6.15 [23]. Realize this using MATLAB only. Obtain the responses with and without turbulence. Use the Dryden model for the turbulence and procedure described here. This model generates moderate turbulence conditions modifying the forward speed, vertical speed, and pitch rate. The dynamic model of the following form [24,25] is considered:

$$\dot{x}_u = [-x_u + y_u k_u \sqrt{\pi / \Delta t}]/t_u \tag{6.42}$$

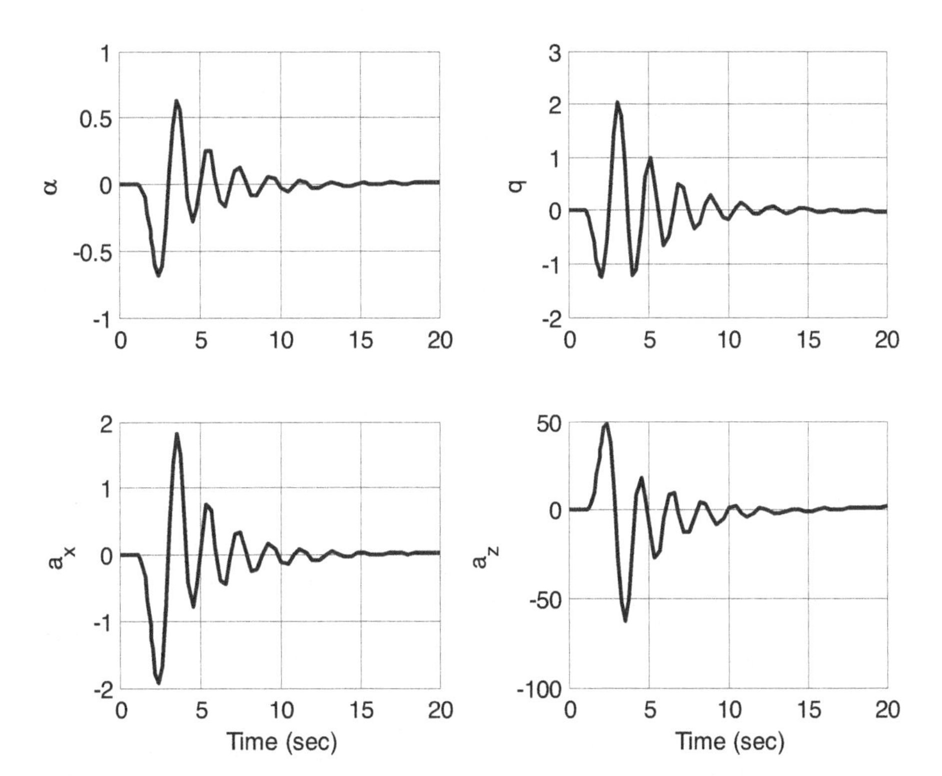

FIGURE 6.14 Time histories of aircraft with a stabilized controller.

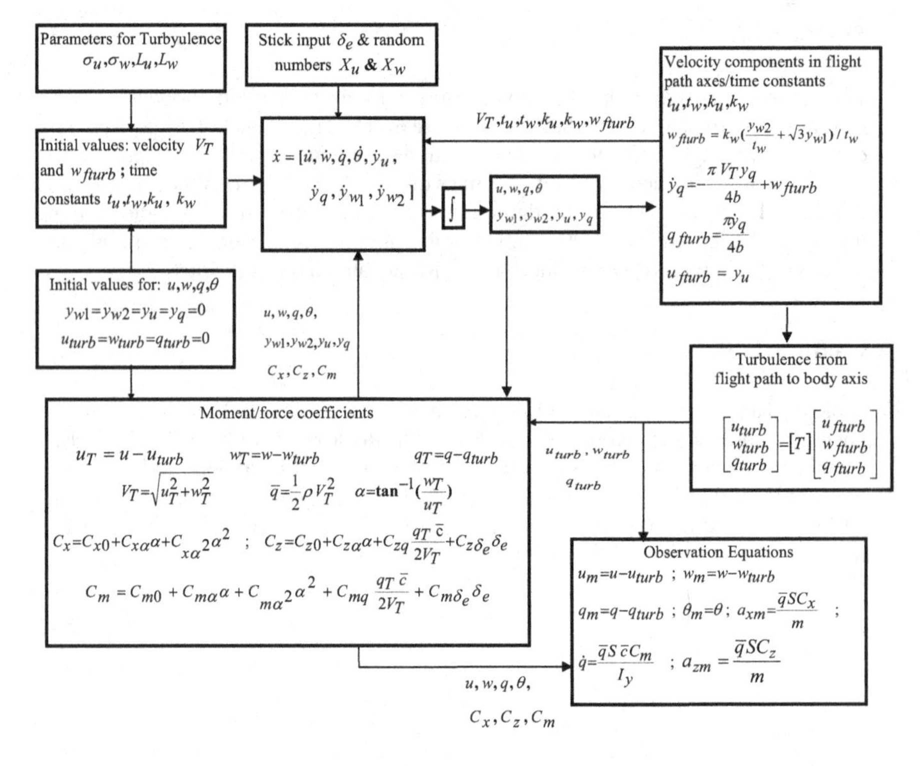

FIGURE 6.15 Block diagram for simulation of aircraft longitudinal motion in turbulence using only MATLAB.

$$\dot{x}_q = -\left[\frac{\pi V_T}{4b}\right] x_q + w_{fturb} \tag{6.43}$$

$$\dot{x}_{w_1} = -\frac{x_{w_2}}{t_w^2} - \frac{2x_{w_1}}{t_w} + y_w \sqrt{\pi / \Delta t} \tag{6.44}$$

Here, y_u and y_w are random numbers used for simulating the random nature of the turbulence; t_u, t_w, k_u and k_w are the time constants as follows:

$$t_u = L_u / V_T;\ t_w = L_w / V_T;\ k_u = \sqrt{(2\sigma_u^2 t_u)/\pi};\ k_w = \sqrt{(2\sigma_w^2 t_w)/\pi} \tag{6.45}$$

$$\text{Here, } V_T = \sqrt{u^2 + w^2};\ \sigma_u = \sigma_w \text{ and } L_u = L_w = 1750 \text{ ft} \tag{6.46}$$

Also, $L = 1750$ ft. and the turbulence intensity is $\sigma = 3$ m/s. This model of the turbulence is incorporated into the flight dynamic state equations, and a 4th order

Runge–Kutta (RK) integration is used to obtain the flight responses of u, w, q, and θ, and the turbulence response variables x_u, x_q, x_{w_2}, and x_{w_1}. Using the procedure outlined in Refs. [24,25], the turbulence in forward and vertical velocity and pitch rate can be obtained:

$$u_{fturb} = x_u;\; w_{fturb} = k_w\left[\frac{x_{w_2}}{t_w} + \sqrt{3}\, x_{w_1}\right]\Big/ t_w \text{ and } q_{fturb} = -\left[\frac{\pi \dot{x}_q}{4b}\right] \tag{6.47}$$

Here, 'b' is the wingspan. As the flight variables u, w, q and θ are computed in the body axis, the quantities u_{fturb}, w_{fturb} and q_{fturb} should be computed in the body axis. The transformation is given as [25]:

$$\begin{bmatrix} u_{turb} \\ w_{turb} \\ q_{turb} \end{bmatrix} = \begin{bmatrix} \cos\alpha & 0 & -\sin\alpha \\ 0 & 1 & 0 \\ \sin\alpha & 0 & \cos\alpha \end{bmatrix} \begin{bmatrix} u_{fturb} \\ w_{fturb} \\ q_{fturb} \end{bmatrix} \tag{6.48}$$

The Dryden mode, (6.48) can be used for simulating the aircraft responses in the presence of atmospheric turbulence. The aircraft's longitudinal responses to turbulence can be obtained using the equations:

$$\begin{aligned} &u_m = u - u_{turb};\; w_m = w - w_{turb};\; q_m = q - q_{turb} \\ &\theta_m = \theta \\ &a_{xm} = \frac{\bar{q} S C_x}{m} \\ &a_{zm} = \frac{\bar{q} S C_z}{m} \\ &\dot{q}_m = \frac{\bar{q} S \bar{c} C_m}{I_y} \end{aligned} \tag{6.49}$$

The foregoing Dryden model for the simulation of turbulence has been used to generate the responses.

Solution 6.3

The blocks and their interconnections are realized using MATLAB and the SW is given in **'ExamplesSolSW/Example 6.3Dryden'**. The aircraft responses with and without turbulence (intensity = 3 m/s) are given in Figure 6.16.

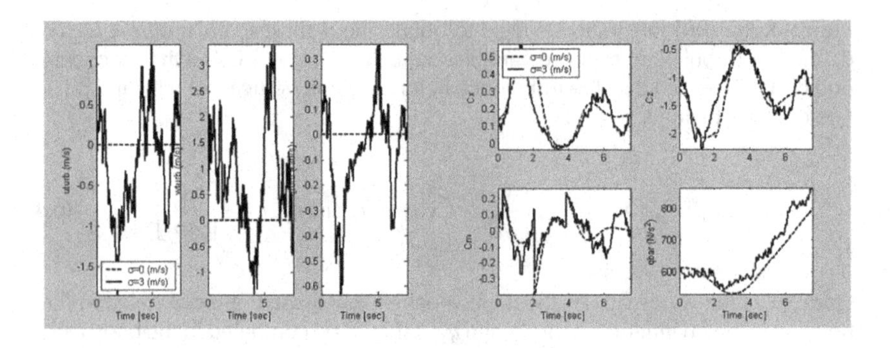

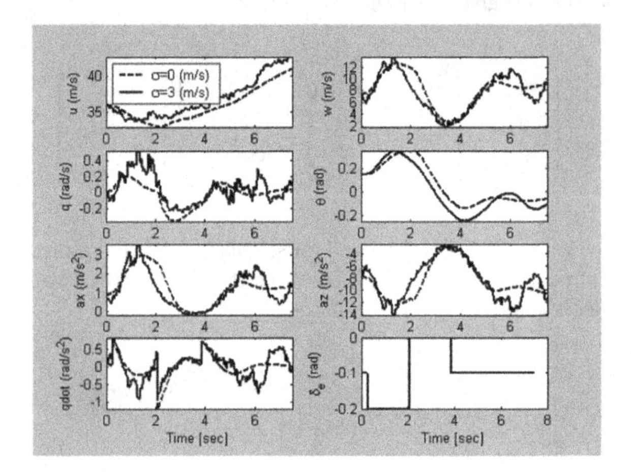

FIGURE 6.16 Time histories of aircraft with and without turbulence.

6.6 REAL-TIME DESKTOP SIMULATOR FOR THE EVALUATION OF FLYING QUALITIES

This simulator avoids extensive coding work and multiple and distributed processors and still achieves a cost-effective desktop solution for handling (and flying) quality evaluation, and the entire application requires only a standard x86-based computing platform provided with Windows operating system using Mathwork's real-time Windows Target (RTWT) toolbox [26]. It does not use any external data acquisition unit for data I/O, and all interfaces are based on a universal serial bus (USB). NALSim has been used for (i) the evaluation of damping factor and natural frequency, (ii) pilot tracking tasks, (iii) real-time AP tuning, and (iv) engine failure studies. An LS optimization-based methodology is also used.

6.6.1 NALSIM Framework

It consists of (i) aircraft model, (ii) out-of-window (OOW) visuals, (iii) avionics displays, (iv) tools for data analysis, and (v) an instructor station application programmer's interface (API) housed in a single x86-based processor. The aircraft model is built using the opensource flight dynamics and control toolbox in SIMULINK10: (i) models for flight dynamics, (ii) aerodynamics, (iii) propulsion, (iv) landing gear, (v) atmosphere, and (vi) sensors; for a transport/fighter aircraft or UAV/MAV simulations.

The AP features are pitch hold, roll hold, altitude select, altitude hold, heading hold, vertical speed hold, nose-up/-down modes, and soft ride modes. The NALSIM uses Mathworks's RTWT kernel for running the auto code of the CL model in real time. The OOW visualization is realized using opensource 3D rendering API, OpenSceneGraph (OSG)-based visualization SW; it requires only C++ and OpenGL SW for programming. Head-up display (HUD) and head-down display are developed using VAPS XT. The data logging and real-time plotting features are developed using VC++ application. The graphical user interface for the instructor station is built using the Microsoft foundation class system framework of visual studio. The VC++ application talks to the MATLAB model using its 'engine' library.

6.6.2 Flying and Handling Quality Evaluation

The simulator has the in-built capability for conducting studies on defining handling quality boundaries for a particular aircraft; the aircraft models are in the form of linear nondimensional derivatives across the flight envelope for a generic fighter/ transport aircraft. By changing mass, inertia, geometry, aerodynamic derivatives, and CG, a designer can evaluate the HQs for the chosen aircraft.

6.6.2.1 Variable Damping and Natural Frequency Features

One can vary appropriate flight mechanic parameters displayed on the instructor station to study the effect of damping and natural frequency. For a learner-test-pilot, an aircraft is trimmed for wings-level conditions; then linear models are generated numerically using central differences; from this linear model, SP/DR damping factor and natural frequencies are calculated and displayed on the instructor station; the user now can change these values for the intended study. With the desired values entered, a nonlinear least square optimization is performed. When optimization converges, the dimensional derivatives are converted into a nondimensional form; the optimization meets the desired damping ratio and natural frequency with good accuracy in conformity with user-entered values.

6.6.2.2 Tracking Tasks

Tracking tasks are conducted to verify that the HQs of an aircraft are satisfactory to perform its intended mission and can be given to a trainee pilot to evaluate the piloting skills. The task is of precision control and is specified in terms of acceptable levels of performance in accomplishing the task itself. The performance results are in the form of pilot comments or an appropriate numerical pilot rating (Cooper–Harper ratings). The desired criteria for the pilot would be to maintain the command bar at the tip of the watermark in the HUD symbology, and the bar is the pilot's target, and the target should be moved according to the user-selected tracking task.

6.6.2.2.1 *Sum of Sines Tracking*

The aim of this task is to expose the phase lag in the pitch-only task. The target's theta command would be formed by summing 7 sine waves, with a randomly appearing frequency-based function that is computed as: $\theta_c = k\sum_{i=1}^{n=7} A_i \sin(\omega_i t)$; $\omega_i = 2\pi\left(\frac{N_i}{63}\right)$ rad/sec. Pilot-in-the-loop (PIL) simulation is performed to track the moving command bar that is driven by θ_c. The gain 'k' is set to achieve the desired task amplitude.

6.6.2.2.2 *Discrete Tracking*

Discrete task consists of a series of steps and ramps; both pitch and roll axes of a command bar are driven by synchronized commands; a pitch error is limited to +3 deg. and roll error is limited to +70 deg. The movement of the command bar in pitch and roll axis is given as follows. The command bar on HUD is driven in the pitch axis as $k\left(\theta_c - \left(\frac{\theta}{\cos(\psi)}\right) + \theta_{\text{bias}}\right)$, and the command bar is driven in roll axis as $k(\phi - \phi_c + \phi_{\text{bias}})$. Here, k is set to achieve the desired task amplitude, and θ_c and ϕ_c are the pitch and roll commands for the target; θ_{bias} and ϕ_{bias} are the aircraft's trim pitch/roll angles.

6.6.2.2.3 *Disturbance Regulation*

This task is computed in the same way as the sum-of-sines task. However, instead of driving a command bar, a command is added to the pilot's stick command. The objective is to maintain wings-level and zero-pitch flight.

6.7 HARDWARE-IN-THE-LOOP-SIMULATION (HILS) FOR A MINI UAV

HILS is an RT simulation platform/system where the output and input signals in an aircraft are maintained as that of a real process/system [27], hence providing an environment to simulate UAV missions that are nearly close to the real environments, yet limited by the accuracy of the 6 DOF model used in the simulation. The HWs that are difficult to simulate such as the actuators, sensors, and embedded controllers are physically placed in the loop. NAL-CSIR has a mini UAV called Slybird with a wing span of 2 m and endurance of 1 h.

6.7.1 A 6DOF Model for SLYBIRD

In order to develop a 6DOF model of the system, one needs data on (i) aerodynamics, (ii) propulsion, (iii) mass, CG locations, inertia, and moment reference point, and (iv) wing aerodynamic chord and wing surface area. The unscaled, 1:1 model of this aircraft was subjected to the wind tunnel tests at HAL's (Bangalore) low-speed wind tunnel to obtain the aerodynamic coefficients which are in the look-up table form. An application rule, Eq. (6.1), has been used for generating the force and moment coefficients for this UAV. The propulsion data are in the form of motor RPM versus thrust. The 6 DOF simulation SW has trimming and linearization features; a nonlinear LS minimization algorithm is implemented to perform a wings-level trim. Linearization is performed using a central difference method and the linear models are generated at the trim point for a particular flight condition which are then used for the control design.

6.7.2 Subsystems of the HILS

We briefly describe the subsystems of HILS.

6.7.2.1 Real-Time Target Machine and Interference

The real-time target machine is a real-time operating system that meets the timing constraints required for real-time applications, which are developed as Simulink models and executed on the target machine for the Slybird UAV, the simulation model which consists of EOMs and sensor models. The HW interface blocks are auto-transmitted to the Target, and this machine TM is connected to the host PC via an Ethernet cable. The TM has the capability to support the interfaces like I2C, SPI, UART, RS232, CAN, PWM, ADC, DAC, and discrete channels. These interfaces are needed to mimic the real scenario in the simulation environment. IMU data is parceled and sent through the I2C interface of the target to the AP board whereas the GPS data are sent using a universal asynchronous receiver/transmitter (UART). A pulse width modulation (PWM) interface is provided to the interface with servo motors. Analog voltage feedback from the servo motors is directly connected to the analog interfaces of the TM. Sensed voltages are mapped into positions for 6DOF simulation. The host PC compiles the 6DOF model and communicates to the TM (via Ethernet), which is a MATLAB real-time kernel; and runs the model in the TM. The I7 single-board computer boots the MATLAB real-time kernel and communicates with Spartan 6 FPGA through the PCIe bus.

6.7.2.2 Autopilot Hardware

NAL-CSIR has designed and developed an indigenous AP board mainly focusing on low weight with reconfiguration capability. The AP electronic hardware is realized using Programmable Systems on Chip. The AP module has onboard (i) 3-axis accelerometers, (ii) 3-axis gyroscope, (iii) 3-axis magnetometer, (iv) static pressure sensor, and (v) temperature sensor. In this AP, the Sensor Suite and Data Logger are integrated, tested, and evaluated on the test platform Slybird.

6.7.2.3 Ground Control Station

The ground-station computer is for direct observation and monitoring purposes, and the onboard computer sends its data to be displayed on the ground-station computer. For this work, an open-source mission planner SW application is used to monitor the telemetry received from a UAV and control its movement by sending waypoints and other commands to the UAV. The SW integrates Google Maps to allow clicking of the waypoint selection and enable the operator to a very simple method to develop a route for the MAV to navigate.

6.7.3 Model-Based Design Framework

In the present work, the main aim is to generate an auto-code for the controller and perform a seamless integration of the onboard AP SW from simulation to onboard implementation. This is achieved using MATLAB/Simulink. The plant model is implemented in Simulink. The estimation and path planning algorithms are also implemented in Simulink around the plant model to constitute a Model in the Loop Simulation. The RTS is carried out in four phases, each with a specific aim: (i) Software in the Loop Simulation (SILS), (ii) Rapid Control Prototyping (RCP), (iii) Processor in the Loop Simulation (PILS), and (iv) Hardware in the Loop Simulation (HILS).

In SILS, compiled AP code is incorporated into the overall simulation and required for the evaluation of auto-code functionality on the designer's desk. The RCP is intended to verify the onboard target before burning the code on the actual target HW; the designs carried out are: (i) the aircraft 6 DOF simulation application will be running in the windows real-time environment called Simulink Desktop Real Time; (ii) the controller simulation application will be running in the xPC target environment known as real-time kernel; and (iii) the data are exchanged by using a user diagram protocol in real time. In PILS, the embedded controller is brought in a loop with the simulation; the steps are: (i) the aircraft model runs in an accelerated mode if one intends to do non-RTS; in order to do RTS, the aircraft model is run on the TM; (ii) the controller runs on the target microcontroller (APHW); (iii) no I/O cards are used, a USB connection is used to exchange the data between the control system and model; and d) the purpose of this simulation is to test that all functionalities of the controller are correctly computed in the target HW.

This HILS had been built to test the missions (i) manual, (ii) stabilized, (iii) return to launch, (iv) loiter, (v) fly by wire, and (vi) auto takeoff, landing, and waypoint navigation.

EPILOG

The simulation of advanced rotorcraft requires high-fidelity mathematical models. The blade element theoretic model possesses good fidelity and accuracy for such applications. However, such models with the required complexity would be slow for real-time considerations. Parallel processing would speed up such computations. The blade element models possess flexibility also. In Ref. [28], an advanced rotorcraft flight simulation model is developed and it is especially suited for implementation on a parallel computer. In Ref. [29], a concept of using flight-recorded data for pilot training via RTS playback of flight maneuvers is developed. It details the requirements for simulation-based flight reconstruction also. A robust KF is utilized as a state-estimator flight-simulator driver. The methodology is validated for UH-60 helicopter nonlinear simulation. A procedure that combined the three steps – identification of derivatives in linear/nonlinear models, comparison of the results to the original database, and update of the database if necessary – is presented in Ref. [30]. Further development of the nonlinear model development can be found in Ref. [31]. In Ref. [32], complete development of a 3DOF motion-based flight simulator is described. Nowadays, well-validated flight simulation/control analysis SW packages are available [3,15,33].

EXERCISES

6.1 It was said in Section 6.2.1 that the aero database is given in the form of lookup tables. Explain the theory of the Table lookup.

6.2 A system of equations with a wide spread of time constants, for the same dynamic system, is known as a stiff system of equations. What problem would it pose for the solution of these equations by the Runge–Kutta method?

6.3 Given the nonlinear state–space equations: $\dot{x}(t) = f(x(t), u(t))$. Linearize this system of equations using Taylor series expansion.

6.4 Obtain the TF, Bode diagram, and step response of the elevator actuator model of Table 6.1. Also, use the booster dynamics: $\frac{2933}{s^2 + 52.83s + 3283.3}$ in conjunction with the actuator TF and obtain the composite TF model, its Bode diagram, and step response.

6.5 Obtain the Bode diagrams for the inner ear models of Eqns. (6.2) and (6.3).

6.6 Represent the basic element operations of an INS in the form of a block diagram.

6.7 Take appropriate SI units and verify the formula dimensionally/unit wise: $Q_o = C_{di} A_o \sqrt{\frac{2\Delta P}{\rho}}$.

6.8 In an actual aircraft, what is the mechanism for trimming? In which way, a trim device would help the pilot?

6.9 What is the trimming in the flight simulation wrt the EOMs/flight dynamics? Explain by taking the state–space Eq. (2.10).

6.10 In a computer simulation of the dynamic equation of an aircraft or other systems, we may have to slow down or obtain a fast solution. For this, scaling is required to be done: $\tau = at$, where τ is the computer time and 'a' is the scale. Explain the significance of this with an example of a second-order system.

REFERENCES

1. Rolfe J. M., and Staples, K. J. (Eds.). *Flight Simulation.* Cambridge University Press, Cambridge, UK, 1986.
2. Stevens, B. L., and Lewis, F. L. *Aircraft Control and Simulation.* John Wiley & Sons, Inc., Hoboken, NJ, 2003.
3. Zipfel, P. Six degrees of freedom modeling and simulation of aerospace vehicle. Notes of a 2-day training program, C-DAC/Zeus Numerix Pvt. Ltd., February 2007.
4. Madhuranath, P. Introduction to flight simulation. In *Aircraft Flight Control and Simulation* (edrs. Chetty S., and Madhuranath P.), NAL Special Publication, SP-9717, National Aerospace Laboratories, Bangalore, 1997.
5. Berry, D. T., and Deets, D. A. Design, development, and utilization of a general purpose airborne simulator. NATO-AGARD Report 529, *Presented at the 28th Meeting of the AGARD Flight Mechanics Panel*, Paris, France 10–11 May 1966.
6. Latha Sree, P. Application formulae for rigid body fighter aircraft mathematical models. Personal Communication. Flight Mechanics and Control Division, National Aerospace Laboratories, Bangalore, June 2007.
7. Maine, R. E., and Iliff, K. W., Identification of dynamic systems – Applications to aircraft. Part 1: The output error approach, AGARD-AG-300, Vol. 3, Pt.1, December 1986.
8. Jategaonkar, R. V. Identification of the aerodynamic model of the DLR research aircraft ATTAS from flight test data, DLR-FB-90-40, July 1990.
9. Rajesh, V. Aircraft modeling in Simulink and autopilot design. NAL PD FC 0416, National Aerospace Laboratories, Bangalore, Dec 2004.
10. Pashilkar, A. A. Subsystem modeling for simulation. In *Aircraft Flight Control and Simulation* (edrs. Chetty S., and Madhuranath P.), NAL Special Publication, SP-9717, National Aerospace Laboratories, Bangalore, 1997.

11. Merritt, H. E. *Hydraulic Control Systems*. John Wiley and Sons Inc., New York, 1967.
12. Johnson, S. A. A simple dynamic engine model for use in a real-time aircraft simulation with thrust vectoring. NASA TM 4240, October 1990.
13. Latha Sree, P. Equilibrium, linearzation and flight mechanics parameters. In *Aircraft Flight Control and Simulation* (edrs. Chetty S., and Madhuranath P.), NAL Special Publication, SP-9717, National Aerospace Laboratories, Bangalore, 1997.
14. Rekha, R., and Madhuranath, P. ALLS- A linearizing link software-links flight simulation to control law design. Project document FC 9007, National Aerospace Laboratories, May 1990.
15. Rauw, M. O. FDC 1.2 – A SIMULINK tool box for flight dynamics and control analysis. Emmaplein 74, 3701, DC Zeist, The Netherlands, 1998.
16. Duke, E. L., Antoniewicz, R. F., and Krambeer, K. D. Derivation and definition of a linear model aircraft model. NASA RP-1207, 1988.
17. Duke, E. L., Antoniewicz, R. F., and Patterson, B. P. User's manual for interactive LINEAR, a FORTRAN program to derive linear aircraft models. NASA TP-2835, 1988.
18. Du Val, R. W., Bruhis, O., Harrison, J. M., and Harding J. W. Flight simulation model validation procedure, a systematic approach. *Vertica*, 13(3), 311–326, 1989.
19. Neville, K. W., and Stephens, A. T. Flight update of aerodynamic math model. AIAA-93-3596-CP, 1993.
20. Pashilkar, A. A., Kamali, C., and Raol, J. R. Direct estimation of nonlinear aerodynamic coefficients. *AIAA Atmospheric Flight Mechanics Conference and Exhibit*, South Carolina, USA, AIAA-2007-6720, 20–23 August, 2007.
21. Kendall, W. N., and Stephens, A. T. Flight update of aerodynamic mathematical model. *AIAA Flight Simulation and Technologies Conference*, Paper No. 3596-CP, Monterey, CA, 1993.
22. Trankle, T. L., and Bachner, S. D. Identification of a nonlinear model of F-14 aircraft. *Journal of Guidance, Control, and Dynamics*, 18(6), 1292–1297, 1995.
23. Raol, J. R., Girija, G., and Singh, J. *Modeling and Parameter Estimation of Dynamic Systems*. IEE Control Series, Vol. 65, IEE, London, UK, 2004.
24. Madhuranath, P. Wind simulation and its integration into the ATTAS Simulator. DFVLR, IB 111-86/21, 1986.
25. Madhuranath, P., and Khare, A. CLASS - closed loop aircraft flight simulation software. NAL PD FC 9207, NAL, Bangalore, October 1992.
26. Kamali, C., Hebbar, A., Vijeesh, T., and Moulidharan, S. Real-time desktop flying qualities evaluation simulator. *Defence Science Journal*, 64(1), 27–32, 2014.
27. Kamali, C. and Jain, S. Hardware in the loop simulation for a mini UAV. *IFAC-PapersOnLine*, 49(1), 700–705, 2016.
28. Sarathy, S., and Murthy, V. R. An advanced parallel rotorcraft flight simulation model: Stability characteristics and handling qualities. *Atmospheric Flight Mechanics Conference*, Monterey, CA, AIAA-93-3618-CP, August 9–11, 1993.
29. Krishnakumar, K., Prasanth, R. K., and Bailey, J. E. Robust flight reconstruction for helicopter simulation and training. *Journal of Aircraft*, 29(3), 421–428, 1992.
30. Rohlf, D. Direct update of a global simulation model with increments via system identification. In *RTO meeting Proceedings 11, System Identification for Integrated Aircraft Development and Flight Testing, The RTO Systems Concepts and Integration Panel (SCI) Symposium*, Madrid, Spain, 5–7 May 1998.
31. Paris, A. C., and Bonner M. Nonlinear model development from flight test data for F/A-18E super hornet. *Journal of Aircraft*, 14(4), 692–702, 2004.
32. Balakrishna, S. Final report on setting up of a three degree of freedom motion based research simulator at NAL. Ref: Aero/Rd-134/100/10/xxxii, NAL, Bangalore, India, 1972–1976.
33. van der Linden, C. A. A. M. *DASMAT-Delft University Aircraft Simulator Model and Analysis Tool*. Delft University Press, The Netherlands, 1998.

7 Flight Test Maneuvers and Database Management

7.1 INTRODUCTION

Flight testing of an aircraft is required for several reasons: (i) for testing of aircraft subsystems in the defined flight test envelope, (ii) for validation of flight control laws (for fly-by-wire inherently unstable/augmented aircraft), autopilot performance, and to prove core and new technologies, (iii) to ascertain the ability of the aircraft to perform conventional and mission-specific maneuvers, (iv) to establish the operating envelope, initial operational certification, and gradual expansion of the flight envelope based on the flight test results of the previous flights, (v) to demonstrate the compliance with Civil Aviation/Military rules/specifications as per their requirements and/or advice, (vi) to generate the data for future work in design and development of a new aircraft or modification of the tested configurations for future upgradation, (vii) to evaluate the ground effects on the low-level flying, (viii) to estimate aerodynamic derivatives for validation of the (predicted/wind tunnel) aero-database used at the design stage of the aircraft, and (ix) to evaluate pilot–aircraft interactions/HQs [1–6]. Flight tests are carried out at various stages of the aircraft modification and for confirmation of the same. The generation and establishment of (updated) aerodynamic database via flight tests is a well-recognized method in an aircraft development program. A reliable and accurate estimation of aerodynamic derivatives from flight test data requires that certain modes of the aircraft are excited properly [7]. For example, it will not be possible to obtain accurate estimates of C_{m_α} and C_{m_q} if the longitudinal SP mode is not sufficiently excited, and hence, the choice of an input form and shape of the signal is very important aspect. From the foregoing, it is very clear that an integrated engineering-system approach is also important for successful accomplishment of the flight tests.

For a flying aircraft, controlling and monitoring its speed and altitude is very important, as one can realize that aircraft controls, input mechanisms, and connected control surfaces are not designed and/or mechanized inherently to do this task quite independently. To perform a certain flight maneuver, most often and perhaps always a complex sequence of movements of a few, if not all, controls (surfaces) is required to be carried out. The concept of aircraft energy and its management is the most important aspect of understanding the requirement of flying an aircraft at a certain altitude and speed. The energy cannot be created (the law of conservation energy), but it can be converted from one form to another. The aircraft energy as such can be put in the form of Quad-E [8]: (i) kinematic energy acquired as a result of flight

DOI: 10.1201/9781003293514-8

speed, (ii) potential energy acquired as a result of the aircraft being at a certain altitude, (iii) fuel's chemical energy, and (iv) thermal energy due to the exhaust gases, where applicable. An appropriate interplay of these energies can be advantageously used to understand flight maneuvers. The airspeed and altitude can be exchanged as a change in mechanical energy. Some important exchanges of energy are as follows [8]: (i) in a climb maneuver, the fuel is used (burnt) to meet the drag force and attain the altitude; (ii) in cruise, there is no maneuver, and hence, the fuel is consumed only to meet the drag resistance, and there is not much change in airspeed and altitude; (iii) in a sudden pull up, the aircraft ascends to a new altitude and the airspeed decreases; (iv) in a sudden pushover maneuver, the altitude is sacrificed and the airspeed is gained, and not much fuel is spent; (v) in an initial portion of the takeoff roll the altitude is not changed but the airspeed is gained; fuel is spent, (vi) in gliding the altitude is sacrificed to meet the drag force; and (vii) in flare, the airspeed is sacrificed to meet the drag force, without much change in the altitude and engine power. It is emphasized here that the altitude and airspeed are very closely interrelated. Reference [8] gives very interesting conversion factors, as a rule of thumb: (i) for a loss of 1 knot speed, one can gain an altitude of 9 ft, and (ii) for a one-ton aircraft, the climb by 6300 ft would spend 1 gallon (~ 4 L) more fuel. The energy exchange guidelines are as follows: (i) if at low altitude and low speed, then there is not much energy, and hence, more power is required to be added to gain the altitude; (ii) if at low altitude and with high speed, the energy might be sufficient, and moderate stick pull might is required; and (iii) *f* at high altitude and with high speed, there is too much energy, and it might be required to reduce the power.

7.2 PLANNING OF FLIGHT TEST MANEUVERS

An aircraft should generally fly in the specified flight envelop defined by the altitude and Mach number (and for any other required configuration parameters). A typical flight envelope is shown in Figure 7.1 [9]. The low speed limit 'a' is stipulated by the maximum lift that can be generated by the aircraft wings (the stall limit). The high speed limit is governed by a constant dynamic pressure contour. One can see that as the speed is reduced from point 'd' to point 'e', the air density is increased, due to the reduction in altitude. This would maintain the dynamic pressure nearly constant. At higher altitudes, the aircraft speed is governed by available maximum thrust. The service ceiling occurs due to the fact that engine cannot produce a certain required rate of climb. The performance limit is due to the low density of air, and this prevents the jet engines and the environmental control systems from sustaining their operations. The temperature limit is due to the kinetic heating of the airframe because of the friction of the air. The loading limit is due to high dynamic pressure to provide a safe margin against the excessive aerodynamic loads.

Flight test planning consists of the following: (i) specification of flight conditions, e.g., flight test points in the envelope, (ii) aircraft configuration, e.g., mass, C.G. (forward, aft), use/deployment of flaps/slats, the undercarriage, i.e., landing gear, is up or down, and carrying of any stores, (iii) axis-specific control inputs, e.g., elevator, aileron, rudder pedal, and (iv) engine conditions, e.g., single or both engines on. It is also important to specify what instruments/sensors are required to monitor/display/

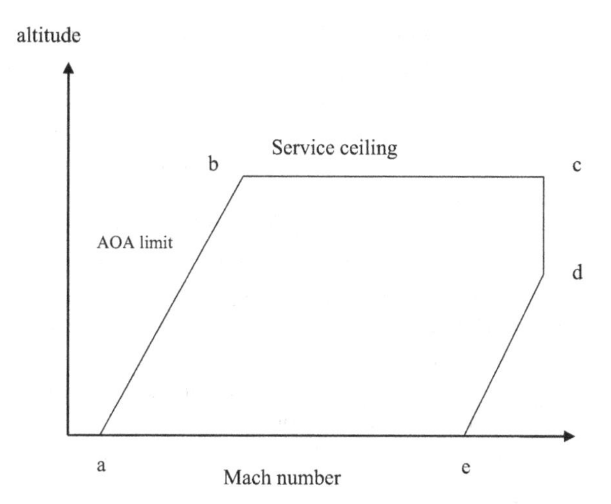

FIGURE 7.1 A typical flight envelope.

collect the flight responses/data that will be used for analysis. The sensor/instrument package should be at the C.G. as far as possible. If the package is not at the C.G., then the offset distance should be known and should be taken into account for correction of the data. The standard sensor/instrument packages should be used for high accuracy and reliability of the data to be gathered. Sampling requirements must be specified well in advance. Flight test data-processing skills are very important and require knowledge of signal processing and estimation methods. A complete flight test matrix consisting of these aspects with the full list of the requirements should be prepared and discussed with the concerned test pilots/flight test engineers. Flight tests exercises are generally carried out in several phases, and often several aircraft versions (called technology demonstrators) are planned to concurrently undergo flight tests. This shortens the overall flight test time to go for certification. However, enough testing should be done in order to get the full confidence in the performance of the aircraft. This would need a gross time of hundreds of hours (that may run from 400h onward depending on the type of the aircraft being tested) of flight testing to scan all the important subsystems (propulsion, landing gears, flaps, airbrakes, electrical systems, cockpit instruments, actuators, movement of control surfaces, structural vibrations, head-up displays, effect of movement of C.G., fuel transfer from one tank to other, etc.) of the aircraft and many configurations (stores/missiles/tanks, etc.). Thus, the flight testing is an iterative exercise and requires a coordinated effort among the test pilots, flight test engineers, design engineers, and subgroups of data analysts. Other important factors to be kept in mind are as follows [1–3,6]: (i) for realizing a simple maneuver, too much time should not be spent (in air), and (ii) the C.G. shift, weight, and flight envelope should be considered at the outset. Properly planned/conducted test experiments save lot of effort, fuel, and time for the flight test engineers, pilots, managers, and the analysts. For the sake of consistency of the results, it is desirable to perform at least 3 repeat runs (maneuvers) at each flight condition. The flight testing of the aircraft and related systems should cover the following areas: (i) air data calibration (very crucial task for feedback control variables), and accurate values of AOA and AOSS are very useful in parameter estimation algorithms; (ii) aircraft loads; (iii) excitation and monitoring of flutter, during the flight tests for system identification experiments, the flutter should not be excited; (iv) high AOA tests and related data generation maneuvers should be done very carefully since safety is involved, and aircraft can enter the stall; (v) preliminary HQ evaluation, and open/closed loops HQ, here many types of tests are involved (see Chapter 10); (vi) stability and control characteristics evaluation; (vi) testing of aircraft/subsystem failure modes/states, e.g., testing of the aircraft when one engine is intentionally failed/switched off; the failure of the starboard engine would cause +ve roll, a transient dip in the rate of yaw tending to +ve value and +ve SS; (vii) engine and drag performance tests; (viii) critical stores testing; and (ix) carefree handling (if applicable), this is useful for the (fully) automatic aircraft.

7.2.1 Flight Test Evaluation of a Transport Aircraft

Guidelines for the flight test evaluation of transport airplanes [2,3] provide an acceptable means of demonstrating confirmation with the applicable airworthiness requirements. Many of these procedures are also applicable to fighters. These methods have evolved over several years of flight testing of several such airplanes in air as well on ground flight simulators and careful analysis of the data gathered. The guidelines are not mandatory; however, it would be definitely advisable to follow them to help compliance with the airworthiness (authorities') requirements for a given transport aircraft undergoing flight testing. Some important aspects of flight testing from aircraft certification point of view are as follows [2,3,8]: (i) effect, if any, of altitude on controllability, stability, trim, and stall characteristics should be evaluated and studied; (ii) performance of the aircraft in terms of stalling speed, takeoff speed, takeoff path, takeoff distance, climb, and landing should be evaluated during flight tests; (iii) controllability, static stability, and maneuverability should be evaluated; (iv) longitudinal and lateral-directional dynamic stability should be demonstrated and evaluated; (v) stall characteristics should be demonstrated/evaluated and stall warning system/software should be evaluated; (vi) critical engine performance should be evaluated; and (vii) vibration, buffeting, and other high speed characteristics should also be evaluated.

7.2.2 Takeoff and Landing Tasks

These tasks are always very critical for any aircraft. Since the aircraft would be carrying some loads, cargo, passengers, etc., for the takeoff task, the required high speed must be reached before the wings can generate the sufficient lift to balance the weight, and hence, it has to travel a long distance on the runway [6,10]. The landing may be required at any time. The aircraft may have to land on a short runway.

7.2.2.1 Approach and Landing Task

The landing-related portion consists of approach, flare, and then landing. Normally, the approach altitude is 16 m (50 ft). This is a very crucial task, for which first the

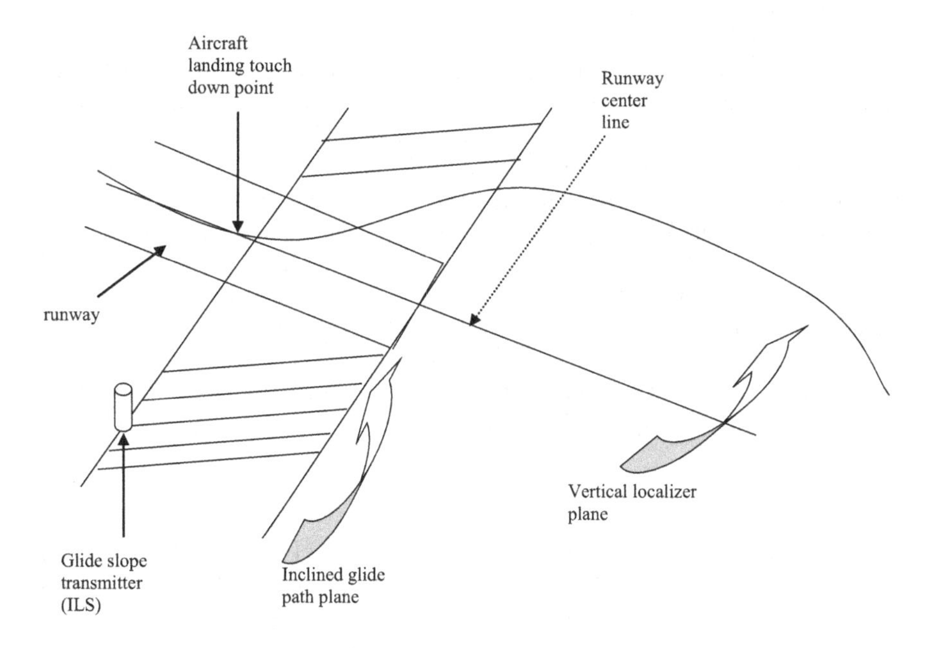

FIGURE 7.2 Landing flight path for an aircraft [6].

aircraft is maneuvered into a certain vertical plane called the 'localizer' plane that is generated electronically by the ILS (Instrument Landing System) [6]. While in a level flight, once intercepted, the nominal path intersects with a second ILS plane called the 'glide path' plane, the base line of which is perpendicular to the runway center line as shown in Figure 7.2. The intersection of the two ILS planes provides the 'glide path' or the 'glide slope'. The pilot concentrates her attention on the primary flight instruments and the ILS deviation indicator during these three sequentially interconnected flight phases: (i) the first interception, (ii) tracking of the localizer, and (iii) getting into the glide path. The pilot decelerates the aircraft, gets into the landing configuration (while lowering the undercarriage), and at the same time, counteracts the atmospheric disturbances and any other deviations. While switching to the outside visual scene, the pilot initiates the landing by gradually rotating the aircraft. The velocity vector virtually changes into horizontal direction. The aircraft is flown with the main (rare) wheels against the runway surface, and a soft and positive touch down is made. The nose wheel is then lowered to make ground contact, and the aircraft is decelerated to the taxi speed. The ground portion of the landing consists of aero/wheel braking and stop. The pilot should be able minimize the distance required to land. A few crucial conditions of the landing are as follows: (i) testing the landing for various weights, (ii) height/elevation, and (iii) the brakes are most important ingredients and must be evaluated for their precise performance. The data/information generally required are as follows: airspeed, pressure altitude, distance along the runway; engine rpm, temperatures/pressure measurements; and wind-related measurements.

7.2.2.2 Take Off Task

The takeoff is also very critical task: (i) the ground acceleration portion and (ii) take-off/climb portion. The pilot uses a suitable method for the first segment, based on the flap settings, control settings, and steering methods. She will accelerate the aircraft sufficiently to obtain sufficient speed and then apply the stick control to obtain the best attitude for takeoff. When the aircraft is airborne, the speed will be controlled to obtain the best angle for the climb. The landing gear would be retracted to reduce the drag effects. For trial fights, the landing gears are kept in down position. The rotation maneuver should be smooth, and the pilot should acquire and maintain the speed for best angle of climb. The takeoff and climbing tasks should be carried out for a few possible climb speeds and deployment of flaps. The pilot should minimize the distance on the runway required to take off. The required data/information are almost similar to that required for landing.

7.2.3 Other Maneuvers

The longitudinal maneuvering implies changing attitude of the aircraft and change of the direction (flight path angle) and magnitude of the velocity vector (airspeed). The pitch attitude variation is achieved by deflecting the (aerodynamic) control surface on the horizontal tail of the aircraft. The magnitude change of the velocity is affected by operating the thrust level of the engines and small change in pitching motion. The direction of the velocity vector is changed by variation of the lift vector though attitude variation. The lateral maneuvering is affected by rolling the aircraft about its longitudinal axis by deflecting ailerons or elevons as the case may be (e.g., for a delta wing aircraft, the elevators and the ailerons are on the rare ends of the wing and are called elevons). The differential movements of end-elevons give rise to rolling moment about the pitch axis of the aircraft.

Aircraft's maneuverability is defined as the ability of an aircraft to change the speed and flight (path) direction. A highly maneuverable aircraft can accelerate or slowdown very quickly. However, a quick maneuver with a short radius would induce high loads on the wings, these being referred as 'g' force. In equilibrium condition, the thrust is equal to the drag force, so that the aircraft is stabilized in the level flight with a constant speed. As per Newton's laws, the aircraft continues to fly with constant speed, unless there is some additional force of disturbance. Now, if the throttle is advanced to obtain full power from the engine, the thrust exceeds the drag force, and this excess force accelerates the aircraft because force = mass × acceleration. The aircraft will speed up due to accelerating force and stabilize when the excess thrust becomes zero, and the aircraft reaches its maximum speed. The importance of the excess thrust is that it can be used either to accelerate the aircraft to a higher speed or to enter a climb at a constant speed or some combination of both. Now, with an increase in speed, if the pilot does not want the aircraft to climb, then she has to lower the elevator (so the inclined/upward force acting on the elevator generates a nose-down moment of the aircraft) by pushing the stick forward.

7.3 SPECIFIC FLIGHT TEST DATA GENERATION AND ANALYSIS ASPECTS

The flight test data have the following roles to play [1–3,6]: (i) to substantiate that the aircraft meets certain specific stability and control parameters; (ii) these data can be correlated with the WT or analytical data, some adjustments based on the flight test data can be made, and these adjusted WT or analytical data can be used to extrapolate the flight data to other regions of the flight envelope where the flight tests are not performed due to limitation of safety and cost; (iii) the data can be used to adjust and substantiate the mathematical models used in the flight simulators, for which the time histories and stability and control derivatives are used for comparison/validation/upgradation of the math models; and (iv) the data can be used for aiding the design of a future aircraft. The flight test data generation and analysis aspects are highlighted next [2,3,8].

7.3.1 LONGITUDINAL AXIS DATA GENERATION

Stabilizer angle to trim the aircraft exercise is carried out to establish trim characteristics, to obtain the effect of configuration or thrust changes on stabilizer trim, and to determine stabilizer effectiveness. The aircraft is trimmed using the stabilizer with zero elevator deflection. The trimmed state is attained when the aircraft can fly hands-off in a steady unaccelerated flight mode. The stabilizer angle required to trim is plotted as a function of lift coefficient for low speeds and Mach number for high speeds for a range of C.G. configurations and thrust settings. Various desired effects are then discerned from these plots.

For a range of flight conditions and configurations, the aircraft is mistrimmed with the stabilizer and retrimmed with the elevator keeping the thrust constant and maintaining the unaccelerated flight. The elevator effectiveness can be obtained by plotting the elevator–stabilizer curves as a function of C.G. In conducting a maneuvering flight, the purpose is: (i) to evaluate the handling characteristics like in pull-ups and windup turns, (ii) to determine pitching moment/lift characteristics, and (iii) to determine elevator hinge moments.

The static (speed) stability test is required to determine aircraft's static stability, and to determine neutral point of the aircraft. The aircraft is trimmed at one speed, and the elevator deflection and stick force required to fly at another speed in a straight flight are measured. Keeping thrust constant, the aircraft is flown at all the speeds.

It is important to investigate controllability and handling characteristics of the aircraft and basic lift and pitching moment characteristics in the high angle of attack range.

The primary interest in the stall exercise is the ability to recover from a stalled condition. Some characteristics related to stall conditions are as follows: (i) It occurs at the critical AOA; (ii) at/after this AOA, the lift coefficient does not increase with an increase in AOA; (iii) this special point on the lift curve also corresponds to a point on the power curve; (iv) if the lift coefficient is small, the aircraft has to fly at a higher speed to support the weight; (v) the lift coefficient has a maximum and the airspeed a minimum – this minimum airspeed is the stalling speed; (vi) normally, it is not possible to sustain the flight below the stalling speed; and (vii) the drag coefficient is high in the stalled regime, and hence, it needs lot of power to maintain the level flight, but with the constant power, the rate of climb decreases. Mainly, there would be an uncommanded and easily recognizable nose-down pitch, and a roll motion that cannot be readily arrested with normal use of the controls. One can recover from the stall by reducing the AOA using normal control by the elevator. The stall maneuvers are flown with varying entry rates. The effect of thrust level is also studied. The $1g$ stall speed is the speed below which $1g$ level flight is impossible. The minimum speed is the lowest measured speed during the stall maneuver. The stall entry rate is defined as the slope of the line between $1.1 \times$minimum speed and minimum speed. During the closing phase of the stall maneuver, the control may be moved forward to maintain a constant deceleration rate and to reduce the large pitch angle changes.

It is well known that ground effect that results from the change of flow field in the presence of the ground can significantly affect the takeoff and landing characteristics of the aircraft. The effect of the ground proximity on the stability and control of the aircraft should be evaluated since it has operational implications and also it is very important for validation of the flight simulators [4,5]. This is because a large part of the pilot training in the simulator is spent on the approach and landing tasks. The simulator models should be checked properly in these flight phases. The aircraft is flown along the runway at progressively lower heights for a range speeds. The height is measured with radar altimeter. The ILS is used for an approach. The autopilot is disengaged prior to entering the ground effect, and the aircraft behavior is observed in the ground effect with hands-off the controls. The aircraft is trimmed for the approach, and when it enters the ground effect, the body angle is kept constant using the elevator control until the touchdown. The data from the ground effect and the reference data from the out-of-ground effect are compared, and the increment in the elevator deflection is found out. Then, the change in the pitching moment is computed due to the ground effect. This change is due to a change in the AOA, a change in the $C_{m\alpha}$, and a change in thrust in order to keep speed constant. Based on this analysis, the update of the simulator models if needed can be made.

The tests for establishing the capability of the elevator to lift the nose wheel off the ground during takeoff phase of the flight are also made. With the forward C.G., the stabilizer is set at the recommended position, and the elevator is fully pulled up some time before the rotation is expected. The rotation speed is determined when the body angle starts increasing, i.e., the pitch rate is positive. The lift-off point is determined when the main gear lifts off the ground.

The dynamic characteristics of the aircraft are assessed by performing the SP and Phugoid maneuvers and generating the data by giving doublet or 3-2-1-1 input commands to pitch stick, as discussed in Section 7.6, Figure 7.3. The SP mode is critical since its period can approach the pilot's reaction time. To determine the stability and control derivatives of the aircraft, the data from these flight test maneuvers are processed in parameter estimation software.

7.3.2 Lateral-Directional Data Generation

The purpose is (i) to establish steady sideslip capability, (ii) to determine the cross-wind landing capability, (iii) to determine the rudder and lateral control hinge

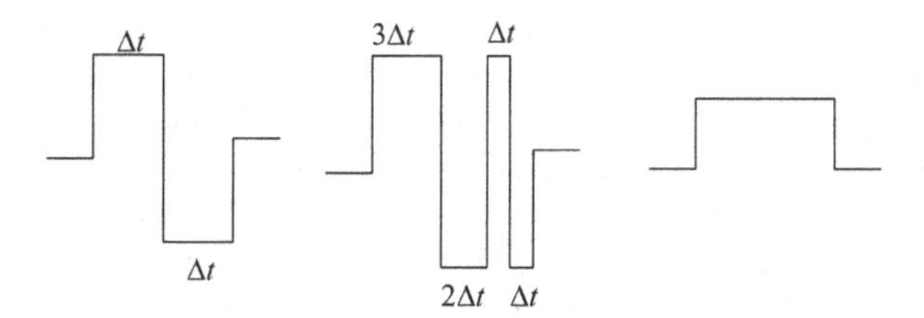

FIGURE 7.3 Specific input types: doublet, 3-2-1-1, and pulse Δt is the unit step width in seconds.

moment, (iv) to determine the rudder effectiveness ($C_{n_{\delta_r}}$), and (v) to determine the directional stability ($C_{n\beta}$). The L-D testing involves two or three axes testing. In a level flight with symmetric thrust, the rudder deflection is given with lateral control such that the flight path is straight, no change in heading, with sideslip and bank angle, for various rudder angles. The maximum values of the sideslip angles are obtained by performing tests at gross weights and aft C.G. The data are used to obtain the relationships between control deflection and various angles: bank angle, sideslip angle, and angle of attack. Maximum values of the rudder deflection, the sideslip angles, wheel deflections, and bank angles are plotted as a function of Mach number and altitude, and the maximum values of the sideslip angles are obtained at heavy weight and aft C.G. The yawing moment produced by the rudder can be computed by knowing the yawing moment produced by the asymmetric thrust and using the rudder to reduce the sideslip angle to zero.

Various dynamic characteristics need also to be established: (i) the DR, (ii) spiral stability, and (iii) lateral/directional rate derivatives. These characteristics are established by generating the flight test data by using doublet/3-2-1-1 inputs and performing the appropriate maneuvers as discussed in Section 7.6. Then, parameter estimation methods are used to determine stability and control derivatives of the aircraft.

7.4 QUALITY OF FLIGHT TEST MANEUVERS

It is very important to keep track of the quality of the flight test maneuvers, because the quality of the estimates obtained and any deductions made from using these in analysis programs depend on the quality of maneuvers [1]. The aim should be to maximize the information content in the data. This requires good planning and experiment design. If the maneuvers are conducted to generate the data for parameter estimation, then specific maneuvers are required to be planned. It may so happen that the maneuvers conducted for other purposes generally do not provide reliable data for parameter estimation exercises. For the latter purpose, very careful consideration of many factors is essential: control inputs, signal magnitudes, sampling rates, and SNR. If the information content in the aircraft responses is not adequate, then these data will not provide good estimates of the aerodynamic derivatives. The SNR should be at least 10 for the most important signals required for the analysis. Also, such data with low SNR are not good to model the plant in general. It is often felt that in order to obtain good estimates of the control effectiveness derivatives, application of one control input at a time would be a good strategy. Other control inputs can be applied in some sequence. Question of sufficient excitation of the modes is of a great importance. Since we are interested in studying the dynamic modes of the vehicle, it is important that the modes are excited properly, and their effects become apparent in the time responses. If the modes are not adequately excited and captured in the response data, then system identification will be ill-conditioned. One should use small magnitude for inputs so that the assumption of linearity is maintained for subsequent analysis. This is mainly due to the fact that aerodynamic derivatives are defined on the assumption of small perturbation theory. The SNR consideration is important for the small-amplitude maneuvers. Typical maneuvers amplitudes for a fight class of aircraft would be (peak): AOA, alpha = 2 deg., AOSS, beta = 1 deg., p, roll rate = 30 deg./s, q, pitch rate = 10 deg./s, r, yaw rate = 5 deg./s, n_z, normal acceleration = 0.25 g and a_y, lateral acceleration = 0.1 g.

The large-amplitude maneuvers will excite nonlinear phenomena/dynamics; however, these might be required for the specific reasons as will be discussed later on. Often and almost always the repeated maneuvers are needed to obtain consistent estimates of the derivatives. If a few maneuvers are bad, then they can be neglected and the overall flight test time is reduced. This would save fuel and test efforts. To the extent, the maneuvers should be conducted in calm atmospheric conditions, unless the purpose is to specifically test the aircraft responses in a turbulent weather.

The command inputs types, if they are simple, then can be applied manually very easily. If not, then the pilot should do a sufficient practice on a real-time flight simulator or on a computer connected to the joy stick. Some input types can be programmed or computerized and applied through a flight test panel. The issue of generating the data for parameter estimation from closed loop responses of an inherently unstable/augmented aircraft is of paramount importance. The generated responses can be often checked on the site, i.e., at the flight test center itself. This will ensure the adequacy of the data for subsequent offline/batch-processing analysis. One should also ensure that the data are adequate and of sufficient length (in time), so that they are suitable for consistency analysis.

7.5 INPUT SIGNALS FOR EXCITING MANEUVERS

Specific command inputs are required to conduct certain maneuvers, then, I/O data from which are used for parameter estimation, and/or drag polar determination.

7.5.1 Design Consideration for Input Signals

The responses of an aircraft to pilot's (or computer's) command signal are the flight data that along with other mandatory data (m, I, dynamic pressure, etc.) are used in a parameter estimation procedure to estimate the aerodynamic derivatives. These responses are generated by exciting certain modes of the aircraft. These dynamic modes contain information on the aerodynamic derivatives implicitly. Thus, it is very essential to choose input test signals with good characteristics so that properly excited dynamic responses are available for further analysis. The test input signal must possess certain desirable features [1,7]: (i) It should have sufficient bandwidth to excite the modes of interest, (ii) the amplitude and form of the signal should be such as to yield good-quality maneuvers, (iii) the amplitude, bandwidth, and slope of the signal should be bounded so that aircraft motion does not violate the assumption of linearity, and (iv) it should be possible to realize (or generate) the signal easily by the pilot. Of course for difficult signals, computer-generated input signals can be used. Different types of input signals that have certain desirable attributes are as follows [7]: (i) a doublet signal excites a band at a higher frequency, the response is not unidirectional, and it is easy to generate by a pilot; (ii) a 3-2-1-1 signal has relatively a broadband spectrum, can effectively excite over a decade of frequency, and it is also easy to generate by a pilot with some practice; (iii) the Schulz input signal is designed by maximizing the trace of Fisher information matrix (FIM), relatively higher frequencies are missing in it, might give relatively large standard deviations compared to next two signals, and are not so easy to generate by a pilot; (iv) the Delft University of Technology (DUT) input signal is designed by minimizing the trace of estimation error covariance matrix P, is a reasonably good signal, and is not so easy to generate by a pilot; and (v) the Mehra input signal is designed by maximizing the determinant of FIM in frequency domain, is a reasonably good signal, and is not so easy to generate by a pilot. For a certain given case study, the 3-2-1-1, DUT and Mehra input signals were found to be of similar efficiency [7]. Many of these signals can only easily be generated by computer or flight test panel. The subject matter of design of the test signals for parameter estimation is relatively complex and requires considerable mathematical sophistication. Hence, only some important aspects of the design procedures are highlighted here. In order to design a multi-step input signal which will excite the modes of an aircraft, one can use Bode diagram approach [7]. This allows one to select the frequencies to be included in the signal based on the identifiability of the derivatives. Let the longitudinal motion be described as follows:

$$\dot{u} + g\theta - X_u u - X_\alpha \cdot \alpha = 0; \dot{\alpha} - q - Z_u u - Z_\alpha \cdot \alpha - Z_{\delta e}\delta_e = 0$$

$$\dot{q} - M_u u - M_\alpha \alpha - M_{\dot{q}} q - M_{\delta e}\delta e = 0$$

The frequency response magnitudes of the terms in the above equations as a function of the input signal frequency could be plotted. Based on this analysis, a derivative is considered identifiable when its term has a magnitude of at least 10% of the largest term's magnitude. This approach is intuitively appealing. There are other approaches based on the following: (i) minimization of the estimation error variance of the parameters by means of an appropriate choice of the input signal characteristics subject to energy constraints, (ii) maximization of trace of the FIM, (iii) minimization of the trace of estimation error covariance matrix P, and (iv) the determinant of FIM.

The required input signal should be generated either manually, by pilot or by some automated technique. In order to obtain accurate signal characteristics, one can use electronics or computer means thereby it may also be possible to automate the generation of the signal. A pregenerated/prerecorded signal can be replayed during the test schedule. While using automatic means for generating and applying the input signal, care should be taken to assure that, in case of any exigency, pilot is able to take over the control of the vehicle. There should be a mechanism to shutoff the automatic application very quickly if so desired and revert back to the manual mode of control.

7.5.2 Specific Input Types

Some specific input signals are shown in Figure 7.3. A doublet stick/rudder pedal input signal excites a band at a relatively higher frequency. It is used to excite longitudinal/lateral-directional short-period modes. It is applied to the aircraft control surface through pilot's pitch-stick, roll-stick, and rudder pedals. At the end of the input, the controls are held constant for some time to permit the natural response of the aircraft to be recorded. When applied to ailerons, it excites the rolling motion that can be analyzed to obtain derivatives for roll damping and aileron control effectiveness. Similar test signal can be used for rudder surface to determine yaw derivatives and rudder effectiveness. If the natural frequency ω_n of the mode to be excited is approximately known, then the duration of the time unit Δt for a doublet can be determined as $\Delta t \cong 1.5/\omega_n$, s.

A 3-2-1-1 type is a series of alternating step inputs with the time duration of the steps satisfying the ratio 3:2:1:1. This input signal has power spread over a wider frequency band compared to the doublet power spectrum. The aircraft SP longitudinal motion can be produced by applying 3-2-1-1 input to the elevator. The time unit Δt needs to be selected appropriately to generate sufficient excitation in the aircraft modes of motion.

A step input signal has energy at relatively lower band of frequencies and is not very suitable for system identification purposes. A longer duration pulse (of 10–15 s) can be applied to the pitch stick/elevator to excite the longitudinal Phugoid motion of the aircraft. The responses should be recorded for sufficient number of cycles before retrimming. From these responses, one can estimate speed-related derivatives and the Phugoid damping and frequency. The doublet inputs tend to excite only a narrow band of frequencies. The pulse inputs have power at low frequencies and therefore are suitable for exciting only low-frequency modes of the system. A combination of various input forms is generally considered the best for proper excitation of the aircraft modes.

7.6 SPECIFIC MANEUVERS FOR AERODYNAMIC MODELING

In this section, many specific maneuvers are described for generating the flight test data that can be further processed in the parameter estimation methods/software for determination of the aerodynamic derivatives of the test aircraft.

7.6.1 Small-Amplitude Maneuvers

The amplitude of the maneuvers is kept small so that the assumption of linearity is not violated and hence simple linear models can be used for parameter estimation.

7.6.1.1 Longitudinal Short-period Maneuver

The SP oscillation is the first one that the test pilot sees/feels after disturbing the airplane from its trim with the pitch control, and the oscillation is damped within about 2 cycles after the completion of initial input. The SP oscillation must be heavily damped both with controls fixed and free. Typical damping ratio would be 0.3–0.9, and the period is 1–4 s. The test for longitudinal dynamic stability is performed by a rapid movement of pulse of the control in a nose-up and nose-down direction at a rate and degree to obtain a short-period response. Dynamic longitudinal stability must be checked at a sufficient number of points in each flight test configuration. In a horizontal-level trimmed flight with constant thrust, a doublet or 3-2-1-1 multistep command is applied to the pitch stick. One usually tries to avoid variations in the lateral-directional axes. Small corrections in the aileron and rudder positions are allowed if necessary. The time width of the input signal is appropriately selected to excite the SP mode of the aircraft. The aircraft is trimmed at the required altitude and speed. Then, the pitch stick is moved in the manner of the specified input signal (either doublet or 3-2-1-1). The mode is expected to be of 2–3 s duration, and the data are recorded for about double the duration to cover the exponentially damped sinusoidal response. One can determine the longitudinal derivatives and also the neutral point of the aircraft from these flight test data. Subsequently, the frequency and damping ratio of the SP mode can be determined using MATLAB program. Typical SP maneuver flight-time histories are shown in Figure 7.4.

7.6.1.2 Phugoid Maneuver

The mode being usually of 40–60 s duration (it could be even up to 120 s), longer data recording is required. A longer duration (10–15 s) pulse command input is applied to the pitch stick keeping the thrust constant. The aircraft is allowed to go through a minimum of one/two complete cycles (for about 2–3 min) of the Phugoid oscillations before retrimming. One can determine the longitudinal Phugoid-related (speed) derivatives of the aircraft from these flight test data. Subsequently, the frequency and damping ratio of this long period mode can be determined using MATLAB program. The Phugoid damping ratio would be 0.03–0.1, and typical maneuver flight-time histories are shown in Figure 7.5.

7.6.1.3 Thrust Input Maneuver

The maneuver is used to determine the effect of a thrust variation on the aircraft motion. Starting from trimmed-level flight, a doublet variation in thrust is applied, and the flight data are recorded.

7.6.1.4 Flaps Input Maneuver

This maneuver can be used to gather information for estimation of flaps effectiveness derivatives. Data are generated by applying a doublet or 3-2-1-1 input to flaps. Other longitudinal controls and thrust are kept constant. Variations in the lateral-directional motion are kept small.

7.6.1.5 Lateral-Directional Maneuvers

Positive static directional stability is the tendency to recover from a skid with the rudder free. Positive static lateral stability is defined as the tendency to raise the low wing in a sideslip with the aileron controls free. Negative static lateral stability is not permitted unless the divergence is gradual and is easily recognizable and controllable by the pilot. For dynamic lateral-directional stability, the oscillation must be positively damped with controls free. It must be controllable with normal skills of use of the primary controls.

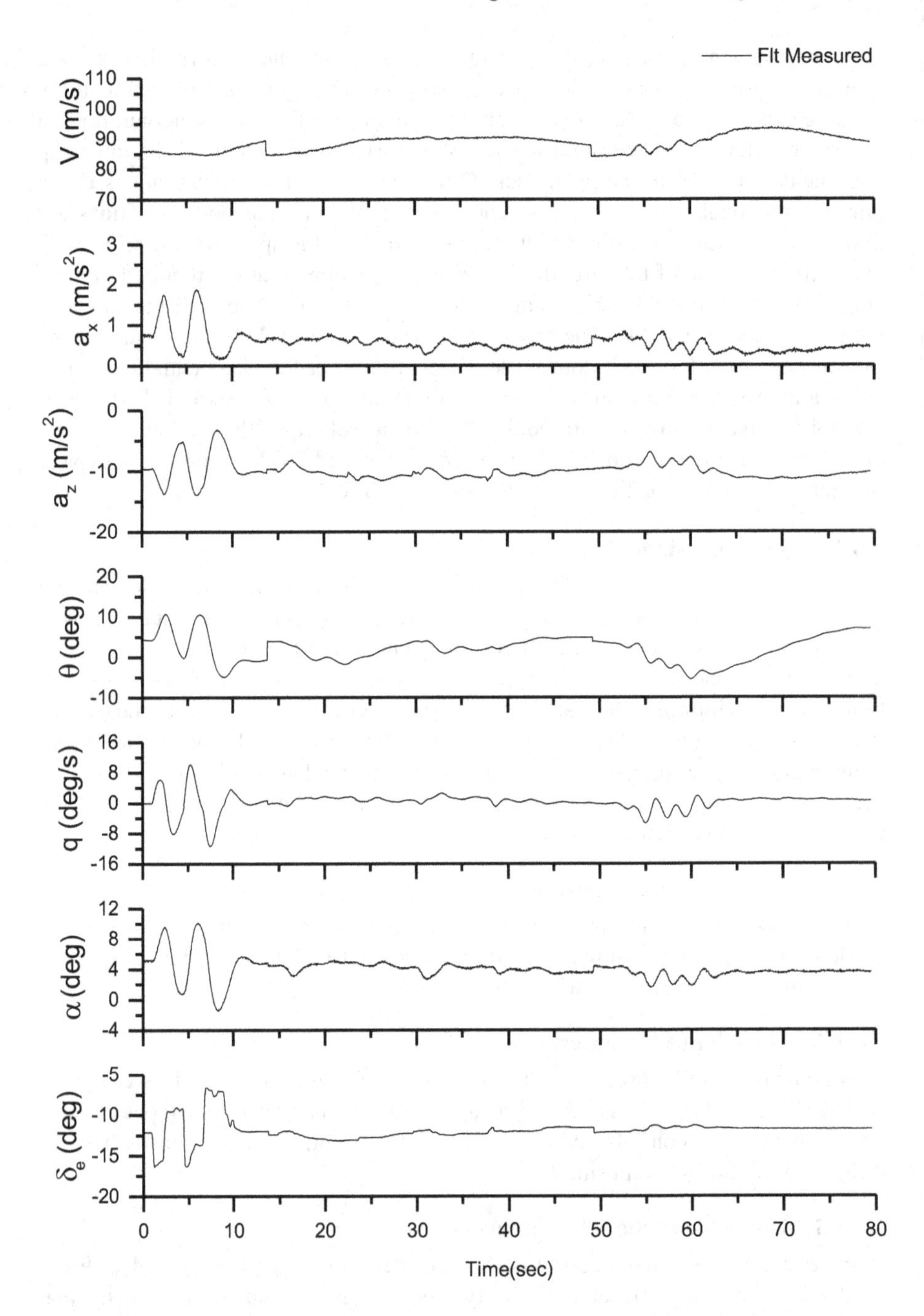

FIGURE 7.4 Typical short-period response in flight.

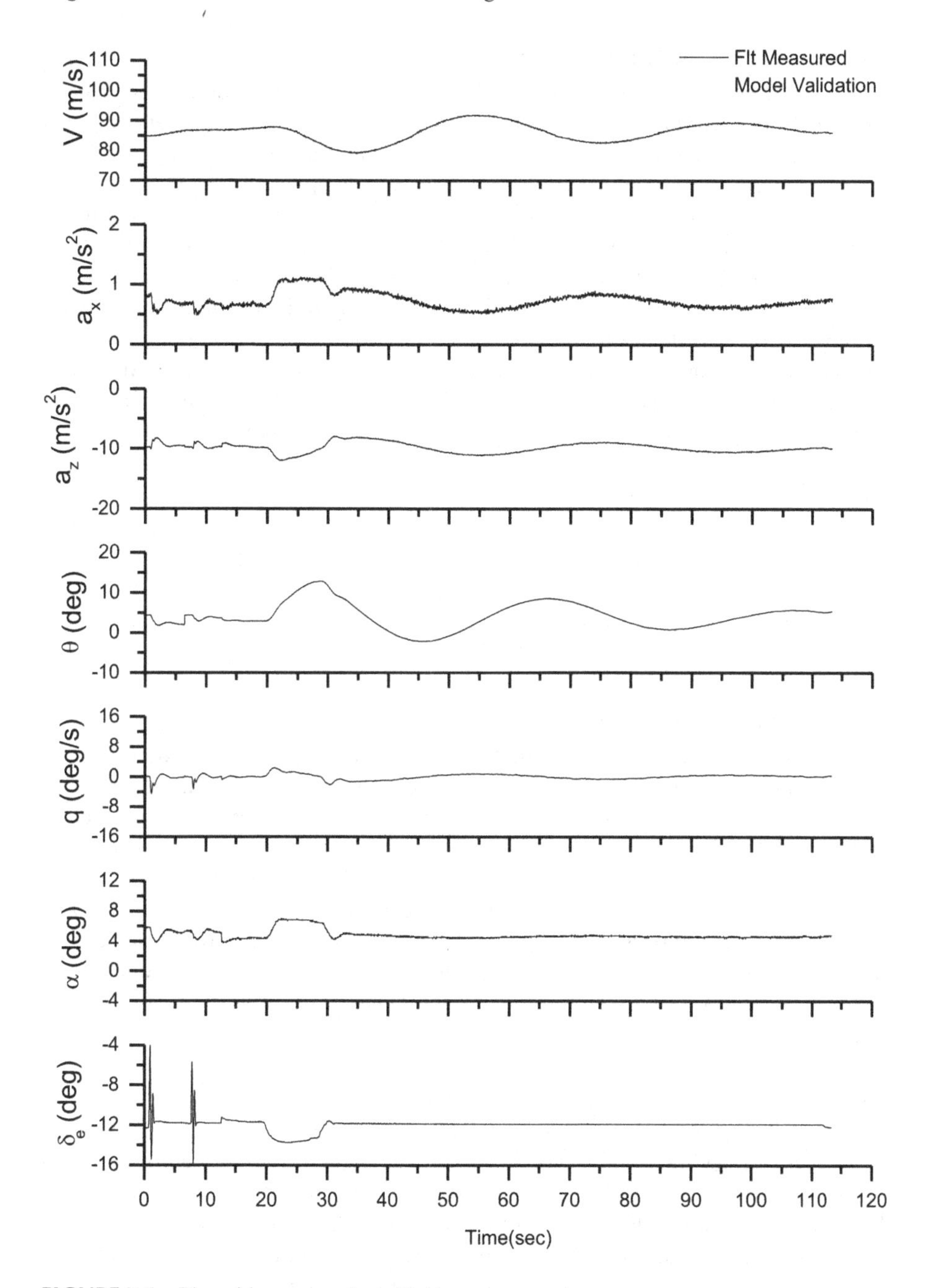

FIGURE 7.5 Phugoid response from flight measurements.

7.6.1.6 Aileron Input Roll Maneuver

This maneuver data is used for estimation of roll damping and aileron effectiveness. Starting from a trimmed horizontal-level flight, a doublet or 3-2-1-1 input command is applied to aileron stick. The time width of the input signal is chosen appropriately to excite dominantly the aircraft rolling motion. Roll damping and frequency are also determined. The roll mode time constant could be 0.1–2.5 s. An aircraft normally has neutral static stability in roll. The maneuver generates bank-to-bank motion that can be used to estimate roll derivatives. The maneuver is initiated with a (short duration) pulse input to aileron in one direction (for about 30 deg. roll), and after few seconds, the aircraft is brought back to horizontal-level position with an input to the aileron in the reverse direction. The process is then repeated in the other direction. At the end of this maneuver, the heading angle should be approximately same as at the beginning. Pure roll-related derivatives L_p, L_{δ_a} can be estimated from these data.

7.6.1.7 Rudder Input Maneuver

This maneuver is used to excite roll-roll motion to estimate yaw derivatives and rudder control effectiveness. Starting from a trimmed-level flight, a doublet or 3-2-1-1 command is applied to rudder pedals keeping the thrust constant. The pulse width of the input signal is appropriately chosen to match the predicted roll-roll frequency.

7.6.1.8 Dutch Roll Maneuver

The purpose is to excite the DR mode of the test aircraft and determine the LD aerodynamic derivatives. Starting from the level trimmed flight (at the chosen flight condition), an aileron stick command is given first and then a command is given to rudder pedals. The command inputs should be either doublet or 3-2-1-1. The thrust is kept constant. The response should be allowed to settle for 5–10 s, and the record should be for about 15–20 s, since the two consecutive responses (due to aileron and rudder inputs) should be analyzed, preferably in one go. The maneuver is actually of a SP kind, and the data generated enable determination of aileron and rudder effectiveness as well as roll/yaw motion-related aerodynamic derivatives $(L_p, L_\beta, N_\beta, N_r)$. Typical DR maneuver flight-time histories are shown in Figure 7.6.

7.6.1.9 Steady Heading Sideslip Maneuver

In a steady heading sideslip (SHSS) maneuver, starting from level flight, rudder pedal is applied and held constant to allow the sideslip angle to build up with minimum variation in speed. As the airplane begins to yaw and enter into a skidding turn, it is banked in the opposite direction by applying aileron deflection in order to maintain a constant heading. As the rudder pedal is further applied, the aileron deflection is adjusted to keep the heading angle constant. The speed variation is marginal. After completing the sideslip maneuver with one rudder pedal, the aircraft is retrimmed and the procedure repeated with the other rudder pedal. This maneuver, along with the DR, can be used to generate the requisite information to assess the aircraft's directional and dihedral stability. Typical SHSS maneuver flight-time histories are shown in Figure 7.7.

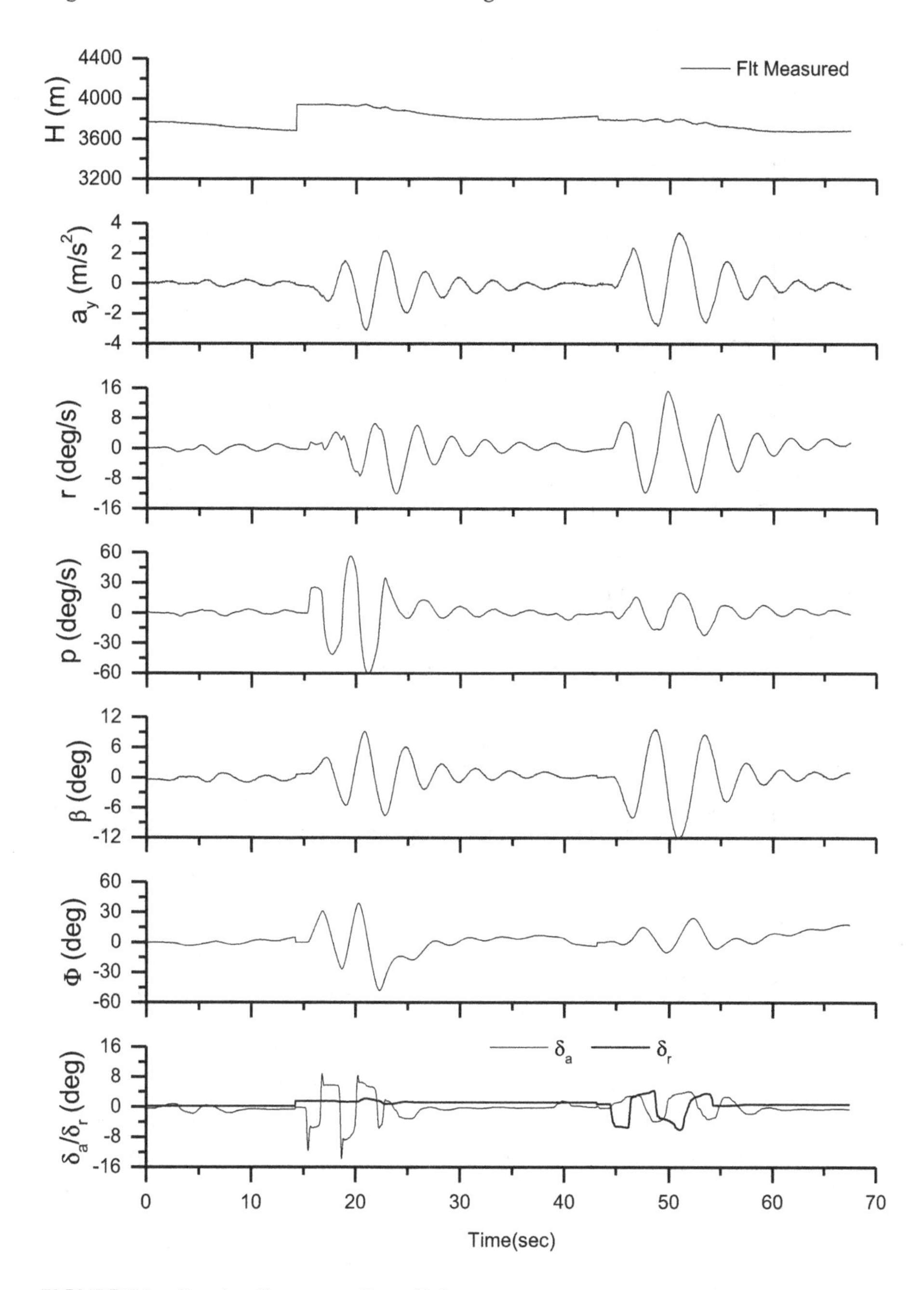

FIGURE 7.6 Dutch roll response from flight measurements.

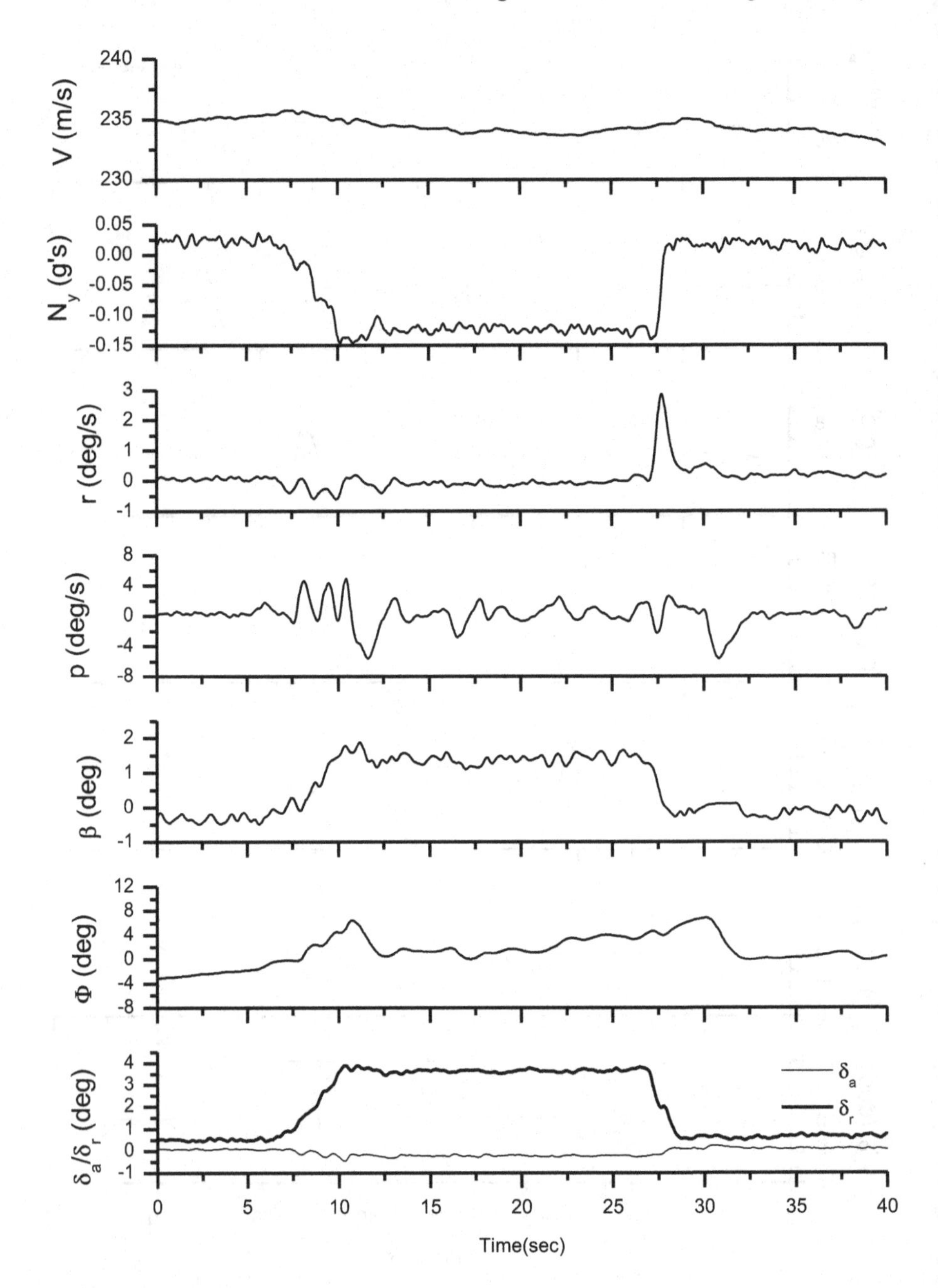

FIGURE 7.7 (a) SHSS response from flight.

(*Continued*)

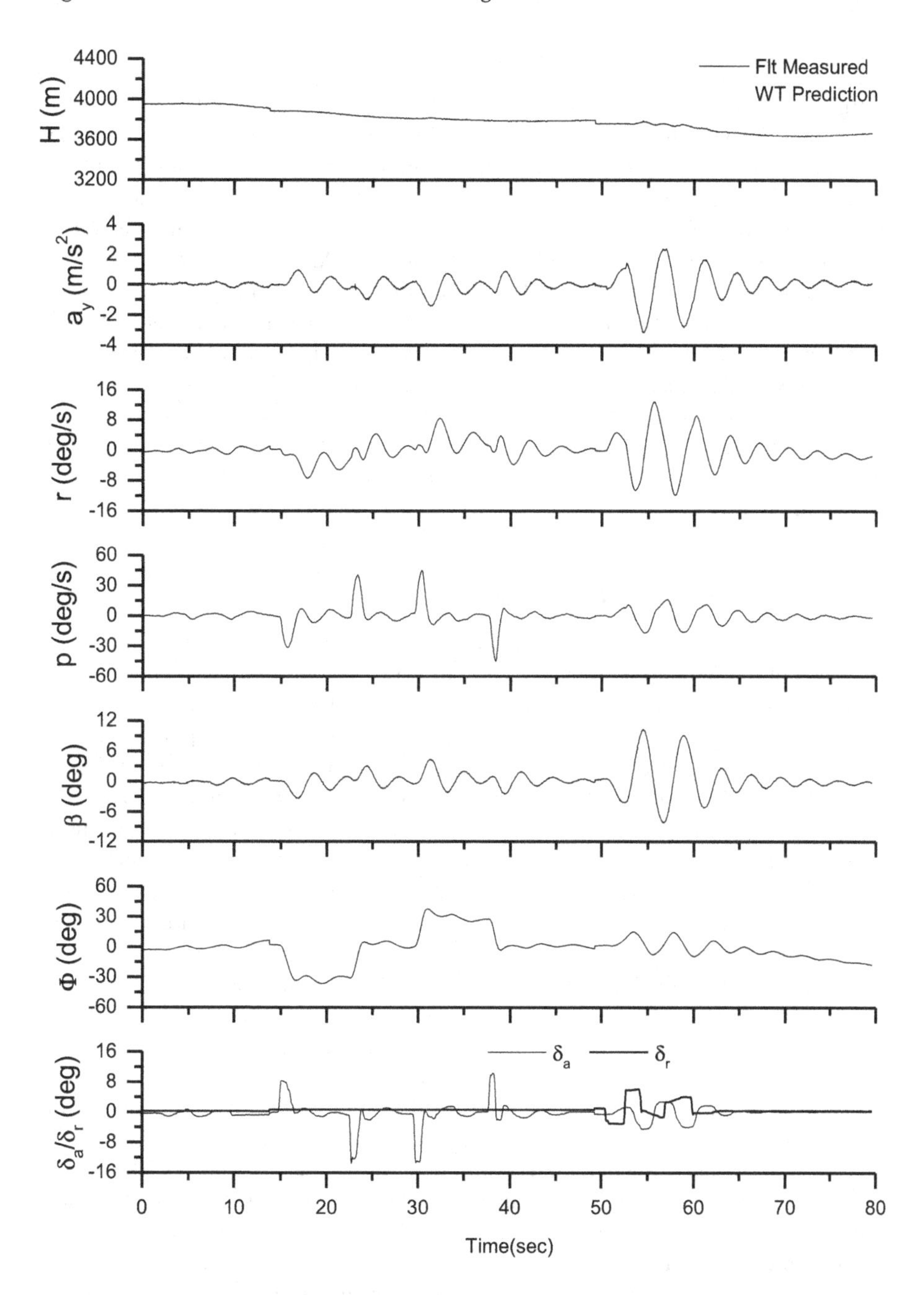

FIGURE 7.7 (*Continued*) (b) Roll and SHSS response from flight measurements.

7.6.2 Large-Amplitude Maneuvers

Large maneuvers could occur during certain regimes perhaps due to loss of stability, damping, or control effectiveness. Also, it may not be possible to trim a given airplane at certain AOAs, and hence, small-amplitude maneuvers are not useful. For such situation, using LAMs and data partitioning, it is possible to generate aerodynamic derivatives over the AOA range covered by the LAM [11]. The method for analyzing these maneuvers consists of dividing the LAM that covers a large AOA range to several bins or subsets, each of which spans a smaller range of the AOA and using a stepwise modified linear regression method to determine the structure of the aerodynamic model within each bin. LAM data in the longitudinal axis are generated by giving doublets and multi-step 3-2-1-1 inputs with different large amplitudes to the pitch stick. The LD characteristics can also be evaluated at HAOA by performing the longitudinal LAMs, that is when large excursions of AOA are made, an L-D maneuver generating command input is given. Typical LAM flight-time histories are shown in Figure 7.8.

7.6.3 A Typical Flight Test Exercise

A basic and the modified version (that had an extra structure on the pylons on the fuselage) of the same basic transport aircraft was flight-tested to evaluate the longitudinal and lateral-directional stability properties, and engine failure performance. Various parameters for the modified aircraft were as follows: wt. up to 39,600 lbs, C.G. range ~ 65.50″ fore, 67″ mid, 71″ aft, (aft of datum, AOD), empty wt. 28,983 lbs, and it acts at 74.33″ (aft of C.G. datum), max speed 200 IAS knots., max altitude 15,000 ft, flap 22 deg., and the autopilot equipment of the aircraft was not used. Instrumentation: for onboard (OB) recording, MARS-2000 tape recorder was to record about 35 parameters; OB PC was used for real-time monitoring of critical flight parameters; the AOA and AOSS vanes were used, the calibration was done inflight, and weighing of the aircraft was done by an electronics system.

Longitudinal stability test: The static stability test was carried out by varying the speed by +/−15 knots from the trimmed speed at 5000 ft and 10,000 ft. There was a reduction in control deflection required for a given altitude/speed change compared to the basic aircraft; the aircraft response to the elevator input was quicker compared to the basic aircraft; the latter indicated a reduction in the longitudinal static stability of the modified aircraft, yet the static stability was +ve; this was borne out by the parameter estimation results (see Chapter 9). The neutral point was obtained as 71.6"AOD; thus, a C.G. limit of 69" was suggested to ensure 2.6″ static margin. This implies that if the C.G. of the aircraft is shifted rearward by movements of loads/weight, then the aircraft would become increasingly less and less stable, and if the C.G. is say 71.6″, then the aircraft would be neutrally stable, and any movement of the C. G. beyond this will render the aircraft statically unstable, and hence dynamically unstable also, such conditions are not permitted for the safety reasons (as per the airworthiness specifications from Civil Aviation Authorities). The Phugoid was neutral to mildly convergent, and the time period was nearly 50 s.

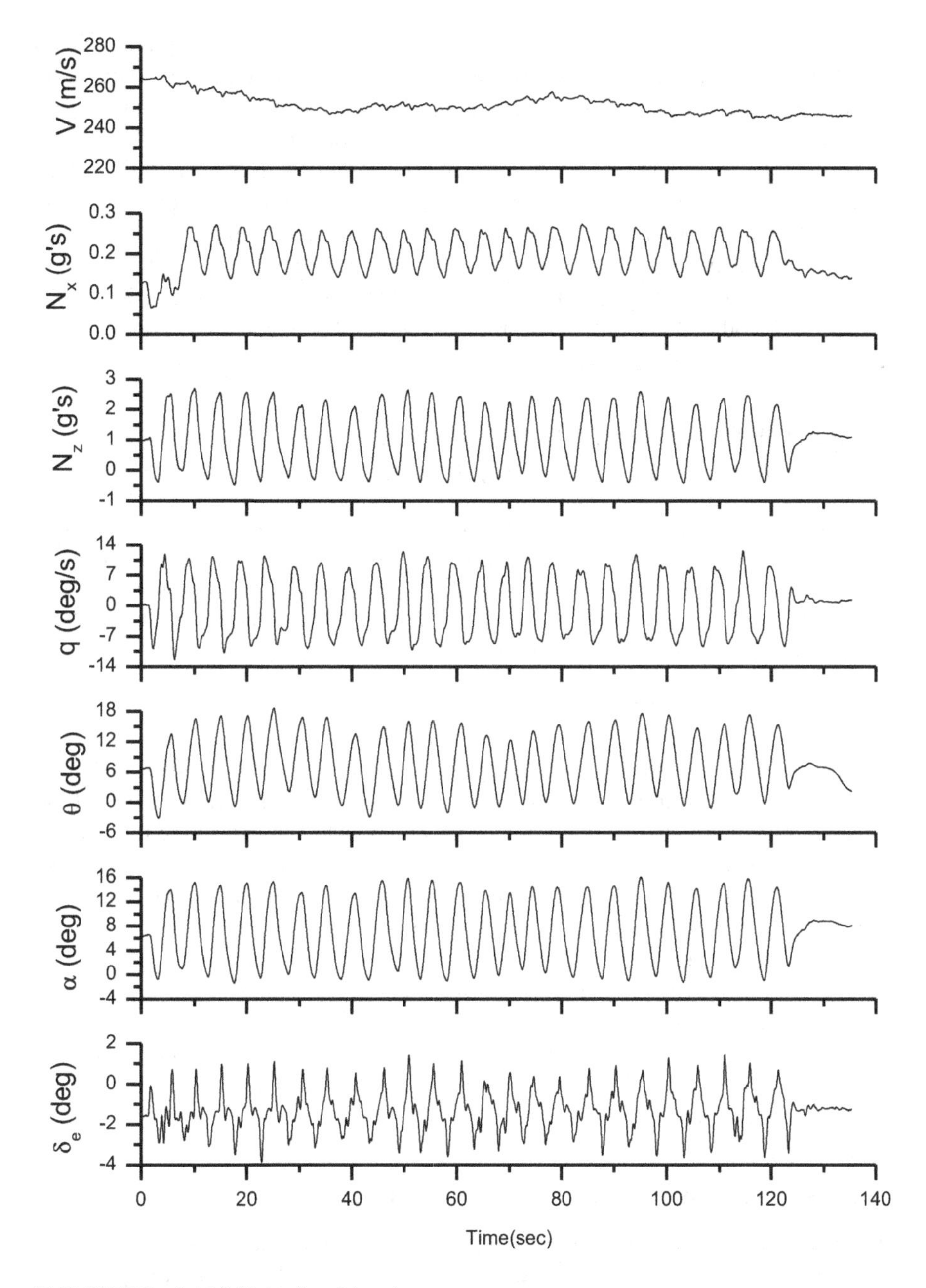

FIGURE 7.8 LAM flight-time histories.

Lateral-Directional test: These tests were carried out through execution of SHSS turns on one control and coordinated turns; the +ve lateral and directional static stability of the aircraft were observed; yet the modified aircraft displayed a tendency of yaw-off and roll-off, that needed a constant correction by the pilot to maintain direction within +/–2 deg.; and it was easy to achieve accuracy, but it added to the pilot's overall workload compared to the basic aircraft flying. The LD dynamic stability was tested by execution of DR maneuver; this indicated high damping in yaw and roll; the modified aircraft was distinctly more stable than the basic one, indicating large effect of the extra structure on this modified aircraft on the LD characteristics; and in the DR, yaw was more predominant than roll, indicating greater increase in damping in roll compared to yaw for the modified aircraft.

Spiral Stability test: The modified aircraft was neutral in a turn to the left and was unstable in a turn to the right; it would lose direction if not corrected for.

HQ: The modified aircraft was stable throughout the explored envelope and did not exhibit any unpleasant flying qualities.

7.7 SPECIFIC DYNAMIC MANEUVERS FOR DETERMINATION OF DRAG POLARS

In order to determine lift/drag performance of an aircraft from flight data, one can plan to conduct the dynamic maneuvers [12,13]. These maneuvers put lesser restrictions on the pitching motion of an aircraft while conducting the maneuver. The classical steady state maneuver is used for drag polar determination by using static weight/lift and drag/thrust balance equation:

$$W = L = \bar{q} S C_{L_{\text{trim}}};\ D = \bar{q} S C_{D_{\text{trim}}};\ T - D - W \sin\gamma = 0$$

When a trimmed aircraft flies in a steady state conditions, the lift is equal to weight and the drag is equal to thrust modulus flight path angle. Since the weight and the thrust are known, the lift and drag forces thus computed can determine the lift and drag coefficients at the trim AOA. Measurement of the dynamic pressure is also important. The aircraft is thus trimmed at required AOA. The total time consumed for these experiments would be much more than that taken by the dynamic maneuvers. The dynamic maneuvers cover a large range of AOA, Mach number, and normal accelerations, and still take flight test time about one-fourth of the static/steady state maneuvers saving time, effort, and fuel for the flight tests. However, it requires to carry out flight data analysis, and using parameter estimation methods, the drag and lift coefficients are estimated. Then, drag coefficients are plotted versus lift coefficients to obtain the drag polars.

7.7.1 Roller Coaster (Pull-up Pushover) Maneuver

This maneuver is used to determine the aircraft drag polars. Starting from a trimmed-level flight (this condition is not really necessary), the pitch stick is first pulled to slowly increase the normal acceleration from 1*g* to 2*g* (at the rate of roughly 0.1 *g*/s) and then returned slowly to the level flight in the same fashion, i.e., to 1*g*. Then, the stick is pushed slowly, causing the acceleration to change from 1*g* to 0*g* at a slow rate and then returned slowly to trimmed-level flight. This is a dynamic maneuver in the vertical plane with load factor varying from 0*g* to 2*g* or up to 4*g* as the case may be. The thrust should be kept constant, and the speed variation should be minimal. The data are recorded at least for about 25–30s in this slow maneuver that covers low angle of attack range (0–10 deg.) and Mach number from 0.4 to 0.9. The speed should be kept constant in the beginning. This maneuver also can be performed up to 4g if found necessary. Typical RC maneuver flight-time histories are shown in Figure 7.9.

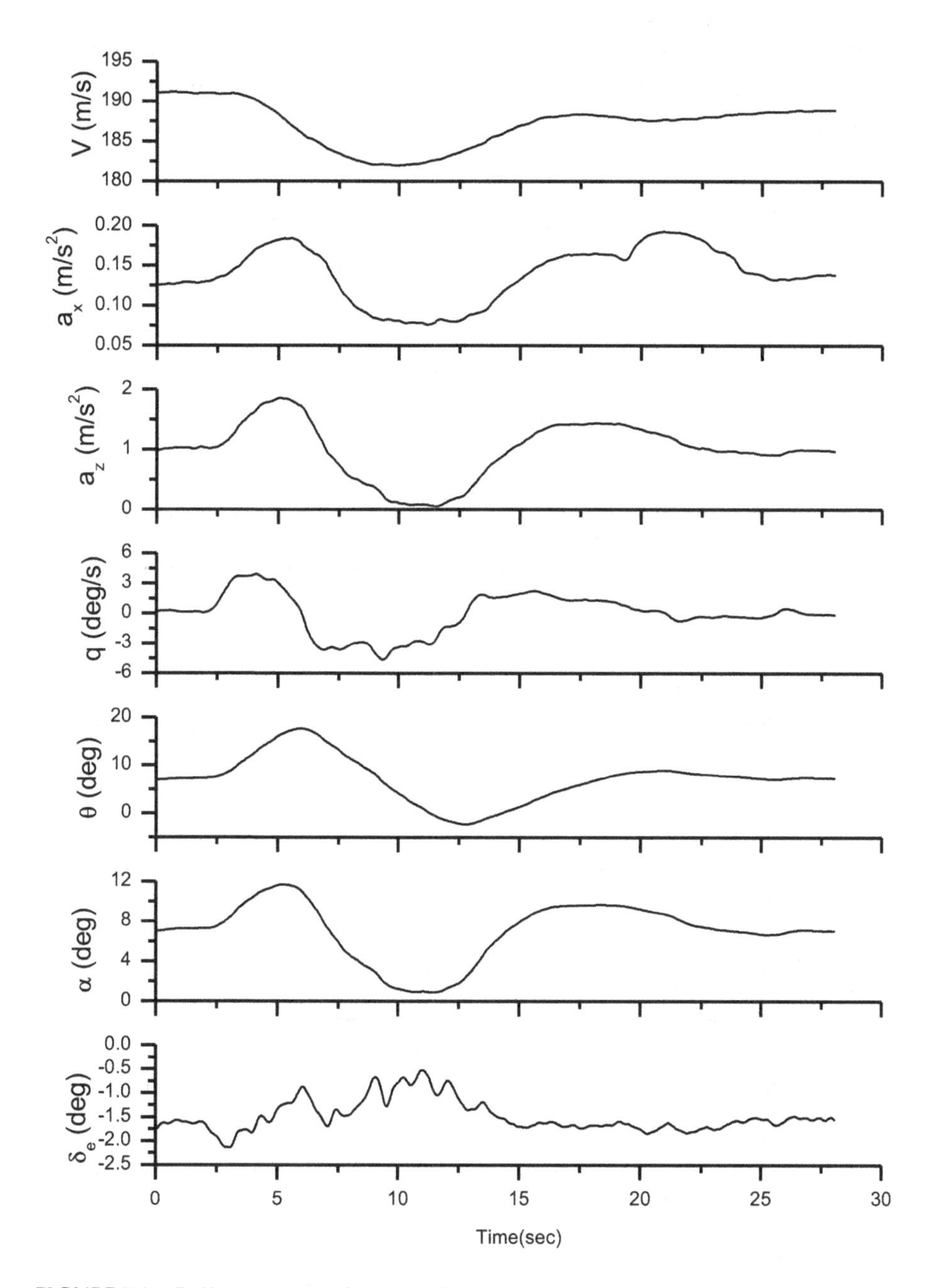

FIGURE 7.9 Roller coaster performance flight maneuver time histories.

7.7.2 Slowdown Maneuver

The purpose of this maneuver is to determine the drag polars in the high AOA range. The maneuver is performed in the vertical plane, and the Mach number range is 0.4 to low speed range. The aircraft is trimmed for the level flight at the chosen flight condition. Then, the elevator stick is pulled slowly to maximum AOA, and then recovery is made from the stall entry. The data are recorded for 25–30 s. The thrust must be kept constant. Typical SD maneuver flight-time histories are shown in Figure 7.10.

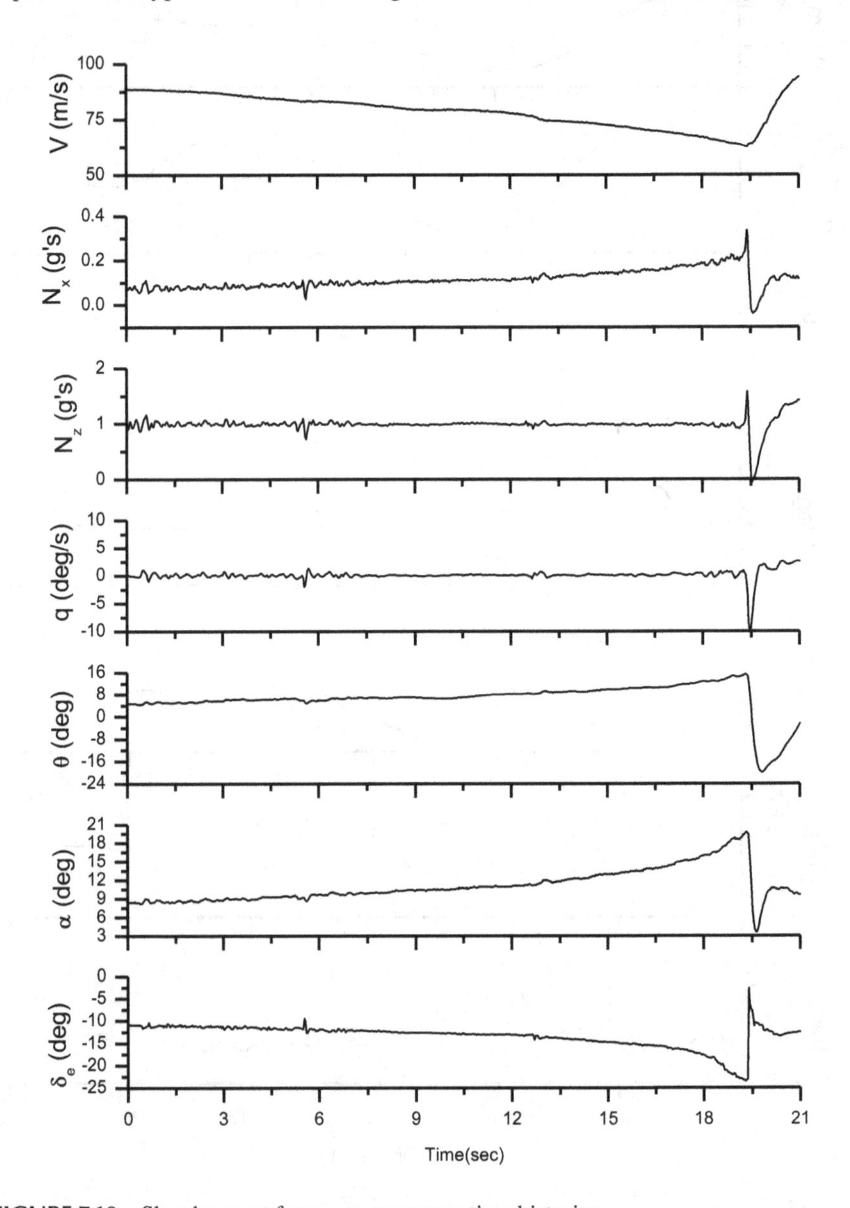

FIGURE 7.10 Slowdown performance maneuver time histories.

7.7.3 Acceleration and Deceleration Maneuver

One of the purposes of this maneuver is to estimate the drag polars at high angles of attack. After trimming the aircraft, the pilot reduces the power to idle and decelerates to the minimum desired speed (a few knots above stall). She/he can hold altitude with only pitch control. Then, the throttle is moved quickly to full power, and the aircraft is allowed to accelerate. On reaching the maximum speed, the power is reduced to idle, and the aircraft decelerates back to the starting trim speed. The maneuver should be performed smoothly. The overall altitude should not change more than 35 m. The power transition should be quite smooth.

7.7.4 Windup Turn Maneuver

In this maneuver, pilot starts banking 2000 ft above the test altitude and then increases the load factor by pulling up slowly from 1*g* to the maximum permissible *g* at 0.5 to 0.8 *g*/s rate. The speed is maintained by reducing the altitude. The throttle deflection is kept constant. This maneuver is useful to get the data for performance above $M > 0.4$ and from 1*g* to max. *g*. Since bank angle is progressively increased from 0 to 80 deg., while pulling the stick, the aircraft starts turning in ever decreasing circles. Actually, the pilot begins with a level turn and increases the bank angle. Since the nose drops due to increasing bank angle, it increases the AOA to maintain speed. This continues until the stall is reached or the '*g*' limit is reached. The WUT is thus a descending spiral and will become tighter and steeper with increasing '*g*'. The aircraft will be in a steep nose-down attitude with high bank angle. Subsequently, recovery to level flight is initiated. The WUT should be performed with smooth increase in '*g*', and speed change should be minimal within 5–10 knots. Typical WUT maneuver flight-time histories are shown in Figure 7.11.

Experience with flight data analysis has shown that no single maneuver, no matter how carefully performed and analyzed, can provide a definitive description of the aircraft motion over the entire envelope or even at a given flight condition in the envelope. Thus, it is always desirable to obtain data from several maneuvers at a single flight condition or a series of maneuvers as the flight condition changes. Often, two or more such maneuvers are analyzed to obtain one set of derivatives. This is more popularly known as multiple maneuver analysis.

7.8 SPECIFIC MANEUVERS FOR ROTORCRAFT

The rotorcraft, unlike fixed-wing aircraft, is more complex with the main rotor playing a key role in defining the overall dynamics of the flight vehicle. In rotorcraft, the off-axis responses cannot be ignored, as these are more often than not of similar magnitude as the on-axis responses. The interference effects from the main and tail rotors, and the presence of vibrations at low frequencies further complicate the matters. A variety of flight test maneuvers are therefore required to generate sufficient information to identify mathematical models with increased reliability for the rotorcraft. Normally, for identification of rotorcraft stability and control characteristics, inputs similar to those for the fixed-wing aircraft, e.g., doublets and 3211, are

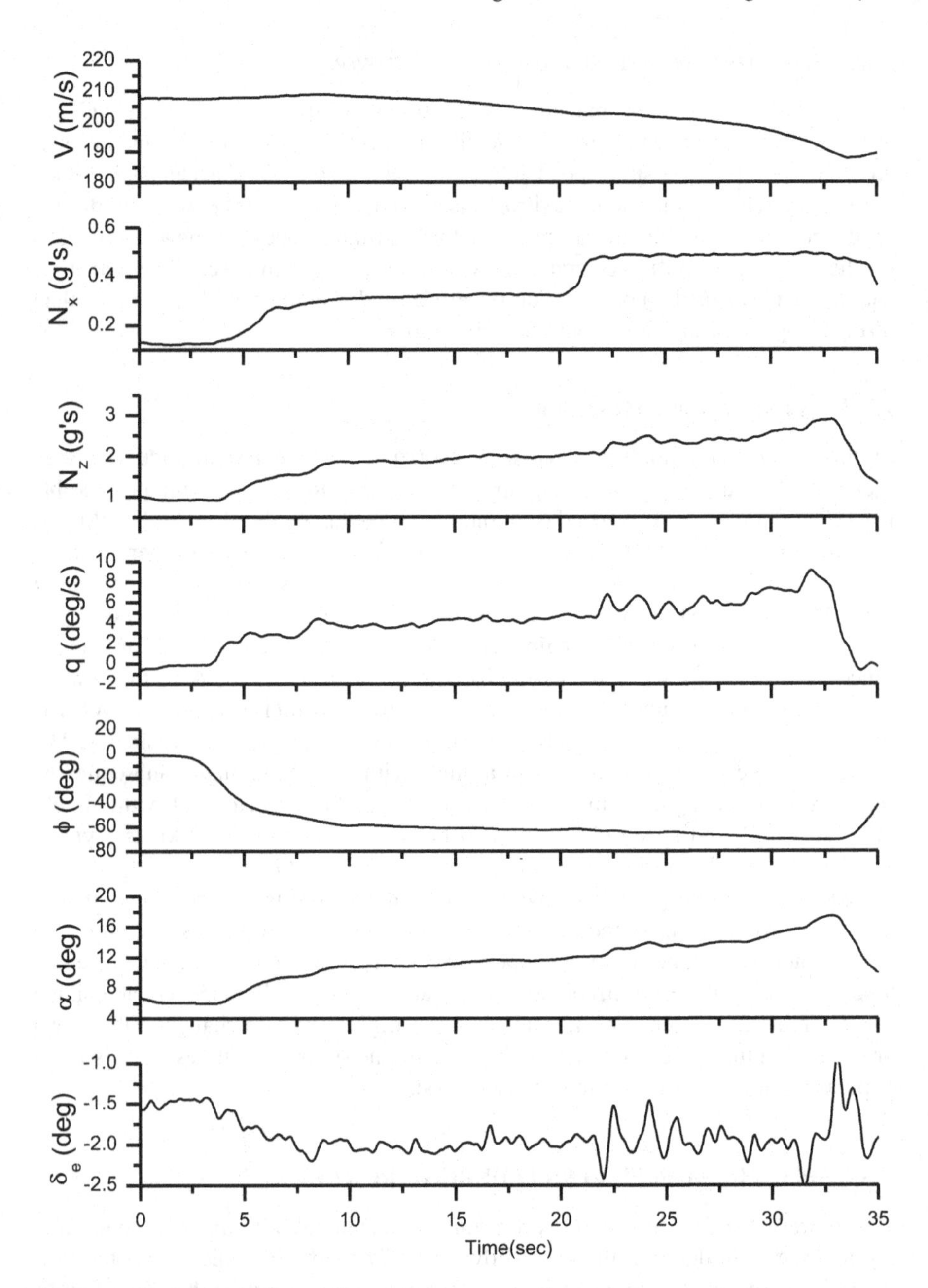

FIGURE 7.11 Windup turn performance maneuver time.

applied to excite the longitudinal Phugoid, pitch, and the lateral DR and spiral modes. Frequency sweep inputs are also used in a good measure to generate flight data for rotorcraft system identification, and an excellent insight into various aspects of rotorcraft identification is available [14].

For rotorcraft handling qualities testing, the ADS-33 stipulates the mission task elements (MTEs) to be performed [15,16]. These are treated as HQ tasks, and mission-oriented maneuvers are flown by pilots who assign subjective pilot ratings using the Cooper-Harper Handling Qualities Rating scale (HQR). The performance of helicopter during piloted mission tasks ultimately defines its HQs. ADS-33E specifies nearly twenty-three MTEs. Execution of these mission maneuvers, which includes maneuvers like Hover, Landing, Hovering Turn, Pirouette, Vertical maneuver, Slalom, Sidestep, Deceleration to Dash, and pull-up/pushover, generates a large amount of extremely valuable data for handling qualities evaluation. A typical rotorcraft flight maneuver is shown in Figure 7.12.

A few of the flight test maneuvers (ADS-33D) required to be performed on a rotorcraft in order to evaluate its HQ performance are described next [17]:

Hover: At a ground speed between 6 and 10 knots, at an altitude of less than 6.1 m, transit to hover, and maintain a precision hover for at least 30 s. This should be done in moderate wind from the most critical direction.

Hovering turn: At an altitude less than 6.1 m, complete a 180 deg. turn from a stabilized hover. With moderate wind from the most critical direction, perform the maneuver in both directions.

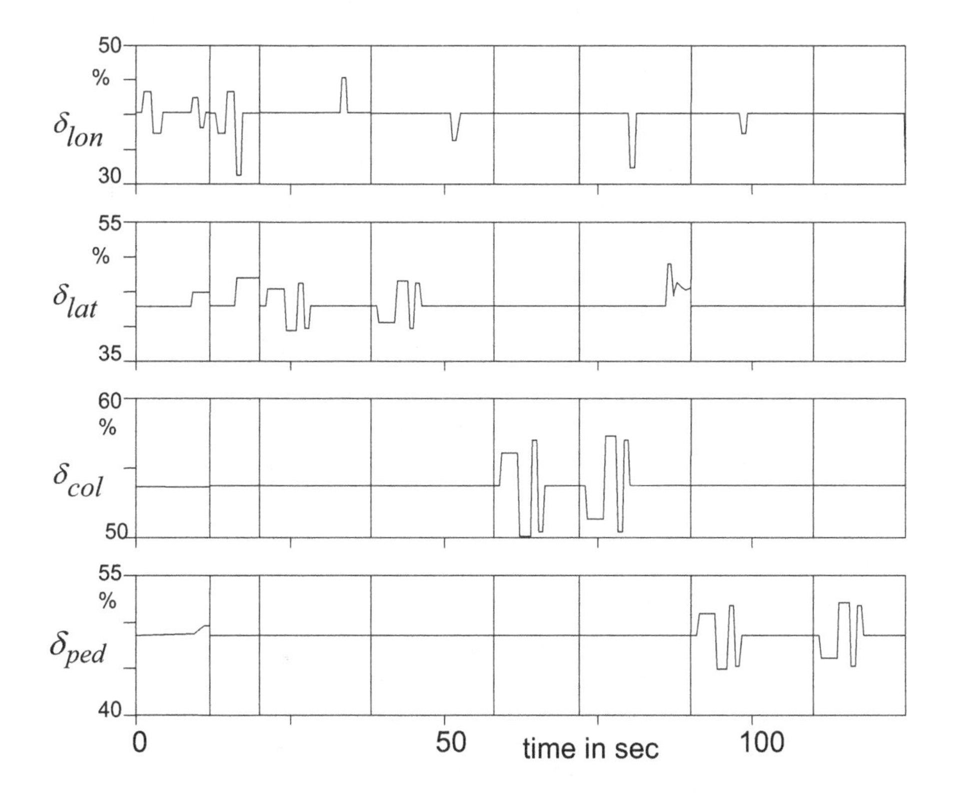

FIGURE 7.12 (a) Control inputs used for rotorcraft data generation.

(Continued)

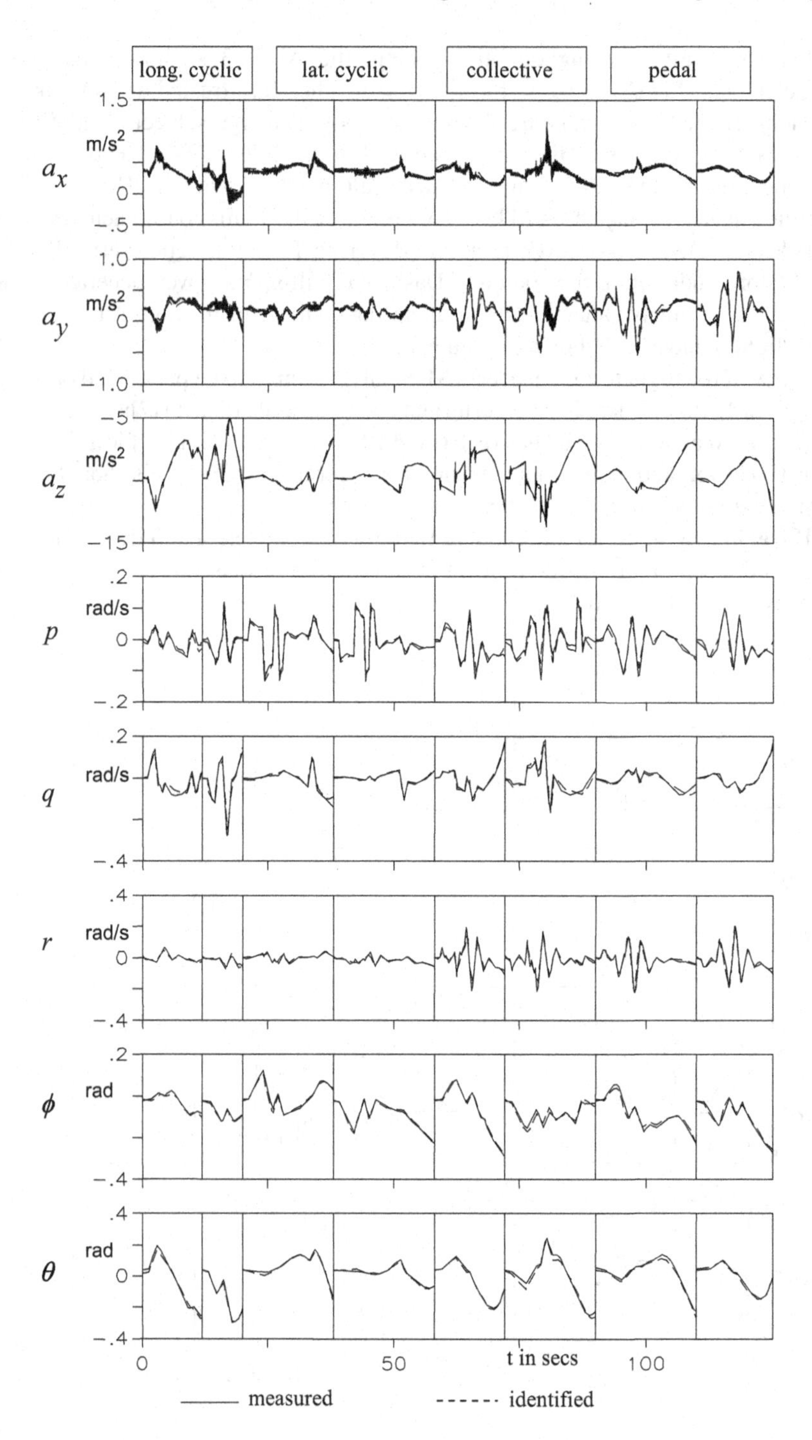

FIGURE 7.12 (*Continued*) (b) Rotorcraft response to the inputs in (a).

Bob-up/Bob-down: Bob-up to a defined reference altitude between 12.2 and 15.2 m from a stabilized hover at 3.05 m. Then, stabilize for at least 2 s. Reestablish the 3.05 m stabilized hover by bob-down. Perform the maneuvers in moderate winds from the critical azimuth.

Acceleration/deceleration: From a stabilized hover, rapidly increase power to nearly maximum and maintain altitude constant with pitch attitude. Accelerate to 50 knots at constant collective. After reaching the target speed, initiate deceleration by aggressively reducing the power and holding altitude constant with pitch attitude, and then stabilize hover.

Sidestep: At a selected altitude below 9.15 m, starting from a stabilized hove, initiate a rapid and aggressive lateral translation, with a bank angle of at least 25 deg., while holding the altitude constant with power. After achieving a lateral velocity within 5 knots of maximum (or 45 knots), initiate deceleration to hover at constant altitude. Establish stabilized hover for 5 s and repeat in the opposite direction.

The foregoing maneuvers are also specified with certain performance/desired parameters in ADS-33D. For example, in 'hover', the altitude should be maintained within +/−0.61 m, the horizontal position should be within +/−0.91 m, and the pilot should be able to stabilize the helicopter in hover within in 3 s.

The pilots need to pay extra attention to avoid exceeding the structural, aerodynamic, or control limits of the helicopter, while flying it to exploit its capability in the complete flight envelope. The piloted simulation evaluations should be made to investigate possible ways of altering and aiding the pilot when close to transmission limits. Aggressive maneuvers can be flown in the simulator and the pilot reaction to various cues (like collective stick force feedback, aural cues, voice cues, and visual HUD cues) can be evaluated. The rotor RPM and rotor transmission limits should be monitored and evaluated. The active flight control approach can effectively eliminate un-commanded limit exceedance in high aggression and large-amplitude maneuvers. The pilot cuing for impending limits can be based on the predictive capabilities of ANNs which can be trained using helicopter flight test load survey database.

7.9 FLIGHT TEST DATABASE MANAGEMENT

It will be a good idea to plan to store the flight test data in a proper format so that one can easily access them as and when required. Also, the results of the analysis should be archived in a proper format. The flight database management system/software (FDBMS) is required to keep track of vast amount of data, the relevant information, and the associated test/parameter estimation results [16,17]. The SW should be interactive, and the user or analyst should be able to enter/retrieve the required data/information with simple queries. The data could be of a huge volume, and it is necessary to manage and store the valuable and large data/information in a suitable way that would enable the analyst select a particular section of data/information for analysis or demonstration as and when required. Also, the parameter estimation results (obtained from these data) should be carefully stored with clear indication of association of these results with the corresponding section of flight data for future references. This can be only achieved with planned and properly defined DBMS. The DBMS will have a number of utility routines like query building, presentation of the results in graphical forms, and plotting of flight data, in addition to usual features like the database creation/manipulation.

7.9.1 Basic Requirements

The basic requirements can be specified as follows [16]:

a. The selection and classification of flight data such that the DBMS is invariant with reference to type of the aircraft, flight test conditions, test maneuvers, aircraft configuration, etc. It should provide queries for an interactive session.
b. Organization of the actual flight data in the form of ASCII files and relevant information/data/parameter estimation results in the data bank. There should be clear indication of association of this information with the corresponding flight data in the ASCII files.
c. Strategy and methodology for quick access to flight data and relevant information and the results with respect to the user requirement.
d. Development of utility routines for generating various graphical outputs. These outputs are used for quick visual checking on the terminal screen. The generation of report quality plots. Documentation of results.

7.9.2 Selection and Classification of Flight Data

For implementation of the DBMS, it is important to create a data bank with a reasonable classification system for easy and quick entry and retrieval of flight data. The flight data bank requires a properly defined data classification system that is generally defined by the user.

7.9.2.1 Classification Based on Type of Maneuvers

Since certain flight tests are conducted with special maneuvers for system identification, it will be easy to identify the flight data for the required maneuvers, if the data are properly classified. The following coding identified the letter code could be useful:

a. Short period and Phugoid	SPP
b. Dutch roll	DRO
c. Roll maneuver	ROL
d. Roller coaster	RLC
e. Windup turn	WUT
f. Level acceleration deceleration	LAD
g. Slowdown	SLD

Similarly, any other maneuver can be coded with three-letter word.

7.9.2.2 Classification Based on Flight Conditions

The maneuvers are performed repeatedly at different flight conditions: Mach number and altitude for general flight testing and/or parameter estimation process. Usually, the maneuvers are performed for different AOA and AOSS. The knowledge of the flight condition is very important for the flight data analysis, and hence, the flight condition details should be included in the data bank to identify the flight data.

7.9.2.3 Classification Based on Aircraft Configuration

The maneuvers are also performed for several aircraft configurations: (i) under carriage is up or down (retracted), (ii) flap position: zero, graded, or full, and (iii) C.G. positions: fore, mid, or aft. Proper classification will need the above configuration information also.

7.9.3 Data Storage and Organization

The aircraft stability and control analysis require flight data, the responses of the aircraft to the chosen input type for a specified maneuver in ASCII format. Also, additional information/data like: maneuver type, flight conditions, mass & inertia, and C.G. position in a form that enables quick storage/modification and retrieval with clear indication of association of this information/data to the ASCII file containing corresponding data. The actual flight data in ASCII files with unique file names can be put in one data bank. The second data bank can contain relevant information/data that are required for analysis/parameter estimation. This data bank can also contain the results of stability/control analysis and parameter estimation. The information in the second data bank is stored as records in different tables that are created using some database engines, e.g., Oracle 8 [18]. The structure of the tables consists of records as rows and fields as columns. The information/data/results particular to one-time segment of the data (i.e., one ASCII file) are stored in different tables as records along with FILE_INDEX. This index is unique and connects the fields between the tables. If the data file is known, then all the information/results corresponding to this data file can be accessed. And this is done by selecting the record in different tables containing the same FILE_INDEX name. For a known FILE_INDEX of a given maneuver at a given flight condition, the corresponding flight data (ASCII file) and related information/data can be identified. With this kind of arrangement, the new data files and the related information/data can be appended to the data banks in respective places as and when need arises.

7.9.4 Flight Test Database in Oracle

Here, the development of a typical DBMS using Oracle 8 (there could be other SW/tools possible [17]) is briefly described [16,18–21]. Its client/server architecture (say, two PCs connected via LAN, local area network) can be used leading to a multiuser access system with a centralized data bank over LAN. The data can be loaded directly into the database tables from ASCII file. This can be done either manually using an interactive menu or via the Oracle Loader. Also, SQL * Loader can be used for loading the data from non-Oracle data source to the Oracle database. The Oracle 8 is a

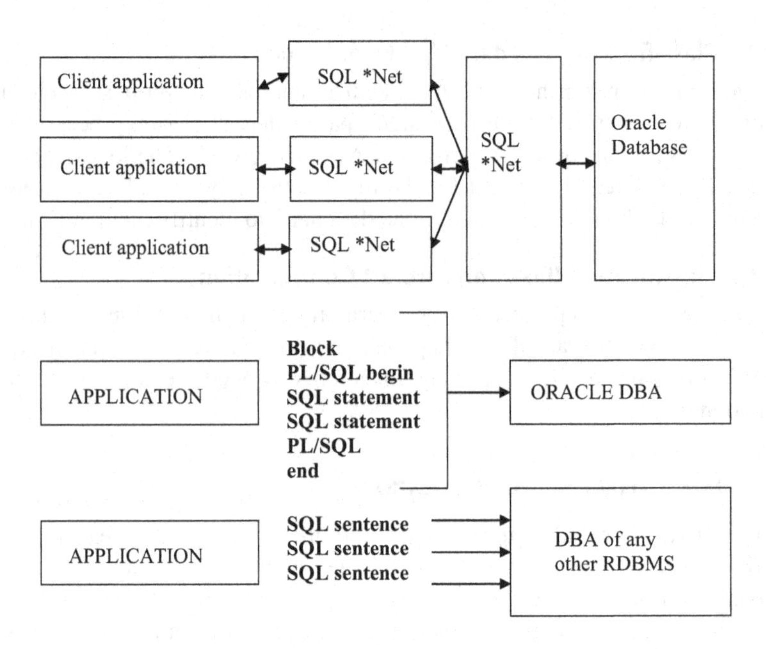

FIGURE 7.13 Communications and code transfer.

component of networking computing architecture, and this feature reduces the cost. Some features of the Oracle 8 are as follows: (i) client/server architecture, (ii) object/relational technology, (iii) concurrent database users, and (iv) SQL * Loader/Oracle Loader facility. These features are briefly described next. A communication and code transfer process in a typical DBMS is depicted in Figure 7.13.

Client/Server Mode: The database system is divided into the following: (i) a client portion or a front end, (ii) a server portion or a back end. The client PC is responsible for controlling the user interface. The server PC is the host to the relational database management system. An application program residing on the client PC interfaces with another SW layer known as SQL *Net. The latter is responsible for communicating requests, and their results between the application program and the relational database management system. The client PCs directly communicate with the database server.

Concurrent database users: The Oracle supports a large number of users at the enterprise level using Net 8, the new version of SQL * Net. The data that are on Windows NT PCs can be accessed by these users.

Object/Relational Technology: This helps in the design of an object-oriented database. This extends the traditional Oracle Relational Database Management System to include object-oriented concepts and structures.

SQL *Loader or Oracle Loader: The tool can be used to move the data from a non-oracle standard source into the oracle database. The data can be loaded from multiple data files and in fixed/delimited formats. The data file containing the actual data and the control file containing the specifications which drive the SQL *Loader session are provided as inputs to the SQL *Loader. The Loader reads text files and places the data in the Oracle database based in the instructions in the control file. The control file provides the information related to the following: (i) names of the input data files, (ii) names of the database tables into which the data are to be uploaded, (iii) the mode in which the data are to be loaded (like replace, appends, or insert), (iv) the correspondence between the data fields and the table columns, and (v) the selection criteria for loading.

A brief note on Oracle 8: It is an Object Relational Database Management System. The objects can be defined as reusable SW codes. These codes are location-independent and perform a specific task on any application. The client/server concept segregates the processing of an application between two systems. One system (PC) performs all activities related to the database server. The other does activities that help the user to interact with the application client. The front-end/client database application also interacts with the database by requesting and receiving information from the database server. The client acts as an interface between the user and the database. It also checks for validation against the data entered by the user. The user front-end tools of Oracle are as follows: SQL *Plus V 8, Oracle Forms 5.0, and Reports 3.0, and of course the later versions. The back-end server manages the database tables among the multiple clients. These clients concurrently request the server for the same data. The server also maintains the integrity across all client applications. It controls the database access and security requirements.

PL/SQL and FORM: The SQL is the natural language of the database administrator. However, it does not have procedural capabilities of looping and branching. It also does not have any conditional checking capabilities. The Oracle has PL/SQL which can be used to create programs for validation and manipulation of table data. It provides the user with all the facilities of a programming environment. The PL/SQL closes the gap between database technology and procedural programming languages. It is a development tool and extends the capability of Oracle's SQL database language. One can insert, delete, update, and retrieve table data. One can use procedural techniques of writing loops and branching to another block of codes. It being an extension of the SQL allows one to use all the SQL data manipulation statements. It also allows the cursor control operations and transaction processing, as well as logical grouping of SQL sentences and passing them to the database administrator as a single block. The entire block of statements can be sent to the RDBMS engine. PL/SQL can be used in SQL*FORMS, and its procedural capabilities can be used for writing complex triggers that would validate data before they are placed in the tables. The applications in PL/SQL are portable to any computer HW and OS, where Oracle is operational. The Oracle Forms Builder provides GUI to design forms. All objects, properties, and triggers can be selected by clicking on an appropriate icon. The tool comprises of the following: (i) Forms Builder, (ii) Forms compiler, and (iii) Forms Runtime. Application built using Oracle Forms Builder would contain the following: (i) Form Module, (ii) Menus, (iii) PL/SQL, (iv) Object Libraries, and (v) Database Objects. The Form Module is a collection of objects: blocks, canvas, frames, items, and the event-based PL/SQL code blocks (these are called triggers in an MS Window). PL/SQL library module is a collection of PL/SQL functions and procedures in a single library file which is attached to a Form/Menu module. The objects in the Form/Menu can access and share the collection of PL/SQL functions and procedures.

An Object Library is used to create, store, maintain, and distribute standard and reusable objects, and create applications by dragging and dropping predefined objects on form. The Database Objects like Stored Procedures, Stored Functions, and Database Triggers are created using appropriate SQL and PL/SQL syntaxes.

Conversion of ASCII file into tables: This can also be done manually by using the built-in option SQL *Loader. This is run on the server machine. A method can be used such that the system will automatically call the subroutines to convert the ASCII file into a text file by copying it from a location to a temporary directory. Then, a table is created with all the required fields and a control file. This file contains the path of the file and the width of all fields. The following code reads each line from file using the file operations with the help of text IO package:

```
LOAD DATA
INFILE <input file name>
DISCARDFILE <disfile>
[APPEND/REPLACE/INSERT]
INTO TABLE <tablename>
(<columnname>
[POSITION <start : end)
<columnname>
[POSITION <(start : end)
...]
)
```

Data storage/manipulation of associated data: This has options like adding/saving of new data, modification/deletion of old data, and viewing of data from the tables. The queries can be given to view the required data.

Data storage: The flight data should be stored in such a way that the analyst should be able to select a particular section of data/information. The user should be able to select the table name from the main screen developed using Developer 2000 product (D2k) – Forms 5.0. From here, the details of the fields with all options can be viewed. The use can enter the new data and click on the SAVE button to save it in the specified field. The subroutines will be activated automatically when the specific buttons are pressed. The data will be saved into the table based on the FILE_INDEX. This is created automatically. The user via a query mode can modify/erase the data, if so desired.

Graphical display: This is the user interface menu for plotting graphs and can be implemented with the help of D2k product – Oracle graphics. One can use the OG.PLL library.

User-defined queries: This can be implemented with the help of D2k product – Reports Builder. Using this screen, the user can select the tables, the field names, the required conditions, etc.

7.9.5 Brief Description of a Typical Program

A typical program of flight database would help perform the following operations/tasks: (i) upload the raw flight data and subsequently the processed flight data, (ii) enter/edit the stability and control (S&C) derivatives, (iii) plot the required graphs of the flight

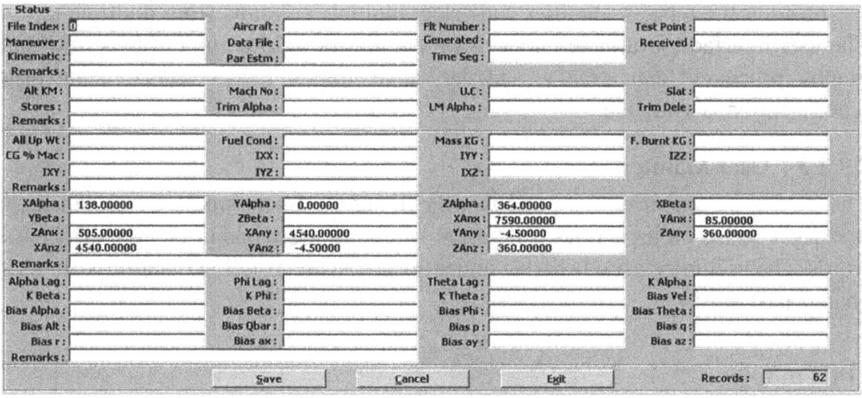

FIGURE 7.14 A typical status screen of the flight DBMS.

data, and (iv) print results based on the selections made. Typical hardware/software requirement would be the following: (i) latest available PCs, (ii) Oracle database server, (iii) Oracle developer suite runtime [22], and (iv) and Windows environment.

Menus would be:

Transactions – to upload raw and processed flight data, to delete the data, to add/edit the stability, and control derivatives
Graphs – would have different types of plotting options
Reports – different types of report options would be available
User maintenance – option to manage users would be available
Help – online help for the flight data management
Exit – exiting from the application

The following description is given mainly with respect to flight database collected from the signal-receiving devices for the purpose of estimation of aerodynamic (S&C) derivatives, and the parameter estimation results thereof. A typical status screen of the flight DBMS is shown in Figure 7.14.

7.9.5.1 Transactions

The signals obtained from the flight maneuvers in ASCII file format can be stored in the data storage device. The FDBM application helps to upload the raw or processed data from the storage device and subsystems. The file index–flight number will be key ID to identify the respective flight/S&C data/parameters. The user interfaces available are as follows: status, bounds on the data, bounds on the longitudinal estimation, longitudinal reference values bounds, lateral estimation bounds, lateral reference values bounds, upload files, and delete transactions.

7.9.5.2 Graphs/Reports

Graph could be plotted wrt the available flight data to view the acquired/stored signal status. It also helps the user decide the scope of values and type of graph to print.

For printing the reports, the user can select multiple columns from multiple tables. The user interfaces available under this menu are as follows: plot graphs, raw data graphs, processed data graphs, and derivative graphs. Under report menu, print reports option is available.

7.9.5.3 User Maintenance

The system administrator can add/modify the users. Users would be provided the privileges based on their roles. The users would have options to change their passwords. The user interfaces available under this menu are as follows: add/modify user, change password.

EPILOGUE

Flight testing of any aircraft is very important aspect from performance evaluation and certification point of view. The energy management concept seems to be very interesting in understanding of various flight test maneuvers. Before performing these maneuvers on the real aircraft, one can carry out the same on flight simulators. The simulator flight data can be studied and analyzed to gain confidence in performing these maneuvers in real flight. The flight database management is very important due to vast amount of flight trajectories that could get accumulated. Keeping track of the data and the related information generated via extensive analysis is very important for successful completion of the flight tasks. Only a few aspects of the flight testing and DBMS have been covered in this chapter. Flight testing is generally not extensively considered in flight mechanics courses. Also, open literature in this area is fairly limited.

APPENDIX 7A AIRCRAFT CERTIFICATION PROCESS AND WEIGHT ANALYSIS

Certification Process: A certification agency (CA) is to be satisfied with the fitness of a newly designed and developed aircraft or a modified one [23]. It is a two-stage process: (i) type certification to establish that the generic-type design meets applicable design and safety requirements, (ii) airworthiness certification to recognize that an aircraft meets any additional operational requirements and is physically airworthy.

Certification process: (i) An application is made to Civil Aviation Authority (CAA) by an approved organization, (ii) a certification team is constituted, (iii) certification basis is provided, (iv) investigations are conducted, (v) based on the results of testing, the team recommends for the type certificate, and (vi) the CAA/CA issues a type certificate.

The certification team consists of experts in the areas of: (i) structures, (ii) engine performance, (iii) systems, (iv) avionics, and (v) flight testing.

The basis of certification: (i) certification specifications, (ii) special conditions, (iii) exemptions, (iv) equivalent safety findings, (v) environmental standards (noise/emissions), and (vi) reversions.

In fact, the process is quite involved one; it would be prudent for the aircraft design agency to involve a small team from the CA from the second phase of their developmental program to familiarize the team with the process of design of the aircraft; and this will help in the certification process.

Weight Analysis: Lot of weight of the aircraft and too far forward will make it difficult to raise the nose of the aircraft; much more aft, and it may be more difficult to recover it from a stall; like a playground seesaw, the weight would exert much more force the farther it is from the fulcrum (where the arm is supported), so a baggage placed in the *rear seat* may be fine, for instance, when the same bag would cause the aircraft to be out of balance in the *rear baggage compartment* [24,25]. The C.G. envelope would show the range of weights/moments (weight times the arm length, its distance from a predefined starting point) that are permissible for a given aircraft plane.

The process is as follows:

1. Determine from the pilot's operating handbook (POH), the empty weight of the aircraft, and determine the weights of people, baggage, and fuel.
2. Multiply each weight by the arm length, the distance from the reference datum to find the moment.
3. Do the summation of all the weights to find the gross weight, and sum all the moments to find the total moment.
4. Divide the total of moments by the gross/total weight to find the C.G. of the aircraft.
5. Locate this total weight and C.G. on the 'C.G. limits' chart in the aircraft's POH to determine if the airplane is within allowable limits.
6. Repeat this process for the expected landing weight, as the C.G. would shift as the fuel is consumed, after the flight phase.

EXERCISES

7.1 What really happens when a short-period mode is excited by using a double input?

7.2 What really happens when a Phugoid mode is excited by giving a pulse input? What a pilot can do to arrest this oscillation?

7.3 What is vertical damping? [8]

7.4 What is roll damping? [8]. Is this natural roll damping very effective at near stall condition?

7.5 Explain why there would be generally three flight path time histories? Which one would not be a straight line path?

7.6 How would you approximately determine the average roll rate from the time history of roll angle?

7.7 How would you approximately compute the yaw rate consistent with the bank angle?

7.8 Give an expression for the total energy of the aircraft at an altitude h and velocity V. Also obtain the expression for the rate of change of this energy, and interpret the resulting terms.

7.9 If the specific excess power of an aircraft is given as $V(X_N - D)/W = dh/dt + \frac{V}{g}\frac{dV}{dt}$, where X is the axial force (in x-direction) and D is the drag force, interpret the expression in terms of the aircraft climb and acceleration performance.

7.10 What is the significance of the total energy and the interplay of its terms from flight maneuvers?
7.11 Give the formula for the energy height of an aircraft when it is at the altitude h meters. Hint: use the concept of specific energy (i.e., per weight).
7.12 Damping is resistance to motion of an aircraft in air. It is characterized as proportional to the rate of the movement. What are these movements?
7.13 In Figure 7.1, why the high speed limit is required?
7.14 Why the speed limit is governed by maximum lift?
7.15 Why it is natural for a pilot to talk about speed in flight and for the experimenter on the ground in terms of angle of attack?
7.16 For a flying airplane, if the drag line is above the thrust line, what kind of moment will be generated?
7.17 What is the condition of the vertical damping at near stall?
7.18 Under what condition of bank angle, the aircraft would enter into spiral dive? What one can do to initially come out of it?
7.19 Why it is feasible to determine the drag variation (drag polar) from the pull-up/push-over (RC) maneuver?
7.20 Why the windup turn maneuver is started at some higher altitude than the specified test altitude?
7.21 Once the yawing moment due to rudder deflection is known, how would you determine the value of the yawing moment due to sideslip?
7.22 What is the acceleration in the direction of the velocity vector called? Give the expression for this acceleration.
7.23 What do the direction and magnitude of the velocity vector signify?
7.24 When flaps are lowered, the lift would increase. The drag force would also increase. What benefit this drag force would provide for landing of the aircraft?
7.25 What is the major division of various maneuvers? Why many of these maneuvers are often classified or called as stability and control (related) maneuvers?

REFERENCES

1. Maine, R. E., and Iliff, K. W. Application of parameter estimation to aircraft stability and control- The output error approach, NASA RP 1168, 1986.
2. Anon. Flight test guide for certification of transport category airplanes. Advisory Circular, FAA, U.S. Department of Transportation, AC 25-7, 1986, US.
3. Carlson, E. F., Galbraith, T., and Rumsey, P. C. New requirements, test techniques and development methods for high fidelity flight simulation of commercial transports. AIAA Paper 80-455, 1980.
4. Roll, L. S., and Koening, D. G. Flight measured ground effect on a low-aspect-ratio ogee wing including a comparison with wind tunnel results. NASA TN D-3431, 1966.
5. Schweikhard, W. A method for in-flight measurement of ground effect on fix-wing aircraft. *Journal of Aircraft*, 4(2), 101–106, 1967.
6. Mooij, H. A. *Criteria for Low Speed Longitudinal Handling Qualities (of Transport Aircraft with Closed-Loop Flight Control Systems.* Martinus Nijhoff Publishers, the Netherlands, 1984.
7. Plaetschke, E., and Schulz, G. Practical input signal design. AGARD Lecture Series LS No. 104, 1979.
8. Anon. http://www.monmouth.com/~jsd/how//htm/.
9. Stevens, B. L., and Lewis, F. L. *Aircraft Control and Simulation.* John Wiley & Sons, Inc., Hoboken, NJ, 2003.
10. Anon. How airplanes fly. http://www.allstar.fiu.edu/aero/.

11. Parameswaran, V., Girija, G., and Raol, J. R. Estimation of parameters from large amplitude maneuvers with partitioned data for aircraft. *AIAA Atmospheric Flight Mechanics Conference*, Austin, USA, August 11–14, 2003.
12. Iliff, K. W. Maximum likelihood estimates of lift and drag characteristics obtained from dynamic aircraft manoeuvres. *Mechanics Testing Conference Proceedings*, pp. 137–150, 1976.
13. Knaus, A. A technique to determine lift and drag polars in flight. *Journal of Aircraft*, 20(7), 587–592, 1983.
14. Hamel, P.G (Ed.). Rotorcraft system identification. AGARD AR-280, September 1991.
15. Ham, J. A., and Metzger, M. Handling qualities testing using the mission oriented requirements of ADS-33C. *Paper presented at the 48th Annual National Forum of The American Helicopter Society*, Washington DC, June 1992.
16. Anon. ADS-33D, Aeronautical Design Standard-handling qualities requirements for military rotorcraft. U. S. Army Aviation and Troop Command, St. Louis, MO, USA, July 1994.
17. Shanthakumaran, P. An industry perspective of future helicopter handling qualities. *Proceedings of International Seminar on Futuristic Aircraft Technologies* (DRDO/Aero. Soc. of India), Bangalore, 3–5 December 1996.
18. Shanthakumar, N., and Ganugapati. Data base management system - developmental aspects. Personal communications, and notes. Flight Mechanics and Control Division, National Aerospace Laboratories, Bangalore, November 2000.
19. Krag, V., Jategaonkar, R. V., Jakel, A., and Wieser, W. Flight data management through data bank IKARUS. DLR-Mitt. 93-14, *Proceedings of DLR-NAL System Identification Conference* at NAL, November 24–25, 1993.
20. Koch, G., and Loney, K. *Oracle 8- The Complete Reference*, McGraw-Hill Education, New York, 1997.
21. Bayross, I. *Oracle Developer 2000- Forms 5.0*, Saraswati Book House, Delhi, 1999.
22. Urman, S. *Oracle 8-PL/SQL Programming*, McGraw Hill, New York, 1997.
23. Muller, R. J. *Oracle developer/2000 – Handbook*, Tata McGraw Hill, New Delhi, 1997.
24. https://www.icao.int/MID/Documents/2019/ACAO-ICAO%20Airworthiness/Session%205%20Part%2021%20%20Aircraft%20Certification%20final.pdf, accessed March 2022.
25. https://www.aopa.org/news-and-media/all-news/2018/september/flight-training-magazine/technique-weight-and-balance.

8 Flight Control

8.1 INTRODUCTION

In aerospace engineering studies, the practice of flight control is a systems discipline. Understanding of the feedback control/systems approach is vital to understanding the flight control theory and practice of piloted, remotely piloted, or even autonomous atmospheric vehicles (viz. fly-by-wire (FBW) aircraft, missiles, rotorcraft, unmanned aerial vehicles (UAVs), and micro-mini aerial vehicles (MAVs)). Even the Wright brothers had appreciation for the fact that the secret to the control of flight was feedback (it could have been a human as a sensor, an actuator, or a controller!); they recognized that the pilot should be able to operate the controls to stabilize, control, and guide the airplane in a desirable way and recognized the need of solving the problem of stability and controllability. In fact, they built their "flyer" as a slightly unstable and controllable one as an engineering experiment. An interesting confluence of theory and practice of automatic feedback control is depicted in Table 8.1 [1]. Applications of the control theory to aerospace (leading to flight control) span four major areas [2]: flight planning, navigation, guidance, and control. In order to build a satisfactory control strategy, adequate mathematical models of the dynamic system to be controlled are required. The control strategy is that of the "feedback" from the output (or/and any inner state) variable to the input variable (added to the pilot input command), and the main idea is that with the information from the output variable, the input variable is suitably altered so that with the new/composite input, the control system's (e.g. aircraft's) response comes as close as possible to the desired output. So, first we discuss some fundamental aspects and concepts of control in next few sections. Then, we deal with the requirements of control and some control strategies in general. Subsequently, we discuss methods of flight control and related performance evaluation aspects.

8.2 CONTROL SYSTEM: A DYNAMIC SYSTEM CONCEPT

A system is something that can be studied as a whole. It is a collection of parts, subsystems, or components that function in such a way in unison as to render the system to carry out an assigned functional role. An airplane is a complex and sophisticated dynamic system in this sense. Systems may consist of subsystems that are interesting in their own right (Chapter 6). They may exist in an environment that consists of other similar systems. Systems are generally understood to have an internal state, inputs from an environment, and methods for manipulating the environment or themselves. Since cause and effect can flow in both directions of a system and environment, interesting systems often possess feedback. The idea behind the dynamic systems theory is studying, understanding, and estimating the long-term behavior of a system that changes with time. The characterization of this behavior consists in

DOI: 10.1201/9781003293514-9

TABLE 8.1
Confluence of aero-control-formative years [1]

Period	Aircraft dynamics theory	Feedback control of aircraft	Feedback control theory
1890	-----	Gyroscope stabilization (Maxim)	-----
1900		Torpedo course control (Obry)	
1910	Study of phugoids (Lanchester) Small perturbation theory (Bryan and Williams)	-----	-----
1920	----- Measurements of derivatives and calculation of motions (Bairstow and Jones); the methods introduced in US (by Hunsaker)	Invention and demo of 2-axis aircraft stabilizer/aerial torpedo (by Elmer & Lawrence Sperry) -----	-----
1930	Measurements/calculations of derivatives (Glauert, Bryant, Irving, Cowley) Full-scale flight tests confirm the theory	-----	----- Study of aircraft under continuous control (Gates, Gamer)
1940	----- The status of theory surveyed (by B M Jones) For a variety of aircraft and conditions, the advances in calculations made Study of response under control (Neumark's) Frequency methods (by Greenberg, Seacord and others)	RAE-Mark IV-Siemens automatic pilot/ development of pneumatic-hydraulic A2-Wiley post-flight ----- Rudder control, missile development (Germany), all-electric and maneuvering automatic pilots, flight of the "Robert R. Lee" ------	Stability of feedback amplifier studied (by Nyquist) Logarithmic plots, sensitivity (by Bode) The techniques applied to servomechanisms (by Harris) Thesis by Hall ---- The new texts ----- Root locus method (by Evans)
1950	----- Lecture by Bollay Volumes by the BuAer-Northrop Improvements in understanding of flight control		

knowing the conditions of a system: (i) the system has a periodic behavior, (ii) the system recurrently returns to a given set, (iii) the system goes to all the possible sets that cover the space of the system, (iv) the system never leaves a given set, and (v) its components interact with each other as desired.

Although there is a clear distinction between the so-called "classical" (frequency domain-based) and "modern" (time domain/optimal control) concepts and theories, we do not want/need to invoke these distinctions here at all. These are matter of the past and bygone era! Since even today we need to and should use available tools or techniques to easily analyze and design systems/control systems. The point is that with so much progress in computing technology (HW memory, speed, HW/SW (hardware/software) parallelization), we can easily use any of the classical or modern tools with much greater ease and flexibility than we could do four decades ago. We believe that we need to use Bode diagrams and transfer functions to have frequency-domain feel and interpretation as well as state-space and time-domain analysis for optimization and direct time history visualization and interpretation of control system performance, and hence, we should use both the approaches in an integrated way for achieving best design with best achievable performance. May be, we should call this a hybrid approach, rather than "classical" or "modern." Also, since even this hybrid approach would further be augmented using the so-called "soft computing" (based on artificial neural networks, fuzzy logic modeling, genetic algorithms, and approximate reasoning), we would then like to consider the entire gamut of control system analysis, design, and validation methods as the "general theory of control systems." The main reason for this is the fact that we should make recourse to all or several such composite, classical, modern, and intelligent approaches and tools for analysis, design, and validation of complex and sophisticated control systems for: FBW aircraft, spacecraft, missiles, UVAs, MAVs, huge powerful computing network systems, and integrated electrical power grids/systems.

8.2.1 Bode Diagrams and Transfer Functions

A dynamic system could be stable or unstable. Even if such a system is stable, its performance might not be satisfactory. So, in general, one can say that all such systems would need some regulation, regulatory mechanism, or control of some variables to improve and enhance the performance of the system. Such a system that functions under partial or full supervision or control of some "control mechanism" (that could be regulatory control, feedback control, or feed forward control) is, in overall sense, called a control system. In general, control system analysis, say for an airplane, can be carried out using frequency-domain and/or time-domain methods. Frequency-domain methods are based on Bode diagrams, root locus, Nichols charts, and transfer functions; and direct time-domain analysis can be carried out using state-space methods [3–5]. Due to easy availability of very high-speed computers, any time-consuming technique can be easily adapted for analysis, design, and evaluation of control systems. The MATLAB/SIMULINK SW tool and its tool kit are very convenient ways for such analysis. These tools can be easily learned and practiced and can save much time and effort in carrying out detailed analysis and validation of control system designs. Transfer function, eqn. (5.2), for a linear system eqn. (5.1), is defined in terms of the Laplace transform (LT). For a function $f(t)$, the LT is defined as follows:

$$L(s) \equiv \int_0^\infty f(t)e^{-st}dt. \tag{8.1}$$

For example, the LT of unit step input u is given as $1/s$, where s is the complex frequency $s = \sigma + j\omega$. We can see that the function $f(t)$ need not be periodic. The inverse LT is defined as follows:

$$f(t) \equiv \frac{1}{2\pi j} \int_{e-j\infty}^{e+j\infty} L(s) e^{st} ds. \tag{8.2}$$

If s in the transfer function (TF) is replaced by $j\omega$, we obtain the complex numbers with respect to frequency. If the input is a sign wave, then we get the steady-state output as a sign wave of the same frequency but with a different magnitude and phase angle difference between output and input waves. The plot of these complex numbers with respect to frequency is called the frequency response. The Bode diagram is a plot of magnitude (or the amplitude ratio) and the phase angle (phase difference between the output and input) of a TF v/s frequency (Figure 5.4). In definition, the Bode plot is the result of the plot of magnitude and phase of poles and zeros of the TF with respect to frequency in logarithmic coordinates. The simplification in Bode plots is due to the fact that in the logarithmic representation, the multiplication and division are replaced by addition and subtraction, respectively. In a filter or control system's Bode diagram, the cutoff frequency is the point where the response is 3 dB down in amplitude from the level of the pass band. Beyond this frequency, the filter will attenuate (the amplitude) at all other frequencies.

8.2.2 Performance: Order, Type of System, and Steady-State Error

Any dynamic control system can be subjected to analysis to see if it meets certain performance expected of it. The performance of the control system is studied in terms of its transient- and steady-state behavior, the first being the response to the initial conditions and the latter being the response when the transients are settled. Since these two requirements are often conflicting, a trade-off is required. The responses of a system are studied in terms of unit step-, unit ramp-, and unit-parabolic inputs. As seen in Chapter 7, the responses of a dynamic system can be obtained by applying short pulse, doublet, 3-2-1-1, and other multi-step inputs for studying the specific behavior of such systems. Often, a steady-state error to standard inputs to the control system is computed. The highest power of s in the denominator of the TF is equal to the order of the highest derivative of the output, and if this highest power is equal to n, the system is of an order "n." A system could have a pole at the origin of multiplicity "N," then it is of type "N." This signifies the number of integrations involved in the open-loop system. If $N=1$, then the system is of type 1; if $N=0$, then it is type zero; and so on. The steady-state error in terms of the gain K of the system is given in Table 8.2. If the type number of the system is increased, the accuracy for the same type of input is increased, but the stability of the system is aggravated progressively.

TABLE 8.2
Steady-state error for types of systems

	Type 0	Type 1	Type 2
Step input $u(t) = 1$	$1/(1+K)$	0	0
Ramp input $u(t) = t$	∞	1/K	0
Acceleration input $u(t) = \frac{1}{2}t^2$	∞	∞	1/K

8.2.3 Stability Criteria

A dynamic system in order to be able to continue to perform its assigned task, for example, flying aircraft, should be stable as such.

8.2.3.1 Static Stability

A linear spring mass system [3] is described (Figure 5.5) as follows:

$$M\ddot{x} + D\dot{x} + Kx = 0 \tag{8.3}$$

Assume that mass M is displaced to the right by a small disturbance x. When the displacement takes place, the spring with its stiffness K (if K is positive) will provide a restoring force proportional to Kx to the mass. The positive restoring force is due to K being positive and it in the direction opposite to the initial movement of the object. The mass will have an initial tendency to move toward the original position. This is the static stability of a dynamic system. If K is negative, then naturally the mass will keep moving in the forward direction since the spring force will be in the direction of x. The mass will further move away from the original equilibrium position. This condition is called the static instability. The differential eqn. (8.3) has two roots:

$$r_1, r_2 = \frac{-D \pm \sqrt{(D^2 - 4KM)}}{2M}; \ \omega_n = \sqrt{\frac{K}{M}}; \ \zeta = \frac{D}{2\sqrt{KM}} \tag{8.4}$$

Now, we see that for ensuring the dynamic stability, the roots should have negative real parts. This is possible if $D > 0$. If $D < 0$ and $K > 0$, then the system becomes dynamically unstable, even though the spring has positive stiffness (assuring static stability). If K is increased further, then $\zeta = \frac{D}{2\sqrt{KM}}$ reduces further. In this case, the excessive static stability leads to a poorly damped system.

8.2.3.2 Routh–Hurwitz Criterion

The characteristic equation of any control system is its denominator equated to zero. The R-H criterion relates to studying the characteristic equation: (i) the control system is dynamically stable if all the roots have negative real parts, (ii) the system is unstable if any root has a positive real part, and (iii) the system is neutrally or marginally (stable/unstable) if one or more roots have pure imaginary values. Let this equation be given as follows:

$$g(s) = a_n s^n + a_{n-1} s^{n-1} + \cdots + a_1 s + a_0. \tag{8.5}$$

Then, the Routh array is constructed, as shown next. The first two rows are obtained from the characteristic equation. The remaining elements of the Routh arrays are calculated from the following expressions:

$$b_{n-1} = \frac{a_{n-1}a_{n-2} - a_n a_{n-3}}{a_{n-1}};\; b_{n-3} = \frac{a_{n-1}a_{n-4} - a_n a_{n-5}}{a_{n-1}};\; \cdots c_{n-1} = \frac{b_{n-1}a_{n-3} - a_{n-1}b_{n-3}}{b_{n-1}}, \tag{8.6}$$

and so on. The number of the sign changes in the elements of the first column of the array is observed (8.7). This number signifies the number of roots with positive real parts. If there are no sign changes, then the system is stable.

$$\left\langle \begin{array}{c|cccc} s^n & a_n & a_{n-2} & a_{n-4} & \cdots \\ s^{n-1} & a_{n-1} & a_{n-3} & a_{n-5} & \cdots \\ s^{n-2} & b_{n-1} & b_{n-3} & b_{n-5} & \cdots \\ s^{n-3} & c_{n-1} & c_{n-3} & c_{n-5} & \cdots \\ \cdot & & & & \\ \cdot & & & & \\ \cdot & & & & \\ s^0 & h_{n-1} & & & \end{array} \right\rangle \tag{8.7}$$

8.2.3.3 Nyquist Criterion

We can examine the frequency response of the $G(s)H(s)$ (or $GH(s)$) loop TF and see if the gain is greater than unity at the phase lag of 180 deg. This is equivalent to studying the condition $GH(j\omega) = -1$ or finding a root of $(GH+1)$ on the s plane $j\omega$ axis, this being the stability boundary. The stability of the system is studied by using the principle of argument advanced by Nyquist. A semicircular contour of an infinite radius is used to enclose the right-hand s plane (RHS). As per this principle, as s traverses the closed contour in a clockwise direction, the increment in the argument of $(1+GH)$ is $2\pi N$. Now, if P is the number of the poles of $GH+1$ and Z is the number of zeros of $GH+1$ inside the Nyquist contour (i.e. the right-hand s plane with an infinite radius), then $N=P-Z$, the number of counterclockwise encirclements of the origin of the s plane. This can be translated to count the encirclements of the so-called critical point $-1+j0$, by the GH contour. The stability criterion is specified as $Z=P-N$. We recognize here that the poles of $1+GH$ are the open-loop poles, and the zeros of $1+GH$ are the closed-loop poles. Thus, the number of the unstable

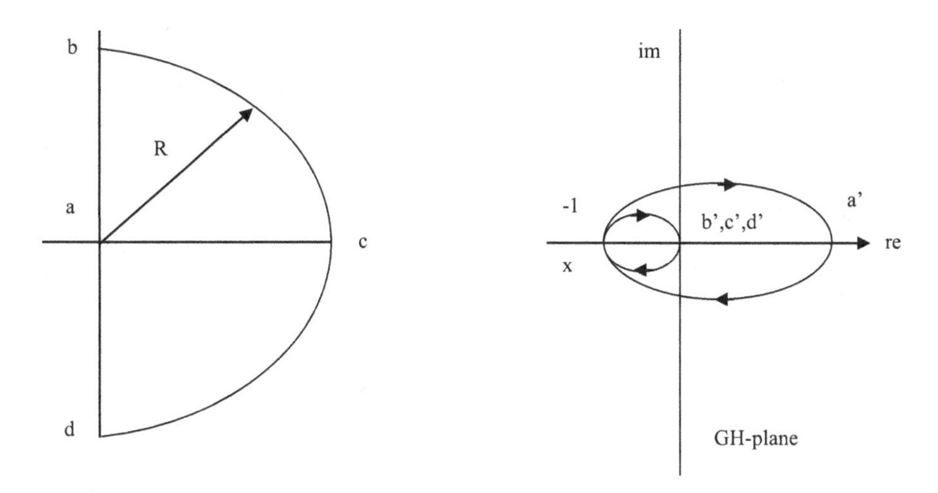

FIGURE 8.1 Nyquist contour (R is infinite radius) and Nyquist plot.

closed-loop poles = number of the unstable open loop – number of the counterclockwise encirclements of the critical point. The steps involved in the determination of the stability of the system using the Nyquist plot are illustrated in Figure 8.1 for the following TF [3]:

$$GH(s) = \frac{60}{(s+1)(s+2)(s+5)}. \tag{8.8}$$

(i) The part "ab" of the Nyquist contour is the $j\omega$ axis from $\omega = 0$ to $\omega = \infty$. This maps to the Nyquist polar from a' to b' (the latter at the origin); (ii) the infinite semicircle "bcd" maps into the origin of the polar plot of GH plane, that is, into points b', c', and d'; and (iii) the negative $j\omega$ axis "da" maps into the curve d'a' in the GH plane. We see that the GH plane Nyquist plot does not enclose the critical point, and hence, $N=0$. Since $P=0$ (there are no poles of the GH in the RHS plane), and hence, $Z=0$, meaning that there are no closed-loop poles in the RHS plane. The closed-loop system is stable.

8.2.3.4 Gain and Phase Margins

We see from the R-H criterion that it gives the absolute stability of the system, that is, whether the system is stable or unstable. The gain/phase margins give the relative stability of the system. They measure the nearness of the open-loop frequency response to the critical point. The gain margin (the *GH* frequency response is plotted from the point where the phase lag is 180 deg.) is the additional gain needed to make the system just unstable, whereas the phase margin (from the point where the gain is one) is the additional phase lag needed to render the system just unstable. The margins for a given system can be easily computed using the MATLAB control system tool box. Gain and phase margins are the criteria specified for design of control systems. What is almost always specified is that the gain margin (of the loop TF $GH(s)$) should be at least 6 dB, and the phase margin should be at least 45 deg.

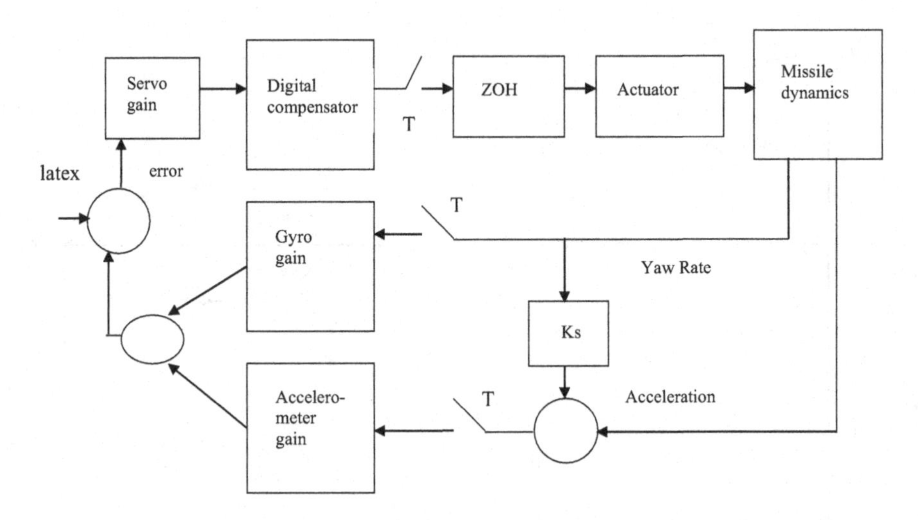

FIGURE 8.2 Schematic of a (missile) sampled data control system. Latex, lateral acceleration demand; ZOH, zero-order hold.

8.3 DIGITAL CONTROL SYSTEM

The digital computers are increasingly being used to implement the control systems and for many related studies. Most modern control systems operate in the digital domain [5]. Control (system) laws are computed via algorithms in a digital computer. One can view the computer-controlled systems as approximation of analog control systems. In such a system, the signal conversion is carried out at t_k, the sampling instant (or time) (Figure 8.2). The system runs open loop in the interval between the A-D and the D-A conversion (analog to digital and digital to analog), and the computational events in the system are synchronized by the real-time clock in the computer. Such a system is a sampled data system, but the events are controlled and monitored by a computer or its dedicated micro-processor (chip). The D-A converter produces a continuous time signal and is done by keeping the control signal constant between the signal conversions. These signals are called the sampled or discrete time signals. Thus, we can use the term sampled data system synonymously for the computer-controlled system. In terms of the system theoretic concept, a computer-controlled system with periodic sampling is a periodic system because if a system is controlled by a clock, the response of the system to an external stimulus will then depend on how the external event is synchronized with the internal clock of the computer system. A digital control system would have several merits: (i) freedom from direct bias and drift of the analog computers, (ii) desired accuracy can be obtained by proper implementation of the control algorithms in the computer, (iii) generally there is no data storage limitation, (iv) time-sharing ability of the digital computer can be used to an advantage for reducing the cost of implementation/computation, (v) effect of quantization and sampling can be modeled and studied in advance of the implementation of the control strategies, (vi) adaptive and reconfiguration control algorithms can be implemented with ease, and (vii) fuzzy logic-based control systems can be easily implemented.

TABLE 8.3
Design approaches and performance specifications [3]

	Bode plots/ frequency responses	Nichols charts/ frequency responses	Root locus	Pole placement method	Time-domain/ optimal[a] control methods
Performance criteria to be specified	Coefficients of steady-state errors/phase margins/ crossover frequency	Coefficients of steady-state errors/ maximum closed-loop frequency response	Coefficients of steady-state errors/location of dominant poles of the closed-loop and root sensitivity	Desired closed-loop TF, poles sensitivities to parameter variations	Integral performance criteria in time domain
Compensation types	Cascade lead, lag, or lead-lag	Cascade lead, lag, or lead-lag	Cascade lead, lag, feedback	More general ones in feed forward, feedback, or both	Varieties of optimization methods used for obtaining the optimum input to the plant

[a] There is a huge body of literature on optimal/modern control methods. Besides these, the approaches based on the quantitative feedback theory, H-infinity, and hybrid (H2 and H-infinity) methods are also used.

8.4 DESIGN COMPENSATION FOR LINEAR CONTROL SYSTEM

A control system might need some adjustments in order that several conflicting requirements can be adequately met. These adjustments are called compensations. A compensator can be inserted either in the forward or feedback path, or in both. In hardware control systems, the compensators are some physical devices like electrical, mechanical, or hydraulic/pneumatic. In digital control systems, wherein the control laws are computer programs and the compensators are also computer programs and perform equivalent tasks as the HW devices. A lead compensator can be used to provide a phase lead between the output and the input. Similarly, the lag and lead lag compensators are used as the case may be. The main idea is to modify the frequency response of the open-loop system/TF such that the performance of the compensated closed-loop system is satisfactory. Table 8.3 gives an overview of various approaches used for a control system design.

8.5 ROOT LOCUS

The method is based on the fact that it is possible to adjust the location of the poles of the closed loop TF by varying the loop gain [4]. The root locus sketches the movement in the s plane of the poles of the closed-loop TF as open-loop gain or some parameter is varied from zero to infinity. Let the characteristic polynomial of the closed-loop system be represented as follows: $KP(s)+Q(s)=0$, with K as some parameter to be varied from zero to infinity to obtain the plots of the roots of this equation leading to the root locus. It can be put in the following form: $1+KP(s)/Q(s)=0$, or $K\frac{P(s)}{Q(s)}=-1=1\angle 180\,\text{deg}$. This will yield the basic conditions to be satisfied by all the points on the root locus: (i) the angle condition –a point on the root locus where the algebraic sum of the angles of vectors drawn to it from the open-loop poles and zeros is an odd multiple of 180 deg., and (ii) the magnitude condition –a point on the root locus where the value of K is given as follows:

$$K=\frac{\text{product of lengths of the vectors from poles}}{\text{product of lengths of the vectors from zeros}} \tag{8.9}$$

The angle condition tells us if any point in the s plane lies on the root locus, and the magnitude condition gives the value of K for which this point will be a root of the characteristic equation. There are certain important properties of the root locus: (i) It is symmetrical about the real-axis of the s plane; (ii) a root locus branch starts from each open-loop pole and terminates at each open-loop zero or infinity; (iii) the number of branches that terminate at infinity is equal to the number of open-loop poles less than the number of open-loop poles; (iv) the sections of the real axis to the left of an odd numbers of poles and zeros are part of the locus, for $K>0$; and (v) if the number of poles and zeros are odd and to the right of a point, then this point is on the root locus.

8.6 AIRCRAFT FLIGHT CONTROL

An atmospheric vehicle can be regarded as being fully characterized by its velocity vector (VV). The time integral of the VV is the path of the vehicle through the space; thus, those qualities of an aircraft that tend to resist any change of its VV, either in its direction or magnitude, or both, constitute its stability. The aircraft's quality of control is governed by the ease with which the VV may be changed. It is the stability that makes possible the maintenance of a steady, un-accelerated flight path, and aircraft's maneuvers are affected by control; the aircraft is provided with reference coordinates for control and stabilization by means of gyroscopic instruments. In general, a small perturbation of the aircraft's motion about the equilibrium (or trimmed) state are considered, and linearized math models of the EOMs (Chapters 3 and 5) are utilized, and since many flight control problems are of very short duration (5–20 s), the coefficient of these equations are regarded as constants (time-invariant), and the TFs (Chapter 5) can be used for the design of control systems (filters/feedback laws). In addition to this, a pilot monitors the aircraft's responses using the cockpit instruments and the view of the window "feel" of the airplane as a qualitative feedback (control).

In the early days of flying, the flight control system (FCS) was mechanical, by means of cables and pulleys, the control surfaces of the aircraft were given the necessary defections to control the aircraft. However, new technologies brought with it the FBW FCS (FBWFCS) in which electrical signals are sent to the control surfaces.

The signals are sent by the flight (control) computer (FC/FCC). There are several merits of this systems: (i) a computer has a much higher reaction velocity than a pilot, (ii) it is not subject to concentration losses and fatigue, (iii) a computer can more accurately know the state the aircraft is in, (iv) computer can handle huge amounts of data better, and (v) it is not necessary to read a small indicator to know, for example, the velocity or the height of the aircraft. However, such an FCS only designed for a certain flight envelope cannot really operate the aircraft anymore when the aircraft is outside of the envelope, and for such situations, pilots are still needed [6].

The FCS of an aircraft generally consists of the following parts: (i) a stability augmentation system (SAS) augments the stability of the aircraft and mostly does this by using the control surfaces to make the aircraft more stable; a good example of a part of the SAS is the phugoid damper (or similarly, the yaw damper), which uses the elevator to reduce the effects of the phugoid – it damp-(en)-s it, and the SAS is always "on" when the aircraft is flying, without which the aircraft is less stable or possibly even unstable; (ii) a control augmentation system (CAS) is a helpful tool for the pilot to control the aircraft, for example, the pilot can tell the CAS to "keep the current heading," the CAS then follows this command, and the pilot does not have to continuously compensate for heading changes himself; and (iii) the automatic control system (ACS) takes things one step further and automatically controls the aircraft, and it does this by calculating, for example, the roll angles of the aircraft that are required to stay on a given flight path, it then makes sure that these roll angles are achieved, and the airplane is controlled automatically. However, the important differences between these systems are given as follows: (i) the SAS is always on, while the other two systems are only on when the pilot needs them, and (ii) in the CAS and ACS, the pilot feels the actions that are performed by the computer. When the computer decides to move a control panel, the stick/pedals of the pilot also move along, making CAS and ACS reversible. The SAS is not reversible since the pilot does not receive feedback, and if the pilot would receive feedback, the only things the pilot would feel are annoying vibrations, and this is, of course, not desirable.

8.6.1 Requirements of Flight Control

What is said regarding general control systems is also applicable to the automatic FCS (AFCS) for an aircraft. The need of the aircraft to fly in clouds and fog and to reduce the pilot workload during flight operation paved the way to the development of devices to stabilize the aircraft by artificial means [7]. In case of an FBW aircraft, artificial stabilization is required because the aircraft is inherently designed to have static instability for performance gains. The automatic maneuver envelope protection takes place due to the limits provided (in the computer) on the responses. Any reasonably desired limit on attitudes, speed, angle of attack (AOA), and "*g*" force can be provided. The relaxed static stability allows to plan smaller tail surfaces on such FBW aircraft, and hence, weight-saving could be up to 10%.

The FCS aids a pilot in controlling the aircraft safely and effectively throughout the entire flight envelope (Figure 7.1). Besides providing the basic stability, if needed, the AFCS in particular provides good handling qualities (Chapter 10). However, fundamentally, the AFCS provides a required stability and improved damping to the basic airframe, if these are inherently poor or insufficient. This aspect could be realized either by an autopilot or a limited authority command and SAS.

Since the response of the AFCS would be much more rapid than that of a human pilot, the effect of the disturbance would not reach a sizeable magnitude. The pilot can detect a change in the pitch attitude of 1 deg. in 0.3 s, and there would be further delay of 0.5 s by the pilot in his decision-making on the amount of the correction required to be applied [8]. However, the autopilot can detect the disturbance of 0.1 deg. in 0.05 s or even less, and then apply an input to overcome the disturbance in 0.1 s, thus gaining considerable advantage in the control response. This also means that less disturbance is seen by the aircraft.

In some case, the AFCS could be a full authority FBW/FCS. Modern high-performance aircraft are designed to have relaxed static stability or are inherently unstable in longitudinal axis. The relaxed static stability concept means that the CG of the airframe is more aft than the conventional designs. If the CG is aft of the neutral point, then the basic aircraft would be statically unstable. This allows reduction of the area of the horizontal tail surface. There will be reduced weight and reduced trim drag. There are certain advantages to be gained by this type of the design. The major reason for this approach is that a more favorable aerodynamic force balance is achieved than the conventional stable airframe configuration. The wing lift is aided by the control surface lift contribution, and for a given AOA, there is a higher lift and reduced drag (for a given lift) for the inherently unstable/relaxed static stability (RSS) airframe configuration. In the conventional aircraft, the balance of momentum is achieved by a downward lift of the elevator. This has a negative effect on the total lift. In the case of the unstable configuration, the moment balance is achieved with the upward lift of the elevator, and hence, the total lift of the aircraft is increased. There are other merits for such a design [9]: (i) about 10%–15% gain (depending upon the configuration) in the maneuver margin, (b) reduction in fuel consumption, and c) an increased climb rate. Since the inherently unstable aircraft as such cannot fly on its own, it needs artificial stabilization. It could have FBW/FCS with either limited or full authority command. The Rafale, gripen, F-22, and Eurofighter Typhoon aircraft are designed unstable.

In particular, FBW/FCS-related control laws have several major functions to perform: (i) SAS for excellent HQs and aircraft maneuverability; (ii) automatic speed and flight trajectory control (via autopilot), thereby reducing pilot workload; (iii) safe operability in all weather conditions; (iv) effective gust load alleviation; (v) an extended service life and enhancement of ride qualities; (vi) performance optimization; and (vii) reconfigurable/restructurable control tasks in the event of sensor/actuator faults or control surface damage.

From the point of view of "flight path," the resulting control problem is to generate adequate deflection of aerodynamic control surfaces or changes in power or thrust to maintain the shape of the flight path and the velocity along this path. In all transport and most other aircraft, autopilot is used to reduce the pilots' workload and to control the motion of the vehicle in pitch, yaw, and roll axes in order to traverse the required trajectory. The main aim is to ensure the stability of the vehicle in the presence of disturbances/forces/moments caused by various sources while maintaining structural integrity. While the autopilot strives to achieve the desired instantaneous

TABLE 8.4
Autopilot modes

Basic modes	Coupled modes	Integrated modes
• Pitch and roll attitude hold • Heading hold	• Heading select/acquire • Altitude hold/select/acquire • Mach hold • Flight path/track/hold/acquire	• Automatic approach and landing • Navigation modes

values, the guidance system aims at achieving the desired end-of-the-flight conditions. The autopilot is the inner control loop, whereas the navigation/guidance is the outer loop. Normally, the BW of the vehicle control loop is 0.2–0.6 Hz, and that for the navigation/guidance is 0.02–0.04 Hz. The guidance system employs the navigation system as sensors to detect the instantaneous velocity/position of the vehicle and generates the guidance commands to reach the desired conditions. Navigation is based on the high-precision measurements of acceleration and high speed/accuracy of the required computations. The autopilot consists of precision angular and angular rate sensors. All the autopilot modes help pilot for effective maneuvers mostly with hands-off. Autopilot could have several types of modes, as shown in Table 8.4 [9].

Due to the fact that some additional information/data are required to perform/conduct certain coupled/integrated modes, the interaction of autopilot with the inertial measurement unit and air data system would be definitely required. The autopilot types are as follows: (i) roll (bank angle position) autopilot holds the aircraft in the level position, (ii) the rate controller stabilizes the aircraft, (iii) heading select and hold autopilot keeps the course (of the aircraft), (iv) vertical speed, (v) airspeed select and hold, (vi) Mach number hold, (vii) pitch attitude hold, (viii) altitude select and hold, (ix) glide slope, (x) approach/flare, (xi) localizer, (xii) runway align, and (xiii) the flight path angle controller is used for climbs. The accelerations autopilots are generally used in missiles, and the altitude hold autopilot alleviates pilot workload. A brief description of a few control aspects is given here [8].

Altitude hold: The pitch attitude-sensing element detects any change in the aircraft attitude. This change will be compensated by the appropriate feedback law. However, changes in vertical displacement cannot be detected by an attitude sensor. So for automatic leveling off at any desired altitude, an altitude hold function is required. The proper sensor is based on a pressure transducer, and it senses a change in altitude as a change in (static) pressure. Correspondingly, the elevator servo actuator is operated to apply elevator control to restore the aircraft to the selected altitude. The altitude hold signal can be obtained from an inertial reference system. This signal is augmented with a barometric pressure-correlated altitude signal (from the altitude sensor) in an air data computer that interfaces with the FCS.

Airspeed hold: An airspeed sensor measures the difference between static and dynamic pressures. The speed error signal is applied to the pitch sensor control channel.

Mach hold: The airspeed hold mode is normally used for the low-altitude flights. The Mach hold is used during the high-altitude phase.

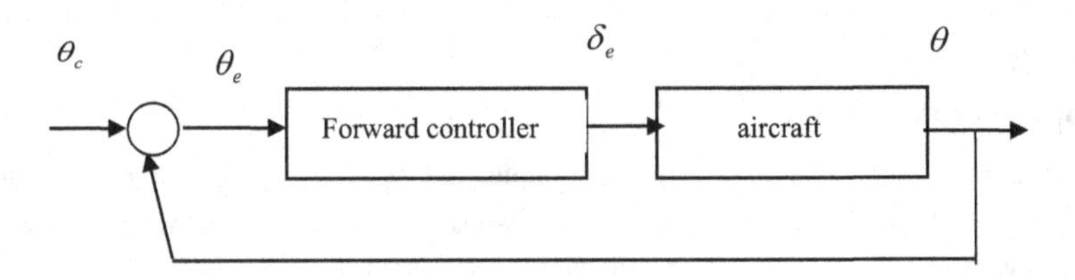

FIGURE 8.3 Attitude stabilization control system.

8.6.2 Stability and Control Augmentation Strategies

An FCS can have different/various roles to play: (i) automatic pilot controller – autopilot, (ii) basic stability augmentation, (iii) guidance control loop, and (iv) FBW/FCS for high-performance fighter or even transport aircraft (like AIRBUS 320). In any case, control deflection would be applied that is proportional to the deviation of the aircraft, say, in the attitude from the reference value (Figure 8.3). New situations and developments paved the way for intense research in flight mechanics and control [7]: (i) increased wing loading, (ii) mass concentration in slender fuselages, (iii) swept/delta wing aircraft, (iv) hydraulic powered controls, and (v) introduction of turbojet engines. These basic changes in design of the aircraft were required to expand the flight envelope of the aircraft and to have enhanced performance of the aircraft [7]. This resulted in deficiency in stability and control characteristics of the aircraft which further gave impetus to the design and development of sophisticated FBW/FCS. In order to handle the large hinge moments of the control surfaces, hydraulic and related control systems came into use, which increasingly "distanced" the pilot from the "feel" of the aerodynamic forces acting on the control surfaces. This further introduced the use of the artificial feel system to aid the pilot in performing the control task. A modern high-performance fighter aircraft would have full-authority FBS/FCS that would cater not only for the basic stabilization task but also do many tasks, which allow for care-free maneuvering of the aircraft and provide overall best aerodynamic efficiency. Any FCS would be based on some control strategies, which are described next [8–10].

With three-axis attitude stabilization, a feasible mechanization of the flight control would be attitude rate and attitude stabilization – attitude hold – when no command is given. This leads to "rate command/attitude hold" system and pilot's task of the stabilization of the aircraft is either eliminated or drastically reduced. The control system strategy based on θ or α and δ_e stabilizes the angular motion of an aircraft. In the (θ, δ_e) strategy, θ is kept constant, and the inherently unstable aircraft is made stable. The attitude response of the aircraft due to gust inputs can be reduced. This helps reduce the pilot's workload. In case of the control strategy based on (α, δ_e), the airflow direction relative to the aircraft is maintained constant. However, the pilot cannot get the "feel" of the AOA since it cannot be visualized from the out-of-window view. Another demerit of this is that AOA measurements could be often very inaccurate. The alternative control strategy based on (n_z, δ_e) would maintain the load factor constant. Over and above the foregoing control strategies, a system based on u and the engine thrust can be used to "stabilize" the airspeed. This strategy, however, has a demerit during takeoff and initial climb phases since maximum thrust is demanded. Yet another control strategy is to use direct-lift (aerodynamic) control surface to affect changes in the flight path angle. The (γ, δ_e) TF is given as follows:

$$\frac{\gamma(s)}{\delta_e(s)} \cong \frac{n_\alpha (g / V)}{s + (g / V) n_\alpha}. \tag{8.10}$$

With smaller value of n_α, there would be a large lag in the variation of γ after the change in pitch attitude has occurred. Also, a small value of n_α requires large AOA changes, and hence, large pitch angle changes. In this situation, limited use of a direct lift control strategy would be useful, due to an increase in drag, thereby restricting the use in the approach and landing flight phases.

The AFCS would mainly consist of sensors, output devices, and onboard digital computers. The sensors measure the relevant parameters and signals and transmit these to the computers. The output devices convert the computed signals to actuator commands. The onboard computers' functions are given as follows [8]: (i) amplification of the signal levels, (ii) integration, (iii) differentiation, and (iv) limiting, shaping, and programming. The control laws are as follows: the control filters, transfer functions, and gain scheduling algorithms are implemented on these computers. In general, the AFCS would be of various types: (i) rate damping systems, (ii) CASs, (iii) autopilots, and (iv) model following control systems. These systems could be (i) duplex, (ii) cross-coupled feedback, (iii) triplex system, or (iv) quadruplex system.

Example 8.1

The state-space model of the short period dynamics of a light transport aircraft are given as follows:

$$\dot{w} = Z_w w + (u_0 + Z_q) q + Z_{\delta_e} \delta_e$$

$$\dot{q} = M_w w + M_q q + M_{\delta_e} \delta_e. \tag{8.11}$$

Assume there is instability in pitch dynamics, and hence, M_w has positive numerical value. The system is stabilized with appropriate feedback from the vertical speed. New state-space equations are given.

Solution 8.1

Let the feedback gain be K units. Then, one can augment the control surface input as $\delta_e = \delta_p + Kw$, where δ_p is the pilot's stick command input. Substituting this for the control surface in the open-loop state-space equations, we get

$$\dot{w} = Z_w w + (u_0 + Z_q) q + Z_{\delta_e} (\delta_p + Kw); \; \dot{q} = M_w w + M_q q + M_{\delta_e} (\delta_p + Kw) \tag{8.12}$$

After simplification, we get

$$\dot{w} = (Z_w + Z_{\delta_e} K)w + (u_0 + Z_q)q + Z_{\delta_e}\delta_p;\ \dot{q} = (M_w + M_{\delta_e} K)w + M_q q + M_{\delta_e}\delta_p \quad (8.13)$$

Since the dimensional moment derivative is augmented with an extra term with K as the feedback gain factor by choosing an appropriate value for K, one can make the system stable. Since the control effectiveness derivative normally has a negative sign, a small value of K might be sufficient to stabilize the system.

Example 8.2

Let the state-space model for a dynamic system be given as follows:

$$x(k+1) = \Phi_k x(k) + Bu(k); \quad y(k) = Hx(k) \quad (8.14)$$

Assume that the system is unstable and the responses/signals are increasing without bound. An appropriate transformation method is used on the unstable data, and the state-space equations are appropriately reformulated.

Solution 8.2

Assume that a suitable parameter "*d*" is available, then the states, input, and output data can be transformed to stable data. The new transformed data would be as follows:

$$\bar{x}(k) = e^{-dk}x(k); \quad \bar{y}(k) = e^{-dk}y(k), \quad \bar{u}(k) = e^{-dk}u(k) \quad (8.15)$$

The previous equations can be written as follows:

$$x(k) = e^{+dk}\bar{x}(k); \quad y(k) = e^{+dk}\bar{y}(k), \quad u(k) = e^{+dk}\bar{u}(k) \quad (8.16)$$

Substituting the previous expressions into the state-space equation, we obtain the expressions for the stabilized data/system:

$$e^{+d(k+1)}\bar{x}(k+1) = \Phi_k e^{+dk}\bar{x}(k) + Be^{+dk}\bar{u}(k); \quad e^{+dk}\bar{y}(k) = He^{+dk}\bar{x}(k) \quad (8.17)$$

After simplifying, we get

$$e^{d}\bar{x}(k+1) = \Phi_k\bar{x}(k) + B\bar{u}(k); \quad \bar{y}(k) = H\bar{x}(k) \quad (8.18)$$

Finally, in terms of the transformed data, the state-space equations would be

$$\bar{x}(k+1) = e^{-d}\Phi_k\bar{x}(k) + e^{-d}B\bar{u}(k); \quad \bar{y}(k) = H\bar{x}(k)$$

$$\bar{x}(k+1) = \Phi_{kd}\bar{x}(k) + B_d\bar{u}(k); \quad \bar{y}(k) = H\bar{x}(k) \quad (8.19)$$

We see that the elements of the transition matrix and the control input vector/matrix are changed. Thus, by appropriately choosing the value of "*d*," the system equations can be stabilized. The growing data of the unstable system can be de-trended and used in system identification/parameter schemes in order to estimate the parameters of the inherently unstable system (Chapter 9).

8.6.3 Performance Requirements and Criteria

The FBW/FCS should achieve (i) the performance robustness in spite of uncertainty in plant, that is, aircraft dynamics, and in the presence of external disturbance, and (ii) good command following. The FCS/control laws should also provide sufficient gain and phase margins and adequate gains' roll-off at high frequencies so that the noise, aircraft model uncertainties, and structural modes have negligible effects on the flight/mission performance of the aircraft [9]. In most cases, the design and development of a flight vehicle, especially for manned aircraft, the specifications of required handling qualities are incorporated in the design cycle. This is more so for design of flight control laws. The time-domain and frequency-domain criteria are specified for handling quality evaluation of the pilot–aircraft interactions (Chapter 10). At times these requirements would be conflicting in nature, and some engineering judgment and insights are needed to arrive at acceptable and satisfactory design and meet the desired goals.

8.6.4 Procedure for the Design and Evaluation of Control Laws

In fact, the design of flight control laws is a multi-disciplinary process in which aerodynamic, structural, propulsive, and control functions are considered together. The modern flight controllers might excite structural modes of the airframe. These modes would interact with the control–actuator dynamics. Also, because of the increasing need to integrate flight controls with engine controls, the interactions between the aerodynamics, propulsive, and structural modes should be investigated and taken into account. Due to the extensive use of composite (material) in the design of the airframe and control surfaces, the aero-elastic coupling would also play a significant role in the design process of flight control laws.

Design and development of an FBW/FCS and autopilot presuppose good understanding of the dynamics of the aircraft and easy availability of the mathematical models of these dynamics over the entire flight envelope (Chapters 3–5). Primarily, such information is available in the form of aero database (Chapter 6) that is usually in the form of look-up tables. These tables are constructed based on the extensive wind tunnel tests carried out on scaled (-down) models of the aircraft during the design/development cycles. The data base would be in terms of aerodynamic coefficients as a function of several independent variables: Mach number, AOA, control surface deflection, store configurations, and any such parameter that would have a major or moderate effect on these coefficients, that is to say, at a given flight condition (altitude, Mach number) and with the specification of other relevant parameters, one should be able to determine the values of these aerodynamic coefficients. These coefficients are used in the EOM to obtain the flight responses, as discussed in Chapter 6. A very close scrutiny of this aero database is made, and the so-called flight mechanics parameters are computed (Chapter 4) [9]. These are in terms of aerodynamic (or stability and control) derivatives and other compound derivatives. As, for example, the maximum roll rate is given as $p_{\max} = -\dfrac{C_{l\delta_a}}{C_{l_p}}\delta_a \dfrac{2V}{b}$. In a combat flight phase operation, the aircraft would have large roll rate. However, a desired value might not be available due to limited control power and limits on the structure. In fact, dimensional linear models (Chapter 5) are obtained at several flight conditions and configurations. These models and the design specifications form a starting point for the design of control laws for the given aircraft. The design specifications could be (i) frequency-domain parameters and/or (ii) time-domain parameters. These performance indices are collected and put in a vector called vector performance index (PI) that is to be optimized. Often the structures of controller blocks are specified. At times, various criteria could be of conflicting nature, and hence, some relaxation/compromise would be required in the performance of the designed system. Also, certain required constraints on the controller gains, actual control gains, and time constants could be specified so that these gains are not unrealizable. This process is the multi-input multi-output (MIMO) pareto-optimal or conditionally optimal control law design procedure. In an interactive design process, the control design engineer also simultaneously looks at the dynamic responses of the closed-loop systems for checking the limits and the shape of the responses to guide the designer in the design process. The entire design process can be almost fully automated, thereby freeing the designer from the tedious design cycles and iterations. This is a very practical design procedure for determination of flight control laws and can form a substantial part of the rapid prototyping formulation and computational paradigm. Hence, in general, the procedure centers around: (i) availability of all the subsystem models that form the entire aircraft close-loop system like, actuators, sensor, anti-aliasing filters, ADC/DACs, quantization errors, and computational delays (Chapter 6); (ii) controller structures; and (iii) closed-loop system performance criteria like, stability margins and conventional control system criteria. The computational delays can be modeled by a second-order lag TF. Often first-order TF of the form $(K/(K+s))$ would suffice as a model for the actuator. The controller structure could be in the form of a state-space or TF.

Subsequently, the handling qualities are evaluated and full nonlinear flight simulation (FS) is carried out. Invariably, the FS is used in the flight control design cycle iterations wherein an engineer (-in-the loop) would fly the simulator and perform several different types of maneuvers and failure modes and give the assessment of the performance of the closed-loop control system (aircraft and controllers). There are several other stages of evaluation of control laws: (i) real-time simulator, (ii) iron bird (and hardware in loop simulator), and (iii) an in-flight simulator (IFS). A comprehensive and iterative procedure [9,10] for design of a FCS is depicted in Figure 8.4.

Example 8.3

The simplified block diagram of an FCS of a high-performance fighter aircraft is given in Figure 6.13. The MATLAB/SIMULINK tool is used to realize the interconnected blocks (Example 6.2) and vary the "destabilizing" gain and study the effects on the output responses.

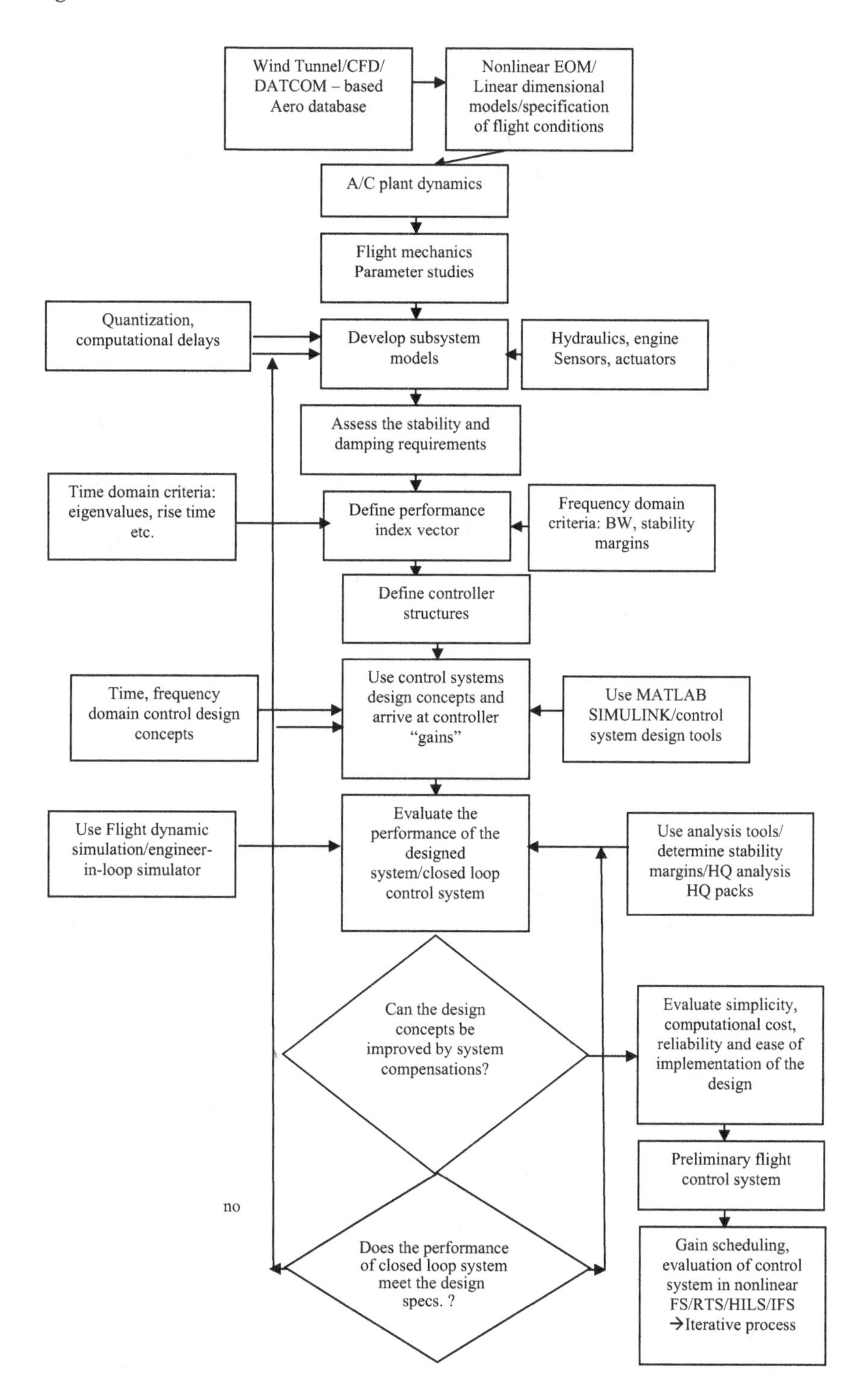

FIGURE 8.4 A typical iterative process for design of flight control.

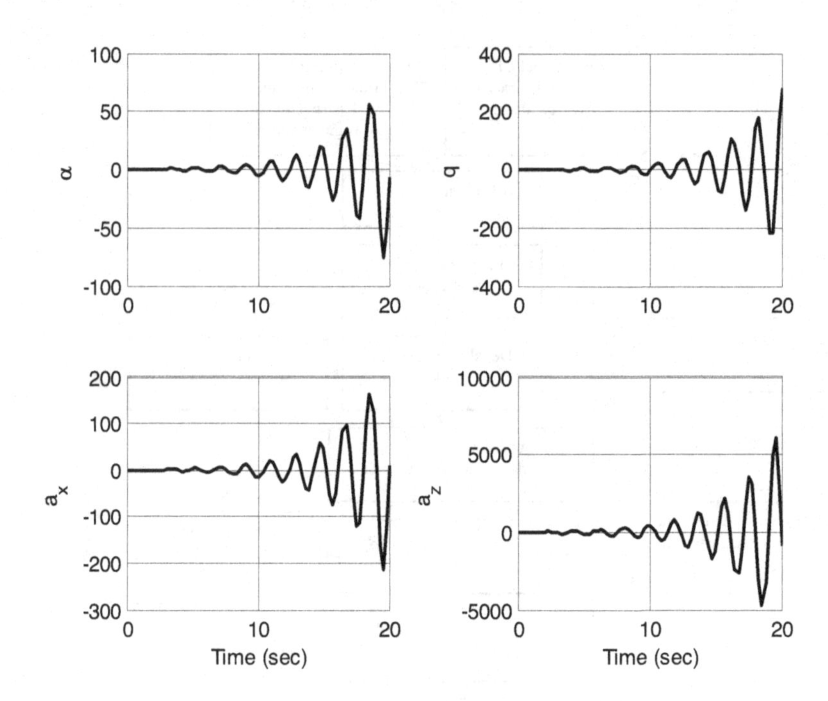

FIGURE 8.5 Time histories of the aircraft with a destabilized controller (Example 8.3).

Solution 8.3

To find the destabilizing gain, one can change each gain value individually, keeping other gains invariant. From the various simulation studies (**ExamplesSolSW/Example8.3ControlSimu**), the gain $k6$ is spotted as a destabilizing gain. If $k6$ is increased, system response becomes destabilizing, as can be noticed from Figure 8.5. The simulation study is carried out with the gain $k6=5$, and all other gains are kept unaltered.

8.7 STABILITY AUGMENTATION SYSTEMS

SAS makes an aircraft more stable: the dynamic stability, whether the eigen-motions do not diverge; the static stability, whether the equilibrium position itself is stable [6].

8.7.1 Dampers: Acquisition of Dynamic Stability

An aircraft can have several eigen-motions, and when certain properties of these natural motions do not comply with the requirements for the safe flight, an SAS is needed, mostly to damp these motions.

8.7.1.1 Yaw Damper

When an aircraft has a low speed at a high altitude, the DR motion of the aircraft deteriorates, and to prevent this, a yaw damper is used [6]. The yaw damper (a feed forward block) gets its feedback input from the yaw rate gyro and the reference signal and then sends a signal to the rudder servo; the rudder control surface is then moved in such a way that the DR is damped much more quickly than usual. The gyros are generally very accurate in low-frequency measurements, but not in high-frequency regions. The gyro's break frequency above which its performance starts decreasing is quite high, higher than any of the important frequencies of the aircraft; hence, it can often be simply modeled as $H(s)=1$; it can be assumed that the gyro is sufficiently accurate. Most actuators are always a bit slow to respond; their outputs lag behind the input. So, the rudder can be modeled as a lag transfer function; the lag time constant depends on the type of actuator. For slow electric actuators, $T_c \sim 0.25$ s; for fast hydraulic actuators, $T_c \sim 0.05$–0.1 s; this time constant or equivalently the servo break frequency is very important. If it turns out to be different than expected, the results can also be very different.

The yaw damper has to reduce the yaw rate but should not always try to keep the yaw rate at zero; then, the pilot will have a hard time to change the heading of the aircraft; thus, a reference yaw rate "r" is also supplied to the system. This yaw rate can be calculated from the desired heading rate by using the following equation [6]:

$$r = \dot{\psi} \cos\theta \cos\phi \tag{8.20}$$

If we do not know r, we can use a washout circuit, which is much less expensive; simply incorporate a washout term in the controller:

$$H_{\text{wo}} = \frac{\tau s}{\tau s + 1} \tag{8.21}$$

This circuit causes the yaw damper to fight less when a yaw rate is continuously present; in other words, the system "adjusts" itself to a new desired yaw rate. The time constant is quite important; for too high values, the pilot will still have to fight the yaw damper; for too low values, the yaw damper itself does not work because the washout circuit simply adjusts too quickly; a good compromise is often that $T_c = 4$ s.

In the yaw damper TF, we have proportional (p, P), integral (I), and derivative (D) actions: if the rise time is to be reduced, we need a proportional action; if the steady-state error needs to be reduced, we add an integral action; and if the transient response is to be reduced (to reduce overshoot), one needs a derivative action; thus, the right values of Kp, KI, and KD can be chosen. Sometimes, the optimal values of these gains differ for a flight phase; then, a gain scheduling can be applied. The gains then depend on certain relevant parameters, like the velocity V and the altitude h.

8.7.1.2 Pitch Damper

When an aircraft flies at a low speed and a high altitude, the short period mode has low damping that is compensated by a pitch damper, which is in many ways similar to the yaw damper, and the set-up is similar. Only this time, the elevators and a pitch rate gyro are used, instead of the rudder and a yaw rate gyro. The gyro TF is unity, as earlier, and the servo TF has a time constant of 0.25 s. The required reference pitch rate is calculated as follows (here n is the load factor) [6]:

$$L = nW = nmg = mg + mUq \rightarrow q = g(n-1)/U \tag{8.22}$$

Alternatively, one can use a washout circuit, for which the time constant can be chosen as 4 s.

Like a yaw damper, the pitch damper has proportional, integral, and derivative actions.

8.7.1.3 Phugoid Damper

This damper is used for adjusting the properties of the Phugoid mode. It is very similar to the previous two dampers; however, this damper uses the measured velocity U as input. Its output is sent to the elevator. The speed sensor is modeled as $H(s)=1$; and the elevator servo is modeled as a first-order lag with its $T_c=0.05$ s. The reference velocity needed is set by the pilot/autopilot or a washout circuit similar to earlier ones can be used; and the proportional, derivative, and integral actions are used for controller actions. Improving the Phugoid property means that the SP motion might become worse.

8.7.2 Feedback-Acquisition of Static Stability

Before an aircraft can be dynamically stable, it should be statically stable: we should have $C_{m\alpha} < 0$; $C_{n\beta} > 0$. Normally, the most aircraft would have this property, but highly maneuverable aircraft would not have this property since less stability generally means more maneuverability.

8.7.2.1 Feedback of AOA

The AOA is used as a feedback parameter. The AOA sensor is modeled as $H(s)=1$. The (canard) servo actuator is modeled as a first order lag with $T_c=0.025$ s [6]. For AOA feedback, usually only a proportional gain K_α is used. By using the mathematical models of the sensor, actuator, and the aircraft, a root locus plot is made, and a suitable value of the gain is chosen; this gain is then used to determine the necessary canard deflection $\Delta\delta_c = K_\alpha \cdot \Delta_\alpha$. A check should be made to see if the canard defection can be achieved since if gust loads can cause a change in AOA of 1 deg. and the maximum canard defection is 25 deg., then K_α should certainly not be larger than 25 or even be close to it.

8.7.2.2 Feedback of Load Factor

It is often hard to measure AOA accurately; instead, load factor feedback can be applied; the value of "n" ($n=L/W$) is used as feedback. The sensor TF is $H(s)=1$. The servo actuator is modeled as a first-order lag with it $T_c=0.025$ sec. We need to derive the TF between the load factor and the canard deflection [6]:

$$\Delta n = \frac{\dot{w}}{g} = \frac{U \tan\dot{\gamma}}{g} \approx \frac{U\dot{\gamma}}{g} = \frac{U\gamma s}{g} \tag{8.23}$$

Then, we get the TF as follows:

$$\frac{n(s)}{\delta_c(s)} \approx \frac{Us}{g}\left(\frac{\theta(s)}{\delta_c(s)} - \frac{\alpha(s)}{\delta_c(s)}\right); \quad \gamma = \theta - \alpha \tag{8.24}$$

Since all the TFs in the right-hand side brackets are known, one can chose appropriate feedback gain for K_n, and one should see that the gain does not require too large a canard deflection. It is often hard to distinguish important accelerations (like the ones caused by turbulence) from unimportant accelerations; hence, good filters need to be used to ensure a useful feedback signal is obtained.

8.7.2.3 Feedback of Sideslip

For lateral stability, feedback of angle of side slip (AOSS) can be used. The SS sensor TF is $H(s) = 1$, and the rudder servo is modeled as a first-order lag with its $T_c = 0.05$ s. The TF between the AOSS and the rudder deflection usually follows from the aircraft mathematical model, and a suitable gain K_β can be chosen for the control system. However, the SS feedback can generate a lateral Phugoid mode, and to compensate for this, another feedback loop is often used, where the roll rate is used as feedback for the ailerons.

8.7.3 Basic Autopilot Systems

Hitherto, we looked at the SASs, and these systems can be seen as the inner loop of the aircraft control. Now, we focus on the outer loop control, called the CAS; when we want to keep (maintain) a certain pitch angle, velocity, roll angle, heading, or something similar, then we use the CAS. With CASs, the pilot's workload is significantly reduced.

8.7.3.1 Longitudinal Autopilots

We focus on the longitudinal autopilot CASs for pitch angle, altitude, airspeed, and climb/descent rate.

8.7.3.1.1 Pitch Attitude Hold

This hold mode prevents a pilot from constantly having to control the pitch attitude; especially in the turbulent air, without the autopilot, it could be very tiring for the pilot. This CAS uses the data from the vertical gyroscope as the feedback input. It then controls the aircraft through the elevators. In actual sense, it sends a signal to the SAS (aircraft+SAS), which uses this reference signal to control the servo. It is assumed that the mathematical model of the aircraft together with the SAS is known. The gyro is again modeled as $H(s) = 1$, and the elevator servo is a first-order lag; if the elevator servo is already modeled in the SAS of the aircraft, then it is not necessary to take it into consideration.

The reference pitch angle needs to be set, which is carried out when the pitch attitude hold mode is activated; in fact, the hold mode usually tries to keep the current pitch angle. The pitch controller block consists of a proportional, an integral, and a derivative action; we need to choose the right gains *Kp*, *KI*, and *KD*. It may happen that with the new gains, the damping ratio of, say, the SP motion has shifted a bit; if it falls outside of the requirements, the SAS of the aircraft needs to be adjusted.

8.7.3.1.2 Altitude Hold

This mode prevents a pilot from constantly having to maintain the aircraft altitude. The feedback input is from the altimeter; the CAS then uses the elevator to control the altitude. For a radar or GPS altimeter, we use $H(s) = 1$; however, for a barometric altimeter, one needs to include a lag. The reference value of the height "h" is set in the model control panel. To control h, we should have an expression for this in the aircraft model. But h is not one of the parameters in the basic state-space model of the aircraft; we need to derive an expression for it [6]:

$$\dot{h} = V\sin\gamma \approx V\gamma \Rightarrow h(s) = \frac{V}{s}\gamma(s) = \frac{V}{s}(\theta(s) - \alpha(s)); \quad \gamma = \theta - \alpha \tag{8.25}$$

If a constant gain is used for the controller, the Phugoid may become unstable; this can happen if a low gain is used, and if it was already lightly damped, to handle this, one could use vertical acceleration feedback, or lead-lag compensation. This hold mode also consists of a proportional, an integral, and a derivative action; an integral action is often not necessary, and one simply uses a proportional derivative (PD) controller.

8.7.3.1.3 Airspeed Hold

This hold mode controller uses the airspeed sensor as input and controls the throttle. For GPS airspeed calculations, we use $H(s) = 1$; and if a pitot-static tube is used, then the sensor is modeled as a first-order lag. Also, the engine servo and the engine itself are modeled as first-order lag TFs. One can also include the engine effects in the state-space model, then add a term $K_{\text{th}}\delta_T$ to the equation for u-dot; this term represents the thrust due to the throttle setting; then, one has to only use the model of the engine servo. The reference/desired value of the velocity V is set at the mode control panel, or it can be derived from the actions of the pilot; if the pilot manually pushes the throttle forward, the computer increases the reference velocity V.

8.7.3.1.4 Climb or Descent Hold

This flight path angle hold mode is similar to the pitch attitude hold mode; here, flight path angle/climb rate is kept constant, and for feedback, γ angle is used. Since we cannot measure it directly, we use $\theta - \alpha$ [6]; the pitch angle can be measured using a gyro, and AOA sensor is used. This hold mode eventually uses the elevators to control γ. The gyro and AOA sensor are both modeled as $H(s) = 1$; one need to take into account the elevator servo since it is already modeled in the SAS of the aircraft. The flight path angle hold mode controller consists of proportional, integral, and derivative actions; however, the derivative action is often not required; the transient behavior is mostly acceptable when adjusting the flight path angle. A steady-state error, however, is more troubling, so integral action is often used.

8.7.3.2 Lateral Autopilots

These deal with roll angle, coordinated roll, and heading control.

8.7.3.2.1 *Roll Angle Hold*

The mode prevents the pilot from constantly having to adjust the roll angle during a turn; it uses the roll angle gyro as sensor and affects the ailerons. The roll angle gyro is modeled as $H(s) = 1$, and the aileron servo is modeled as a first-order lag. The reference roll angle is defined on the control panel. The reduced model for the roll angle is used [6]:

$$\frac{\phi(s)}{\delta_a(s)} = \frac{-L_{\delta_a}}{s(s - L_p)} \tag{8.26}$$

It can happen that the DR mode becomes unstable in the full model, whereas the reduced model used does not show this.

8.7.3.2.2 *Coordinated Roll Angle Hold*

This mode is an extension of the roll angle hold mode and also tries to ensure that the AOSS is equal to zero, resulting in a coordinated turn, thus giving the aircraft less drag and the passengers more comfort. This hold mode uses the SS sensor as a feedback signal, in addition to the roll angle gyro that was already used in the roll angle hold model, and then sends a signal to the rudder, in addition to the signal to the aileron that was already present. The SS sensor is modeled as $H(s) = 1$. One does not need to model the rudder servo as this is already incorporated in the inner-loop SAS, to be more precise, in the yaw damper. The SS angle β that is used as reference input is always zero; one does not want any SS in a coordinated turn. For AOSS measurement, one can use a vane-type SS sensor, but this signal is easily distorted due to some aerodynamic effects. Instead of this signal, one can use the lateral acceleration [6]:

$$\text{Force} = \text{mass} \times \text{acc.} \Rightarrow mA_Y = Y = C_Y \frac{1}{2}\rho V^2 S \approx C_{Y_\beta} \beta \frac{1}{2}\rho V^2 S$$

$$\Rightarrow \beta = \frac{2mA_Y}{C_{Y_\beta} \rho V^2 S} \tag{8.27}$$

In equation (8.27), we have neglected the effects of p, r, δ_a, δ_r on C_Y.

8.7.3.2.3 *Control of Heading Angle*

This is carried out by giving the aircraft a roll angle; it sends a signal to the coordinated roll angle hold mode, telling it which roll angle the aircraft should attain, then this roll angle is maintained until the desired heading is achieved. As a sensor, it uses the directional gyro, modeled as $H(s) = 1$. Its output affects the ailerons since the system controls the roll angle hold mode. The reference angle is set by the pilot on the mode control panel. In the aircraft model, if we do not have ψ as one of the state parameters, we use the following equation [6]:

$$\dot{\psi} = q\frac{\sin\phi}{\cos\theta} + r\frac{\cos\phi}{\cos\theta} \tag{8.28}$$

We assume that when $q=0$, there is no pitching during the turn, that the pitch angle is constant, and that the roll angle is small, then we have

$$\dot{\psi} = \frac{r}{\cos\theta}, \quad \text{or} \quad \psi = \frac{r}{s\cos\theta} \tag{8.29}$$

One can use the relation $\psi = \frac{g}{Us}\phi$.

8.7.4 Navigational Autopilot Systems

These are more advanced autopilot systems, and the aircraft is not only going to hold a certain parameter; instead, it is going to fly on its own and that it is navigating a glide slope, automatically flaring during landing, following a localizer or following a very high frequency (VHF) omnidirectional radio range (VOR) beacon.

8.7.4.1 Longitudinal Autopilot

8.7.4.1.1 Glide Slope Hold

In this mode, the aircraft automatically follows a glide slope. It reduces the pilot workload and is more accurate than when the pilot follows the glide slope on his own. The glide slope antenna is positioned at the aircraft CG and measures the glide slope error angle Γ (the sensor is modeled as $H(s)=1$), then the CG of the aircraft is driven along the glide slope. To achieve this, the aircraft is kept on the glide slope using pitch angle control, and the airspeed is controlled using the auto throttle; thus, it is assumed that the pitch and airspeed controls are already present. Let "*d*" be the deviation from the glide slope, then we have [6]

$$\dot{d} = V\sin(\gamma + 3^\circ) \approx V(\gamma + 3^\circ)\frac{\pi}{180^\circ} \Rightarrow d(s) = \frac{V}{s}\frac{\pi}{180^\circ}L\{\gamma + 3^\circ\} \tag{8.30}$$

In equation (8.30), $L\{.\}$ is the LT. We need a feedback signal, but we cannot measure "*d*"; hence, we use the error angle Γ, which is related to "*d*" as follows:

$$\Gamma \approx \sin(\Gamma) = \frac{d}{R}\frac{180^\circ}{\pi} \tag{8.31}$$

Here, R is the slant range. Based on the measured error angle Γ, which should be kept at zero, we can compute a desired pitch angle, then pass this angle on to the pitch attitude control system. The required pitch angle is computed using a glide slope coupler:

$$H_{\text{coupler}}(s) = K_c\left(1+\frac{W}{s}\right) \tag{8.32}$$

The coupler gain K_c is to be chosen such that the closed loop performance is acceptable. The weighing constant W is to cope with turbulence, and it is chosen as $W=0.1$. We also need to know the TF between flight path angle and the pitch angle [6]:

$$\frac{\gamma(s)}{\theta(s)} = 1-\frac{\alpha(s)}{\theta(s)} = 1-\frac{\alpha(s)/\delta_e(s)}{\theta(s)/\delta_e(s)} = 1-\frac{N_\alpha(s)}{N_\theta(s)} \tag{8.33}$$

If the slant range R changes, the property of the control system changes. In fact, if the gain K_c remains constant, then the closer the aircraft, the worse the performance. We can use some gain scheduling, we let K_c depend on the distance measured by the distance measuring equipment (DME) beacon, or even simpler but less accurate, let it depend on time; finally, one can also add a lead-lag compensator to the control system. The autopilot should be made robust; if certain parameters change, the autopilot should still work within a good performance range; the parameters that are subject to change in the real world are the aircraft CG location, weight, speed, and presence or intensity of turbulence; the autopilot should be able to cope with these variations.

8.7.4.1.2 Automatic Flare Mode

Getting the right vertical velocity v (not V, total velocity) on landing is difficult, it should not be too high; hard landings ($\dot{h} \le -6$ ft/s) are challenging for both the landing gear and the passengers; and too soft landings ($\dot{h} \approx 0$) are also undesirable as there will be floatation of the aircraft [6]. Ideally, we will have a firm landing with $\dot{h} \le -2$ to 3 ft / s. The relationship between the normal velocity and the vertical velocity during landing is usually $\dot{h} = -V\sin 3°$. The faster an aircraft flies, the harder the touchdown; this can be a problem for aircrafts with a low minimum speed, which thus have to fly fast. Therefore, such aircraft usually is flare right before touching down; it pull up its nose. By doing this, the airplane follows the so-called flare path. This path starts at the height h_{flare}. It ends by touching down 1100 ft. further than the point where the glide slope ends, which is where the glide slope antenna is positioned.

The aircraft is kept on the flare path by the pitch angle control system, which needs to have some input; for this, we approximate the flare path by

$$h = h_{\text{flare}}e^{-t/\tau} \tag{8.34}$$

We need constants h_{flare} and τ, which depend on the time t_d between the start of the flare and touchdown. First, we get the horizontal distance that the aircraft travels during the flare maneuver [6]:

$$Vt_{td} = 1100 + \frac{h_{\text{flare}}}{\tan 3^\circ} \tag{8.35}$$

From equation (8.35), we can derive the height h_{flare}. Now, we differentiate eqn. (8.34) as follows:

$$\dot{h} = -\frac{h_{\text{flare}}}{\tau} e^{-t/\tau} = -\frac{h}{\tau} \Rightarrow h_{\text{flare}} = -\tau \dot{h}_{|h_{\text{flare}}} \tag{8.36}$$

Now, we need the vertical velocity $\dot{h}_{|h_{\text{flare}}}$ (at the flare height); this can be found using $\dot{h} = -V \sin 3^\circ$, and once we know τ, we will have the control law for the automatic flare mode $\dot{h} = -\frac{h}{\tau}$; but we need to see if the aircraft stays at the correct altitude/height, h. We know the vertical speed of the aircraft (h-dot), which can be measured, and we also know the desired vertical airspeed from the control law; based on the difference, we can compute a desired pitch angle by using the coupler of eqn. (8.32), then the desired pitch angle is passed onto the pitch attitude control system. We need a model of the aircraft to see the effect of changes in the pitch angel to the change in the vertical velocity:

$$\dot{h} \approx V\gamma \Rightarrow \frac{\dot{h}(s)}{\theta(s)} = \frac{\gamma(s)}{\theta(s)} V \tag{8.37}$$

We have already derived an expression for $\frac{\gamma(s)}{\theta(s)}$ (8.33). For the automatic flare mode to work, an accurate altitude measurement system is required, and a radar altimeter is usually sufficiently accurate. In fact, when it is used, the system is so precise that aircraft always lands on exactly the same spot; this often results in runway damage at that point; to prevent this, a Monte Carlo scheme is used; the flare mode system now chooses a random point inside a certain acceptable box. It then makes sure that the aircraft touches down at that point.

8.7.4.2 Lateral Autopilot

Here, we discuss the localizer hold mode and the VOR hold mode.

8.7.4.2.1 Localizer Hold Mode

During an instrument landing, a pilot needs to follow the instrument landing system (ILS) localizer, but the localizer hold mode can perform this much more accurately, and it reduces the pilot workload. We assume that the aircraft CG follows the localizer beam center line and that the localizer error angle λ is sensed by the on-board localizer receiver. The aircraft is then kept on the center line using the heading angle,

ψ controller, which is assumed to be present. The deviation from the intended path is denoted as "d," then we have [6]

$$\dot{d}(s) = V\sin(\psi(s) - \psi_{\text{ref}}(s)) \approx V(\psi(s) - \psi_{\text{ref}}(s)) \Rightarrow d(s) = \frac{V}{s}(\psi(s) - \psi_{\text{ref}}(s)) \quad (8.38)$$

The ψ_{ref} is the desired heading angle and is the heading angle of the runway. For the feedback, the localizer error angle λ is used and is given by eqn. (8.30):

$$\lambda \approx \sin\lambda = \frac{d}{R}\frac{180^\circ}{\pi} \quad (8.39)$$

Based on the measured error angle λ that must be kept at zero, we compute a desired heading angle ψ, which is passed on to the heading angle controller. The desired heading is calculated using a coupler TF given by eqn. (8.32). When the slant range R becomes too small, a dynamic instability may occur, so the gain K_c needs to depend on the slant range R; or alternatively, a compensating network $H_{\text{comp}}(s)$ needs to be added. But, the localizer does not have to work for slant ranges smaller than $R=1$ nm because the localizer antenna is at the end of the runway, while the aircraft already touches down near the start of the runway.

8.7.4.2.2 VOR Hold

The VOR hold mode tries to follow a certain VOR radial. The principle of following the VOR radial is similar to that of following the ILS localizer path; the VOR error angle λ is used as feedback, and it be kept at zero. The VOR transmitter has a bandwidth of 360 deg. (range), whereas the ILS localizer only has a bandwidth (range) of 5 deg. since the localizer works only when the aircraft is more or less in line with the runway [6]. Also, the range of possible slant ranges R is much different for the VOR; the maximum range of a VOR beacon is roughly 200 nm; when an aircraft is flying at 6000 ft, the slant range simply cannot become less than 6000 ft (~1 nm). So, aircraft hardly ever comes closer than 1 nm to a VOR beacon. When an aircraft is above the VOR, it does not receive a signal, it is in the so-called cone of silence, and the VOR hold mode system should be able to cope with that.

8.8 FLIGHT CONTROL DESIGN EXAMPLES

In this section, we present three examples of control system design (implemented in MATLAB/SIMULINK), the MATLAB codes for which are provided with the book, and are re-scripted and re-run, and the figures, plots, and the results are re-generated by running the appropriate codes by the authors of this book.

8.8.1 DESIGNING A HIGH AOA PITCH MODE CONTROL

Example 8FC.1: This controller is designed to permit an aircraft to operate at a high AOA with minimal pilot workload. A controller is also included that is a

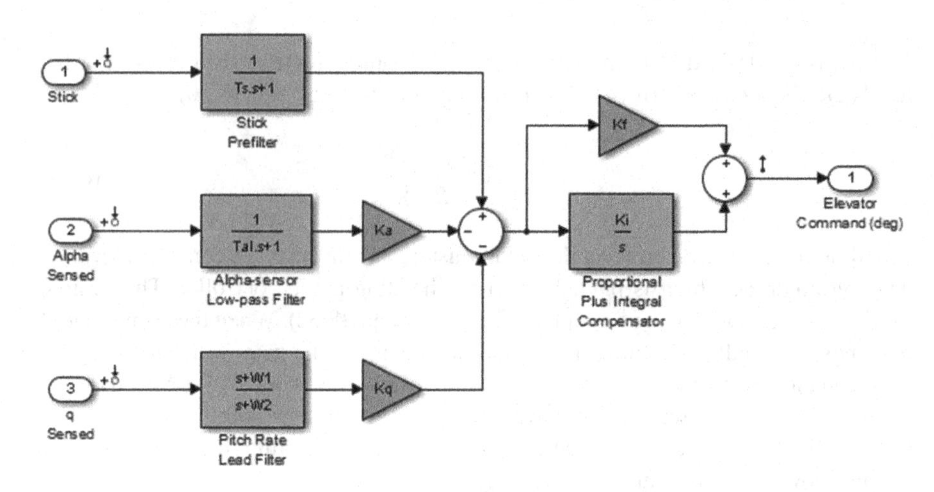

FIGURE 8.6 Autopilot scheme for HAOA pitch control [10]. HAOA, high angle of attack.

discrete implementation of the analog design that is similar to the algorithm that would go into an on-board flight computer [11]. The results can be generated by running the code **ExamplesSolSW/Example 8FC.1/haoapitchcontrol.m**. Figure 8.6 shows the controller block diagram [11]. One can linearize the model using Simulink Control Design SW, and to interactively linearize the model, one can use the **Model Linearizer** app; to open the app. in the Simulink Editor (on the **Apps** tab, under **Control Systems)** click **Model Linearizer; Or one** can programmatically linearize the model using the linearize function. There are three types of LTI objects to develop a linear model: state-space (SS), TF, and zero-pole-gain (ZPG) objects. When the code **haoapitchcontrol.m** is run, the results of trimming and linearization (and many other subsequent numerical results) will appear; these are included in the comments of the code itself.

Zero-order hold discretized controller: The LTI object generated can be used to design the digital autopilot. The analog system is coded into the LTI object called contap (CONtinuous AutoPilot). A digital autopilot is also created by using a zero-order hold ($T=0.1$ s). The discrete object maintains the type: ss, tf, or zpk. From Figure 8.7 it is seen that the systems do not match in phase from 3 rad/s to the half sample frequency (the vertical black line) for the pilot stick input and the AOA.

Tustin Bilinear transformation: Next, Tustin transformation with $T=0.1$ s is used; still the systems do not match (Figure 8.8). When $T=0.05$ s is used, we see from Figure 8.9 that the responses do match very well; with the new time constant, the discrete controller is able to track the performance better than with $T=0.1$ s.

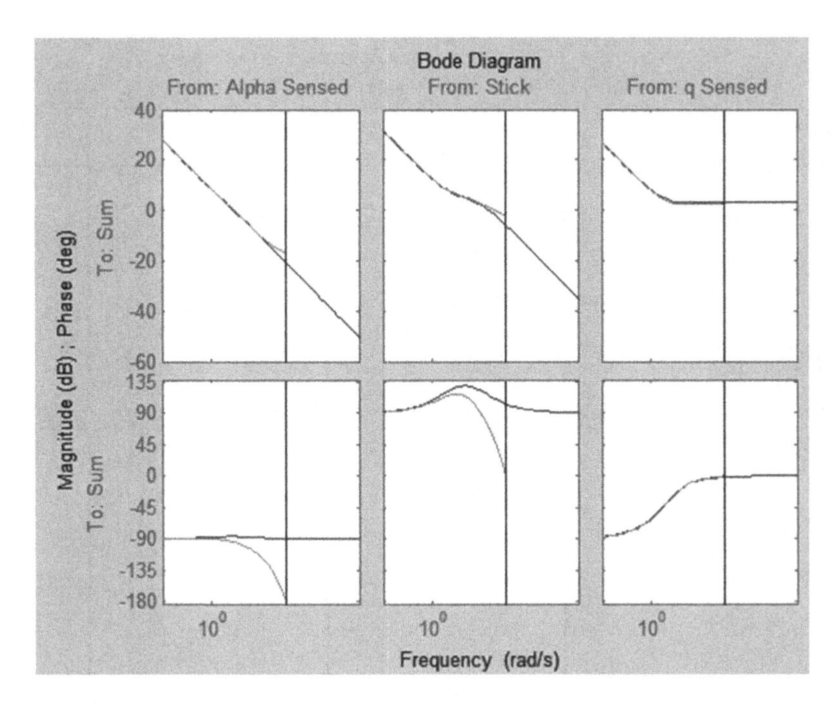

FIGURE 8.7 Comparison of analog and ZOH controllers. ZOH, zero-order hold.

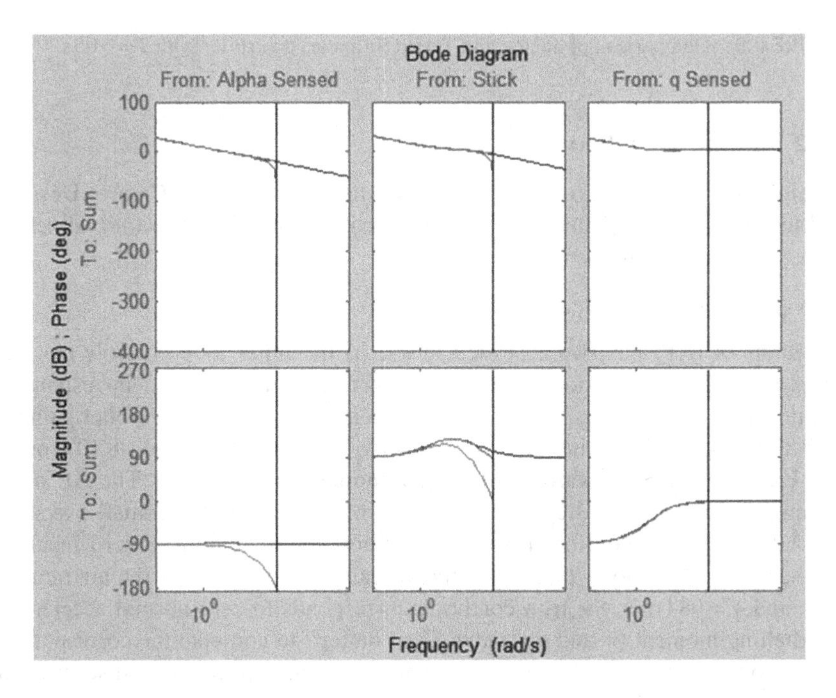

FIGURE 8.8 Comparison of analog and Tustin (bilinear) controller with $T=0.1$ s.

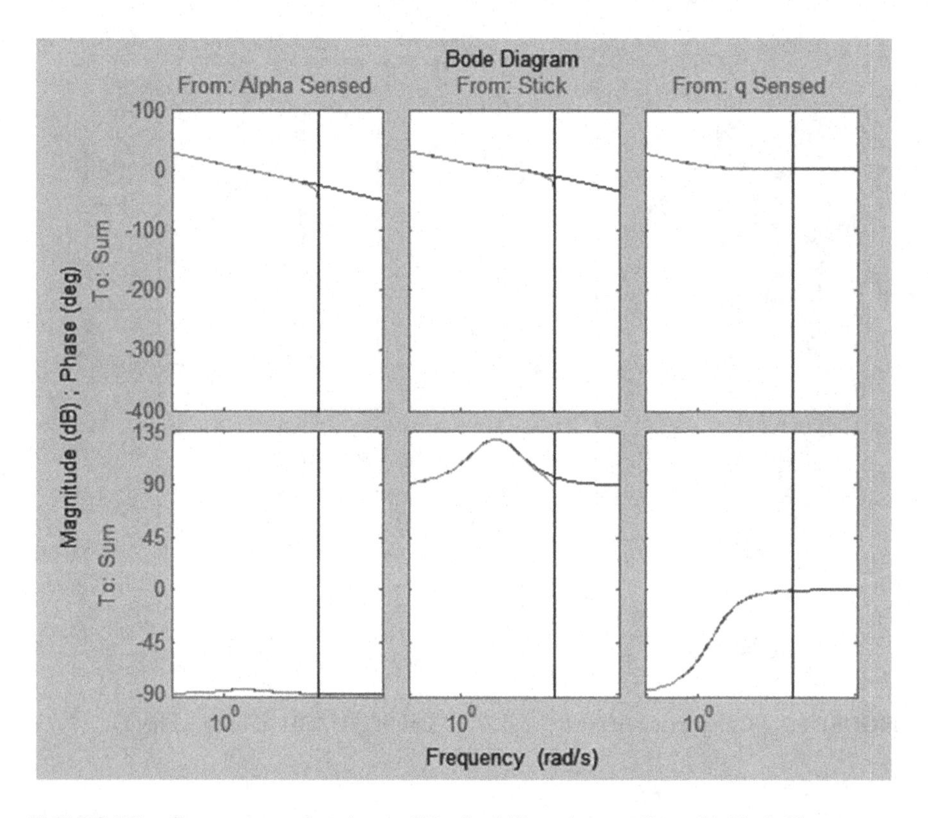

FIGURE 8.9 Comparison of analog and Tustin (bilinear) controller with $T = 0.05$ s.

8.8.2 Tuning of a Two-Loop Autopilot

Example 8FC.2: In this example, it is shown how the Simulink Control Design is used to tune a two-loop autopilot that would control the pitch rate (q) and vertical acceleration of an airframe [12].

8.8.2.1 Model of Airframe Autopilot

The autopilot (AP) has two cascaded loops; (i) the inner loop controls the pitch rate q, and (ii) the outer loop controls the vertical acceleration a_z; in response to the pilot stick command a_{zref}; the tunable elements are (i) the PI controller gains("a_z control" block), and (ii) the pitch-rate gain ("q gain" block), Figure 8.10 (run the code **ExamplesSolSW/Example 8FC.2/twoloopautopilot.m**) [12]. The AP should be tuned to respond to a step command a_{zref} in about 1 s with minimal overshoot. The AP gains are tuned for one flight condition corresponding to zero incidence and the speed of 984 m/s. To analyze the airframe dynamics, trim the airframe for $\alpha = 0$, and $V = 984$ m/s; the trim condition corresponds to zero normal acceleration and pitching moment (w and q steady). Use "findop" to compute the corresponding closed-loop operating condition, and a "delta trim" input port is added so that "findop" can adjust the fin deflection to produce the desired equilibrium of forces and

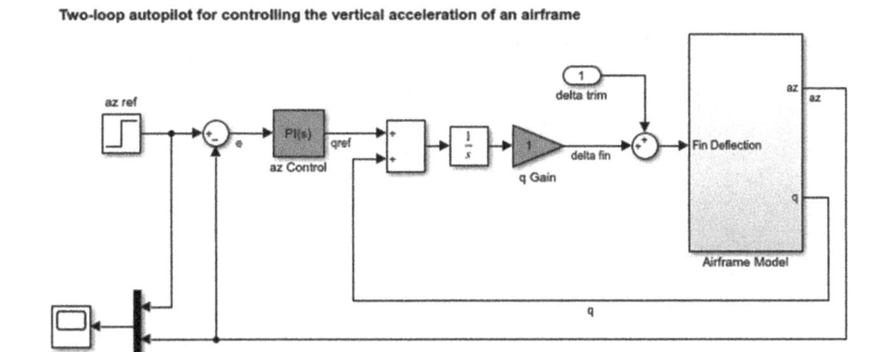

FIGURE 8.10 Block diagram of the two-loop autopilot tunable AP. (https://in.mathworks.com/help/slcontrol/ug/tuning-of-a-two-loop-autopilot.html, the figure is re-generated by running the code twoloopautopilot.m; accessed January 2022.)

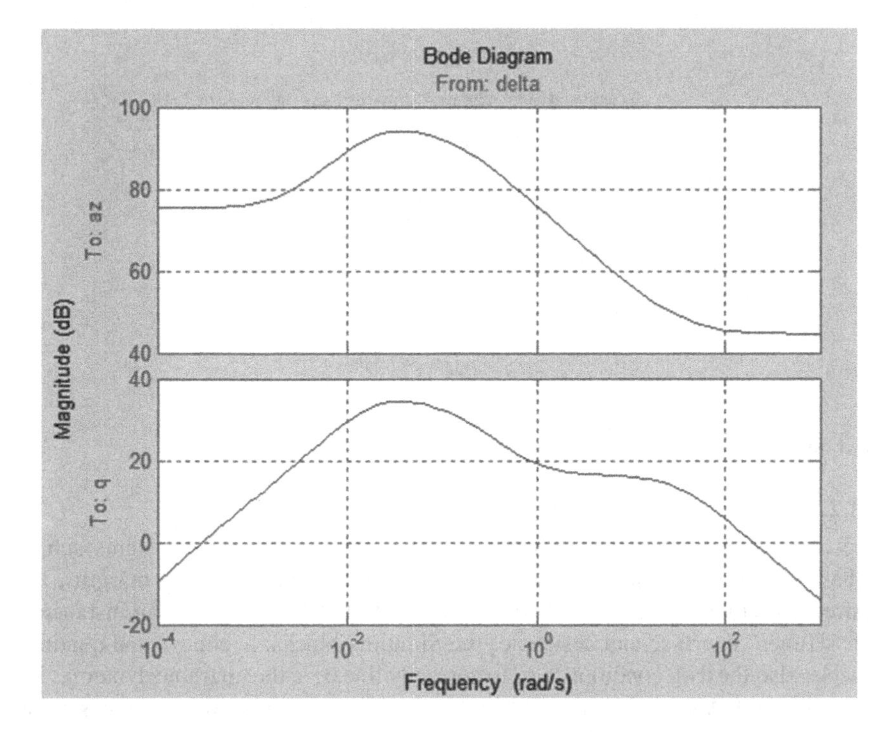

FIGURE 8.11 Two-loop autopilot tuning: the gains from the fin deflection to q, a_z.

moments. Linearize the "Airframe Model" block for the computed trim condition "op" and plot the gains from the fin deflection delta to a_z and q, Figure 8.11; note that the airframe model has an unstable pole: −0.0320, −0.0255, **0.1253**, −29.4685, obtained by pole(G).

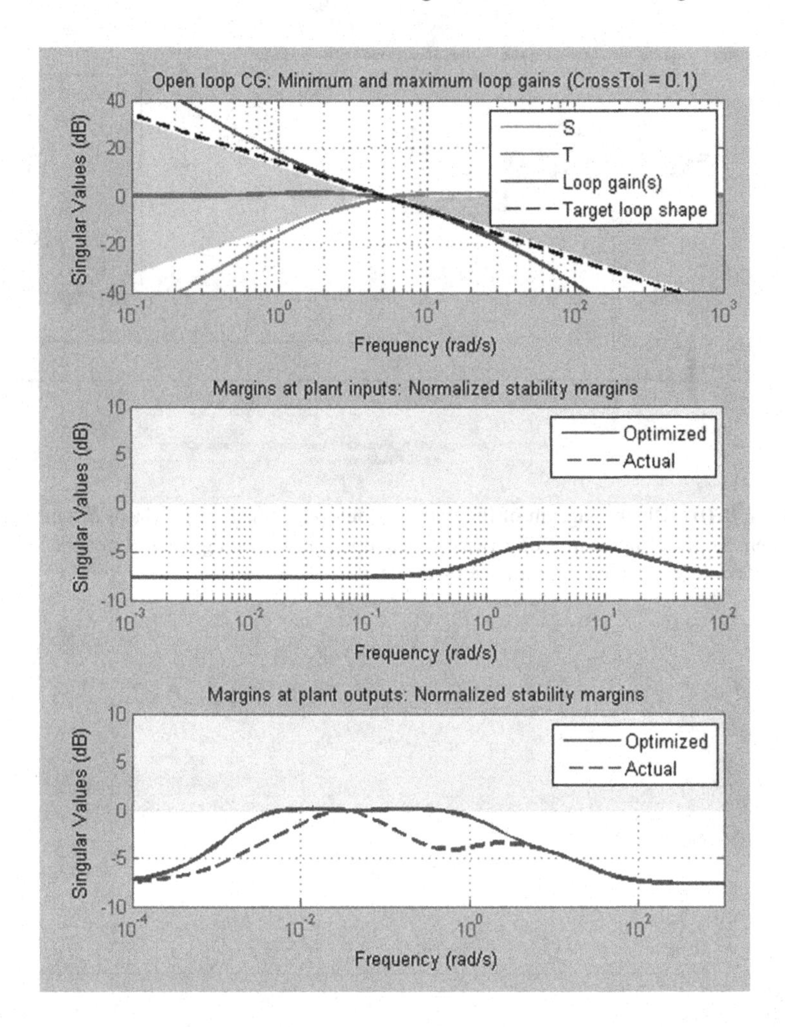

FIGURE 8.12 Requirements of the control met.

8.8.2.2 Tuning with Looptune

One can use the "looptune" function to tune multi-loop control systems subject to basic requirements such as integral (I) action, adequate stability margins, and desired BW; to apply "looptune" to the autopilot model, one creates an instance of the "slTuner" interface and designates the Simulink blocks: a_z control and q gain as tunable; also the trim condition "op" to correctly linearize the airframe dynamics has to be specified. Also, mark the reference, control, and measurement signals as points of interest for tuning/analysis. Then, tune the control system parameters to meet the 1 s response time specification; in the frequency domain, this roughly corresponds to a gain crossover frequency $w_c = 5$ rad/s for the open-loop response at the plant input δ_{fin}. The final peak gain = 1.01, the number of iterations = 72; the requirements are normalized so a final value near 1 means that these are met; this is confirmed by Figure 8.12, the top graph confirms that the open-loop response has integral action

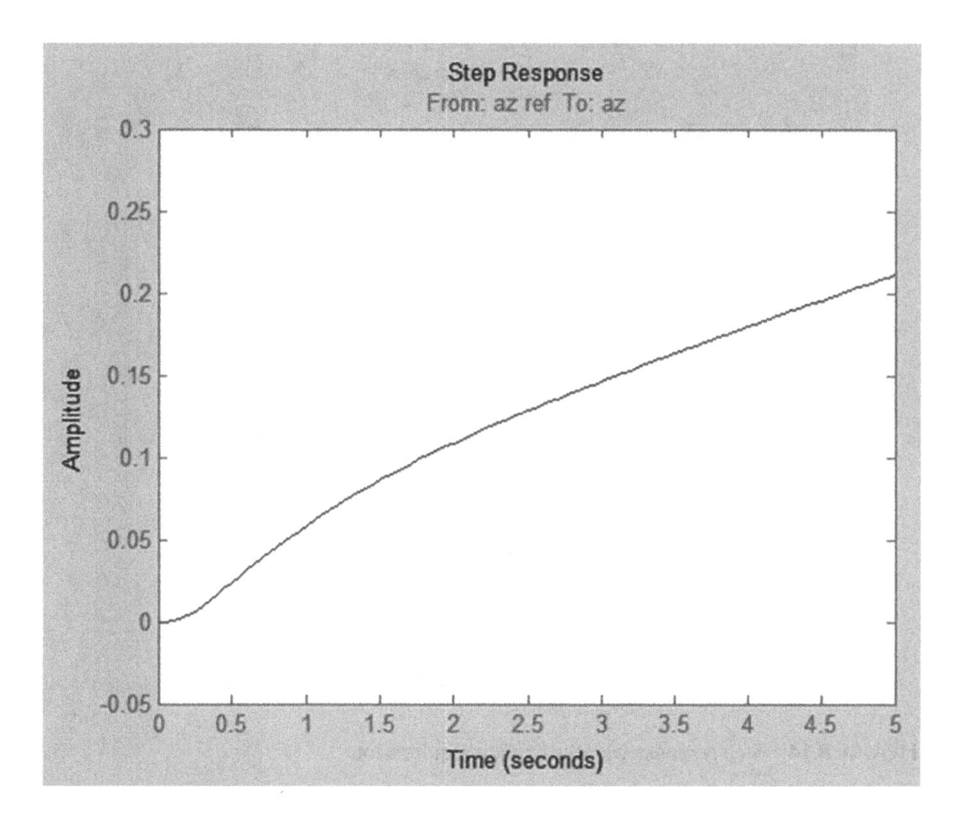

FIGURE 8.13 Step response.

and the desired gain crossover frequency is $w_c = 5\,\text{rad/s}$, while the next two graphs show that the MIMO stability margins are satisfactory (the optimized curve/s should remain below the yellow/gray shade-bound). Also, check the response from the step command a_{zref} to the vertical acceleration a_z, which does not track a_{zref} despite the presence of an integrator in the loop (Figure 8.13); this is because the feedback loop acts on the two variables a_z and q and it is not specified. which one should track a_{zref}.

8.8.2.3 Addition of a Tracking Requirement

An explicit requirement that a_z should follow the step command a_{zref} with a 1-s response time is now added, and the gain crossover requirement is relaxed to the interval (3,12 rad/s) to let the tuner find the appropriate gain crossover frequency; the final peak gain is 1.23, with 54 iterations. The step response is now satisfactory (Figure 8.14). Next, the disturbance rejection characteristics is checked by looking at the responses from a disturbance entering at the plant input (Figures 8.15 and 8.16). "showBlockValue" is used to see the tuned values of the PI controller and inner-loop gain. If this design is satisfactory, "writeBlockValue" is used to apply the tuned values to the Simulink model, and the tuned controller is simulated using Simulink.

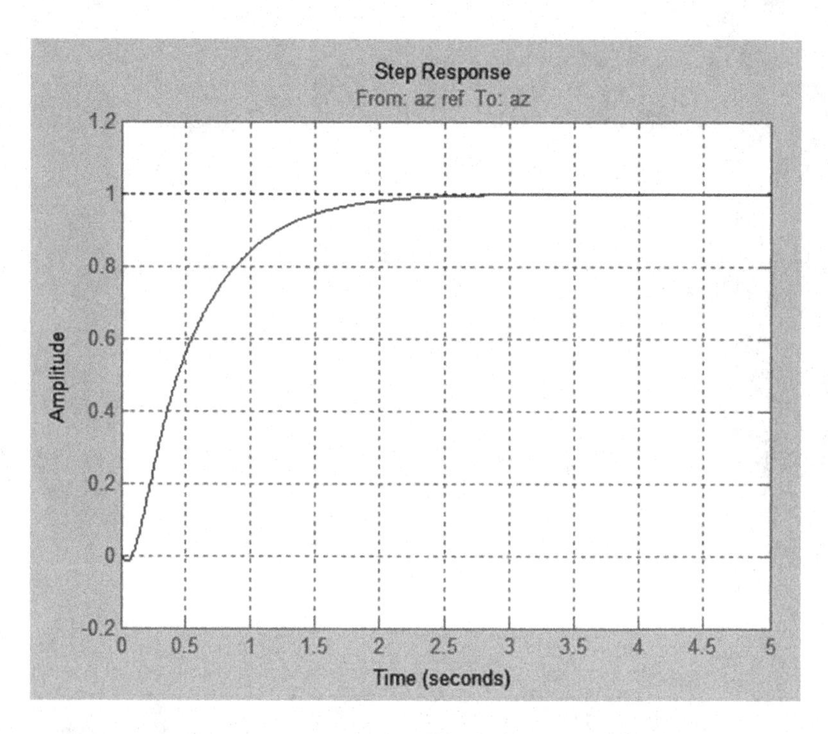

FIGURE 8.14 Step response with the explicit requirement.

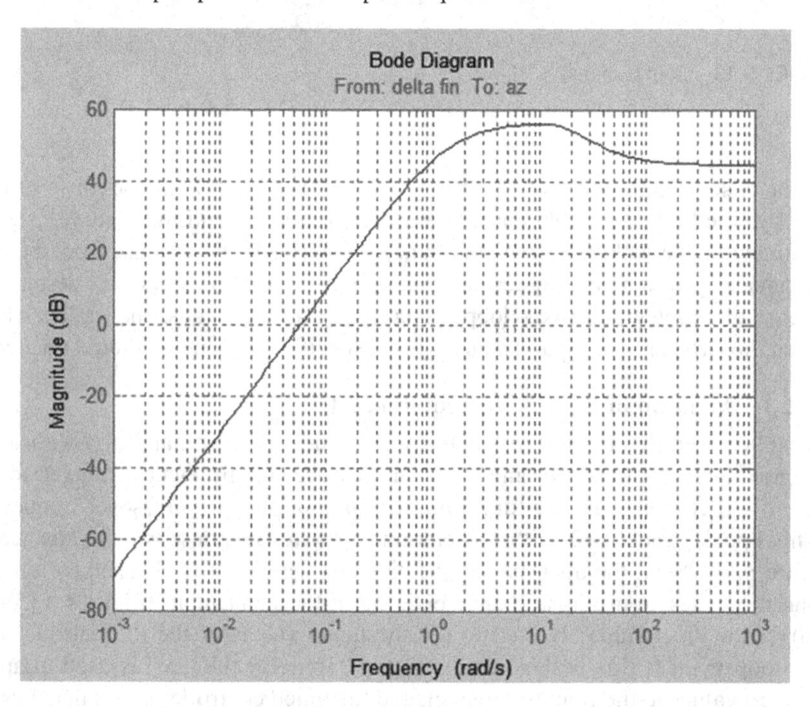

FIGURE 8.15 Bode diagram: from fin command to vertical normal acceleration.

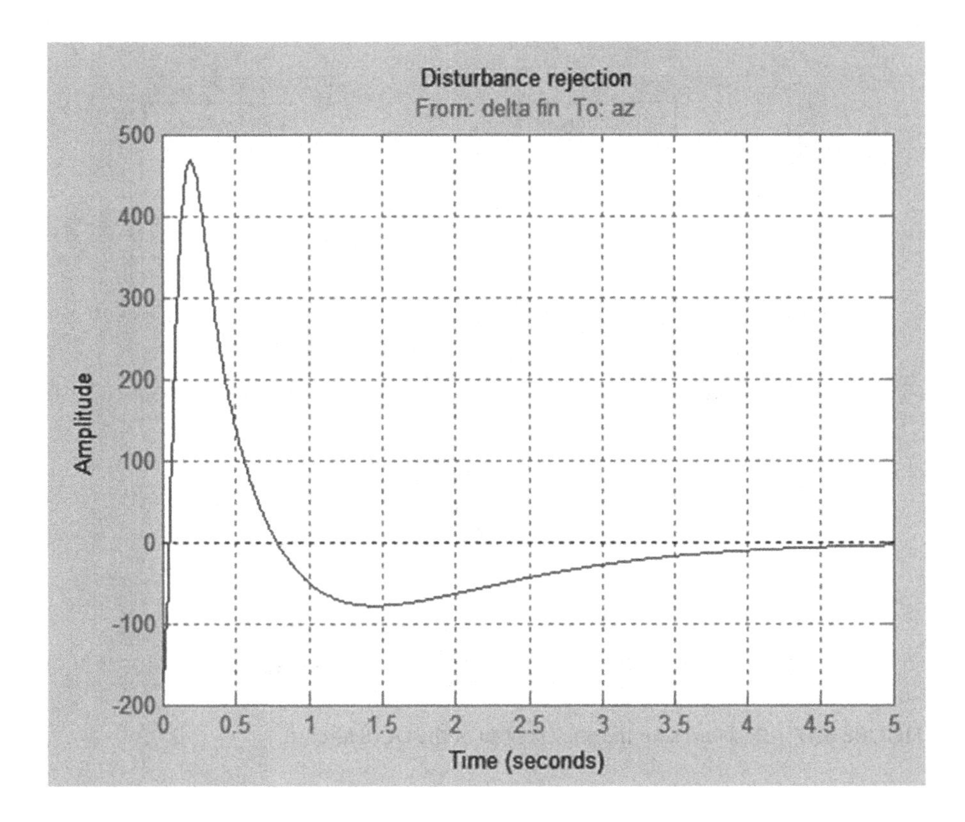

FIGURE 8.16 Disturbance rejection: from delta fin command to vertical normal acceleration.

8.8.3 DC-8 Aircraft Pitch Attitude Control

Example FC.3: The aircraft model is specified in the MATLAB script **ExamplesSolSW/Example 8FC.3/DCFC1PAC.m**, and the control results are generated by running the same program. (DC8FC1Long is called by the DC8FC1PAC.) The linearized 4-state mathematical model for the longitudinal axis of the aircraft at a certain flight condition is given as follows [13]:

$$\begin{bmatrix} \dot{u} \\ \dot{\alpha} \\ \dot{\theta} \\ \dot{q} \end{bmatrix} = \begin{bmatrix} -0.0291 & 15.3476 & -32.2000 & 0 \\ -0.0010 & -0.6277 & 0 & 1 \\ 0 & 0 & 0 & 1 \\ 0.0003 & -1.9592 & 0 & -1.053 \end{bmatrix} \begin{bmatrix} u \\ \alpha \\ \theta \\ q \end{bmatrix} + \begin{bmatrix} 0 \\ -0.0418 \\ 0 \\ -1.3391 \end{bmatrix} \delta_e;$$

$$C = \text{eye}(4,4), \text{ and } D = []. \tag{8.40}$$

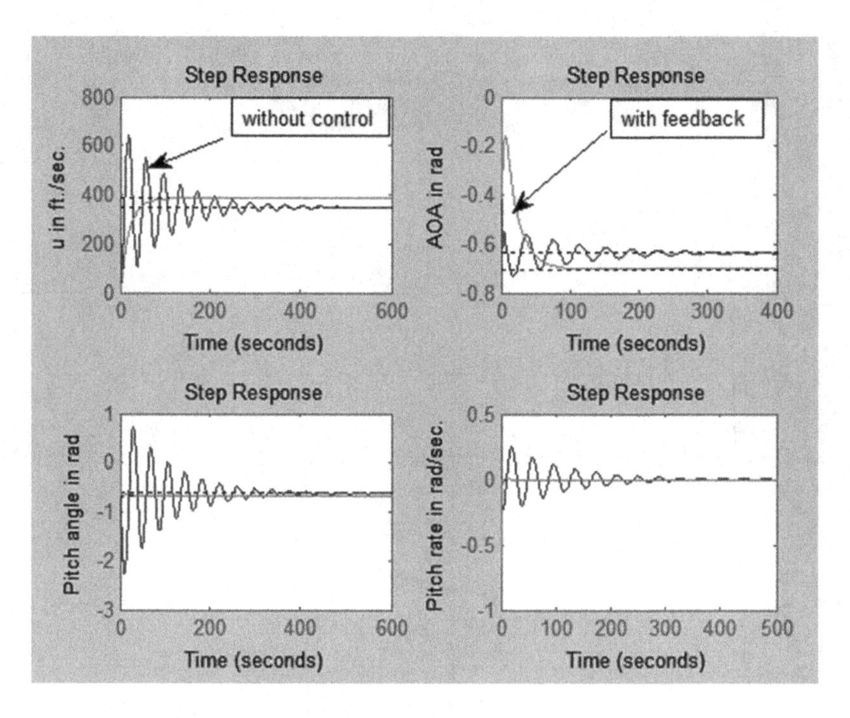

FIGURE 8.17 Responses for the pitch control of the DC-8 aircraft.

The open loop eigenvalues are found to be as follows:

eig(a): $\begin{bmatrix} -0.8450+1.3823i \\ -0.8450-1.3823i \\ -0.0099+0.1631i \\ -0.0099-0.1631i \end{bmatrix}$, which seem to be weakly placed so far as the dynamic stability is concerned; see the second pair of the EVs. Hence, with pitch damper (pitch rate feedback) and the PI controller $\{-20/(s+20)\}$, the closed loop eigenvalues are found to be as follows:

eig(acl): $\begin{bmatrix} -17.8995+0.0000i \\ -1.6559+2.0856i \\ -1.6559-2.0856i \\ -0.2246+0.1929i \\ -0.2246-0.1929i \\ -0.0492+0.0000i \end{bmatrix}$, which seem to be much improved. Figure 8.17 shows the step responses of the four states of the aircraft before the pitch attitude

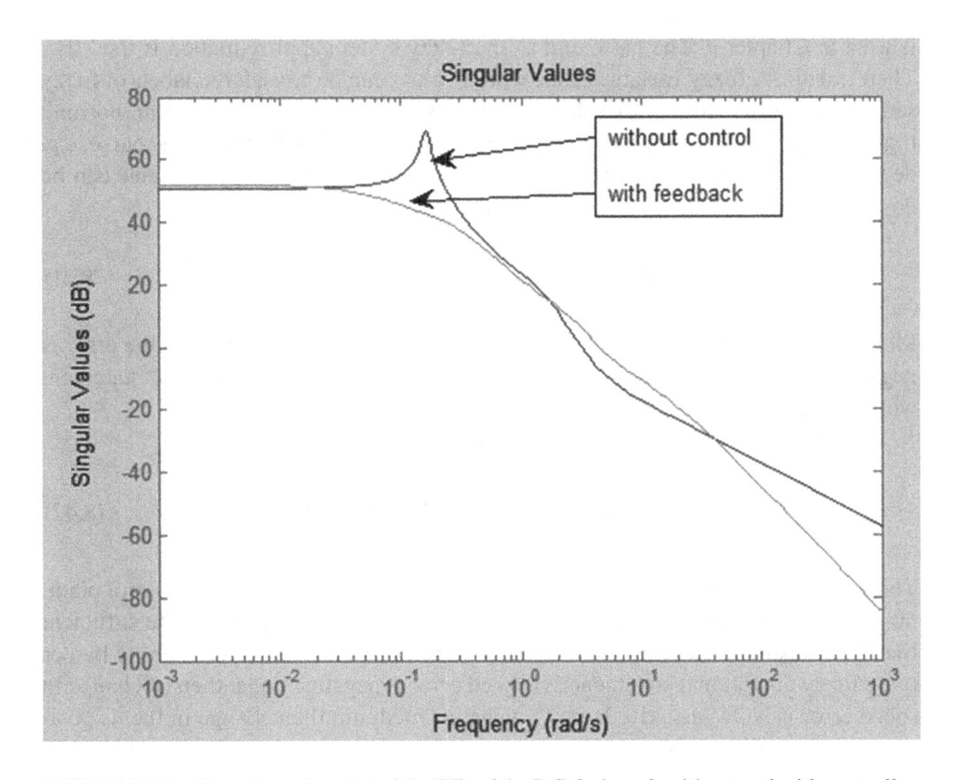

FIGURE 8.18 Singular value plot of the TFs of the DC-8 aircraft without and with controller.

control is applied and after the control augmentation; a very satisfactory design has been obtained, and this is further ascertained by the sigma plot (Figure 8.18) of the TFs of the dynamics of the DC-8 aircraft without and with the pitch damper controller.

8.9 FUZZY LOGIC CONTROL

Conventional control systems are not regarded as the intelligent systems. If, in the conventional control strategy, some logic is used, then it can be a start of the intelligent control system. Fuzzy logic provides this possibility. The fuzzy logic-based controller is suitable to keep the output variables between the specified limits and also to keep the control actuation, that is, the control input and the related variables, between limits. In fact, the fuzzy logic deals with vagueness, rather than uncertainty. The fuzzy logic in control would be useful if the system dynamics are slow and/or nonlinear, when models of the system are not available, and competent human operators, to derive the expert rules, are available [14]. Thus, the rule-based fuzzy system can model any continuous function or system, and the quality of fuzzy approximation would depend on the quality of rules. These rules can be formed by the experts, and if required, the ANNs can be used to learn the rules from the empirical data. The latter systems are called the adaptive neuro-fuzzy inference system (ANFIS), which is used in Chapter 9. The basic unit of the fuzzy system/approximation is the "If... Then"... rule. A fuzzy variable is one whose values can be considered labels of fuzzy sets: temperature→fuzzy variable→linguistic values such as low, medium, normal, high, and very high, leading to membership values (on the universe of discourse, e.g. degree Celsius). The dependence of a linguistic variable on another variable can be described by means of a fuzzy conditional statement:

$$R: \text{If S1 (is true), then S2 (is true) Or S1} \rightarrow \text{S2.} \tag{8.41}$$

More specifically, (i) if the load is small, then torque is very high; (ii) if the error is negative large, then output is negative large. The composite conditional statement would be as follows:

R1: If S1 then (If S2 then S3) is equivalent to:

$$R1: \text{If S1 then R2 AND R2}: \text{If S2 then S3.} \tag{8.42}$$

The number of rules could be large (30), or for more complex process control plant, 60–80 rules might be required [14]. For small tasks, 5–10 rules might be sufficient like for a washing machine. Then, a fuzzy algorithm is formed by combining two or three fuzzy conditional statements: If speed error is negative large, then (if change in speed error is NOT (negative large Or negative medium) then change in fuel is positive large), …, Or, …, Or, …

The knowledge necessary to control a plant is usually expressed as a set of linguistic rules of the form: If (cause) then (effect). These are the rules that the new operators are trained to control a plant, and the set of rules constitute the knowledge base of the system. However, all the rules necessary to control a plant might not be elicited or even known. It is therefore essential to use some techniques capable of inferring the control action from available rules.

Forward fuzzy reasoning (generalized modus ponens):

Premise 1: x is A',
Premise 2: If x is A then y is B,
Then the Consequence: y is B'.

This is the forward data-driven inference, used in all fuzzy controllers, that is, given the cause infer the effect. The directly related to backward goal-driven inference mechanism, that is, infer the cause that leads to a particular effect, is as follows:

Generalized Modus Tollens:
Premise 1: y is B',
Premise 2: If x is A then y is B,
Then the consequence: x is A'.

Reasons for the use of fuzzy control are as follows: (i) for a complex system → math model is hard to obtain, (ii) fuzzy control is a model-free approach, (iii) human experts can provide linguistic descriptions about the system and control instructions, and (iv) fuzzy controllers provide a systematic way to incorporate the knowledge of human experts. It reduces the design iterations and simplifies the design complexity. It can reduce the hardware cost and improve the control performance, especially for the nonlinear control systems. The assumptions involved in a fuzzy logic control system are as follows [14]: (i) plant is observable and controllable; (ii) expert linguistic rules are available/or formulated using engineering common sense, intuition, or an analytical model; (iii) a solution exists; (iv) look for a good enough (approximate reasoning) solution and not necessarily the optimum one; (v) desire to design a controller to the best of our knowledge and within an acceptable precision range; and (vi) the problem of stability/optimality could remain an open issue. The heuristic knowledge-based control does not require deep knowledge of the controlled plant, and this approach is popular in the industry and manufacturing environment, where such knowledge is lacking. It is case-dependent and does not resolve the issue of overall system's stability and performance. The model-based fuzzy control (MBFC) combines the fuzzy logic and theory of modern control, especially the known model of the plant can be used, and this approach is useful for control of high-speed trains, in robotics, helicopters, and flight control. In the classical gain scheduling control, the control gains are computed as a function of some variable, for example, the dynamic pressure at flight conditions. The fuzzy gain scheduling (FGS) is a special case of the MBFC and uses linguistic rules and fuzzy reasoning to determine the control laws at various flight conditions. The issues of stability, pole placement, and closed loop dynamic behavior are resolved by using the conventional and modern control approaches.

The Takagi and Sugeno [14] controller is defined by a set of fuzzy rules. These rules specify a relation between the present state of the dynamic system to its model and the corresponding control law with a general rule→R: If (state) Then (fuzzy plant/process model) AND (fuzzy control law).

Let the dynamic system be given as follows:

$$\dot{x}(t) = f(x,u);\ x_0 \tag{8.43}$$

The fuzzy state variables for the discrete case is defined as follows:

$$\phi X_{ij} = \sum \mu_{\phi X_{ij}}(x)/x \tag{8.44}$$

Each element of the crisp state variable is fuzzified, and in each fuzzy region of fuzzy state space, the local process model is defined as follows:

$$R_s^i : \text{if } x = \phi x^i \text{ then } \dot{x}^i = f_i(x^i, u^i) \tag{8.45}$$

One example of fuzzy process rules is given in Example 8.4. The process rules can be given in terms of the elements of the crisp process state. The process is specified by the state space with degrees of fulfillment of the local models of the process employing Mamdani rule:

$$\begin{aligned} \dot{x}^i &= \mu_s^i(x) f_i(x,u) \\ &= \min\left(\mu_{\phi x 1}^i(x_1),\ldots,\mu_{\phi x n}^i(x_n)\right) f_i(x,u) \end{aligned} \tag{8.46}$$

The fuzzy open-loop model is given as follows:

$$\dot{x} = \sum w_s^i(x) f_i(x,u) \tag{8.47}$$

Here, $w_s^i(x) = \dfrac{\mu_s^i(x)}{\sum \mu_s^i}$ is the normalized degrees of fulfillment. Now, the fuzzy control law can be defined as follows:

$$R_s^i : \text{if } x = \phi x^i \text{ then } u = g_i(x) \tag{8.48}$$

In a similar manner as the development of the fuzzy open-loop/process model, the control law is defined as follows:

$$u = \sum w_c^j(x) g_j(x) \tag{8.49}$$

Here, w is the control weights.

An aircraft's dynamic characteristics would change with the Mach number and/or altitude (or dynamic pressure). The gain scheduling is used for selecting appropriate filters' (feedback controllers) characteristics (gain and time constants) so that the performance of the aircraft/control system is acceptable, despite the change in its dynamics. The concept is illustrated by the following example:

Example 8.4

Use the system process rules as follows [14]:

$$\begin{aligned} &R^0 : \text{if } x_d = 0, \text{ then } \dot{x} = f_1(x,u) = -0.4x + 0.4u \\ &R^1 : \text{if } x_d = 1, \text{ then } \dot{x} = f_2(x,u) = -2.5x + 2.5u \end{aligned} \tag{8.50}$$

Then,

i. obtain step responses of these two systems/process rules and compare the results;
ii. use the fuzzy gain scheduling controller concept with the following rules:

$$\begin{aligned} &R^0 : \text{if } x_d = 0, \text{ then } u_1 = g_1(x,x_d) = k_1(x - x_d) \\ &R^1 : \text{if } x_d = 1, \text{ then } u_2 = g_2(x,x_d) = k_2(x - x_d) \end{aligned} \tag{8.51}$$

Use the state feedback gains $k_1 = 0.5$ and $k_2 = 0.92$. Then, the overall fuzzy process model is given by the weighted sum:

$$\dot{x} = w_1\{-0.4(x - x_d) + 0.4u\} + w_2\{-2.5(x - x_d) + 2.5u\} \tag{8.52}$$

Design the above fuzzy gain scheduling and obtain the responses of the final system.

iii. compare the responses of the designed system with the system: $\dot{x} = -0.2x + 0.2u$

iv. plot all the results: step responses and Bode diagrams of the open-loop and closed-loop systems.

Solution 8.4

The process time constants are 2.5s and 0.4s, respectively. The step responses of these processes are shown in Figure 8.19 (**ExamplesSolSW/Example 8.4FuzzyControl**). It is obvious from the plots that the second process rule signifies the faster process. The fading of the slow plant dynamics to the fast one is

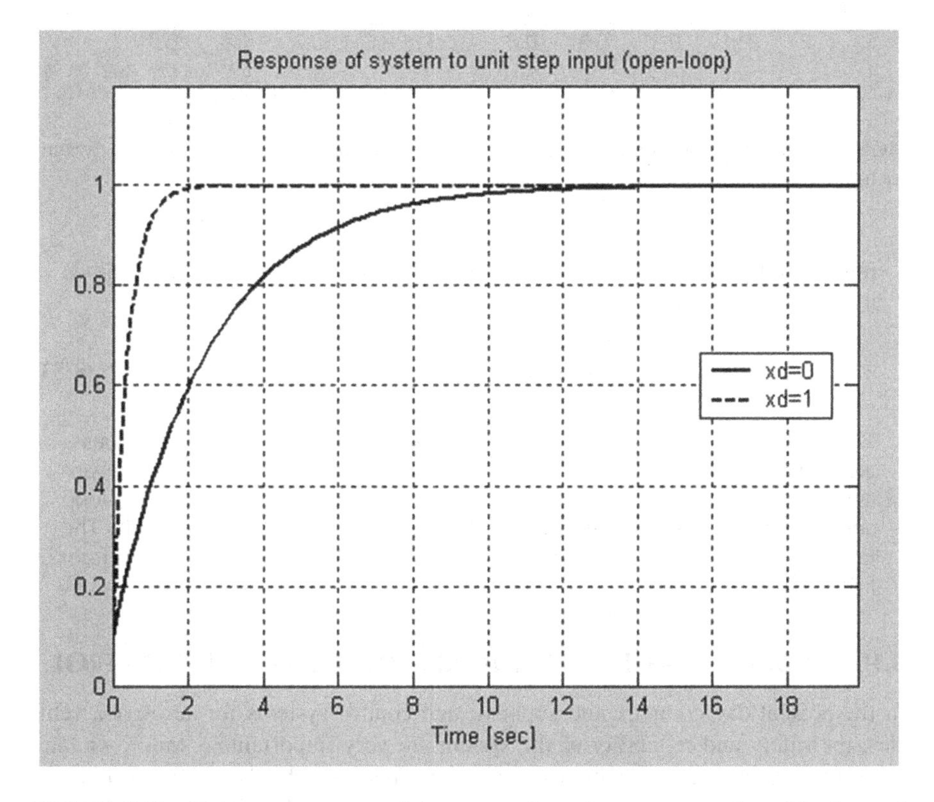

FIGURE 8.19 Unit step responses of the system for two processes: at low altitude $x_d = 0$ (slow dynamics, -); at high altitude x_d.

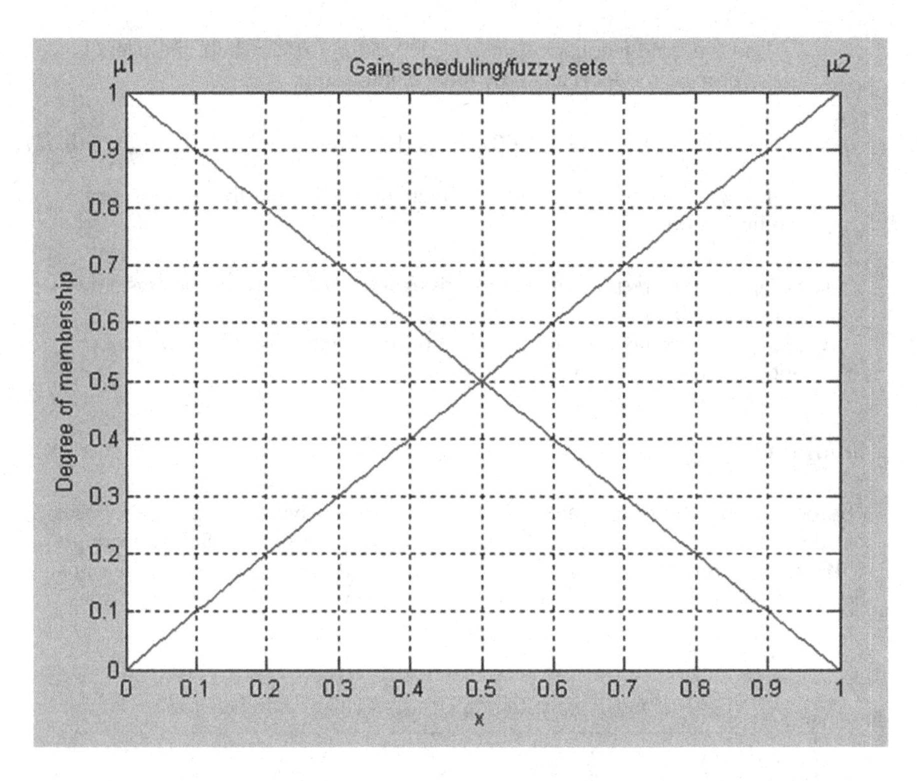

FIGURE 8.20 Gain scheduling fuzzy membership functions: transition from slow dynamics to fast dynamics as the altitude increases.

represented by the fuzzy membership functions plotted in Figure 8.20, as altitude increases. The fuzzy control law is given as follows:

$$u = 0.5 w_1 (x - x_d) + 0.92 w_2 (x - x_d) \tag{8.53}$$

The fuzzy controller is realized, as discussed in Section 8.6, and the responses are shown in Figure 8.21. The comparison of the time responses with the invariant system is shown in Figure 8.22. The frequency responses for the open-loop and closed-loop systems are shown in Figures 8.23 and 8.24, respectively. The fuzzy controller meets the requirements at both the ends, as can be seen from Figure 8.22.

8.10 FAULT MANAGEMENT AND RECONFIGURATION CONTROL

In the present-day complex and sophisticated control systems for aerospace vehicles, the safety and reliability of the system are very important. A failure or fault could be in a hardware system or in software. This could be due to one or more of these: (i) design error, (ii) damage, (iii) deterioration, (iv) EM interference, and (v) external disturbance of high magnitude. A fault could be transient, intermittent, or

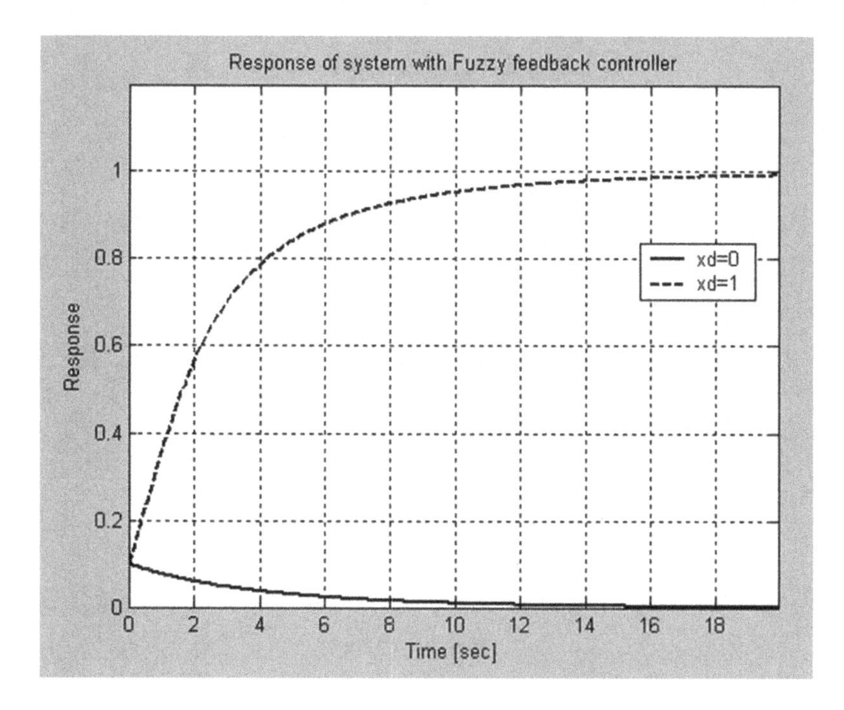

FIGURE 8.21 Unit step responses of the system for two processes: at low altitude $x_d=0$ (slow dynamics, -); at high altitude $x_d=1$ (fast dynamics, --) with fuzzy controllers.

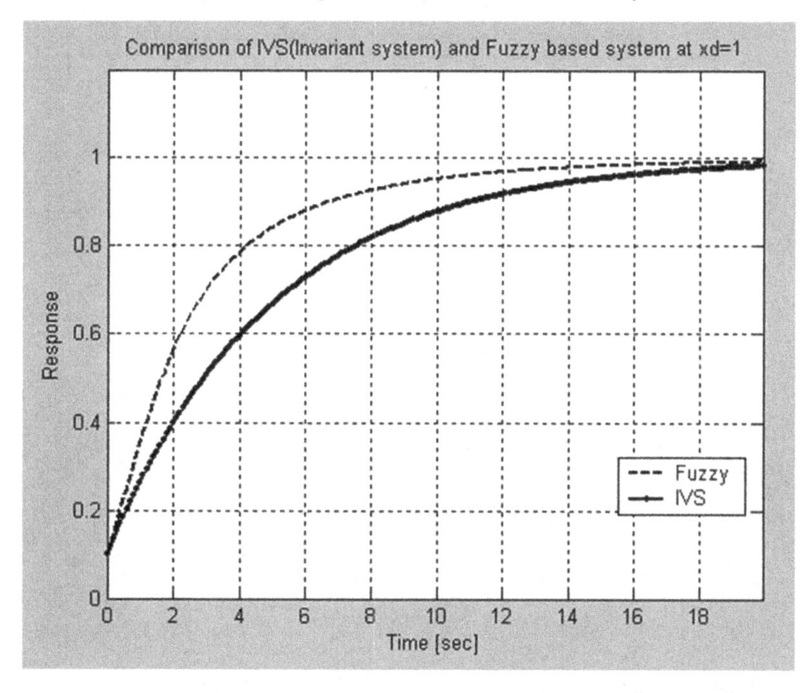

FIGURE 8.22 Unit step responses of the fuzzy controlled (--) and the invariant system (-).

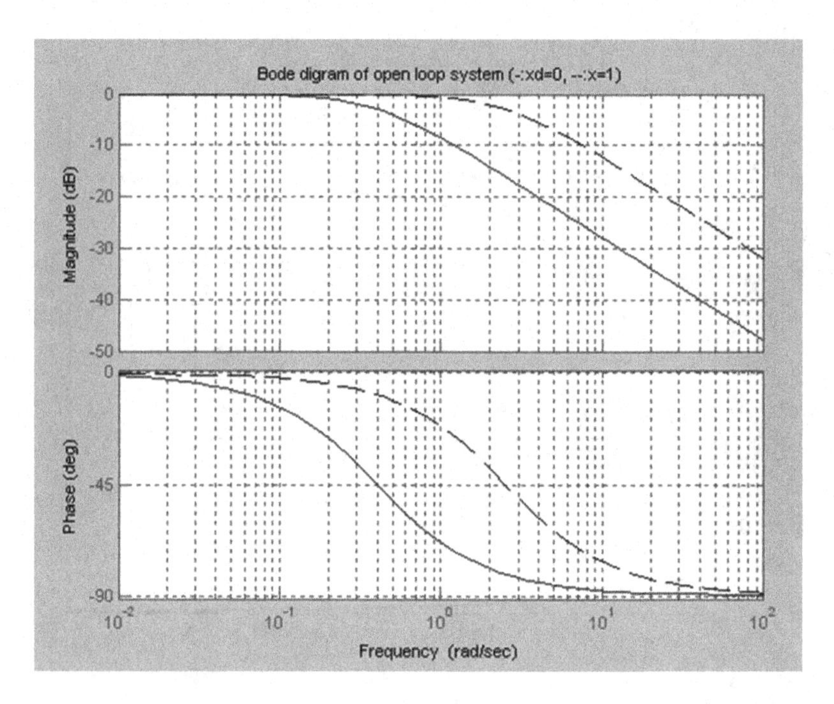

FIGURE 8.23 Frequency responses of the system for two processes: at low altitude $x_d=0$ (- slow dynamics); at high altitude $x_d=1$ (-- fast dynamics).

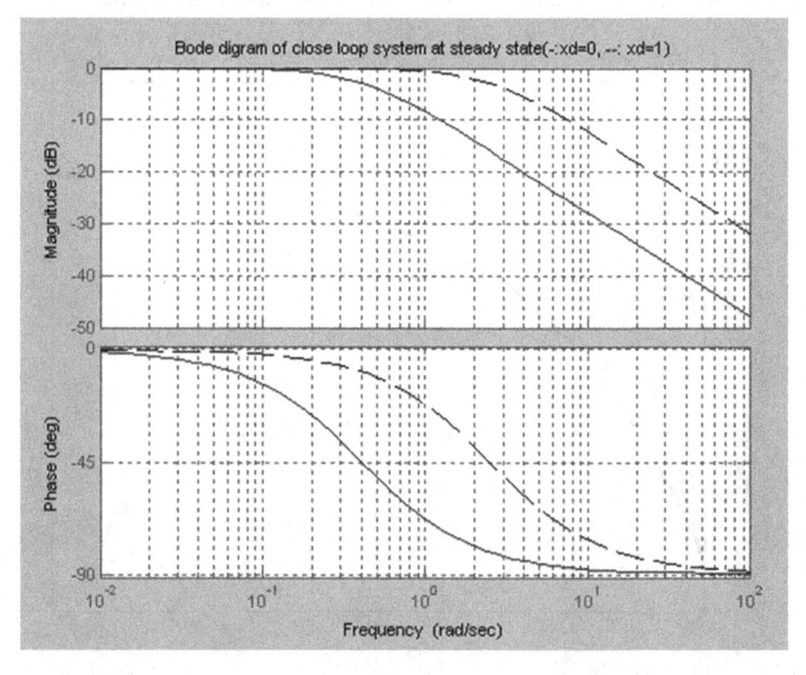

FIGURE 8.24 Closed-loop frequency responses of the system for two processes: at low altitude $x_d=0$ (- slow dynamics); at high altitude $x_d=1$ (-- fast dynamics).

permanent [15]. Hence, the importance of a fault-tolerant (FTR) control system in safety critical systems need not be overemphasized. The technology of fault detection, identification, and isolation in the context of a reconfigurable control system is a rapidly developing field in aerospace engineering. Due to more demand of autonomous navigation, guidance, and control of aerospace vehicles, the subject assumes a greater significance than the basic control system. For ensuring safe flight operation of an aircraft, proper functioning of sensors, actuators, and control surfaces should be assured. The aircraft's FCSs or its subsystems should be FTR to failures. Alternatively, the AFCS should be adaptive, and the aircraft should be able to recover from the effects of such failures. For this fault detection, identification, isolation, and subsequent reconfiguration, if needed, should be carried out in the near real-time mode for it to be really effective and useful.

8.10.1 Models for Faults

The most crucial aspect in the study and analysis of the FTR/reconfiguration control system is specification of models for various states (two) of the systems behavior [16]. There are several methods and approaches based on linear and nonlinear dynamic mathematical models for analysis and design of FTR control systems. Different faults would exhibit distinct characteristics, and hence, naturally, a multiple model approach offers a logical solution for dealing with multiple types of faults in sensors, actuators, and system dynamics. So, it would be a good idea to represent each fault with a separate model, thereby enhancing the overall handling capacity of various types of faults. Based on the requirements of the system performance, multiple reference models can be defined. Once these models are defined, the FTR control system can be synthesized using the model reference approach.

Let the dynamic system be described as follows:

$$\dot{x}(t) = (A + \Delta a_j)x(t) + (B + \Delta b_j)u(t) + w(t) = A_j x(t) + B_j u(t) + w(t) \quad (8.54)$$

$$z(t) = C_j x(t) + v(t);\ j = 0,\ldots,N-1. \quad (8.55)$$

Eqn. (8.55) represents the system under normal and $(N-1)$ failure modes. $\Delta a_j, \Delta b_j$ and Δc_j define the fault-induced changes in the system dynamics, actuators, and sensors, respectively. $j=0$ is a normal and non-faulty condition. We need to specify the characteristics of the system under each fault. For this purpose, one can use "performance-reduced reference model." Let a reference model of the system under a normal situation be given as follows:

$$\dot{x}^m = A^m x^m + B^m r;\ y^m = C^m x^m \quad (8.56)$$

When a fault occurs, the eigenvalues of performance-reduced reference model would shift toward the imaginary axis, reflecting the loss of performance.

8.10.2 Aircraft FTR Control System

A typical FTR control system would consist of actuators, sensors, computer systems, data acquisition, and often mechanical, hydraulic, pneumatic, electronic devices, and systems [16–18]. A FTR FCS (FTFCS) is expected to perform the task of failure detection, identification, and accommodation of actuator and general failures. The FTFCS is also useful for UAVs and MAVs. In a reconfigurable FTFCS, a full set of possible failures for different control surfaces and other probable faults is predefined. The required changes in control gains are computed offline in advance and stored in onboard computer (OBC) for subsequent online usage. This approach requires extensive design to consider all possible failure modes, and hence, large storage space in OBC is required. In a restructurable FTFCS, the required control laws are restructured online based on some identified system/fault parameters, especially damage to a wing or control surface. Artificial neural networks (ANNs) and fuzzy logic find a good use for this type FTFCS. For FTFCSs, two aspects are important: a) detection and identification of actuator fault and b) failure accommodation. For an FBW aircraft, FTR capacity requires an increase in the number of independent control surfaces. This selection depends on (i) control effectiveness ($C_{m\delta}$), (ii) increased aircraft complexity and cost, (iii) weight constraint, (iv) aerodynamic drag force, and (v) aircraft type. When any type of control surface failure occurs, the coupling between longitudinal and lateral-directional dynamics is of a great concern since this might lead to loss of stability of the vehicle. In many of these studies, an accurate post-failure aerodynamic model for simulation is required. From eqn. (4. 6), we can guess that damage of a control surface would cause instantaneous changes in its aerodynamic characteristics. Neglecting the effect on the axial forces, the net effect of the damage of the control surface would be normally the reduction in the relative normal force coefficient.

The sensors and actuators play a vital role in the control mechanisms of various aerospace vehicles/systems. The problem arises when a particular sensor stops working properly due to an unknown internal fault or component impairment or any catastrophic disturbance from an external source. In that case, the whole architecture of the control system becomes hampered. There comes the urgent requirement of a model that detects the failed sensor as quickly as possible and reconfigures the damaged architecture so that the controlling mechanism continues to deliver desired results without any alterations. Timely detection and adjusting the system to cope with these uncertainties are highly desirable. This process of identification and reconfiguration is known as sensor fault detection, isolation, and accommodation (SFDIA). The two most popular methods for SFDIA are model- and non-model-based approaches. These techniques are based on analysis of residuals between measured and predicted plant states. In the model-based technique, the algorithms commonly used are robust Kalman filter [18], multiple model adaptive estimator, and interacting multiple model). In the non-model-based approach, neural network and fuzzy logic are used to predict the plant responses and, in conjunction with the measured states, are subsequently utilized to detect and isolate the sensor fault. The other potential cause to hamper aircraft performance is its faulty actuator. The process of detection, isolation, and reconfiguration of actuator faults is known as actuator fault detection, identification and accommodation (AFDIA). In this case, the strategy is to pinpoint the cause of failure and then appropriately reconfigure the FCS such that impaired aircraft can be made flyable.

Figures 8.25 and 8.26 show the architectures used for fault detection, isolation, and reconfiguration for the sensor and actuator, respectively. For data simulation, the longitudinal dynamics of the aircraft control system are considered. The state-space formulation of longitudinal motion of aircraft in the continuous time domain is as follows:

$$\dot{x} = Ax + Bu_c + Gw_n \tag{8.57}$$

Here, $x = [u, w, q, \theta]$ are the physical variables corresponding to aircraft longitudinal motion. The w_n is zero mean white Gaussian process noise. The equation used to model the sensor is given as follows:

$$Z_m(k) = Hx + v \tag{8.58}$$

The observation matrix H is an identity matrix having dimension 4×4. The measurements can be obtained with a sampling time interval T equal to 0.01 s.

8.10.2.1 Sensor Fault Detection Scheme

In order to evaluate the scheme, primarily only one kind of sensor fault can be considered, that is, shift of mean of measurement noise. For fault detection, isolation, and accommodation, the extended Kalman filter (EKF) in the discrete time domain is mainly used. Sensor fault detection is performed by checking the change in the mean of time series of innovation sequences computed over a window length M (usually $M = 20$ would suffice). The hypothesis used for fault detection is as follows:

$$\mathrm{H}_0\ (\text{no fault}) : \beta(k) \le \chi_\alpha^2;\ \mathrm{H}_1\ (\text{fault}) : \beta(k) > \chi_\alpha^2 \tag{8.59}$$

Here, χ_α^2 is a threshold obtained from the chi-square table for a confidence probability $\alpha = 0.95$ (present case) and degree of freedom $M_s = M$ * number of measurement channels. The statistical function $\beta(k)$ is computed using the following equation:

$$\beta(k) = \sum_{j=k-M+1}^{k} v^T(k) S^{-1}(k) v(k) \tag{8.60}$$

Once the hypothesis H_1 is satisfied, the next step is to find the source of fault, that is, which measurement channel is malfunctioning? Assuming that there are total of "m" channels, the algorithm to isolate the faulty channel is given as follows:

$$\mathrm{H}_{0,\mathrm{ch}}\ (\text{no fault in 'ch' channel}) : \beta_{\mathrm{ch}}(k) \le \chi_{\alpha,\mathrm{ch}}^2 \tag{8.61}$$

$$\mathrm{H}_{1,\mathrm{ch}}\ (\text{faulty 'ch' channel}) : \beta_{\mathrm{ch}}(k) > \chi_{\alpha,\mathrm{ch}}^2 \tag{8.62}$$

Here, "ch" is the channel number varying from 1 to "m," $\chi^2_{\alpha,\text{ch}}$ is a threshold obtained from the chi-square table for a confidence probability $\alpha = 0.95$, and degree of freedom $M_{s,\text{ch}} = M - 1$. The statistical function $\beta_{\text{ch}}(k)$ is computed using the following equation:

$$\beta_{\text{ch}}(k) = \sum_{j=k-M+1}^{k} \left[\left(\nu(j,\text{ch}) \left(\sqrt{S(j,\text{ch},\text{ch})} \right)^{-1} \right) - \frac{1}{M} \sum_{j=k-M+1}^{k} \nu(j,\text{ch}) \left(\sqrt{S(j,\text{ch},\text{ch})} \right)^{-1} \right]^2 \tag{8.63}$$

In the present case, sensor fault is accommodated by not using them in the measurement update stage of EKF. This is achieved by resetting the all the elements (corresponding to faulty channel) of the observation matrix "H" and measurement "Z_m" to zero values.

Example 8.5

Realize the sensor fault detection scheme of Figure 8.25 in MATLAB based on the foregoing theory. The following matrices A, B, and C [10] can be used for generation of the data:

$$A = \begin{bmatrix} -0.033 & 0.0001 & 0 & -9.81 \\ 0.168 & -0.3870 & 260 & 0 \\ 0.005 & -0.0064 & -0.55 & 0 \\ 0 & 0 & 1 & 0 \end{bmatrix};$$

$$B = \begin{bmatrix} 0.45 \\ -5.18 \\ -0.91 \\ 0.00 \end{bmatrix}; G = \begin{bmatrix} 1 & 0 & 0 & 0 \\ 0 & 1 & 0 & 0 \\ 0 & 0 & 1 & 0 \\ 0 & 0 & 0 & 1 \end{bmatrix} \tag{8.64}$$

Solution 8.5

The data simulation is carried out using eqns. (8.26) and (8.27) to generate true and measured states for aircraft longitudinal motion (**ExamplesSolSW/Example 8.5KF**). Each element of "Q" is kept at 1.0e-8, whereas "R" under the normal sensor operation is kept at diag$[0.1296 \quad 0.0009 \quad 0.0225 \quad 0.010]$. In case of the SFDIA scheme, the total number of data points simulated is kept to 100, and for AFDIA, 2000 data points are simulated. The control input "u_c" is kept at a constant value of 8. In order to first evaluate the SFDIA scheme, a fault is introduced in the third channel (i.e. in pitch rate) measurement at the 31st second by adding a constant bias of 0.4 deg./s to that channel. Figure 8.27 shows the comparison of actual and faulty measurements of pitch rate for 100 data points. The fault detection is first carried out by comparing the statistical function $\beta(k)$ with the threshold χ^2_α selected using the chi-square table. The threshold is kept to 101.88 for 20 degrees of freedom (DOF) and with a confidence level of 0.95. Figure 8.28 shows

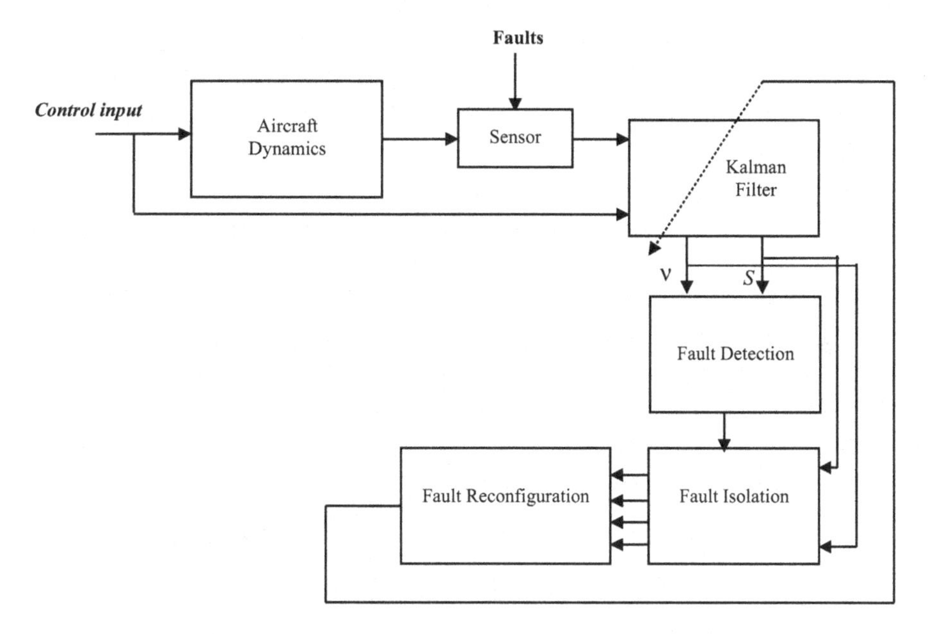

FIGURE 8.25 Schematic for SFDIA. SFDIA, sensor fault detection, isolation, and accommodation.

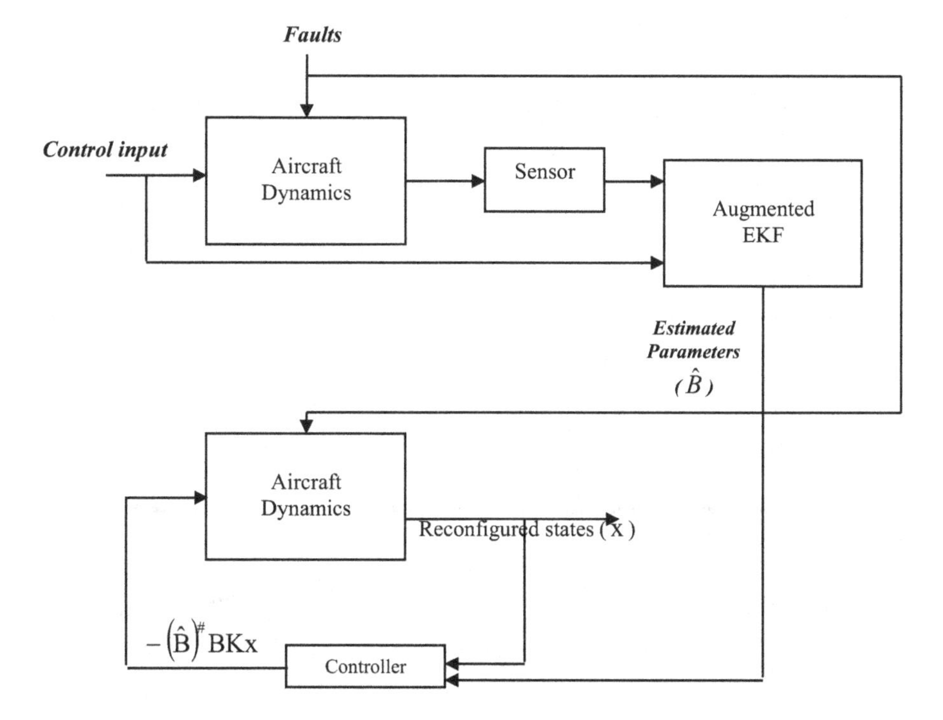

FIGURE 8.26 Schematic for AFDIA (#, pseudo-inverse// B, true parameters // K, feedback gain for unimpaired aircraft).

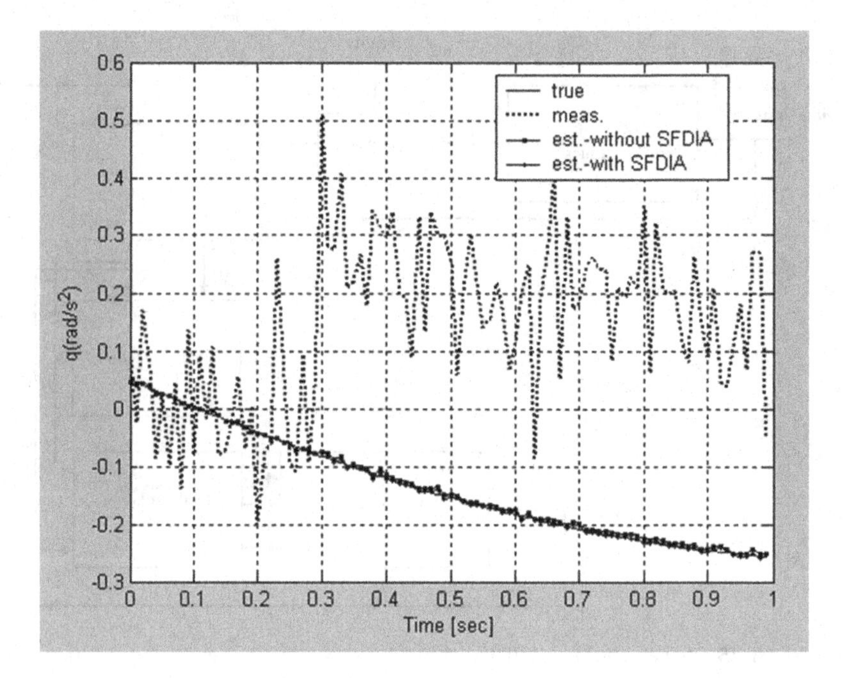

FIGURE 8.27 Comparison of actual and faulty pitch rate measurements.

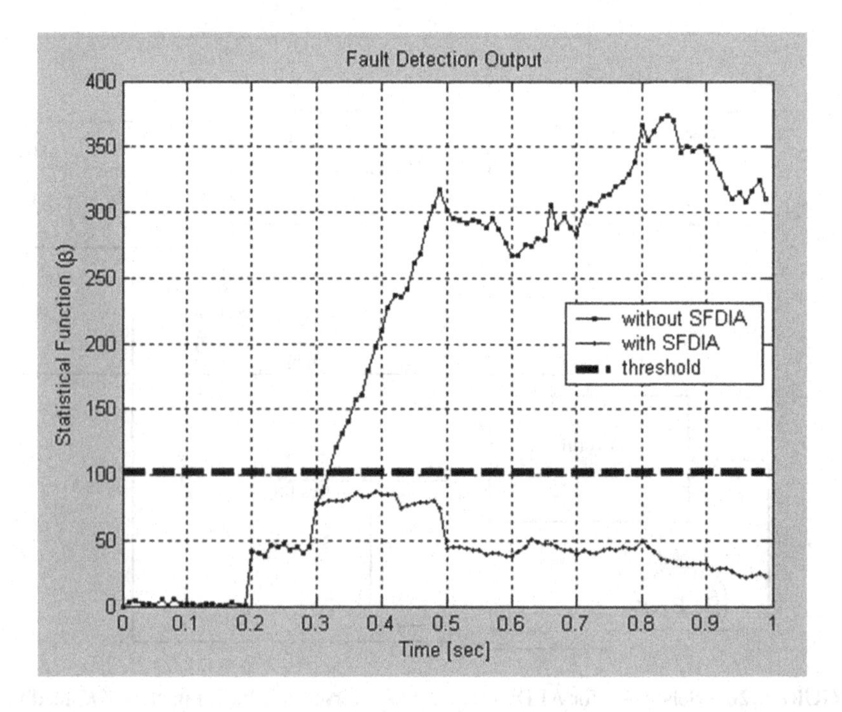

FIGURE 8.28 Comparison of β's without/with reconfiguration – fault detection.

the comparison of $\beta(k)$ with the threshold χ^2_α for the entire set of data points. It can be observed from the figure that fault is detected with a delay of 4 scans (0.04 s). The fault detection automatically triggers the isolation algorithm to locate the faulty channel, and once the fault is located, the accommodation algorithm bypasses the faulty measurements used in state update of the EKF. Figure 8.13 also compares the $\beta(k)$ computed for faulty and reconfigured cases. It is clear that after fault reconfiguration, $\beta(k)$ never exceeds the threshold value, thereby indicating the proper handling of the faulty channel through the algorithm. The outcome of fault isolation in terms of comparison of statistical function computed for each measurement channel and compared with a threshold value of 31.4 for 19 DOF and with a confidence level of 0.95 was that, except for the third channel, $\beta(k)$ never exceeded the threshold for the remaining channels, thereby indicating the source of the fault occurrence. $\beta(k)$ goes below threshold immediately after the isolation and reconfiguration. Figure 8.14 depicts the innovation sequences along with the theoretical bounds of $\sqrt{S_{ii}}$, where S is the innovation covariance matrix and term "i" indicates the index of diagonal elements. It seen from Figure 8.29 that for a faulty channel (3rd), the innovation sequence exceeds the bound if not reconfigured, whereas it is forced to zero after reconfiguration. The estimated states were found to be close to the true states, thereby confirming satisfactory performance of the scheme.

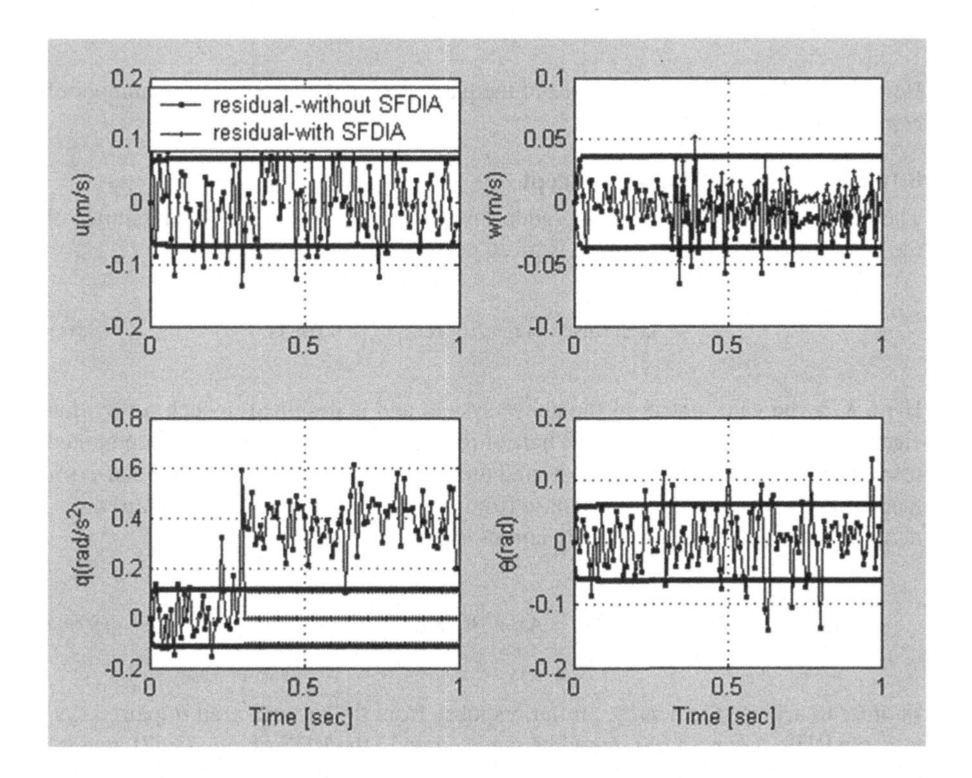

FIGURE 8.29 Comparison of residuals without/with reconfiguration – fault accommodation.

8.10.2.2 Actuator Fault Detection Scheme

The actuator fault detection scheme is shown in Figure 8.26. The actuator fault can be modeled by multiplying the appropriate component of the control vector "B" (eqn. (8.26)) by factor of effectiveness. The 50% loss in effectiveness means the faulty value is half of the true value. After introduction of fault in some of the components of the control vector "B," the measurements should be generated using eqns. (8.26) and (8.27). The EKF should be reformulated as far as the plant model is concerned by augmenting the plant state with the states pertaining to components of "B" that have to be estimated. The reformulated plant model used in the filter is given as follows:

$$\tilde{x}_a = A_a \hat{x}_a \tag{8.65}$$

Here, the augmented state x_a consists of $\hat{x}_a = \begin{bmatrix} \hat{x} & \hat{B} \end{bmatrix}$, and the augmented matrix A_a for a single control input is given as follows:

$$A_a = \begin{bmatrix} [A]_{n \times n} & [u * I]_{n \times m} \\ [0]_{m \times n} & [0]_{m \times m} \end{bmatrix} \tag{8.66}$$

Here, "n" is the number of variables of the plant state and "m" is the total number of components of vector "B."

8.10.2.3 Reconfiguration Concept

The technique used is known as pseudo-inverse or control mixer. The dynamics of the unimpaired close-loop system is given as follows:

$$\dot{x} = Ax + Bu_c = Ax + B(-Kx) = (A - BK)x \tag{8.67}$$

Here, K is the gain matrix of the above system and is designed in such a way that eigenvalues of $(A - BK)$ are in left half of the s plane. The gain K can be computed using MATLAB® function "LQR ()" and tuning its parameters by the trial-and-error method till it gets the desired response from the system. In case of the impaired system, the reconfigured close-loop dynamics is given as follows:

$$\dot{x} = Ax + Bu_c = Ax + \hat{B}(-Kx) = (A - \hat{B}K_i)x \tag{8.68}$$

In order to achieve the nearby similar response from the reconfigured impaired system, the RHS of eqn. (8.68) should be made equal to the RHS of eqn. (8.67), that is,

$$A - BK = A - \hat{B}K_i; \ \ BK = \hat{B}K_i \tag{8.69}$$

The gain matrix for the impaired system can be computed from eqn. (8.35):

$$K_i = \left(\hat{B}\right)^{\#} BK \tag{8.70}$$

Here, # is the pseudo-inverse of control vector parameters of "B" estimated using the augmented EKF.

Example 8.6

Realize the actuator fault detection scheme of Figure 8.26 in MATLAB based on the foregoing theory.

Solution 8.6

In the present case, "m" equals "n," that is, 4; hence, A_a will be as follows:

$$A_a = \begin{bmatrix} \begin{bmatrix} -0.033 & 0.0001 & 0 & -9.81 \\ 0.168 & -0.3870 & 260 & 0 \\ 0.005 & -0.0064 & -0.55 & 0 \\ 0 & 0 & 1 & 0 \end{bmatrix} & \begin{bmatrix} u & 0 & 0 & 0 \\ 0 & u & 0 & 0 \\ 0 & 0 & u & 0 \\ 0 & 0 & 0 & u \end{bmatrix} \\ \begin{bmatrix} 0 & 0 & 0 & 0 \\ 0 & 0 & 0 & 0 \\ 0 & 0 & 0 & 0 \\ 0 & 0 & 0 & 0 \end{bmatrix} & \begin{bmatrix} 0 & 0 & 0 & 0 \\ 0 & 0 & 0 & 0 \\ 0 & 0 & 0 & 0 \\ 0 & 0 & 0 & 0 \end{bmatrix} \end{bmatrix} \tag{8.71}$$

The rest of the equations used in the augmented EKF are same as those used in the scheme of Example 8.5.

The present script is in **ExamplesSolSW/Example 8.6AFDIA**. The estimated value of the control vector is subsequently used for reconfiguration of the impaired aircraft. For evaluation of this scheme, the fault in the actuator is introduced at the 100th iteration by changing the third element of the control input vector "B" from its actual value of −0.91 to −0.45, that is, nearly 50% loss in effectiveness. Figure 8.30 shows the states corresponding to longitudinal motion of the aircraft for unimpaired and impaired configurations of the same aircraft. It can be seen that the states for both the configurations are comparable up to the first 5 s, and thereafter, a clear difference in magnitude of state values can be seen. The measurements are generated using the state of the impaired aircraft and subsequently used to recursively estimate the elements of the input control matrix "B" using the scheme. For the estimation of "B" as an augmented state with the state pertaining to longitudinal motion, the EKF is used with its initial state kept at zero, and accordingly, state initial error covariance is computed. The tuning parameters such as "Q" and "R" are assumed to be known precisely. Figure 8.31 illustrates the estimated values of "B" compared with true values. The "$B(1)$" converges to the true value as the number of data increases. In case of faulty element $B(3)$, initially, the estimated value starts from zero and then converges to true value, and when fault occurs, it slowly tries to converge to the faulty value. These estimated values are subsequently used in control reconfiguration to make the performance

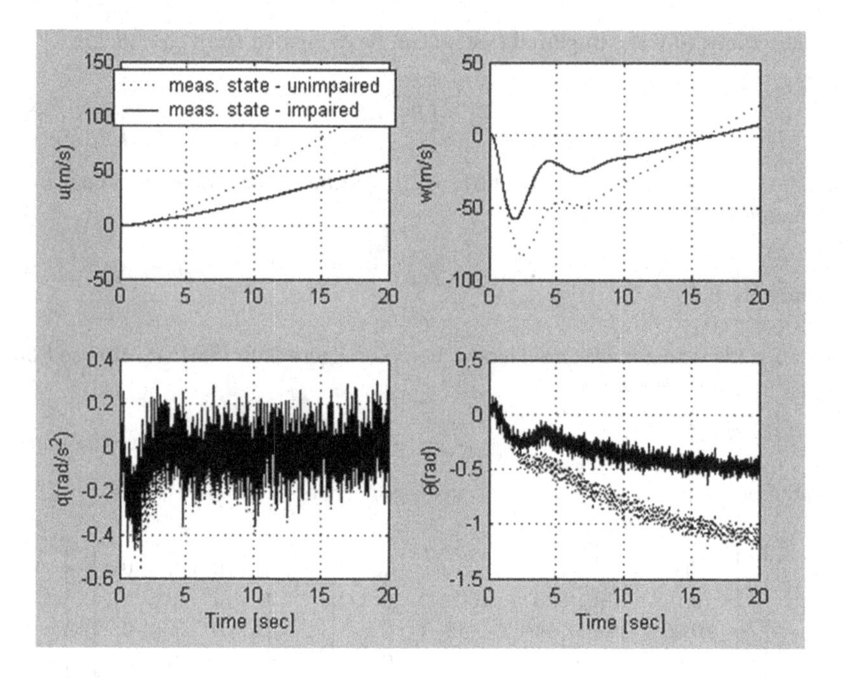

FIGURE 8.30 Comparison of measured states of unimpaired and impaired aircrafts.

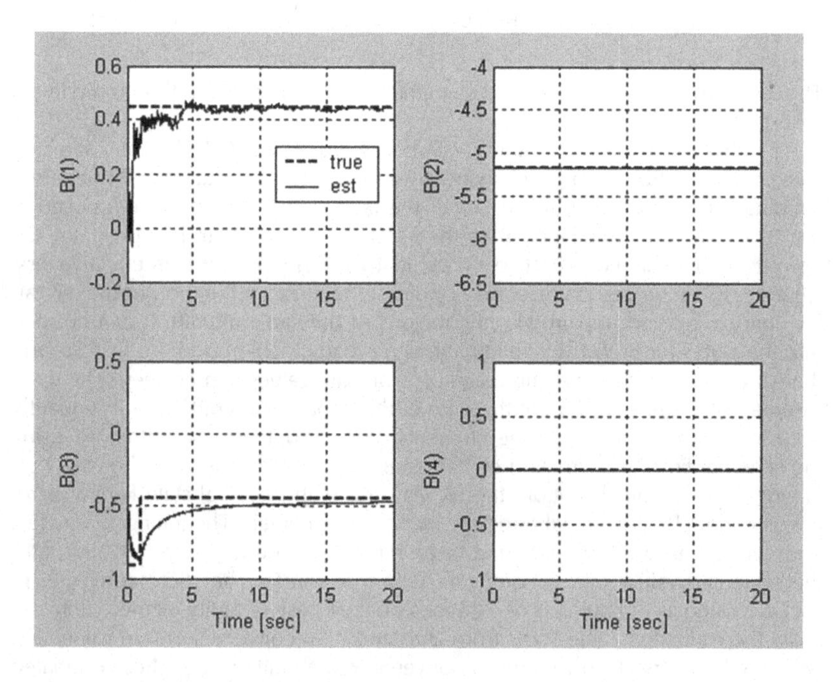

FIGURE 8.31 Comparison of true and estimated elements of the control input matrix "*B*."

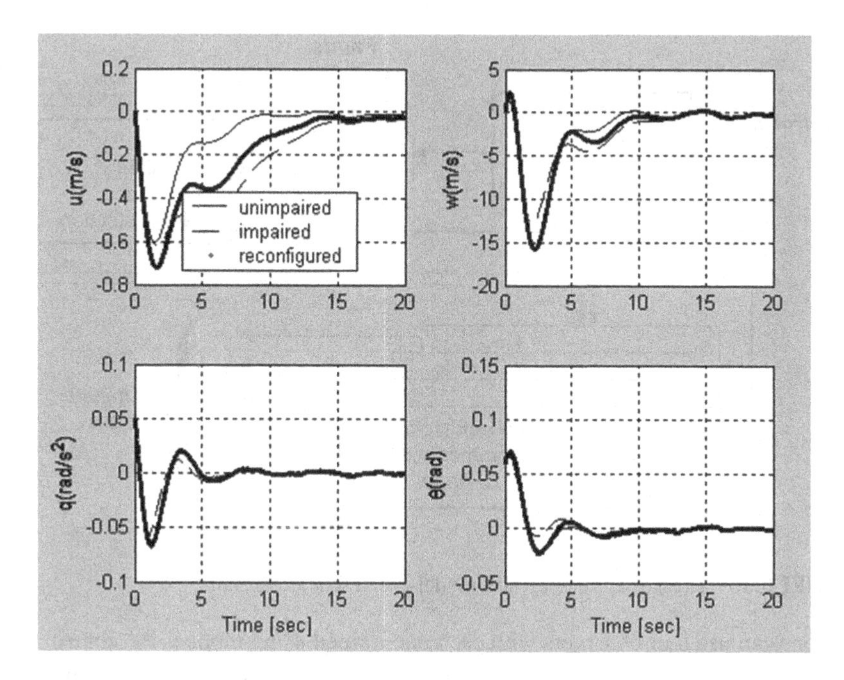

FIGURE 8.32 Comparison of states of unimpaired, impaired, and reconfigured states of an aircraft.

of the impaired aircraft close to unimpaired one. Under the scheme, first, the unimpaired aircraft response with state feedback and without control input (i.e. $u_c = 0$) is realized by designing the feedback gain matrix "K", eqn. (8.36), using the "LQR" method (in MATLAB). During the analysis, it is found that with feedback gain $K = \begin{bmatrix} 0.0886 & 0.0048 & -0.8906 & -2.1274 \end{bmatrix}$, closed-loop or controlled response of the longitudinal state tends to be zero, as required. The parameters, such as feedback gain "K" for the unimpaired aircraft, estimated "B" along with true B and A, are used to reconfigure the feedback gain for the impaired aircraft. Figure 8.32 compares the closed-loop response of the unimpaired, impaired without reconfiguration, and reconfigured impaired aircraft. It is observed from the plots that the response of reconfigured state converges to that of unimpaired one, thereby indicating reconfiguration of the faulty system. However, much work for refinement remains to be pursued.

8.10.2.4 Non-Model-Based Approach

A non-model-based approach for sensor fault detection, isolation, and accommodation is shown in Figure 8.33. The scheme utilizes the ANFIS. The MATLAB-based functions "GENFIS" and "ANFIS" can be used to build the state observers to estimate the state pertaining to longitudinal motion of the aircraft considered. The first part consists of state observer for each measurement channel for fault detection and isolation [19]. Each observer is trained using inputs as the control input (u_c) and measurement (e.g. pitch rate) and the output as corresponding true state. For the detection purpose, thresholds are computed using the Monte–Carlo simulation of random noises (one can use 1000 runs). The second part is for fault reconfiguration,

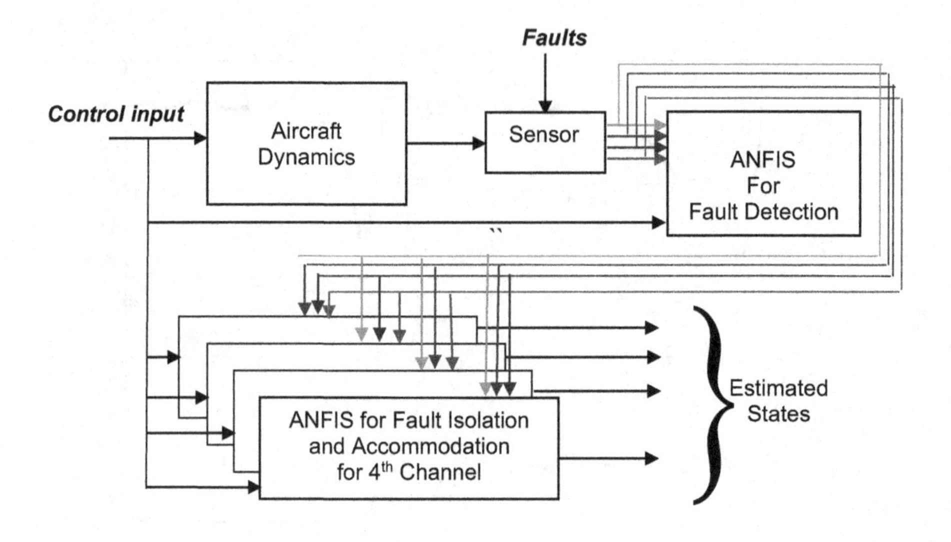

FIGURE 8.33 Block diagram of a non-model-based scheme for fault.

and one can use four observers with each one trained using input as the control input (u_c) and the measurements from three channels (excluding a channel for which state is estimated) and the output is the estimated state corresponding to the measurement of the excluded channel. For example, for faulty pitch rate (q) accommodation, inputs to observer are u_c, w, u, and θ.

Example 8.7

Realize the non-model-based sensor fault detection scheme of Figure 8.33 in MATLAB based on the foregoing theory. Use the same state-space models used in Examples 8.7 and 8.8.

Solution 8.7

The simulation is carried out as per Example 8.5, and the present script is in **ExamplesSolSW/Example 8.7ANFIS**. The minimum and maximum values of the noises were computed for each run, and then averaged value of all the runs was considered the threshold for detection purpose. The sensor fault is introduced only to a forward velocity "*u*" at 31st data point by adding a constant bias of 4 m/s. The fault was detected around 32nd data point, and most of the time, the fault was in the first channel only. In between it went to other values but only for a single time-point, and the reason for this could be a false alarm. Figure 8.34 illustrates reconfigured estimated states compared with measured (including faulty "*u*"), true, and reconfigured estimate states. In the case of faulty measurement, it is observed that overall, the estimated value closely matches with the true, except at some points where the scheme is affected perhaps due to a false alarm. Figure 8.35 shows the residuals compared with the threshold used in fault detection and isolation for each channel. It is observed for the faulty channel (i.e. "*u*") that after reconfiguration, error lies within the threshold.

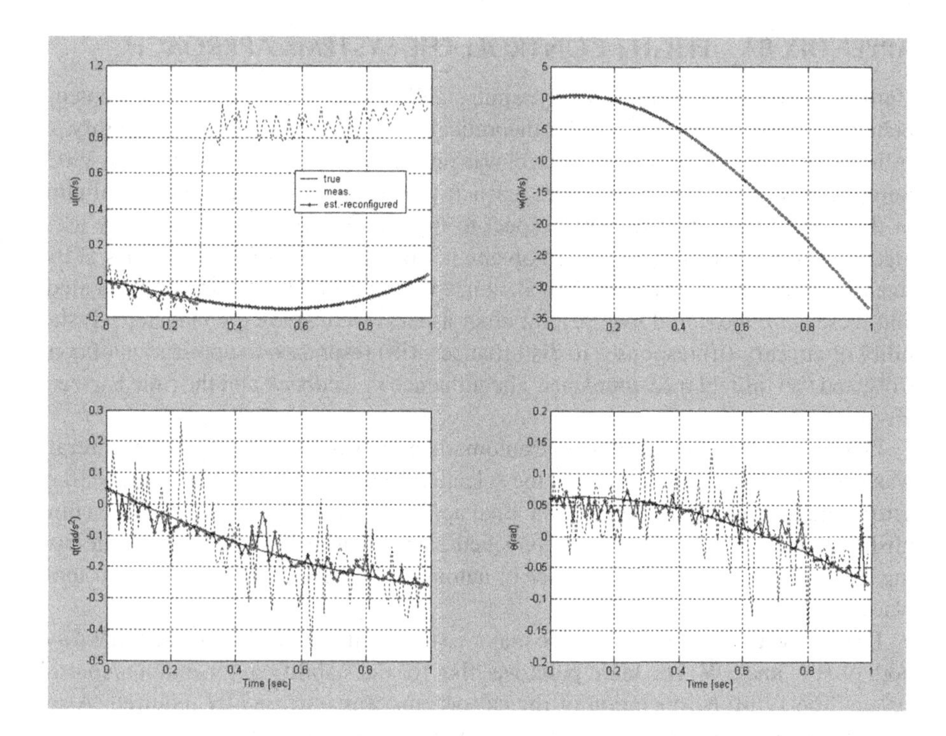

FIGURE 8.34 Comparison of true, measured, and estimated states (reconfigured).

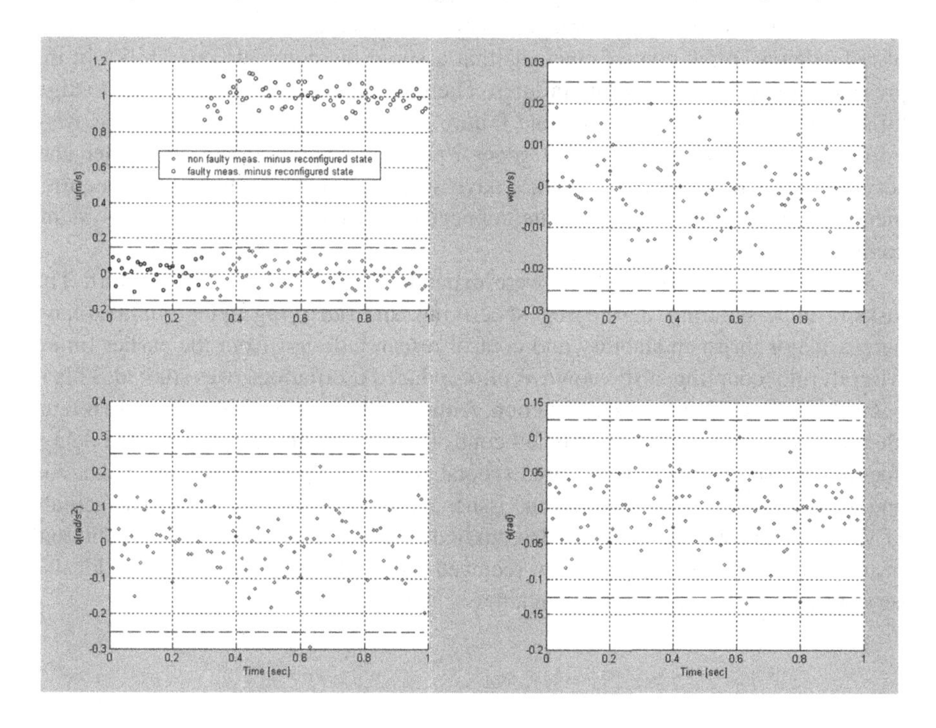

FIGURE 8.35 Residuals compared with threshold.

APPENDIX 8A FLIGHT CONTROL: THE SYSTEMS APPROACH

Until 1901, the ideas of aircraft's dynamic stability and flight control were contemplated [1]. Also, Routh advanced a theoretical background for inherent stability. A fully stable glider under manual control was invented by the Wright brothers in 1902. Subsequently, the aircraft engineers learned to achieve a desirable bare minimum of three-axis static stability with respect to the relative wind, for which they used algebra and the rule of thumb based on empirical data. Until 1931, the studies of the linearized motions were accomplished, concept of stability derivatives was invented, and these were measured too. Several other aspects were studied/conducted: (i) stability of aircraft, (ii) responses to disturbances, (iii) responses to application of controls, and (iv) inflight measurements. The influence of feedback and the time lag were also studied.

During the period 1930–1956, the automatic pilot was discovered, and the aircraft was under automatic control. Also, the selection of radio station, course, speed, flap setting, landing gear position, and the final application of wheel brakes were accomplished from a program stored on punched cards. During this period, an interesting confluence of theory and practice of automatic feedback control of aircraft took place.

During the period 1956–1981, the major issues dealt were of safety and reliability, both of HW and SW. The main point was that for the stability augmentation/control system, the full-time operation of the subsystems was intrinsically required. Also, since augmenting systems operate in series with the pilot's inputs, safety is very important. Also, the full-authority, fail-operational (reliable) stability augmentation systems were accomplished over a period of time, and this required to provide a level of redundancy in the control channels that assured the complete operability in the presence of several independent failures. There was a transition from analog to digital systems, and the control laws and failure detection/management functions were implemented in SW for redundant general purpose computers, and the latter also served the functionalities of guidance, navigation, and other control (GNC) requirements; such a GNC is a highly interconnected and interactive composite system/computer.

Since 1981, the FBW systems were expanded to the commercial aircraft. The tasks of understanding, developing, specifying, and satisfying flying qualities have been a major thrust in stability and control research/design from the earlier times. Aircraft pilot coupling, also known as pilot-induced oscillations, was studied. This is better named as aircraft pilot interaction. Although the original aircraft might be stable, one with pilot and the augmented control system might not be. Also, the delays due to actuator rate limiting required special attention. Many aspects of verification and validation of critical SW and the issues of health monitoring and management of various control systems were also studied. Many inputs provided by traditional instruments and sensors were now received from GPS signals, which could also serve as part of the analytical redundancy scheme in the aircraft.

APPENDIX 8B ASPECTS OF FLY-BY-WIRE FLIGHT CONTROL DESIGN

For flight control, basically we require the control of forces and moments acting on the aircraft and then can control the accelerations and, hence, the velocities; and a flight control system (FCS) accomplishes this by aircraft control surfaces [20]. In earlier times, the mechanical linkages were used between the pilot's controls (in cockpit) and aircraft's control surfaces. The earlier FCS has moved to the present-day FBW control systems, with dependence on the digital computing, and the mechanical linkages have been replaced by the electrical signaling of the motion commands; hence, it is called the fly-by-wire (FBW) system. The major benefits of the FBW technology usages are as follows: (i) the aircraft system's characteristics can be tailored at each point in the flight envelope; (ii) the complex algorithms for various functionalities (redundancy management and health monitoring) can be implemented on the aircraft (flight) computer; (iii) allows "care-free-handling" of the AOA control and suppression of AOSS; this allows the protection against the stall, and departure; (iv) the HQs can be optimized across the flight envelope; (v) the aircraft agility can be improved, and this will improve the fuselage aiming to enhance the target capture and evasive maneuvering; (vi) improved lift/drag ratio, by controlling the unstable airframe, to increase the turning capability; (vii) reduced drag due to optimized trim controls; (viii) reconfiguration ability to allow continuation of missions following failures or battle damage; (ix) advanced autopilot that reduces the pilot's workload; and (x) reduction in weight, maintenance cost, and mechanical complexity.

The important thing to consider for the design of an FBW control system is the aircraft's flight envelope (Section 7.2). For the design of an FCS, a grid of operating points is selected and then localized controllers (control laws) are obtained; these points are minimized by taking into account the dynamic pressure within the structure of the control laws, and yet the design cases could become thousands in number; these local designs are integrated by gain scheduling. These designs are evaluated and tuned by time- and frequency-domain methods for satisfactory aircraft handling and design robustness. For gain scheduling, the information is used from the air-date system, which might be triplex/quadraplex redundancy system. The information from the air data system is augmented with that from the aircraft's inertial sensors.

For the control law, design several aspects need to be integrated and balanced: (i) basic airframe configuration; (ii) aircraft stores; (iii) uncertainties in the wind tunnel data; (iv) variation of aircraft mass, inertia, and CG; (v) possible symmetric/asymmetric combination of stores and their release; (vi) fuel state; (vii) high lift devices; (viii) airbrakes; (ix) wing sweep; (x) performance schedules; (xi) power plant interface; (xii) reversionary modes; (xiii) undercarriage operations; (xiv) ground handling; (xv) effect of nonlinear aerodynamics; (xvi) reduction of wing and tail effectiveness; (xvii) effect of transition from subsonic to supersonic flight; and (xviii) shock induced flow separations and effects of air compressibility.

For the implementation of FBW-FCS, several inter-disciplinary areas that need to be taken into account are (i) equipment specifications, (ii) required levels of functionalities, (iii) system qualification process, (iv) validation of the modes assumed for the control law design, (v) guarantee of required levels of reliability and integrity, (vi) appropriate levels of multiplexing, (vii) redundancy management, (viii) built-in-test capability, (ix) advanced sensors and actuators, (x) digital computing with its interfaces, (xi) lags and delays introduced by HW components, and (xii) filtering required to reduce, within the control loops, the responses from the flexible airframe; this might again introduce additional lags into the loops.

Finally, the flight control laws are evaluated on piloted simulators, and ground and flight testing of the aircraft (see Chapters 6 and 7). The HQs are also evaluated by gathering the data from these simulation runs and flight testing (see Chapter 10).

APPENDIX 8C MISSILE CONTROL METHODS

One of the tasks of the guidance system is to detect whether the missile is flying too high or too low, or too much to the right or left. The system measures these deviational errors and sends appropriate commands to the control system to reduce these errors to zero. The task of the control system is to maneuver the missile quickly and efficiently as a result of these signals. The guided missiles usually have one or two axes of symmetry. The aileron deflection is given as $\frac{1}{4}(\delta_1+\delta_2+\delta_3+\delta_4)$, or any other combination as per the available surfaces, and can act even differentially. The effects produced by the deflections would be considered as per the sign conventions adopted for the given missile control surfaces and the guidance system. There are several control methods possible for a missile, and for the Cartesian approach, two main groups are (i) aerodynamic and (ii) thrust vector control. In aerodynamic group, we have subgroups: (i) real control surface, (ii) moving wing, (iii) canards, (iv) freely rolling, (v) roll position control, and (vi) roll rate control. In thrust vector control, we have (i) gimballed motors, (ii) ball and socket or flexible nozzles, (iii) spoiler vanes, and (iv) secondary fluid or gas injection. In the second major approach, the twist and steer, we have aerodynamic method that use (i) rear control surfaces, (ii) moving wings, and (iii) canards. With Cartesian control, ideally one would like the missile to remain in the same roll orientation as at launch during the whole flight. Missile is not designed like an airplane, and there is no tendency to remain in the same roll orientation; it will tend to roll due to (i) asymmetrical loading of the lifting and control surfaces (in supersonic flight), which occurs when pitch and yaw incidences occur simultaneously and are not equal, and (ii) atmospheric disturbances.

EPILOGUE

In this chapter, we have introduced some fundamental aspects that are very important for design and analysis of flight control and reconfiguration control systems. The robustness issues, feedback properties, and classical and modern synthesis for multivariable feedback control design are surveyed in Ref. [21–23]. The evaluation of the FCSs using the stability margins and low-order equivalent systems is discussed in Ref. [24]. Applications of ANNs and artificial intelligence to sensor fault management, restructurable control systems, and decision and control are dealt with in Ref. [25–27]. In Ref. [28], the authors develop the physical insights into the relation between the singular values and the aircraft and controller dynamics by using the so-called literal approximate factors and provided a procedure for the determination of the most important parameter uncertainties for an aircraft and controllers. The idea is to obtain approximate relations for transfer function poles and zeros literally in terms of the stability and control derivatives of the atmospheric vehicle. In Ref. [29], the authors deal with the application of online parameter estimation for the restructurable control system. In Ref. [30], the authors discuss the issues and applications of evolutionary algorithms and hybrid neural and fuzzy control schemes. Some examples of flight control design using MATLAB are given.

EXERCISES

8.1 Draw the block diagram for the closed-loop system of Example 8.1.

8.2 How the parameter "*d*" should be chosen in the transformation method for the unstable state-space system?

8.3 If the positive real root is "*a*," what is the time to double the response of a system?

8.4 What is signified by the conditional optimality in control/estimation context?

8.5 What kind of feedback will help primarily improve the damping in a control system?

8.6 If the latex demand is in volts (Figure 8.2), what would be the units of gyro gain, accelerometer gain, servo gain, and *K*. What is the significance of *K* correction?

8.7 Can you guess on possible error performance criteria in the evaluating/design of control system (performance)?

8.8 How would you augment the control system to compensate for deficiency in damping in pitch?

8.9 An attitude control system of a missile is given as

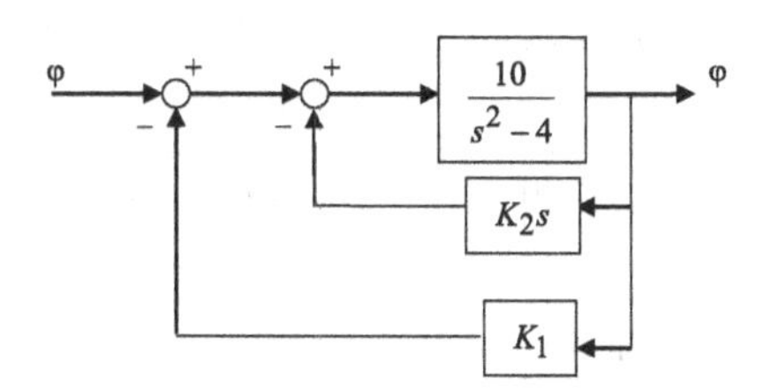

Simplify the block diagram and obtain the values of K_1 and K_2 so that the closed-loop system natural frequency is 20 rad/s and damping ratio is 0.7?

8.10 Show the force/moment balance for unstable aircraft by explicit geometry of an aircraft elevator and disposition of lift forces.

8.11 Refer to the block diagram of the attitude control system of Exercise 8.9. What are the inner and outer control loop and the feedback signals?

8.12 What are general limitations in sensor systems?

8.13 What are primary aspects of an artificial intelligent-reconfigurable control system?

REFERENCES

1. McRuer D., and Graham D. Flight control century: Triumphs of the systems approach. *Journal of Guidance, Control, and Dynamics*, 27(2), 161–173, 2004.
2. Bryson Jr., A. E. New concepts in control theory, 1959–1984. *Journal of Guidance, Control, and Dynamics*, 8(2), 417–425, 1985.
3. Sinha, N. K. *Control Systems.* Holt, Rinchart, and Winston Inc., New York, 1988.

4. Ogata, K. *Modern Control Engineering*, 2nd Edn. Prentice Hall, Hoboken, NJ, 1990.
5. Rabiner, L. R., and Rader, C. M. (eds.) *Digital Signal Processing, Vols. I and II*. IEEE Press, New York, 1975.
6. Anon. Automatic Flight Control Summary-Aerostudents. http://www.aerostudents.com/courses/automatics-flight-control/automaticFlightControlFullVersion.pdf; accessed January 2022.
7. Mooij, H. A. *Criteria For Low Speed Longitudinal Handling Qualities (Of Transport Aircraft With Closed-Loop Flight Control Systems)*. Martinus Nijhoff Publishers, the Netherlands, 1984.
8. Pallett, E. H. J., and Shawn, C. *Automatic Flight Control*, 4th Edn. Blackwell Scientific Publications, Oxford, 1993.
9. Chetty, S., and Madhuranath, P. (eds.). Aircraft flight control and simulation, NAL SP 9717, NAL-UNI lecture series No. 10, National Aerospace Laboratories, Bangalore, 26–29 August 1997.
10. Nelson, R. C. *Flight Stability and Automatic Control*, 2nd Edn. McGraw Hill, International Editions, New York, 1998.
11. https://in.mathworks.com/help/simulink/slref/designing-a-high-angle-of-attack-pitch-mode-control.html, accessed January 2022.
12. https://in.mathworks.com/help/control/ug/tuning-of-gain-scheduled-three-loop-autopilot.html, accessed January 2022.
13. Schmidt, D. K. *Modern Flight Dynamics*. McGraw-Hill, New York, 2011.
14. King, R. E. *Computational Intelligence in Control Engineering*. Marcel Dekker, New York, 1999.
15. Hajiyev, Ch., and Caliskan, F. *Fault Diagnosis and Reconfiguration in Flight Control Systems*. Kluwer Academic Publishers, Boston, MA, 2003.
16. Jiang, J., and Zhang, Y. Accepting performance degradation in fault-tolerant control system design. *IEEE Transactions on Control Systems Technology*, 14(20), 284–292, 2006.
17. Hajiyev, Ch., and Caliskan, F., Integrated sensor/actuator FDI and reconfigurable control and fault-tolerant flight control system design. *The Aeronautical Journal*, 105, 525–533, 2001.
18. Hajiyev, Ch., and Caliskan, F. Sensor/actuator fault diagnosis based on statistical analysis of innovation sequence and robust Kalman filtering. *Aerospace Science and Technology*, 4(6), 415–422, 2000.
19. Nagarajan, S., Shanmugam, J., and Rangaswamy, T. R. Sensor fault detection in a satellite launcher using adaptive neuro-fuzzy observer. *Journal of Aerospace Sciences and Technology, AeSI*, 59, 41–49, 2007.
20. Fielding, C. The design of fly-by-wire flight control systems. *The Aeronautical Journal*, 105, 543–549, 2001.
21. Lehtomaki, N. A., Sandell, Jr., N. R., and Athans, M. Robustness results in linear- quadratic Gaussian based multivariable control designs. *IEEE Transactions on Automatic Control*, Ac-26(1), 75–93, 1981.
22. Doyle, J. C., and Stein, G. Multivariable feedback design: Concepts for a classical/modern synthesis. *IEEE Transactions on Automatic Control*, Ac-26(1), 4–16, 1981.
23. Safonov, M. G., Laub, A. J., and Hartmann, G. L. Feedback properties of multivariable systems: The role and use of the return difference matrix. *IEEE Transactions on Automatic Control*, Ac-26(1), 47–65, 1981.
24. Anderson, M. R., Rabin, U. H., and Vincent, J. H. Evaluation methods for complex flight control systems. AIAA paper no. 89–3503-CP, 667–675, 1989.
25. Napolitano, M. R., Neppach, C., Casdorph, V., and Naylor, S. Neural-network-based scheme for sensor failure detection, identification and accommodation, *Journal of Guidance, Control, and Dynamics*, 18(6), 1280–1286, 1995.

26. Napolitano, M. R., Naylor, S., Neppach, C., and Casdorph, V. On-line learning direct neuro- controllers for restructurable control systems. *Journal of Guidance, Control, and Dynamics*, 18(1), 170–176, 1995.
27. Rauch, H. E. A control engineer's use of artificial intelligence. *Control Engineering Practice*, 6, 249–258, 1998.
28. McRuer, D. T., Myers, T. T., and Thompson, P. M. Literal singular-value-based flight control system design techniques. *Journal of Guidance*, 12(6), 913–919, 1989.
29. Napolitano, M. R., Song, Y., and Seanor, B. On-line parameter estimation for restructurable flight control systems. *Aircraft Design*, 4, 19–50, 2001.
30. Fleming, P. J., and Purshouse, R. C. Evolutionary algorithms in control systems engineering: A survey. *Control Engineering Practice*, 10, 1223–1241, 2002.

9 System Identification and Parameter Estimation for Aircraft

9.1 INTRODUCTION

The system identification and parameter estimation process (SIPEP), as one of the important approaches to mathematical modeling, provides a powerful tool for systematic analysis and optimization of industrial processes, thereby reducing losses and saving the cost of production. The SIPEP is a highly matured technology and can be looked upon as a data-dependent model-building process. For aerospace applications, it can be advantageously utilized to aid the iterative control law design/FS cycles as well as in certification of atmospheric vehicles. SID refers to determination of adequate mathematical model structure based on the physics of the problem and analysis of available data using an optimization criterion so as to minimize the sum of the squares of errors between the responses of the postulated mathematical model and the real system. The computational procedure is generally iterative and requires engineering judgment and use of objective model selection criteria [1]. Parameter estimation is also utilized in the SID procedure, and is regarded as a special case of SI procedure and also of Kalman filtering methods. Parameter estimation refers to explicit determination of numerical values of the unknown parameters of the postulated state-space mathematical model, or any type of the model. Basic principles are same as in SID, but in many cases, the model structure selection procedure may not be needed due to well-defined structure available and used from the physics of the system (Chapters 3–5), for example, aircraft parameter estimation (APE). The SID/parameter estimation process for a flight vehicle is illustrated in Figure 9.1.

The SIPEP utilizes techniques and principles from several established branches of mathematics, science, and engineering: (i) general and applied mathematics for definition of mathematical models, (ii) statistical and probability theory for interpretation of results and definition of cost functions/criteria, (iii) optimization methods and theory of model reduction to have optimal solutions, (iv) numerical techniques for solution of differential equations, matrix inversion, (v) system and control theory for state-space and transfer function modeling, (vi) signal processing for FFT, spectral density evaluation, and filtering of signals, (vii) linear algebra for vector/matrix and related theories, and (viii) information theory for interpretation of results and definition of criteria [2]. More often, the configuration of the SIPEP used for a particular application is governed by the state-of-the-art development in the foregoing. So a continual upgradation may be required in order to extract benefits from the supportive disciplines. The development of the SIPEP takes place at various stages and levels as an independent R&D effort. In initial stages, a considerable trial and

DOI: 10.1201/9781003293514-10

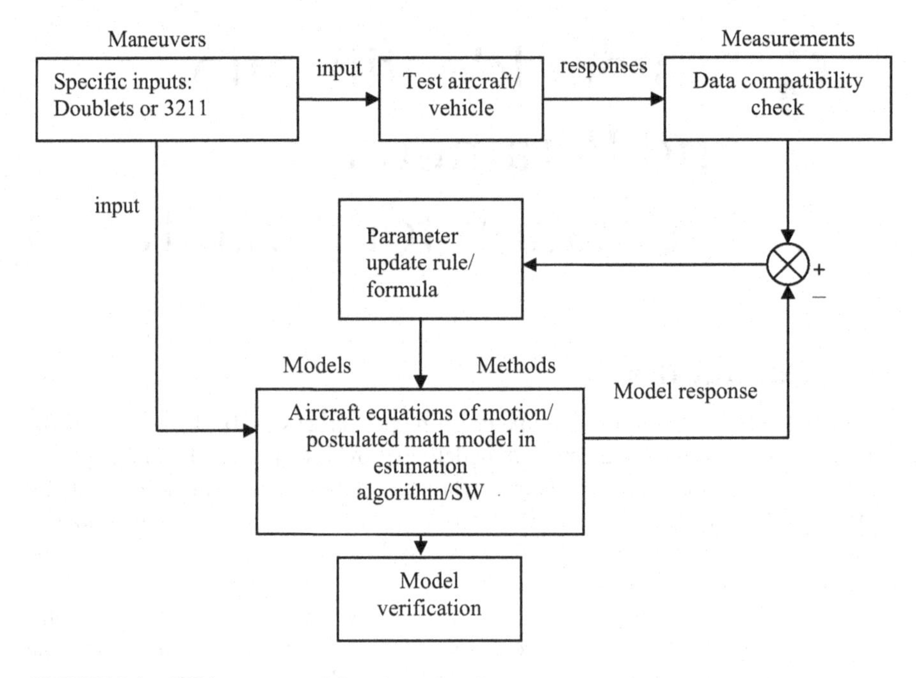

FIGURE 9.1 SID/parameter estimation procedure.

error approach is used with simulated and real data to arrive at the most appropriate methods. No single method is perfectly suitable for all kinds of the problems studied/ encountered.

Typical major developments have been in: (i) implementation and use of time-series analysis methods for modeling of dynamic systems (e.g., modeling of human operator's activity in a compensatory control task in a flight research simulator) [3], (ii) enhancement of output error method-based algorithm for parameter estimation of inherently unstable/augmented control system (e.g., unstable fly-by-wire (FBW) aircraft), development of analysis techniques for parameter estimation of unstable dynamic systems [1], (iii) development of factorization-based extended Kalman filtering programs for flight path reconstruction and data compatibility checking of flight test data [4], (iv) development of serial/parallel schemes for implementation of genetic algorithms (GAs) with an aim to using for parameter estimation [1], (v) development of new architectures for parameter estimation using recurrent neural networks and fast algorithms for training of feed-forward neural networks (FFNNs) [1], and (vi) development of filter-error-based program for parameter estimation of stable/unstable systems with data corrupted by colored noise (turbulence) [5]. The expansive scope of the SIPEP encompasses other related areas as follows: (i) Online real-time SID and parameter estimation technology poses a challenging task of arriving at synergism of robust, accurate, and stable computational algorithms, efficient programming, choice of suitable cost-effective and yet-fast computers (serial or parallel), and thorough validation under realistic test conditions to guarantee overall trouble-free operation in real-life practical environment;

certain time-varying properties of systems can be tracked by using online real-time SID-cum-parameter estimation for use in adaptive control [6]; (ii) incorporation of the so-called 'soft computing' into the traditional/conventional SIPEP, here, ANNs, fuzzy logic-based concepts, and GAs can be effectively used to develop expert system-based SIPEPs; (iii) use of SID procedures for near real-time determination of gain and phase margins of a FBY aircraft during flight testing exercises; (iv) GPS receiver signals can be used in conjunction with the existing sensor-measurement data to improve the accuracy of estimates, for example, flight path reconstruction using GPS signals, air data calibration exercises [7]; and (v) multi-sensor data fusion technology utilizes Kalman and information-based filtering algorithms for kinematic fusion of the state vector from individual sensor updates, and SIPEP can play a significant role in this field also [8].

Several criteria are used to judge the 'goodness' of the estimator/estimates: Cramer Rao bounds of the estimates, correlation coefficients among the estimates, determinant of the covariance matrix of the residuals, plausibility of the estimates based on physical understanding of the dynamic system, comparison of the estimates with those of nearly similar systems or estimates independently obtained by other methods (analytical or other parameter estimation methods), and model predictive capability. The time–history match is a necessary but not the sufficient condition. It is quite possible that the response match would be good, but some parameters could be unrealistic due to the following: (i) deficient model used for the estimation, and (ii) not all the modes of the system might have been sufficiently excited. One way to circumvent this problem is to add a priori information about the parameter in question, or by adding a constraint in the cost function, with a proper sign (constraint) on the parameter. One more approach is to fix such parameters at a priori value, which could have been determined by some other means or available independently from another source from the system.

A simple and commonly followed procedure [9] for SIPEP (see Figure 9.1) has the following steps: (i) data gathering/recording from the planned experiments on the process/plant/system, which presupposes proper experiment planning, design of input signal, test conditions, etc.; (ii) data preprocessing to remove spikes/extraneous noise, kinematic consistency checking for APE; (iii) postulation of appropriate model structure (Chapters 3–5); (iv) selection of criteria for model structure determination, usually data-dependent (Chapter 6 of [1]); (v) estimation of parameters of the postulated model using suitable parameter estimation method [1]; (vi) performance evaluation of the identified model/parameters using goodness-of-fit criteria; (vii) model validation across the validation data sets from the same experiment; (viii) revisit the postulated model and refine the model if needed, may be with additional data, to improve models; and (ix) interpretation of the results and specification of the confidence in the derived results from SIPEP.

9.2 SYSTEM IDENTIFICATION

In general, SID is a process of determination of model structure, (often the model order, i.e. the number of coefficients in an equation or a TF), and numerical values of these coefficients and model validation based on empirical data of a given dynamic system. This process is iterative in nature requiring engineering judgment at various stages of the procedure. A priori knowledge of the system is often used in selecting an appropriate model structure (state-space, transfer function).

9.2.1 Time-Series and Regression Model Identification

The parameters of an autoregressive (AR) and LS models can be estimated using LS method [1]. Assumption of the identifiability of the coefficients of the postulated models is presupposed. The general time-series model is given as:

$$z = \frac{B(q^{-1})}{A(q^{-1})}u + \frac{C(q^{-1})}{A(q^{-1})}e \tag{9.1}$$

$$\begin{aligned} A(q^{-1}) &= 1 + a_1 q^{-1} + \cdots + a_n q^{-n} \\ B(q^{-1}) &= b_0 + b_1 q^{-1} + \cdots + b_n q^{-m} \\ C(q^{-1}) &= 1 + c_1 q^{-1} + \cdots + c_n q^{-p} \end{aligned} \tag{9.2}$$

The LS model is given when the polynomial 'C' is neglected in the above equations. The 'z' is the time-series/output data, 'u' the input to the model/process and 'e' the equation error or noise representation. The equation error, Figure 9.2, for the LS model is defined as:

$$e(k) = A(q^{-1})z(k) - B(q^{-1})u(k) \tag{9.3}$$

Since, the data $u(k)$ and $z(k)$ are available from the conducted experiments, the above equations can be put in the form:

$$z = H\beta + e \tag{9.4}$$

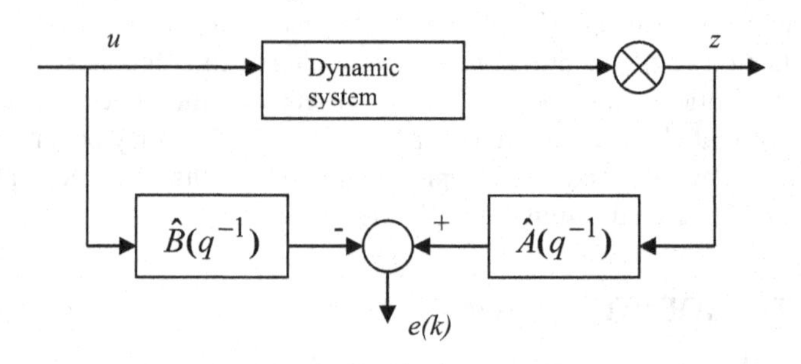

FIGURE 9.2 Equation error formulation.

Here, $z = \{z(n+1), z(n+2), \ldots z(n+N)\}^T$ with

$$H = \begin{bmatrix} -z(n) & & -z(1) & u(n) & & u(1) \\ -z(n+1) & & -z(2) & u(n+1) & & u(2) \\ \vdots & & \vdots & \vdots & & \\ -z(N+n-1) & \cdots & -z(n) & u(N+n-1) & \cdots & u(N) \end{bmatrix} \quad (9.5)$$

N = number of the data used. For example with $n = 3$ and $m = 1$, we have

$$e(k) = z(k) + a_1 z(k-1) + a_2 z(k-2) + a_3 z(k-3) - b_0 u(k) - b_1 u(k-1) \quad (9.6)$$

$$z(k) = \begin{bmatrix} -z(k-1) & -z(k-2) & -z(k-3) \end{bmatrix} \begin{bmatrix} a_1 \\ a_2 \\ a_3 \end{bmatrix} + \begin{bmatrix} u(k) & u(k-1) \end{bmatrix} \begin{bmatrix} b_0 \\ b_1 \end{bmatrix} + e(k)$$

$$z(k+1) = \begin{bmatrix} -z(k) & -z(k-1) & -z(k-2) \end{bmatrix} \begin{bmatrix} a_1 \\ a_2 \\ a_3 \end{bmatrix} + \begin{bmatrix} u(k+1) & u(k) \end{bmatrix} \begin{bmatrix} b_0 \\ b_1 \end{bmatrix} + e(k+1)$$

and so on, thereby yielding, in collective form the eqn. (9.4), with appropriate equivalence. Using LS method, we get the estimates of the parameters as:

$$\hat{\beta} = \left\{\hat{a}_1, \ldots, \hat{a}_n \,\vdots\, \hat{b}_1, \ldots, \hat{b}_m\right\} = (H^T H)^{-1} H^T z \quad (9.7)$$

The coefficients of time-series models can be estimated using the SID Toolbox of MATLAB. An example of human operator modeling using the time-series approach is given in Chapter 10. The covariance matrix of parameter estimation error is given as:

$$\text{cov}(\beta - \hat{\beta}) \approx \sigma_r^2 (H^T H)^{-1} \quad (9.8)$$

Here, σ_r^2 is estimation-residual variance.

In APE, a general form of the model to be identified occurs as follows:

$$y(t) = \beta_0 + \beta_1 x_1(t) + \cdots + \beta_{n-1} x_{n-1}(t) + e(t) \tag{9.9}$$

In (9.9), the time history $y(t)$ is assumed to be available. Actually, depending upon the problem at hand, the variable $y(t)$ would not be necessarily the states of the dynamic system; in fact, some intermediate steps might be required to compute y from $\hat{x}$. This is true for APE problem, and the intermediate computations will involve all the known constants and variables like x_i and y. It is necessary to determine which parameters should be retained in the model and estimated. This problem is handled using model order determination criteria and least-square (LS) method for parameter estimation. In APE problem, often $y(t)$ represents the aerodynamic coefficients (Chapter 4) and β the aerodynamic derivatives.

9.2.2 Comparison of Several Model Order Criteria

The crucial aspect of time-series modeling is that of selection of model structure (AR, MA, ARMA, or LS) and the number of coefficients for fitting this model to the time-series data. There are several model order selection criteria available in the open literature [1,10]. Selection of a reliable and efficient test criterion could be difficult, since most criteria are sensitive to process statistical properties which are often unknown. In the absence of *a priori* knowledge, any system that is generating time-series output data/signals could be represented by more popular AR or a LS model structures. These structures represent a general nth-order discrete linear time invariant system affected by random noise. The problem of model order determination is to assign a model dimension to adequately represent the unknown system, and the model selection procedure involves selecting a model structure and its complexity. Often a model structure can be ascertained based on the knowledge of the physics of the system and for certain processes; if physics is not well understood, then a black box approach can be used, leading to trial-and-error iterative procedure. In many situations, some knowledge about the system or the process would be available. Several model selection criteria arising out of different, but sometimes related principles of goodness-of-fit and statistical measures are possible. The criteria based on fit error of estimation, number of model parameters/coefficients, and whiteness of residuals are given in Table 9.1.

The test based on whiteness of residuals is widely used to check whether the residuals of fit are a white noise sequence. Autocorrelation-based whiteness of residuals test (ACWRT) is performed as shown in Table 9.1. Here, it is assumed that $r(k)$ is a zero-mean sequence. The autocorrelations must lie in the band $\pm 1.96/\sqrt{N}$ at least for 95% of the times for the null hypothesis, and usually the normalized ratio is used: $\hat{R}_{rr}(\tau)/\hat{R}_{rr}(0)$. The autocorrelations tend to be impulse function if the residuals are uncorrelated, and hence, at other times, their values must be too small, within the above-prescribed bounds. For input-/output-type models (ARMA, transfer function/LS), the process of cancellation of zeros with poles provides a model with less complexity. The numerator and denominator polynomials are factored, and cancellation, if any, then becomes obvious, but a subjective judgment is involved, since the cancellation might not be perfect. Often, eigenvalues are used for model reduction and selection. The eigenvalues

TABLE 9.1
Comparison of Some Model Order Selection Criteria

Criteria based on the concept of fit error	Fit error criterion (FEC1)	$\text{FEC1} = \dfrac{1/N\sum_{k=1}^{N}[z_k - z_k(\hat{\beta}_1)]^2}{1/N\sum_{k=1}^{N}[z_k - z_k(\hat{\beta}_2)]^2}$ N = no. of data points	If FEC1 < 1, select the model with $\hat{\beta}_1$. If FEC1 > 1, select the model with $\hat{\beta}_2$.
	Fit error criterion (FEC2)/prediction fit error (PFE) criterion	$\text{FEC2} = \dfrac{1/N\sum_{k=1}^{N}[z_k - z_k(\hat{\beta})]^2}{1/N\sum_{k=1}^{N} z_k^2}$	Insignificant change in the value of FEC2 determines the order of the model by locating the knee of the curve FEC2 v/s model order. PFE = FEC2 × 100.
	Residual sum of squares (RSS)	$\text{RSS} = \sum_{k=1}^{N}[z_k - \hat{z}_k(\hat{\beta})]^2$	With new parameter in the model, there should be large reduction in the RSS.
	Deterministic fit error (DFE)	$\text{DFE} = z - \dfrac{\hat{B}(q^{-1})}{\hat{A}(q^{-1})}u$	Useful for models of output/input types.
Criteria based on fit error and number of model parameters	FPE	$\text{FPE} = \sigma_r^2(N,\hat{\beta})\dfrac{N+n+1}{N-n-1}$ σ_r^2 = variance of the residuals.	A minimum is sought with respect to increasing n.
	Akaike's information criterion (AIC)	AIC = −2 ln(maximum likelihood) + 2 (number of independent parameters in the model) or $\text{AIC} = -2\ln(L) + 2n$ $\text{AIC}(n) = N\ln\sigma_r^2 + 2n$ for AR models; n is the no. of estimated parameters	If the two models are equally likely ($L_1 \approx L_2$), then the one with less number of parameters is chosen again according to the principle of parsimony. If the number of parameters increases, the AIC also increases, and hence less preferable the model.
Tests based on whiteness of residuals	ACWRT	$\hat{R}_{rr}(\tau) = 1/N\sum_{k=\tau}^{N} r(k)\,r(k-\tau)$ Estimate the autocorrelation function $R_{rr}(\tau)$ of residual sequence $r(k)$, for lag $\tau = 1, 2, \ldots, \tau_{max}$	

TABLE 9.2
The Human Operator's LS Model from the Tracking Experiments

Criterion/Order	$n=1$	2	3	4	5
% FEC2	3.13	0.03	0.03	0.046	0.08
% DFE	94.07	17.70	19.17	72.7	–

cannot be meaningfully determined for a covariance matrix if input/output units differ. Dimensional inconsistencies can be removed by appropriate normalization of the covariance matrices used in computation of model order criteria.

The detailed modeling using time-series and transfer function analysis are covered in Ref. [10], wherein three applications of model order determination were considered. The data sets were derived from (i) a simulated second-order system, (ii) experiments with human operator in a fixed-base simulator, and (iii) forces on a scaled-down model of an aircraft in a wind tunnel exposed to mildly turbulent flows. For the simulated system, the AR model identification was carried out using LS method, and it was found that several model order criteria, studied in the foregoing, provided sharp and consistent decisions. Next, the time-series data for human operator's responses were derived from a compensatory manual control tracking experiment on a fixed-base research flight simulator. Before recording the data, the operator was allowed to reach his steady performance in the tracking task. The LS/AR models were fitted to the data, and several model order determination criteria were used and studied. 500 data points sampled at 50 ms were used for SID. A fifth-/sixth-order AR model for human activity in the tracking task was found suitable, whereas LS model of second order was found to be also suitable for the same data sets as can be seen from Table 9.2.

From the discrete Bode diagrams obtained for the LS models, it was found that adequate amplitude ratio (plot vs. frequency) was obtained for the second-order TF. Thus, the AR pilot model differs from the LS pilot model in model order because the LS model has distinct input/output structure and hence the data information is well captured by the numerator part also. This is not so for the AR model, and hence a longer (large order) model is required. Estimation of pitch damping derivatives was attempted using random flow fluctuations inherent in the tunnel flow, wherein an aircraft's scaled physical model was mounted on a single DOF flexure having a dominant second-order response. The excitation to the model was inaccessible and not measurable, being the random flow fluctuation of the tunnel itself. The AR models were fitted using 1000 sample data. Since response was known to be of second order, the natural frequency was determined by evaluating the spectra using a frequency transformation of the discrete AR models. The estimated natural frequency stabilized for AR(n), $n \geq 10$.

Based on the experience gained, using several model order criteria, the following working rule is considered adequate for selection of the model order to fit varieties of experimental data. For order determination: (i) evaluate entropy criterion (AR only), (ii) evaluate final prediction error (FPE), (iii) perform F-test, and (iv) check for pole-zero cancellations (for input/output model), and for model validation: (i) see time–history prediction, (ii) test residuals for whiteness, and (iii) do the cross-validation.

9.2.3 Transfer Function Models from Real Flight Data

It is often advisable to fit a TF to the real flight data to quickly assess the dynamic characteristics of the aircraft. The TF and the Bode plots can be easily obtained by using the SID toolbox of the MATLAB. The main principle is that an output/input mathematical form like the LS model is fitted to the discrete-time data, and the Bode diagram from this model is determined [10]. Another approach is to fit a TF model to the frequency response data (the amplitude v/s frequency and the phase v/s frequency data) by a LS method [3]. From this TF, again the Bode plot can be obtained. These two approaches are available in the MATLAB toolboxes. Basically, it is a SID/parameter estimation problem, and the criterion for the minimization would be the same as in time domain, except that here we are dealing with the frequency domain data instead of the time histories. One can fit a delta operator TF to the continuous time data of a dynamic system [11]. The 'arx' model of second order was fitted to the real flight data (of a transport aircraft) using the following sequence in MATLAB: Data=iddata(q, ele, 0.4); model=arx(Data, [2 1 0]);

$A(q^{-1}) = 1 - 1.981q^{-1} + 0.9898q^{-2}$; $B(q^{-1}) = -0.002198$. The discrete time Bode diagram was then obtained by dbode([-0.002148], [1-1.981 0.9898], 0.4). Figure 9.3 shows the Bode diagram of this pitch rate-elevator control input TF.

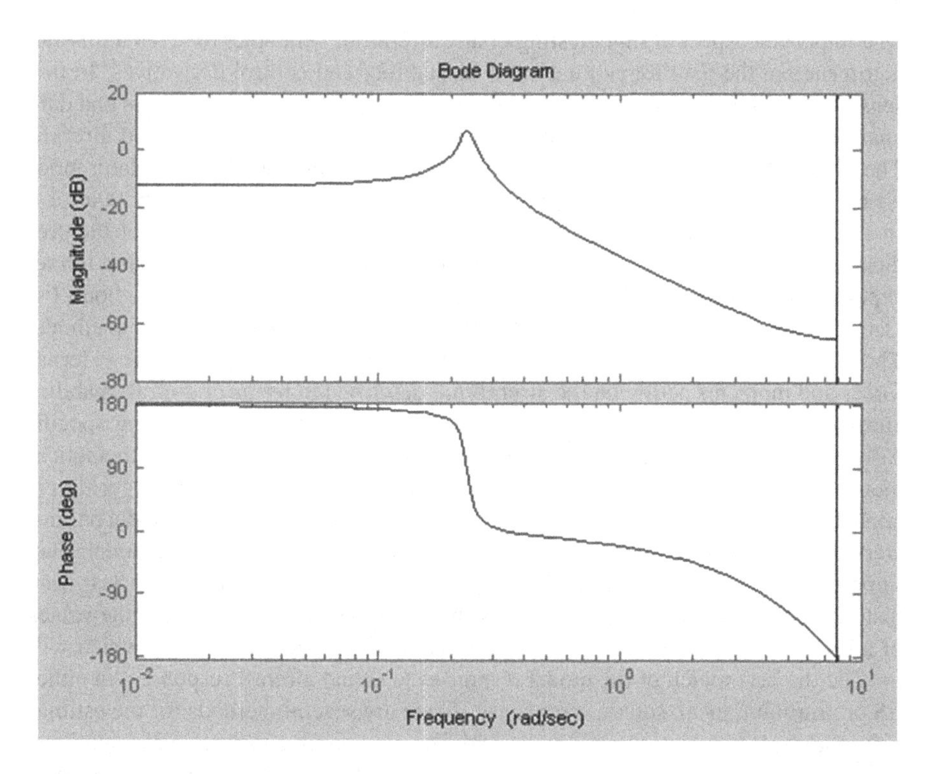

FIGURE 9.3 Frequency response of the pitch rate to elevator TF estimated from real flight data of a transport aircraft.

9.2.4 Expert Systems for System Identification

Although the SID procedure is well established, there is some scope for improvement in keep trying until a good enough or best model is obtained. A typical exercise would need a dozen steps. It should be possible to build an expert system that utilizes a set of rules that attempts to mimic the way in which an experienced SID expert proceeds while still using the ID toolbox of MATLAB. The segment interconnect diagnostic expert system (SIDES) should be able to run without any intervention from an analyst. The SIDES would start its iterative journey once driven by the data and rules. The SIDES would have the following features: (i) an efficient sampling device with appropriate filters to select the data at lower sampling intervals without loss of information, and (ii) a set of permissible model structures. A typical expert system has the steps [12]: (i) parsimonious walk through the model set defined by the analyst, (ii) selection of best models based on a quality index, and (iii) modification of the sampling interval. The first step avoids the examination and comparison of all the models in the model set. The quality index reflects the best trade-off among many validation tools. The rules are goal-oriented, and every set of rules is organized around a particular goal. Each rule would contain a conditional expression of one or more patterns which would be followed by an unconditional action.

9.3 AIRCRAFT PARAMETER ESTIMATION

One important aspect of flight testing of any aircraft or helicopter (or even a missile) is to generate the data for estimation of its stability and control derivatives. In this sense, the parameter estimation is an important tool for flight test engineers and data analysts to determine the aerodynamic characteristics of new and untested aircraft. The flight-estimated derivatives are very useful in updating the flight simulator models (of aircraft), improving the flight-control laws and evaluating handling qualities. In addition, the flight-determined derivatives (FDDs) help in validation of the predicted derivatives, which are based on one or more of the following: (i) wind tunnel experiments on scaled models of aircraft and subsequent data reduction methods, (ii) Data Compendium (DATCOM) methods, and (iii) some analytical/CFD methods. The aircraft dynamics are modeled by a set of differential equations. The external forces and moments acting on the aircraft are described in terms of aircraft stability and control derivatives, which are treated as unknown parameters. Using specifically designed control inputs, the responses of the test aircraft and the mathematical model are obtained and compared. An appropriate parameter estimation algorithm is applied to minimize the error (in responses) by iteratively adjusting the model parameters. The key elements such as maneuvers, measurements, methods, and models play important role in APE (Figure 9.1). Assume that a set of measured flight test time histories are given, then the parameter estimation problem is to determine the values of a given set of parameters in the system equations/postulated model, which will provide the best match of the model responses to actual aircraft responses, in either LS or minimum mean square error sense. There are several methods for the estimation of aircraft parameters, and the basic differences are due to certain assumptions, choice of optimal criteria, formulation of cost function that reflects the existence of external disturbances, and the presence of measurement noise in the data.

9.3.1 Maneuvers, Measurements, and Mathematical Models

A detailed description of maneuvers specifically to be performed for generating data that are suitable for APE was presented in Chapter 7. The first major step in APE is the definition and planning of flight test experiment and subsequent data acquisition. This primarily addresses the issue of specification of flight test maneuvers, and obtaining measurements of control surface deflections, airspeed, sideslip, angle-of-attack, angular velocities, linear and angular accelerations, and attitude angles. Also, definition of flight conditions, aircraft configuration, sensors, fuel consumption for estimation of aircraft, C.G. location, weight, and inertias is required. Detailed information of these must be sought before the analysis of the flight data.

The accuracy of estimated derivatives depends on the quality of measurements that are subjected to systematic and random errors. It is essential to evaluate the quality of the measured data and rectify these measurements before using for parameter estimation. These evaluations include consideration of factors like the frequency content of the input signal, sampling rate, signal amplitude, and signal-to-noise ratio. Some quantities may be difficult to measure directly, this being true for helicopters because of their high vibration levels. As a first step, the raw data are inspected for obvious gross errors and imperfections, such as incorrect signs of measured variables, missing points, spikes, and extensive noise. The corrections are made in data to account for errors resulting from sensor location (C. G. offset and flow corrections) and time skew errors. The procedure for evaluating data quality and correction is the kinematic consistency checking, since the aircraft states are related by a set of differential equations, one checks for consistency among the kinematic quantities. As an example, the measured roll and pitch attitudes should match with those reconstructed from the rate measurements with great accuracy. The process ensures that the data are consistent with the basic underlying kinematic models; since the aircraft is flying, it must be according to the kinematics of the aircraft, but the sensors could have gone wrong in sensing/generating the data. The data compatibility check provides the estimates of the bias parameters and scale factors in the measured data. An accurate determination of these errors is very important for accurate estimation of the aerodynamic derivatives from flight data.

The mathematical models used in APE are discussed in Chapters 3–5, and specifically simple longitudinal and lateral-directional models are discussed in Chapter 5. If the modes of aircraft dynamics are not properly excited by the chosen input, then an attempt to identify too many parameters from limited amount of data might fail or yield estimates with reduced accuracy.

The response of an aircraft to pilot's (or computer's) command signal are the flight data that along with other mandatory data (m, I, and dynamic pressure, etc.) are used in a parameter estimation procedure in order to estimate the aerodynamic derivatives. These responses are generated by exciting certain modes of the aircraft. These modes are dynamic modes, and they contain information on the aerodynamic derivatives implicitly. It is very essential to choose input test signals with good characteristics, so that properly excited dynamic responses are available for further analysis. The test input signal must possess certain desirable features: (i) It should have sufficient bandwidth to be able to excite the modes of interest; (ii) the amplitude and the form of the signal should be such as to yield good-quality maneuver; (iii) the amplitude, bandwidth, and slope of the signal should be bounded so that aircraft motion does not violate the assumption of linearity; and (iv) it should be possible to realize (generate) the signal easily by the pilot. Of course for complicated signals, computer simulation or prerecording can be used. In order to design a multistep input signal which will excite the modes of an aircraft, one can use Bode diagram approach [13]. This allows one to select the frequencies to be included in the signal based on the identifiability of the derivatives.

9.3.2 Parameter Estimation Methods

Several useful techniques are available in the literature for estimation of parameters [1] from flight data, selection of which is governed by the complexity of the mathematical model, *a priori* knowledge about the system, and information on the noise characteristics in measured data. In general, the estimation technique provides the estimated values of the parameters along with their accuracies, in the form of standard errors or variances. In this section, some widely used methods are briefly discussed.

9.3.2.1 Equation Error Method

In the equation error method (EEM), the measurements of the state variables and their derivatives are assumed to be available/measureable and used. The method is computationally very simple and is a one-shot procedure. Since the measured states and their derivatives are used, the method would yield biased estimates, and hence, it could be used as a startup procedure for other methods. It minimizes a quadratic cost function of the error in the state equations to estimate the parameters. The EEM is applicable to linear as well as linear-in-parameter systems. The EEM can be applied to unstable systems because it does not involve any numerical integration of the system equations that would otherwise cause divergence of certain state variables. Thus, the utilization of measured states and state derivatives for estimation in the algorithm enables estimation of the parameters of unstable systems. If a system is described by the state equation

$$\dot{x} = Ax + Bu; \quad x(0) = x_0 \tag{9.10}$$

then, the EE can be written as

$$e(k) = \dot{x}_m - Ax_m - Bu_m \tag{9.11}$$

Here, x_m is the measured state, and the subscript m denotes 'measured'. Parameter estimates can be obtained by minimizing the EE w.r.t. β (β being the unknown parameter vector). Eqn. (9.10) can be written as

$$e(k) = \dot{x}_m - A_a x_{am}; \; A_a = \begin{bmatrix} A & B \end{bmatrix}; \; x_{am} = \begin{bmatrix} x_m \\ u_m \end{bmatrix} \tag{9.12}$$

The cost function and the estimator are given by

$$J(\beta) = \frac{1}{2}\sum_{k=1}^{N}\left[\dot{x}_m(k) - A_a x_{am}(k)\right]^T \left[\dot{x}_m(k) - A_a x_{am}(k)\right] \tag{9.13}$$

$$\hat{A}_a = \dot{x}_m (x_{am}^T)(x_{am}\, x_{am}^T)^{-1} \tag{9.14}$$

Example 9.1

The equation error formulation for parameter estimation of an aircraft is illustrated with one such state equation here. Let the moment equation be given:

$$\dot{q} = M_\alpha \alpha + M_q q + M_{\delta_e}\delta_e \tag{9.15}$$

Then, moment coefficients of Eqn. (9.15) are determined from the system of linear equations given by (Eqn. (9.15) is multiplied in turn by α, q and δ_e)

$$\sum \dot{q}\alpha = M_\alpha \sum \alpha^2 + M_q \sum q\alpha + M_{\delta_e} \sum \delta_e \alpha \tag{9.16}$$

$$\sum \dot{q}q = M_\alpha \sum \alpha q + M_q \sum q^2 + M_{\delta_e} \sum \delta_e q$$

$$\sum \dot{q}\delta_e = M_\alpha \sum \alpha\delta_e + M_q \sum q\delta_e + M_{\delta_e} \sum \delta_e^{\,2}$$

Here, Σ is the summation over the data points $(k = 1,\ldots,N)$ of α, q and δ_e signals. Combining the terms, we get

$$\begin{bmatrix} \sum \dot{q}\alpha \\ \sum \dot{q}q \\ \sum \dot{q}\delta_e \end{bmatrix} = \begin{bmatrix} \sum \alpha^2 & \sum q\alpha & \sum \delta_e\alpha \\ \sum \alpha q & \sum q^2 & \sum \delta_e q \\ \sum \alpha\delta_e & \sum q\delta_e & \sum \delta_e^2 \end{bmatrix} \begin{bmatrix} M_\alpha \\ M_q \\ M_{\delta_e} \end{bmatrix} \tag{9.17}$$

The above formulation can be expressed in a compact form as $Y = X\beta$, and the EE is formulated as $e = Y - X\beta$ keeping in mind that there will be modeling and estimation errors combined in 'e'. It is presumed that measurements of $\dot{q}, \alpha, q$ and δ_e are available, then the EE estimates of the parameters can be obtained by using Eqns. (9.7) and (9.14).

9.3.2.2 Maximum-Likelihood Output Error Method

The ML process, invoking the probabilistic aspects of random variables (measurement noises/errors), defines a method by which one can obtain estimates of the parameters. The idea is that these parameters are most likely to produce the model responses, which would closely match the measurements. A likelihood function, similar to a probability density function (PDF), is defined when measurements are used. This likelihood function is maximized to obtain the estimates of the parameters of the dynamic system. The output error method is a maximum-likelihood estimator which accounts only for measurement noise and not process noise. The main idea is to define a function of the data and the unknown parameters, and this function is called the likelihood function. The parameter estimates are those values, which maximize this function. If $\beta_1, \beta_2, \ldots, \beta_r$ are the unknown parameters of a system and $z_1, z_2, \ldots, z_n$ the measurements of the true values $y_1, y_2, \ldots, y_n$, then these true values could be made function of these unknown parameters as $y_i = f_i(\beta_1, \beta_2, \ldots, \beta_r)$.

If z is a random variable with PDF $p(z, \beta)$, then to estimate 'β' from z, choose the value of 'β' that maximizes the likelihood function $L(z, \beta) = p(z, \beta)$. Thus, the problem of parameter estimation is reduced to maximization of a real function called the likelihood function (of the parameter 'β' and the data z). In essence, the 'p' becomes 'L' when the measurements are obtained and used in 'p'. The parameter 'β', which makes this function most probable (to have yielded these measurements), is called the ML estimate. Such a likelihood function is given as

$$p(z|\beta) = p(z_1, z_2, \ldots, z_n | \beta_1, \beta_2, \ldots, \beta_r)$$

$$= \frac{1}{(2\pi)^{n/2} \sigma_1, \sigma_2, \ldots, \sigma_n} \exp\left[\sum_{i=1}^{n} -1/2 \frac{(z_i - y_i(\beta))^2}{\sigma_i^2}\right] \tag{9.18}$$

The main point in any estimator is to determine or predict the error made in the estimates relative to the true parameters, although the true parameters are unknown in the real sense. Only some statistical indicators for the errors can be worked out. The Cramer-Rao (CR) lower bound is, perhaps, the best measure for such errors. The likelihood function is defined as

$$L(z|\beta) = \log p(z|\beta) \tag{9.19}$$

The likelihood differential equation is obtained as

$$\frac{\partial}{\partial \beta} L(z|\beta) = L'(z|\hat{\beta}) = \frac{p'}{p}(z|\hat{\beta}) = 0 \tag{9.20}$$

The equation is non-linear in $\hat{\beta}$, and a first-order approximation by Taylor's series expansion is used to obtain $\hat{\beta}$:

$$L'(z|\beta_0 + \Delta\beta) = L'(z|\beta_0) + L''(z|\beta_0)\Delta\beta = 0 \tag{9.21}$$

The increment in 'β' is obtained as

$$\Delta\beta = \frac{L'(z|\beta_0)}{-L''(z|\beta_0)} = -\left(L''(z/\beta_0)\right)^{-1} L'(z/\beta_0) \tag{9.22}$$

The likelihood-related partials can be evaluated when the details of the dynamic systems are specified. In a general sense, the expected value of the denominator of Eqn. (9.22) is defined as the information matrix:

$$I_m(\beta) = E\{-L''(z|\beta)\} \tag{9.23}$$

From (9.23), we see that if there is large information content in the data, then $|L''|$ tends to be large, and the uncertainty in estimate $\hat{\beta}$ is small (9.22). The so-called CR inequality provides a lower bound to the variance of an unbiased estimator. If $\beta_e(z)$ is any estimator of 'β' based on the measurement z and $\bar{\beta}_e(z) = E\{\beta_e(z)\}$, the expectation of the estimate, then we have as the CR inequality, for unbiased estimator:

$$\sigma_{\beta e}^2 \geq \left(I_m(\beta)\right)^{-1} \tag{9.24}$$

For unbiased efficient estimator, $\sigma_{\beta e}^2 = I_m^{-1}(\beta)$. The inverse of the information matrix, for certain special cases, is the covariance matrix, and hence, we have the theoretical expression for the variance of the estimator. Thus, the actual variance in the estimator, for an efficient estimator, would be at least equal to the predicted variance, whereas for other cases, it could be greater but not lesser than the predicted value. The predicted value provides the lower bound.

9.3.2.3 Maximum-Likelihood Estimation for Dynamic System

Let a linear dynamical system be described as

$$\dot{x}(t) = Ax(t) + Bu(t) \tag{9.25}$$

$$y(t) = Hx(t) \tag{9.26}$$

$$z(k) = y(k) + v(k) \tag{9.27}$$

In many applications, the actual systems are of continuous time, but the measurements would be available as discrete samples, with

$$E\{v(k)\} = 0; \quad E\{v(k)v^T(l)\} = R\delta_{kl} \tag{9.28}$$

It is assumed that the measurement noise is zero mean and white Gaussian random variable with covariance matrix R, allowing us to use the Gaussian probability density concept for deriving the maximum-likelihood estimator. We have the likelihood function as

$$p(z(1),z(2),\ldots,z(N)\,|\,\beta,R) = \prod_{k=1}^{N} p(z(k)\,|\,\beta,R)$$

$$= \left((2\pi)^m |R|\right)^{-N/2} \exp\left[-\frac{1}{2}\sum_{k=1}^{N}[z(k)-y(k)]^T R^{-1}[z(k)-y(k)]\right] \tag{9.29}$$

The parameter vector 'β' is obtained by maximizing the likelihood function with respect to 'β' by minimizing the negative log-likelihood function given as

$$L = -\log\, p(z\,|\,\beta,R) = \frac{1}{2}\sum_{k=1}^{N}[z(k)-y(k)]^T R^{-1}[z(k)-y(k)]] + N/2\log|R| + \text{const} \tag{9.30}$$

The R can be estimated as

$$\hat{R} = 1/N\sum_{k=1}^{N}[z(k)-\hat{y}(k)][z(k)-\hat{y}(k)]^T \tag{9.31}$$

When the estimated value R is substituted in the likelihood function, the minimization of the cost function w.r.t 'β' results in

$$\partial L/\partial\beta = -\sum_{k}(\partial y(\beta)/\partial\beta)^T R^{-1}(z-y(\beta)) = 0 \tag{9.32}$$

This set is a system of nonlinear equations, and an iterative solution can be obtained by quasi-linearization method, known as modified Newton-Raphson or Gauss Newton method. We expand

$$y(\beta) = y(\beta_0 + \Delta\beta) \tag{9.33}$$

$$\text{as } y(\beta) = y(\beta_0) + \frac{\partial y(\beta)}{\partial\beta}\Delta\beta \tag{9.34}$$

A version of the quasi-linearization is used for obtaining a workable solution for the output error method. Substituting this approximation in Eqn. (9.34), we get

$$\left[\sum_{k}\left[\frac{\partial y(\beta)}{\partial\beta}\right]^T R^{-1}\frac{\partial y(\beta)}{\partial\beta}\right]\Delta\beta = \sum_{k}\left[\frac{\partial y(\beta)}{\partial\beta}\right]^T R^{-1}(z-y) \tag{9.35}$$

Finally we have

$$\Delta\beta = \left\{ \sum_k \left[\frac{\partial y(\beta)}{\partial \beta} \right]^T R^{-1} \frac{\partial y(\beta)}{\partial \beta} \right\}^{-1} \left\{ \sum_k \left[\frac{\partial y(\beta)}{\partial \beta} \right]^T R^{-1}(z-y) \right\} \tag{9.36}$$

Thus, ML estimate is obtained as

$$\hat{\beta}_{\text{new}} = \hat{\beta}_{\text{old}} + \Delta\beta \tag{9.37}$$

The CR bound is a primary criterion for evaluating accuracy of the estimated parameters. Maximum-likelihood method (MLE) gives the measure of this parameter accuracy without any extra computation. The information matrix is computed as

$$(I_m)_{ij} = E\left\{ -\frac{\partial^2 \log p(z|\beta)}{\partial \beta_i \partial \beta_j} \right\} = \sum_{k=1}^{N} \left(\frac{\partial y(k)}{\partial \beta_i} \right)^T R^{-1} \frac{\partial y(k)}{\partial \beta_j} \tag{9.38}$$

The diagonal elements of the inverse of the information matrix give the individual covariance, and the square roots of these elements are measures of the standard deviations called the CR bounds (CRBs). The OEM/MLE can also be applied to any nonlinear system. Computational and accuracy aspects are further discussed in Ref. [1].

9.3.2.4 Filtering Methods

Kalman filtering has evolved to a very high state-of-the-art method for recursive state estimation of dynamic systems [14]. It has generated worldwide extensive applications to aerospace system and target tracking problems. It has an intuitively appealing state-space formulation, and it gives algorithms that can be easily implemented on digital computers, since the filter is essentially a numerical algorithm and is the optimal state observer for the linear systems. It being a model-based approach uses the system's model in the filter:

$$x(k+1) = \phi x(k) + Bu(k) + Gw(k); \; z(k) = Hx(k) + v(k) \tag{9.39}$$

The process noise w is a white Gaussian sequence with zero mean and covariance matrix Q, the measurement noise v is a white Gaussian noise sequence with zero mean, and covariance matrix R and ϕ is $(n \times n)$ transition matrix that propagates the states from k to $k+1$. Given the mathematical model of the dynamic system, statistics Q and R of the noise processes, the noisy measurement data, and the input, the KF obtains the optimal estimate of the state, x, of the system. It is assumed that the values of the elements of ϕ, B, and H are known.

9.3.2.4.1 Discrete Time Filtering Algorithm

Let the initial state estimate at k move to $k+1$. At this stage, a new measurement is available, and it contains information regarding the state as per Eqn. (9.39). Intuitively, the idea is to incorporate the measurement into the data-filtering process and obtain a refined estimate of the state. The algorithm is given as mentioned later.

9.3.2.4.2 State/Covariance Evolution or Propagation

$$\tilde{x}(k+1) = \phi\hat{x}(k) \quad (9.40)$$

$$\tilde{P}(k+1) = \phi\hat{P}(k)\phi^T + GQG^T \quad (9.41)$$

9.3.2.4.3 Measurement Data/Covariance Filtering/Update

$$r(k+1) = z(k+1) - H\tilde{x}(k+1) \quad (9.42)$$

$$K = \tilde{P}H^T(H\tilde{P}H^T + R)^{-1} \quad (9.43)$$

$$\hat{x}(k+1) = \tilde{x}(k+1) + K\,r(k+1) \quad (9.44)$$

$$\hat{P} = (I - KH)\tilde{P} \quad (9.45)$$

The matrix $S = H\tilde{P}H^T + R$ is the covariance matrix of the residuals/innovations. The actual residuals can be computed from Eqn. (9.42), and these can be compared with standard deviations obtained by taking the square root of the diagonal elements of S. The performance of the filter can be evaluated by checking (i) the whiteness of the residuals and (ii) comparing the computed covariance with the theoretical covariance obtained from the covariance equations of the filter. The test (i) signifies that the residuals being white, no information is left out to be utilized in the filter. The test (ii) signifies that the computed covariance from the data matches the filter predictions/theoretical estimates of the covariance. Thus, a proper tuning has been achieved. The Kalman filter could diverge due to many reasons [15]: (i) modeling errors due to the use of highly approximated model of the nonlinear system, (ii) choice of incorrect a priori statistics (P, Q, R), and (iii) finite word length computation. For (iv), a factorization-based filtering method should be used, or the filter should be implemented on a computer with large word length.

9.3.2.4.4 Extended Kalman Filtering Algorithm

Many real-life dynamic systems are nonlinear, and estimation of states of such systems is often required. The nonlinear system can be expressed as

$$\dot{x}(t) = f[x(t), u(t), \Theta] \quad (9.46)$$

$$y(t) = h[x(t), u(t), \Theta] \quad (9.47)$$

$$z(k) = y(k) + v(k) \quad (9.48)$$

The f and h are general nonlinear vector-valued functions. The Θ is the vector of unknown parameters, $\Theta = [x_0, b_u, b_y, \beta]$, x_0 as values of the state variables at time $t = 0$, b_u represent the bias parameters in control inputs, b_y represents the biases in model responses y, and β as the parameters in the mathematical model of the system.

We need to linearize the nonlinear functions f and h and apply the KF with proper modifications to these linearized models. The linearizations will be around previous/current best state estimates that are more likely to represent the truth. Simultaneous estimation of states and parameters is achieved by augmenting the state vector with unknown parameters, as additional states, and then using the filtering algorithm with the augmented system non-linear model. The new augmented state vector is given as

$$x_a^T = \left[x^T \; \Theta^T\right] \tag{9.49}$$

$$\dot{x}_a = \begin{bmatrix} f(x_a,u,t) \\ 0 \end{bmatrix} + \begin{bmatrix} G \\ 0 \end{bmatrix} w(t) \tag{9.50}$$

$$\dot{x}_a = f_a(x_a,u,t) + G_a w(t) \tag{9.51}$$

$$y(t) = h_a\left(x_a,u,t\right) \tag{9.52}$$

$$z(k) = y(k) + v(k), \;\; k = 1,\ldots N \tag{9.53}$$

$$f_a^T(t) = \left[f^T \quad 0^T\right]; \; G_a^T = \left[G^T \quad 0^T\right] \tag{9.54}$$

The linearized system matrices are defined as

$$A(k) = \left.\frac{\delta f_a}{\delta x_a}\right|_{x_a=\hat{x}_a(k),\, u=u(k)} \tag{9.55}$$

$$H(k) = \left.\frac{\delta h_a}{\delta x_a}\right|_{x_a=\tilde{x}_a(k),\, u=u(k)} \tag{9.56}$$

The state transition matrix is given by

$$\phi(k) = \exp\left[-A(k)\,\Delta t\right] \text{where } \Delta t = t_{k+1} - t_k \tag{9.57}$$

We notice the time-varying nature of A, H, and ϕ, since they are evaluated at current state estimate, which varies with time k.

9.3.2.4.5 Time Propagation

The states are propagated from the present state to the next time instant. The predicted state is given by

$$\tilde{x}_a(k+1) = \hat{x}_a(k) + \int_{t_k}^{t_{k+1}} f_a[\hat{x}_a(t),u(k),t]\,dt \tag{9.58}$$

The covariance matrix for state error propagates from instant k to $k+1$ as

$$\tilde{P}(k+1) = \phi(k)\hat{P}(k)\phi^T(k) + G_a(k)QG_a^T(k) \tag{9.59}$$

9.3.2.4.6 Measurement Update

The EKF updates the predicted estimates by incorporating the new measurements as follows:

$$\hat{x}_a(k+1) = \tilde{x}_a(k+1) + K(k+1)\{z_m(k+1) - h_a[\tilde{x}_a(k+1), u(k+1), t]\} \tag{9.60}$$

The covariance matrix is updated using the Kalman gain and the linearized measurement matrix. The Kalman gain is given by

$$K(k+1) = \tilde{P}(k+1)H^T(k+1)[H(k+1)\tilde{P}(k+1)H^T(k+1) + R]^{-1} \tag{9.61}$$

The posteriori covariance matrix is given as

$$\hat{P}(k+1) = [I - K(k+1)H(k+1)]\tilde{P}(k+1) \tag{9.62}$$

A sophisticated filter error method accounts for process and measurement noise using a Kalman filter in the structure of the maximum-likelihood/output error method. The details of this method are given in Ref. [1].

9.3.3 Parameter Estimation Approaches for Inherently Unstable, Augmented Aircraft

In certain practical applications, it is required to estimate the parameters of the open-loop system (which might be inherently unstable) from the data generated when the system is operating in a closed loop. The feedback causes correlations between the input and output variables and identifiability problems. In the augmented system, the measured responses would not display the modes of the system adequately since the feedback is generating controlled responses. The estimation problem complexity is more when the basic system is unstable because the integration of the state model could lead to numerical divergence and the measured data would be usually corrupted by process and measurement noises. The EKF-UD approach can be used for parameter estimation of unstable systems because of the inherent stabilization properties of the filters [1]. The OEM poses some difficulties when applied to highly unstable systems since the numerical integration of the unstable state model equations would lead to divergence. One can provide artificial stabilization in the model (in the software algorithm) used for parameter estimation resulting in feedback-in-model method. This requires a good engineering judgment. Another way to circumvent this problem is to use measured states in the estimation leading to the so-called stabilized output error method (SOEM) [1]. The filter error method is the most general approach to parameter estimation problem. The filter error method treats the errors arising from data correlation as process noise that is suitably accounted by KF part of the filter error method. The detailed exposition of the filter error method is given in Refs. [1,5].

9.3.3.1 Stabilized Output Error Methods

There are two major approaches of the SOEMs. In the equation-decoupling method, the system state matrix is decoupled: (i) only diagonal elements pertaining to each of the integrated states, supposed to be the stable part, and (ii) the off-diagonal elements associated with the measured states. Thus, the state equations are decoupled, and the unstable system is changed to a stable one. This stabilization of the output error method by means of measured states would prevent the divergence of the integrated states. The measured states are used, and the input vector u is augmented with the measured states x_m to give

$$\dot{x} = A_d x + \begin{bmatrix} B & A_{od} \end{bmatrix} \begin{bmatrix} \delta \\ x_m \end{bmatrix} \tag{9.63}$$

The integrated stable variables are present only in the A_d part, and all the off-diagonal variables have measured states. This renders each differential equation to be integrated independently of the others.

In the regression analysis SOEM, the measured states are used with those parameters in the state matrix that are responsible for instability in the system and integrated states are used with the remaining parameters. The matrix A is put into two parts: (i) one containing that part of matrix A that has parameters that do not contribute to instability and (ii) the other having the parameters that contribute to system instability. The partitioned system has the form of

$$\dot{x} = A_s x + \begin{bmatrix} B & A_{us} \end{bmatrix} \begin{bmatrix} \delta \\ x_m \end{bmatrix} \tag{9.64}$$

The integrated states are used for stable part of the system matrix and measured states for the parameters contributing to the unstable part of the system.

Thus, the SOEMs seem to fall in between EEMs that use the measured states, and output error methods, and can be said to belong to a class of mixed equation error-output error methods. The output error method does not work directly for unstable systems because numerical integration of the system equations causes divergence. For the SOEMs, since the measured states (obtained from the unstable system operating in closed loop) are stable, their use in the estimation process attempts to prevent the divergence. The parameter estimation of basic unstable system directly in a manner similar to that of OEM for a stable plant is accomplished, and this has been established by the asymptotic analysis of SOEMs when applied to unstable systems [1].

9.4 DETERMINATION OF STABILITY AND CONTROL DERIVATIVES FROM FLIGHT DATA – CASE STUDIES

In this section, several case studies are presented with some results from real data analysis in the sanitized forms. The flight tests have been conducted by certain flight test agencies, and certain experiments/exercises were carried out with specific intentions of estimation of stability and control derivatives and/or performance characteristics. To the extent possible, the experiments were conducted very carefully, data were gathered with adequate sampling rates and where necessary the data compatibility analysis was carried out, and the corrected data were used for parameter estimation. In a few cases, the data generated from the flight simulator have been used. For the sake of brevity, CR uncertainty bounds are not given, and the aerodynamic derivatives are based on only a few data sets, where the repeat runs are not available, due to cost and time constraints in conducting several flight tests. Despite this, the results are found to be reasonably good and are representative of the characteristics of the specified aircraft or vehicle. Since most of the data analyzed for these case studies are real flight data, it gives good feel and experience of the application of flight data analysis procedures and parameter estimation methods. Wherever applicable the multiple maneuver analysis and model validation were carried out. Often, the aerodynamic derivatives are obtained as a function of AOA or Mach number. The plots of these derivatives wrt these independent parameters would be very useful. These plots would depict the trends, linear or nonlinear, and thus, reveal the specific characteristic variation of the derivatives. Also, it would be a good practice to plot these derivatives wrt, say AOA, such that any graphical difference in the variation or any discerning effects while comparing the values with other derivatives shows the correct correlations in the percentage differences and the graphical distances of these variations. For this, the minimum and maximum values of all derivatives are determined, and then the y-axis scale is chosen by multiplying these extreme values with the same factor, say 1.5; this factor remains the same for all the derivatives. With this method, any graphical difference (measured in a 'distance unit') for any derivative will correspond to the same % difference in the numerical values. In literature, some practitioners do not use this scaling procedure and just compare and present their results as they are and this often confuses, since the graphical 'distance' does not match uniformly with the % difference for all the parameters, and give a wrong judgment on the nearness and farness of the matching of the FDDs and the theoretical/predicted derivatives.

One of the earliest exercises of flight testing with an aim to utilize parameter estimation program was conducted by a flight testing agency in 1985–86 on a fighter-trainer MiG 21-xxx/2270 test aircraft with an objective to determine stability and control derivatives with a final aim to study the effect of a vortex plate (VP) that was attached to the wing's leading edge. The geometric data were: Wt. = 7675 kg, mac = 4 m. The flight conditions were: Mach No. = 0.5, 0.6, and 0.7; Alt. = 5000 m, AOA range < 5 deg. The configurations were: (i) BV – big capacity saddle tank with the VP (mass variation was 6626–7100 kg); (ii) B – without the saddle tank (mass variation was 6785–7020 kg); (iii) SV – standard tank with the plate (mass variation was 7737–7971 kg); and (iv) S – without the plate (mass variation was 7076–7900 kg). The C. G. variation was not much. Variation in the dynamic pressure (q) was: for BV-9661 to 18477 N/m^2 and for B-8467 to 20801 N/m^2. The SP and Phugoid modes were excited for generation of the data. The aircraft was fully instrumented, and the data were recorded on the photo trace recorder for 2 min for each test maneuver; finally, the useable data were extracted by graphic expansion-cum-digitization, and stored/used for parameter estimation. The observables used were: forward acceleration, normal acceleration, angular rates, AOA, flap angle, tail plane position, and AOSS. For SP mode, the sampling time used was 20 ms, and second-order linear model was used; for the Phugoid, the sampling time was 200 ms, and three-state and two-observable model was used for analysis. The results obtained were: (i) the $C_{m\alpha}$ was converted to the reference position as follows: $C_{m\alpha}|_{\text{ref}} = C_{m\alpha}|_{\text{any C.G.}}\ (x_{\text{ref}} - x_{\text{cp}}) \cdot C_{L\alpha}$; (ii) NP and MP shifts with Mach number; and (iii) the short period and Phugoid characteristics (the data for Phugoid for the conditions B and BV were not available). The neutral point shift with the VP was assessed with the formula: $\text{N.P.}(\%\text{mac}) = x_{\text{ref}}(\%\text{mac}) - C_{m\alpha} / C_{L\alpha}$; the wind tunnel result was: for S-37%, and for SV-33%; the VP reduces the static margin by 3–4 units or by 10.81%. The Phugoid natural frequency was found to be 0.07, and the damping ratio was from 0.004 to 0.111. The main inferences were as follows: (i) VP reduces the subsonic trim drag, (ii) reduces maneuver margin, (iii) increases SP damping at low Mach, and (iv) there is no appreciable change in the lift curve slope. An interesting observation on the quality of the data generated was as follows: Out of 50 sets, only 30 sets were good, and out of 30 sets, only 23 were successfully analyzed; hence, overall quality of the data generation/utilization was only 46%; despite this, reasonable and plausible inferences were obtained from this flight test exercise, and this shows how difficult it was to do such experimental analysis in the mid-eighties.

9.4.1 Fighter Aircraft FA1

This is a modern high-performance fighter aircraft fully instrumented for the SIPEP experiments. The maneuvers conducted were as follows: SP, LD, roller coaster (RC), slow down (SD), and windup turn (WUT). The latter three maneuvers are the dynamic maneuvers for determination of drag polars from the flight test data using parameter estimation methods. The idea behind this aspect is that these maneuvers save flight test time compared to the steady-state methods (Chapter 7). The RC maneuver data were generated for Mach range 0.6–0.8 for 1 *g* excursion of acceleration covering 0–9 deg. AOA. The SD maneuver is a low-speed maneuver at about 0.4 Mach, and the AOA range covered is 9–19 deg. The WUT involves a Mach 0.6–0.8 steep turn of decreasing radius with '*g*' force varying from 1–8 *g*. The AOA range is 8–19 deg. The WUT is a combined longitudinal/lateral maneuver. The various tests were performed at 1.5, 3, and 6 km altitudes. The signals were sampled at 16/32 samples/s. Where applicable the sideslip angle correction factors wrt altitude and Mach were used. Various math models used for the data analysis are given in Ref. [16]. The thrust data were used as input to the mathematical model. For drag polar estimation, the model-based approach was used as against the static/steady state one, since maneuvers performed were dynamic. This shifts the burden from flying aircraft at several AOA/Mach numbers (to obtain the static force balance) to the offline analysis of the dynamic performance maneuvers. Thus, OEM method can be used with equal ease to determine stability and control derivatives as well as drag polars (by estimating lift and drag coefficients as a function Mach number/AOA) from the dynamic

TABLE 9.3
Neutral Point (for FA1) as % of MAC

Mach Number (H=10 kft)	Reference Value (Manufacture's)	FD Value
0.6	34.2	33
0.7	34.0	33.8
0.8	33.8	34

maneuvers (RC, SD, and WUT). Most of the (longitudinal and L-D) FDDs as well as the drag polars compared well with the manufacturer's data. The flight-determined longitudinal HQ met the Level 1 requirements. The flight-determined SP natural frequency varied from 2.3 to 3.65 rad/s, and the damping ratio was 0.3. The DR roll frequency was determined as 2.5–3.4 rad/s. The neutral point was also determined from the real flight data, and the results are shown in Table 9.3.

9.4.2 Fighter Aircraft FA2

An unstable aircraft cannot fly without the aid of flight controller, thereby rendering the entire system as a closed-loop control system. However, the estimation of derivatives of the basic unstable aircraft (open-loop system) is of primary interest for many applications in the field of flight mechanics, especially for the validation of the math models used for control law design. The accuracy of the estimated derivatives in such cases is adversely affected by the feedback from the flight controller that tends to produce highly damped responses and correlated input/output variables. Using the filter error program, it was shown that better estimation can be achieved if modeling errors arising from data correlation are treated as process noise. Parameter estimation of an augmented aircraft equipped with a controller was carried out using output error and filter error methods. It was shown that the feedback signals from the controller and the aileron–rudder interconnect operation cause correlations between the input/output variables that degrade the accuracy of the parameter estimates. The filter error method was found to yield reliable parameter estimates, while the aircraft derivatives estimated from output error method did not compare well with the reference derivative values. Parameter estimation results from data generated from a research simulator for an unstable augmented aircraft are plotted in Figure 9.4. Due to model compensation

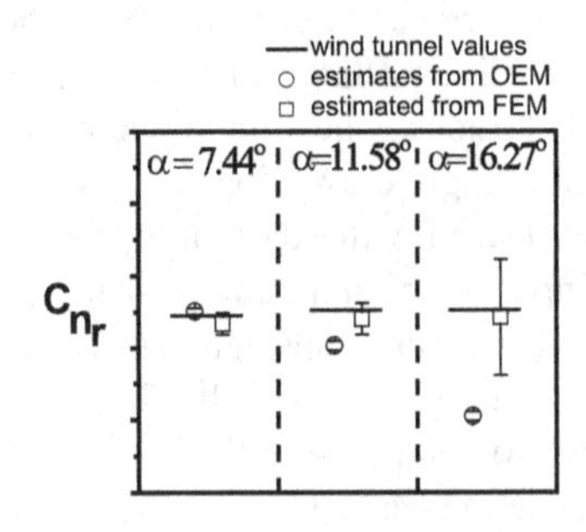

FIGURE 9.4 Parameter estimation using FEM and OEM for an unstable/augmented aircraft FA2.

ability of FEM, the derivatives estimated show better match with wind tunnel values compared to the derivatives estimated from OEM [17].

9.4.3 Basic and Modified Transport Aircraft

A program to modify a transport aircraft to carry some electronics payloads was taken up by an aircraft testing agency. It was important to establish the incremental effects on the dynamic derivatives due to the modification of/on the basic aircraft. The basic aircraft had a medium range with two Rolls Royce Dart turboprop engines. It had manually operated controls for elevators, rudder, and starboard ailerons. Other characteristics were: MAC = 2.69 m, span = 30 m, dihedral = +7 deg., aspect ratio (b^2/S) = 11.96 m. The engine (R-Da. 7/Mk. 531) modeling: static thrust = propeller thrust + small jet thrust – Ram drag – feather drag (the latter is small drag when propeller is at feather); Ram drag is an intake drag proportional to the intake airflow; the dynamic response of the engine was assumed to be of second order with natural frequency of 1 rad/s, damping ratio 0.52. On the basic aircraft, an external structure (7.2 m × 1.5 m) was mounted on the pair of pylons fixed to the fuselage, converting it to the modified aircraft. It was expected that the modified a/c would have some characteristics altered compared to the basic one. Hence, it was very important to predict its characteristics and HQs. Several planned maneuvers were performed on this transport aircraft: DR, steady turn, sideslip wings level, sideslip steady heading, roll performance, longitudinal stability, SP/Ph, stall, and RC maneuvers. The dynamic data were acquired at the rate of 32 samples/s and were corrected using kinematic consistency checking and subsequently used for parameter estimation. The math models used are given in Ref. [18]. The results of parameter estimation from these flight test data using the maximum-likelihood output error are discussed here along with some results of the analytical studies also carried out by the same agencies.

Analytical studies of the modified aircraft: The $C_{m\alpha}$ (−ve) magnitude reduces from −1.71 to −1.3 as C. G. moves aft from 22% to 29% of MAC, at $M = 0.29$, whereas the C_{m_q} is not affected. There was no marked effect on $C_{y\beta}, C_{y_p}, C_{y_r}, C_{l\beta}, C_{l_p}, C_{l_r}$ derivatives of the C. G. The $C_{n\beta}$ +ve magnitude decreases, and DR frequency decreases, less +ve moment is generated. There is no marked effect on C_{n_r}; there are no effects on RS time to half, and SM time to double. Dynamic stick-fixed characteristics are as follows: SP frequency slightly decreased with increase in weight of the aircraft from 30,000 to 42,000 lbs, and the DR frequency also decreased; other time constants remained same. Some more observations of the analytical studies are as follows (the accuracy of which is within 10%–15%): (i) the additional vertical structure of the modified aircraft acts as lifting surface and generates additional side force when aircraft is sideslipping, yawing, or rolling; (ii) because of the longitudinal and vertical separation between point of action of these additional surfaces, these forces give rise to yaw and roll moments, hence the magnitudes of all the rolling and yawing derivatives would increase due to the presence of these structures on the basic aircraft; (iii) however, the reduced dynamic pressure (due to the large wake) behind the structure would tend to decrease the aerodynamic forces on the vertical tail; thus, the direct and interference effects of the unit on the lateral stability derivatives could be of opposite nature than the usual effects, and the net effect would depend on the relative contributory magnitudes of these direct and interference effects; and (iv) the magnitudes of some derivatives would increase and for some other would decrease for the modified aircraft over the basic aircraft. Hence, further verification and validation could be based on the analysis of the real flight test data.

FDD results: The results of parameter estimation using the flight test data in the MLE software are given in Table 9.4 for the specified flight condition/s; here moment of inertia (for the modified version) used is for example for the wt. = 36,000 lbs, and C. G. position 333.7″ from the nose (25.6% of mac) would be: I_{xx} = 277,644 kg-m^2 (204,780 Slug.ft^2 × 1.35582), I_{yy} = 265,078 kg-m^2, I_{zz} = 515,266 kg-m^2, and I_{xz} = −19,140 kg-m^2.

It is inferred from the data presented along with the corresponding comments that for the most important (and especially for the moment related) derivatives, the observations from the FDDs, analytical, and the Pilot's comments nearly agree. This further establishes the importance of the role played by the parameter estimation method in flight test data analyses.

Some L-D characteristics of the modified aircraft are given in Table 9.5. The damping-in-roll and the static stability of the modified a/c were quite improved compared to the basic a/c. However, the DR frequency and damping were a bit poor for the modified a/c compared to the basic a/c. The L-D static stability was improved, and this could be due to the pylons providing additional vertical surface above the fuselage (similar to the vertical tail). The predicted longitudinal and L-D HQs were well validated by the flight (test data)-derived HQs for both the aircrafts [18]; for the most flight conditions, the HQs were found to be Level 1 for both the versions of the aircraft on the short-period limits chart: SP frequency v/s $\frac{n_z}{\alpha}$; for basic, it is 6.26, and for the modified, it is 8.94 g's/rad, and $\frac{n_z}{\alpha} = -(V/g)Z_w$.

A new procedure for estimation of neutral point is applied to the basic and the modified configurations [19]. The experiments were conducted at three C.G. points. In the classical method of the neutral point estimation, the lift coefficient was determined from the airspeed and the weight data. The results (on the average) of neutral and maneuver points (as % of mean aerodynamic cord) estimated by using parameter estimation method are also shown in Table 9.4. The new method is expected to be highly reliable due to the fact that the data used are corrected by kinematic consistency and the longitudinal static and maneuver stability of the aircraft are extracted from the short-period dynamic response parameters M_α and the natural frequency ω_n^2 which vary consistently with C.G. location. We see that the static stability of the modified aircraft is reduced compared to the basic aircraft; however, the damping is increased as seen from Table 9.4.

9.4.4 Trainer Aircraft

The basic jet trainer is a two-seater low-wing monoplane aircraft. The flight tests were carried out by Air Force Test Pilots School and Aircraft Systems and Testing Establishment at 4570 and 6700 m altitudes on this aircraft. The sampling interval for data acquisition was 31.25 ms. The flight-determined values are as follows: $C_{m_\alpha} = -0.5$, $C_{m_q} = -15$, and $C_{L_\alpha} = 4.9$ (WT value is 4.5). The latter derivative was fairly constant since the lift curve was fairly and monotonically increasing up to

TABLE 9.4
Preliminary Observations of the FDDs (for flight tests; FC1: Mach = 0.22; Test Alt. = 10,000′; flap 0 deg.)

	Basic C.G. ~19.5% mac	Modified (Avg.), C.G. ~ 21% mac ([a]FC2)	Increase/ Decrease Relative to the Basic Aircraft	Analytical for Modified Unit Mach = 0.29; C.G.~22% mac Wt. 42,000 lbs.	Effect (Expected) Due to Pylons on the Basic Aircraft
$C_{y\beta}$	−0.474	−1.3 (−1.4)	Small/+ve values undesirable	−1.44	DR damping up ([c]the same)
$C_{l\beta}$	−0.031	−0.115 (−0.131)	Small −ve values desirable	−0.16	DR damping down
$C_{n\beta}$	0.043	0.073 (0.101)	DR frequency up ([c]LD static stability increased)	0.137	DR frequency up
C_{l_p}	−0.239	−0.55 (−0.604)	Helps increase in RS mode	−0.692	RS damping up ([c]the same)
C_{n_p}	−0.058	−0.215	Small −ve or +ve values desirable	–	DR damping down
C_{l_r}	0.073	0.11	Small +ve values desirable	–	Spiral mode ([c]SM dive/conv)
C_{n_r}	−0.087	−0.182 (−0.121)	Large −ve values desirable, yaw damping up	−0.092	DR damping up ([c]the same)
Control surface	–	–	Some increased in −ve or +ve mag.	Not so for all	–
$C_{m\alpha}$ (ω rad/s)	−1.48 (2.1)	−0.65 (1.7)	Static stability down/ undesirable	−1.71	SP frequency down ([c]the same)
C_{m_q} (ς)	−38 (0.5)	−45 (0.75)	SP damping up desirable	–	SP damping up ([c]the same)
bNP:					
CM (%mac)	50.75	22	Decreased		
MLE (%mac)	47	33	Decreased		
bMP:					
CM (%mac)	–	–	–		
MLE (%mac)	56	38	Decreased		

[a] Flt. Condition, *FC2:* Mach No. 0.29; C. G. 21% mac (66.5″ AOD); Wt. 38,000 lbs.; Alt. 5000′.
[b] CM: Conventional method; MLE: maximum-likelihood method; NP: neutral point; MP: maneuver point.
[c] Test Pilot's observations/remarks.

TABLE 9.5
Some L-D Characteristics of the Modified Aircraft

Mach No./ Altitude (ft)	DR Mode	Roll Subsistence (/s)	Spiral Convergence/ Divergence(/s)	DR ω rad/s	ς
0.202/5000	0.32+/−j1.15	−2.24	0.026	1.2	0.268
0.235/10,000	0.27+/−j1.9	−2.27	−0.014	1.23	0.198
0.319/10,000	0.318+/−j1.58	−3.1	−0.004	1.61	0.22
		RS $t_{1/2}$	**SM $t_{1/2}$ or t_2**	DR time period	DR ς
FC2	–	0.17	130 ($t_{1/2}$) 350 (t_2)	3.62	0.184
Analyticala	–	0.2	88 (t_2)	3.82	0.14

[a] Wt. 36,000 lbs.; Alt. 12,000′.

AOA of 12 deg. In the range of AOA = 8–12 deg., the SP frequency was 1.5 rad/s, and at low AOA, it was 3 rad/s, whereas the SP damping ratio was fairly constant at the average value of 0.45 [20]. Another jet trainer version has the following characteristics determined from the specific flight tests conducted (at M = 0.5, 0.6, 0.7, and H = 5000 m/ C.G. = 27.5% MAC): NP = 32.5% of MAC; SP frequency = 2.5 rad/s; SP damping ratio = 0.4; the Ph frequency = 0.08; and the damping ratio = 0.11.

9.4.5 Light Canard Research Aircraft

The LCRA is an all-composite aircraft with canard and is based on the Rutan long-EZ design. It is a two-seater a/c with pusher propeller and has a tricycle landing gear with retractable nose wheel. The wings have moderate sweeps, and the canard has trailing edge flaps that function as the elevators. The LCRA has two rudders. The ailerons are situated in the wings. The SP and DR maneuvers were carried out at Alt. = 1.53 and 2.74 km and at three speeds 65, 85, and 105 knots at each altitude (1 nautical mile, nm = 1852 m, 1 knot = 1 nm/h = 1.852 km/h; 1 knot = 1852/3600 = 0.51444 m/s). Table 9.6 shows the comparison of the analytical (values which were almost at constant values) and the average values of FDDs of the LCRA for longitudinal and L-D axes. The scatter and the CR bounds for most of the estimated derivatives were reasonably low [21]. Also, the SP natural frequency and damping ratio were fairly constant over the AOA range, and hence, average values are given in Table 9.6. The '*' indicates that there is a slight trend of the damping-in-roll derivative (C_{l_p}) with AOA and signifies the reduction in the damping with increasing AOA, whereas the analytical value is constant. This shows that the flight data provide additional information and hence enhanced confidence in the dynamic characteristics of the vehicle. This exercise shows that for this light and small aircraft, the aerodynamics are fairly linear wrt AOA. It also shows that though the FDDs are not far away from the analytical values, the pitch and directional static stability are reduced in flight. However, the dihedral static stability is increased.

TABLE 9.6
The Comparison of LCRA Derivatives

Derivative (/rad)	Analytical	FDD
$C_{L\alpha}$	6.2	5.6
C_{m_q}	−14.75	−12.0
$C_{m\alpha}$ (AFT CG)	−0.75	−0.65
$C_{m\alpha}$ (Fore CG)	−1.2	−1.1
$C_{m\delta_e}$	0.7	0.75
ω_n rad/s	—	3.6
ς_{SP}	—	0.4
$C_{l\beta}$	−0.1	−0.17
C_{l_p}	−0.65	−0.48(*)
$C_{n\beta}$	0.11	0.05
ω_{DR}	-	2.5
ς_{DR}	-	0.1

9.4.6 Helicopter

The flight tests on a helicopter consisted of 14 sets of data runs at the uniform sampling rate of 8 samples/s. The longitudinal modes were excited by applying a doublet input command and a 3-2-1-1 like input for the longitudinal cyclic control, and lateral maneuver was excited. The longitudinal derivatives are shown in Table 9.7. Since the reference values were not available, no comparison could be made with the FDDs. Hence, the approach used was to analyze the individual data sets and also use the corresponding data in a multiple maneuver analysis mode. The latter involves the concatenation of the time histories of the repeat runs (obtained at the same flight conditions) and use in the OEM such that average kind of the results are obtained. Since the results from both these approaches were almost identical, except for a few cases, and the estimates had low CR bounds, the FDDs were believed to be of acceptable values. In most of the APE applications, the forces and moments acting on the

TABLE 9.7
The Helicopter Derivatives (Longitudinal)

Derivative	FDD-Multiple Maneuver Run Analysis (3–4 Runs Together)
X_w	0.025
Z_w	−0.645
M_w	0.0127
M_q	−1.115
$X_{\delta_{lon}}$	1.71
$M_{\delta_{lon}}$	−2.6

TABLE 9.8
The Helicopter Derivatives (Lateral)

	Without Inclusion of Cross-Coupled Derivatives	Cross-Coupled Derivatives Included for Estimation
N_u	–	−0.0104
N_v	0.0407	0.0331
N_w	–	−0.0038
N_p	0.1953	0.3889
N_q	–	0.0079
N_r	−0.9953	−0.1142
N_{lat}	−2.5362	−2.1073
N_{ped}	0.6397	1.7560

aircraft are approximated by terms in the Taylor's series expansion resulting in a model that is linear in parameters. However, when the aerodynamic characteristics are rapidly changing, mathematical models having more number of parameters and cross-derivatives may be required. Table 9.8 gives the lateral FDDs of the helicopter estimated with and without cross-derivatives. The improvement in the lateral velocity component and yaw rate response match is observed when cross-coupled derivatives N_u, N_w, and N_q are included in the yawing moment equation during estimation. With cross-coupled derivatives included, estimated values of other yaw derivatives, particularly N_r, register a change resulting in improved match. The time–history match with flight data improved when the model included some cross-coupled derivatives [22]. This aspect is important for the flight data analysis of a helicopter. The search for obtaining adequate aerodynamic models that can satisfactorily represent the helicopter characteristics can be further pursued using the SMLR and the model error estimation (MEE) method.

9.4.7 AGARD Standard Model

In this study, the OEM is used for estimation of the aerodynamic derivatives of AGARD standard ballistic model (SBM) from the data available in the open literature, and the hybrid model with aerodynamic forces represented in the wind axis and the moments in the body axis coordinate systems was used [23]. The SBM is composed of a blunt nose cone with a cylindrical mid-section followed by a 10 deg. flare. This axis-symmetric, un-instrumented model is solid in construction. The free-light tests were performed at Mach No. 2.0 in the NASA Ames pressurized ballistic range facility. The model, as it flew the down range, was photographed by 22 fully instrumented orthogonal shadowgraph stations. The translational and angular motions were acquired in Earth-fixed orthogonal coordinates based on analysis of the film records. The NASA method was based on LS/Gaussian LS differential approach for parameter estimation. The V, α, β and the body axis models were used. The available data for V, x, y, z, θ, ϕ in a short flight duration were not uniformly sampled, and hence,

TABLE 9.9
The Comparison of AGARD SBM derivatives

Derivative (/rad)	NASA	OEM (NAL) V, α, β	OEM (NAL) body axis
C_D	1.24	1.21	1.2
C_{m_q}	−83	−73.7	−72.7
$C_{m\alpha}$	−1.26	−1.28	−1.28
$C_{L\alpha}$	3.6	3.2	4.4
$C_{n\beta}$	—	1.28	1.28

a cubic spline interpolation was utilized to generate uniform data sets. The flow angle trajectories were generated using $\alpha = \theta + \tan^{-1}(dz/dx)$ and $\beta = -\phi + \tan^{-1}(dy/dx)$. Table 9.9 shows comparison of some of the aerodynamic derivatives.

Due to axis-symmetry property of the AGARD SBM, it is seen that the static stability derivatives in longitudinal and lateral axes are the same except the polarity.

9.4.8 Dynamic Wind Tunnel Experiments

The dynamic wind tunnel (DWT, CSIR-NAL) experiments on the scaled (-down) model of a delta-wing a/c were conducted using single degree of freedom in pitch, yaw, and roll. The DWT facility is a 1.2 m × 1.2 m low-speed (20–60 m/s) tunnel of open jet-induced draft type. The FRP a/c model was supported at the back on a low-friction bearing assembly for the roll motion experiments. The DC motors drove the elevons of the a/c model. In the DWT experiment, the a/c model is free to move about a gimbal having translational and rotary degrees of freedom. The experiments are of semi-free-flying type as against the conventional/static wind tunnel experiments. The miniature (incidence), rate, and acceleration sensors were mounted inside the hollow model of the aircraft to pick up the dynamic responses to the given inputs. The control surface movements were measured using precision potentiometers. The angular accelerations were sensed using a set of strain gauge-type accelerometers. The test inputs were applied to servo-controlled (control-) surfaces to excite the modes of the a/c model. The aerodynamic derivatives were estimated from the motion responses of the model by using OEM, and the results were compared with other wind tunnel test methods [24] (Table 9.10).

TABLE 9.10
The DWT Experiment Results

Derivative/rad	$C_{m\alpha}$	$C_{m\delta_e}$	$C_{l\beta}$	$C_{n\beta}$
DWT (NAL)	−0.11	−0.29	−0.1	0.9
Static WT	−0.095	−0.3	−0.11	1.0

TABLE 9.11
The Iron Bird Experiment Results

Derivative/rad	$C_{m\alpha}$	$C_{m\delta_e}$ (1/deg.)	$C_{L\alpha}$	C_{m_q}
NLM	0.0079	−0.0071	2.611	−1.218
FS	0.0058	−0.0067	2.66	−1.129
IB	0.0073	−0.0067	2.6692	−1.037

9.4.9 Iron Bird Results

An attempt of analysis of the short-period maneuver data generated from an iron bird (IB) for a fighter aircraft was made for the flight condition: Mach number = 0.41, H = 3 km, under carriage down, alpha trim = 7.69 deg. A SP maneuver was excited with a pitch stick command input doublet. It so happened that the pulse width was 2.8 s, the sampling rate was 40 samples/s, and 320 data points were available for parameter estimation. The estimation results are compared with those from FS and the non-dimensional linear model (NLM) parameters in Table 9.11. The results seem satisfactory.

It is seen that for this flight condition, the aircraft is almost statically neutrally stable or unstable. The pitch damping is also very poor. It must be emphasized here that the open-loop dynamics of this are unstable, and the aircraft operates in a closed loop with control system.

9.5 APPROACHES FOR DETERMINATION OF DRAG POLARS FROM FLIGHT DATA

The estimation of drag polars (drag coefficient versus lift coefficient) is an important aspect in any aircraft development/flight test program. The drag polar data are required to assess actual performance capability of the aircraft from flight test data. Dynamic flight test maneuver of various types [25,26] can be performed on an aircraft and the data analyzed using OEM for a stable aircraft and SOEMs for unstable/augmented aircraft. Dynamic maneuvers like RC, SD, and WUT are most suitable for drag polar determination. The model-based and non-model-based approaches are suitable for determination of drag polars, and these are linked in Figure 9.5 [27]. The estimation before modeling method can be used for determination of the structure of the aerodynamic model to be used in the model-based approach.

9.5.1 Model-Based Approach for Determination of Drag Polar

In this method, an explicit aerodynamic model for the lift and drag coefficients is formulated next.

State model:

$$\dot{V} = -\frac{\bar{q}S}{m}C_D + \frac{F_e}{m}\cos(\alpha+\sigma_T) + g\sin(\alpha-\theta);\ \dot{\alpha} = -\frac{\bar{q}S}{mV}C_L - \frac{F_e}{mV}\sin(\alpha+\sigma_T)$$

$$+q+\frac{g}{V}\cos(\alpha-\theta)$$

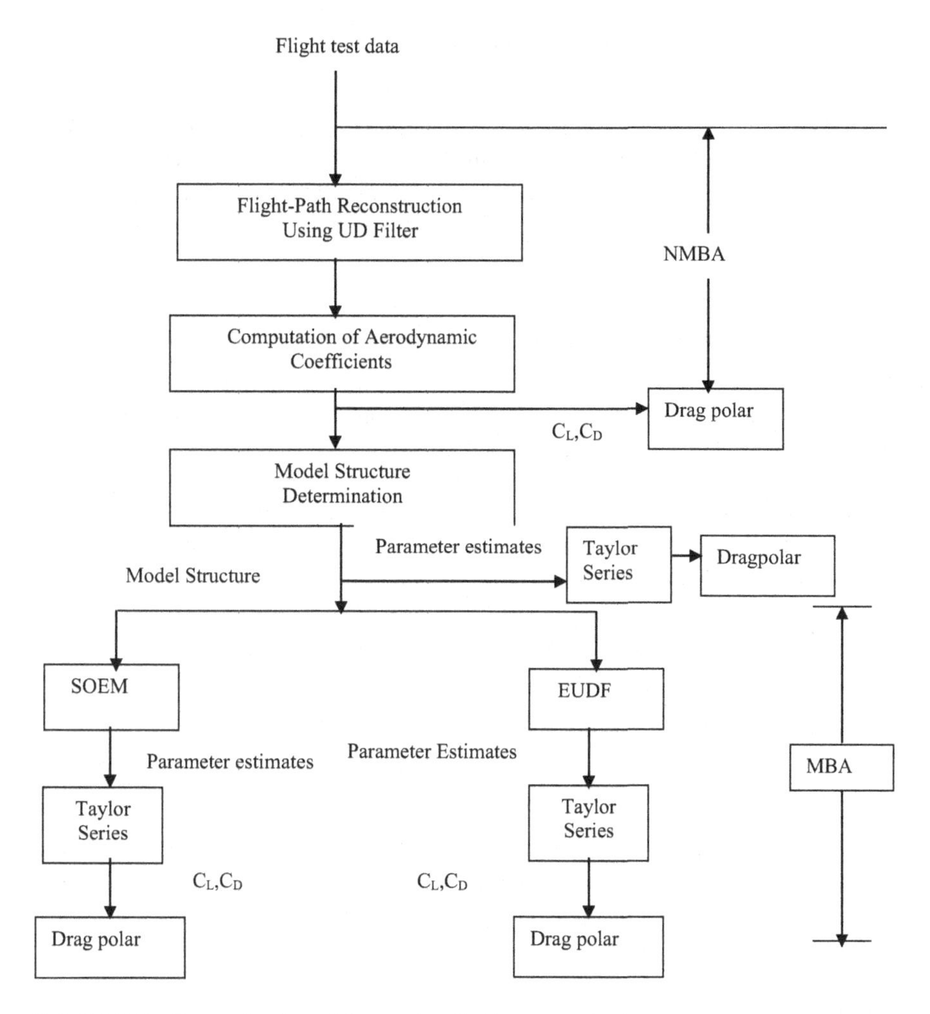

FIGURE 9.5 Determination of drag polars linking of methods.

$$\dot{\theta} = q \tag{9.65}$$

Here, the C_L and C_D are modeled as

$$C_L = C_{L_o} + C_{LV}\frac{V}{u_o} + C_{L\alpha}\alpha + C_{L_q}\frac{q\bar{c}}{2u_o} + C_{L\delta_e}\delta_e;$$

$$C_D = C_{D_o} + C_{DV}\frac{V}{u_o} + C_{D\alpha}\alpha + C_{D_{\alpha^2}}\alpha^2 + C_{D_q}\frac{q\bar{c}}{2u_o} + C_{D\delta_e}\delta_e \tag{9.66}$$

Observation model:

$$V_m = V;\ \alpha_m = \alpha;\ \theta_m = \theta;\ a_{x_m} = \frac{\bar{q}S}{m}(C_X) + \frac{F_e}{m}\cos\sigma_T;\ a_{z_m} = \frac{\bar{q}S}{m}(C_Z) - \frac{F_e}{m}\sin\sigma_T$$

$$C_Z = -C_L \cos\alpha - C_D \sin\alpha; \; C_X = C_L \sin\alpha - C_D \cos\alpha \tag{9.67}$$

The aerodynamic derivatives in the above equations could be estimated using OEM for stable aircraft (SOEM for unstable aircraft) or using extended UD filter. In the EKF/EUD filter, the aerodynamic derivatives in Eqn. (9.66) would form part of the augmented state model. The estimated C_L and C_D are then used to generate the drag polars [16,25–27].

9.5.2 Non-Model-Based Approach for Drag Polar Determination

This method does not require an explicit aerodynamic model to be formulated. The determination of drag polars is accomplished using the two steps: (i) suboptimal smoothed states of an aircraft are obtained using the EKF smoother [1,14]. Scale factors and bias errors in the sensors are estimated using the data compatibility checking procedure, and ii) the aerodynamic lift and drag coefficients are computed using the corrected measurements (from Step (i)) of the forward and normal accelerations using the following relations:

$$C_x = \frac{m}{\bar{q}S}\left(a_x - \frac{F_e}{m}\cos\sigma_T\right); \; C_z = \frac{m}{\bar{q}S}\left(a_z + \frac{F_e}{m}\sin\sigma_T\right) \tag{9.68}$$

The lift and drag coefficients are computed from C_x and C_z using

$$C_L = -C_Z \cos\alpha + C_X \sin\alpha; \, C_D = -C_X \cos\alpha - C_Z \sin\alpha \tag{9.69}$$

C_D versus C_L is plotted to obtain the drag polar. The first step could be accomplished using the state and measurement models for kinematic consistency. Typical results of this case study are shown in Table 9.12.

TABLE 9.12
Fit Errors (%) for Lift and Drag Coefficients – Parameter Estimation (Alt. = 8 km)

RC Maneuver

Estimation Method	Set-1 Mach No. = 0.6		Set-2 Mach No. = 0.7		Set-3 Mach No. = 0.8		Set-4 Mach No. = 1.0	
	C_L	C_D	C_L	C_D	C_L	C_D	C_L	C_D
EUDF-NMBA	0.0216	0.1586	0.0201	0.1112	0.0190	0.1036	0.0442	0.4275
EUDF-MBA	0.2597	0.9108	0.3940	0.6660	0.4939	1.0506	0.5439	0.3659
SOEM	0.2198	0.7540	0.3747	0.5979	0.4746	1.0009	0.4973	0.2996

WUT Maneuver

	Set-1 Mach No. = 0.6		Set-2 Mach No. = 0.7		Set-3 Mach No. = 0.8		Set-4 Mach No. = 1.0	
EUDF-NMBA	0.0411	0.5960	0.0543	0.4930	0.0624	0.4588	0.0423	0.2261
EUDF-MBA	0.6193	0.9564	0.4524	1.1780	0.3922	0.8220	0.3995	0.7903
SOEM	0.3871	0.8862	0.4811	1.2402	0.5078	0.8579	0.2487	0.4741

The indicated maneuvers (Chapter 7) were performed on a supersonic fighter aircraft at several flight conditions. We can easily see from Table 9.12 that the non-model-based (EUDF-NMBA) approach gives better fit to the data and hence accurate determination of the drag polars from flight data. A NMBA could be preferred over a model-based approach, as it requires less computation time and still gives accurate results for drag polars from flight data, and is a potential method for real-time online determination of drag polars.

9.6 ANALYSIS OF LARGE-AMPLITUDE MANEUVER DATA

The usual procedure for APE is to perturb it slightly from its trim position by deflecting one or more of its control surfaces and gathering flight data from such small-amplitude maneuvers. However, in practice, it may not be possible to trim the aircraft at all at certain test points in the flight envelope. Under such conditions, large-amplitude maneuvers (LAMs) have proved to be more useful in determining the aerodynamic derivatives. The approach is not valid in the region where stability and control derivatives vary rapidly. The results for the LAM are generated for the fighter aircraft FA2 for the conditions: (i) simulated longitudinal LAM data at two flight conditions and analyzed using the SMLR method [1] without partitioning – here, entire data set is handled as a single LAM, (ii) the data as in (i) but analyzed using the model error method (MEM) [1], and (iii) the real flight data (*) analyzed using the previous two approaches [28]. The results are shown in Table 9.13.

The other method consists of partitioning the LAM that covers a large angle-of-attack (AOA) range into several bins, each of which spans a smaller range of AOA [29]. These data would not be contiguous, and therefore, regression method is employed to estimate derivatives. This approach was also used for the analysis of the data mentioned earlier, and the detailed results can be found in Ref. [28].

For LD data generation from LAMs (in an RTS/real-time simulator), the modes were excited in roll and yaw axes by giving aileron and rudder inputs superimposed on a steadily increasing AOA due to slow but steady increase in elevator deflection. Normal practice is to estimate linear derivative models but, if necessary, SMLR approach can be used to determine model structure with higher-order terms for better representation of aircraft dynamics. A pilot flew an RTS (Chapter 6) and persistently excited the roll and yaw motions at increasing AOA in pull-up. The data were generated for Mach numbers 0.4 and 0.6 with AOA variation between 2 and 20 deg. [30].

TABLE 9.13
Longitudinal LAM Data Results

Trim AOA (deg.)	C_{m_α}			C_{m_q}			$C_{m_{\delta_e}}$		
	Ref.	SMLR	MEM	Ref.	SMLR	MEM	Ref.	SMLR	MEM
11.90	0.077	0.071	0.081	−1.173	−1.295	−1.252	−0.391	−0.375	−0.379
4.3	−0.014	−0.01	−0.020	−1.32	−1.19	−1.16	−0.415	−0.398	−0.411
11.55*	0.077	0.058	0.061	−1.173	−0.992	−0.922	−0.391	−0.394	−0.399

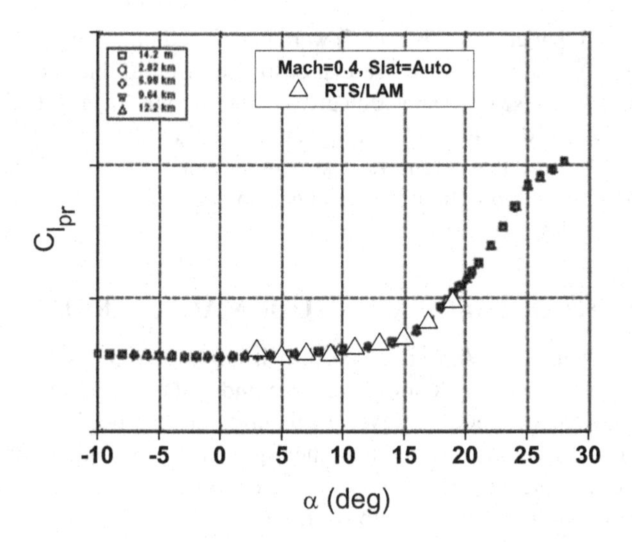

FIGURE 9.6 A compound derivative from RTS.

Significant variation in yaw during lateral stick inputs, possibly caused by aileron–rudder interconnect, was seen in the time–history plots. Regression and data partitioning techniques were used for estimation of the LD derivatives. For each Mach, the data were partitioned into bins, where each bin corresponds to a mean value of AOA: At Mach 0.4, the total variation in AOA is from 2 to 20 deg. The data from 2 to 4 deg. are put into one group, the data from 4 to 6 deg. are put into another group, and so on until all the data are accounted for. The data in a group would have resulted from different portions of a single maneuver and/or from certain portions of several other maneuvers. These data in each group were analyzed using regression. Each bin should have sufficient number of data points for successful estimation of derivatives and should have sufficient variations in the aircraft motion variables so that accurate estimation of derivatives is possible. Several criteria, for example, the number of data points in each bin, the CR bounds of the estimated derivatives, and the % fit error, were evaluated to check the quality of the estimates. The derivatives $C_{n\beta}$, $C_{l\beta}$, $C_{n\delta_a}$ and $C_{l_{pr}}$ were estimated and cross-plotted with the reference derivatives. The results were very encouraging. One compound derivative (for the flight condition as Mach number = 0.6 and with auto slat) is: $C_{l_{pr}} = C_{l_p} + C_{l_r} \tan\alpha$ is shown in Figure 9.6 [30]. This RTS exercise established that the approach if followed would work for the data gathered from the real flight tests. Using this approach, it is possible to determine the aerodynamic derivatives at those flight conditions that cannot be covered during 1-*g* trim flight.

9.7 GLOBAL NONLINEAR ANALYTICAL MODELING

The point math models are valid for small ranges in AOA and sideslip at a fixed Mach number; however, many applications require knowledge of the aerodynamics characteristics for large AOA range. Also, at certain flight conditions, it may not be possible to trim the aircraft and obtain the point models from the flight data. Thus, the global models obtained from the wind tunnel and flight data can constitute a total aerodynamic database that would be a good source for validation of flight-control performance, flight envelope expansion, and validation of wind tunnel aero-database [31,32]. This aero-database can be updated if required based on the results from flight data analysis. A global model with gradient and large range of validity can replace many local models in the flight envelope. In addition, the partial derivatives of the global analytical model would provide point models (at trim conditions) that can be used in control system design. Multivariate orthogonal functions are very useful for global modeling of aerodynamic coefficients [31]. A minimum predicted squared error (PSE) criterion [1] can be used to determine which orthogonal functions should be included in the global model. The other criteria used are the sum of the conventional mean squared error and an over-fit penalty [33]. Each orthogonal function included in the global model can be decomposed into an ordinary polynomial in independent variables. Thus, the global model can assume a form of a truncated power series expansion providing an insight into the physical relationship between the aerodynamic coefficients and independent variables. The orthogonal functions approach also allows for an easy upgrading of the model with the newly acquired flight data. Alternatively, the spline functions can be used to fit the FDDs, padded with linear model derivative values obtained from wind tunnel data to determine global models of force and moment coefficients as functions of trim AOA and Mach number [34]. A spline is a smooth piecewise polynomial of degree m, with the function values and derivatives agreeing at the points where the piecewise polynomials join. These points are called 'knots' defined by the value of their projection on to the plane or axis of independent variables. For the polynomial splines of degree m, a single variable function $f(\alpha)$ can be approximated in the interval $\alpha \in [\alpha_0, \alpha_{\max}]$:

$$S_m(\alpha) = \sum_{h=0}^{m} C_h \alpha^h + \sum_{i=1}^{k} D_i (\alpha - \alpha_i)_+^m \text{ with} \tag{9.70}$$

$$\begin{aligned}(\alpha - \alpha_i)_+^m &= (\alpha - \alpha_i)^m \quad && \text{if } (\alpha \geq \alpha_i) \\ &= 0 && \text{if } (\alpha < \alpha_i)\end{aligned}$$

The $\alpha_1, \alpha_2, \ldots, \alpha_k$ are knots obeying the condition, $\alpha_0 < \alpha_1 < \alpha_2 < \cdots < \alpha_k < \alpha_{\max}$ and the coefficients C_h and D_i are constants. A function of two variables $f(\alpha, \beta)$ can be approximated in the range, $\alpha \in [\alpha_0, \alpha_{\max}]$ and $\beta \in [\beta_0, \beta_{\max}]$, using polynomial splines in two variables. First-order splines are generally adequate for modeling the non-dimensional derivatives of an aircraft. It is assumed that the spline polynomial of an aerodynamic derivative C, that is a function of α and Mach number, can be expressed as follows:

$$\begin{aligned}C(\alpha, M) = {} & a_{00} + a_{01}\alpha + a_{10}M + a_{11}\alpha M + \sum_{i=1}^{k} b_i (\alpha - \alpha_i)^1 + \sum_{j=1}^{l} c_j (M - M_j)^1 \\ & + \sum_{i=1}^{k} \sum_{j=1}^{l} d_{ij} (\alpha - \alpha_i)^1 (M - M_j)^1\end{aligned} \tag{9.71}$$

Here, the constants $a_{00},\ldots a_{11}$, b_i, c_j, and d_{ij} are to be determined. A higher order may be used if the degree of nonlinearity in the data is believed to be high. The LS regression matrix computed using the values of the NDDs and the independent variables α and M is orthogonalized using the Gram-Schmidt procedure. The minimum PSE metric is used determine the model structure, and the global model parameters are estimated using the LS method.

The global models of the non-dimensional derivatives of a high-performance aircraft were identified using splines technique and validated over an AOA range of 0–22 deg. and Mach range of 0.2–0.9 [33]. The GM of C_{m_α} from splines comprises of 25 terms with the knots placed at every 1 deg. α from 0 to 22 deg., and the Mach knots were placed at intervals of 0.1 between 0.2 and 0.9 Mach. According to Eqn. (9.70), as α increases, each spline function $(\alpha - \alpha_i)$ contributes lesser than the previous function allowing for different values of derivative to be computed for different values of α. The static validation of the GMs is done by comparing them with several point values. The results of model fitting with splines are shown in Figures 9.7–9.9. The GMs were validated dynamically by using data from the 6 DOF non-linear simulations.

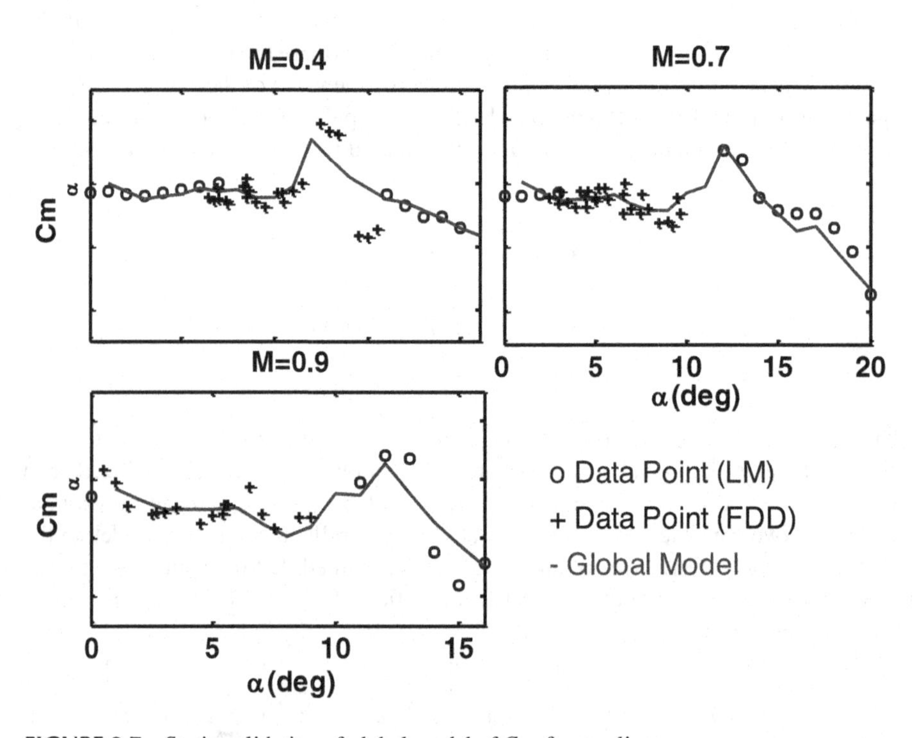

FIGURE 9.7 Static validation of global model of C_{m_α} from splines.

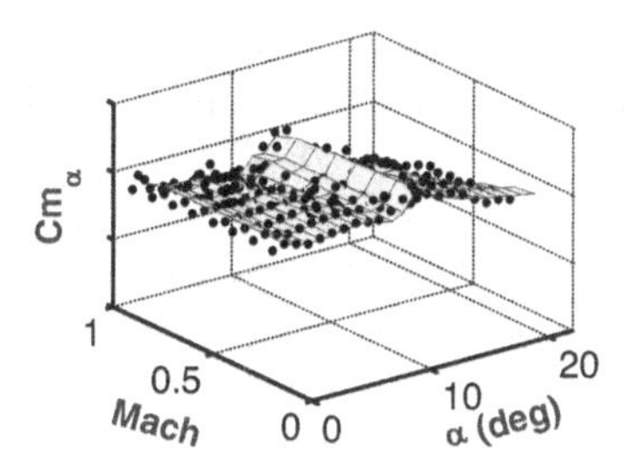

FIGURE 9.8 Surface plot of global model of $C_{m\alpha}$ from splines dotted with point values.

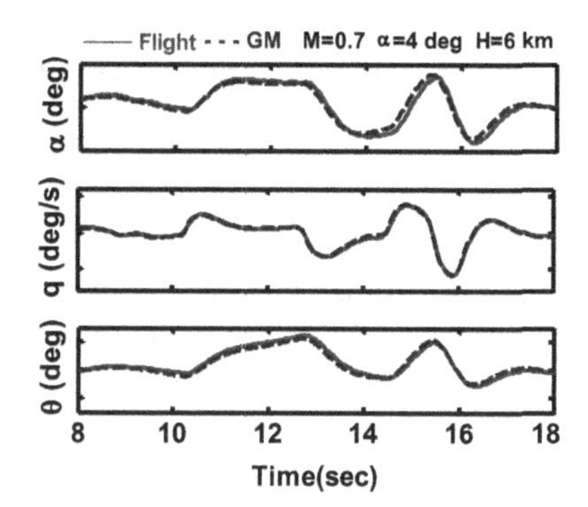

FIGURE 9.9 Dynamic validation of global model of $C_{m\alpha}$ from spline using 6 DOF simulation.

9.8 FUZZY KALMAN FILTER FOR STATE ESTIMATION

It is essential to obtain accurate estimation of target states (position, velocity, and acceleration) from the noisy measurements originating from single or multiple sensors. Kalman filter is a suitable algorithm for such applications. In case of multiple sources, either single Kalman filter can be used by fusing the measurements at data level or one can use state vector fusion. The accuracy of estimated/fused states depends upon the following: (i) how precisely the target and measurement models are known and (ii) tuning parameters such as process noise covariance matrix 'Q' and measurement noise covariance matrix 'R' that basically decide the bandwidth of the filter. In many situations, either mathematical models are not known accurately or are difficult to model. In practice, modeling errors are partially compensated by tuning Q, by using trial-and-error or some heuristic approach. A fuzzy Kalman filter (FKF) is found suitable in such cases, and it is investigated here. The performances of KF and FKF are compared with adaptive Kalman filter (AKF) in which process noise covariance 'Q' is computed online using sliding window method.

FL is a multi-value logic used to model any events or conditions that are not precisely defined or unknown. In the FL-based system, we use the following: (i) membership function (fuzzification) – it converts the input/output crisp values to corresponding membership grades indicating its belongingness to respective fuzzy set, (ii) rule base consisting of IF-THEN rules, (iii) fuzzy implications used to map the fuzzified input to an appropriate fuzzified output, (iv) aggregation used to combine the output fuzzy sets (single-output fuzzy set for every rule fired) to single fuzzy set, and (v) defuzzification to convert the aggregated output fuzzy set from its fuzzified values to equivalent crisp values (Figure C.1 of Appendix C). In a KF, since the innovation sequence is the difference between sensor-measurement and predicted values (based on filter's internal model), this mismatch can be used to perform the required adaptation using fuzzy logic rules. The advantages derived from the use of the fuzzy technique are the simplicity of the approach and the possibility of accommodating the heuristic knowledge about the phenomenon. This aspect is accommodated in Eqn. (9.44) as given by

$$\hat{X}(k+1/k+1) = \tilde{X}(k+1/k) + KC(k+1) \tag{9.72}$$

Here, $C(k+1)$ is the fuzzy correlation variable (FCV) [35] and is a nonlinear function of the innovations vector e. It is assumed that target motion in each axis is independent. The FCV consists of two inputs (i.e., e_x and $\dot{e}_x$) and single output $c_x(k+1)$, where $\dot{e}_x$ is computed by

$$\dot{e}_x = \frac{e_x(k+1) - e_x(k)}{T} \tag{9.73}$$

Here, T is the sampling time interval in seconds. In any fuzzy inference system (FIS), fuzzy implication provides mapping between input and output fuzzy sets. Basically, a fuzzy IF-THEN rule is interpreted as a fuzzy implication. The *antecedent* membership functions would define the fuzzy values for inputs e_x and $\dot{e}_x$. The labels used in linguistic variables to define membership functions are as follows: LN (large negative), MN (medium negative), SN (small negative), ZE (zero error), SP (small positive), MP (medium positive), and LP (large positive). The rules for the inference in FIS are created based on the past experiences and intuitions. For example, one such rule is

$$\text{IF } e_x \text{ is LP AND } \dot{e}_x \text{ is LP THEN } c_x \text{ is LP} \tag{9.74}$$

This rule is created based on the fact that having e_x and $\dot{e}_x$ with large positive values indicates an increase in innovation sequence at faster rate. The future value of e_x (and therefore $\dot{e}_x$) can be reduced by increasing the present value of c_x (a function of $\approx Z - H\tilde{X}$) with a large magnitude. Table 9.14 summarizes the 49 rules [36] used to implement FCV. Output c_x at any instant of time can be computed using the inputs e_x and $\dot{e}_x$, input membership functions, rules mentioned in Table 9.14, fuzzy inference engine, aggregator, and defuzzification.

The properties of FIS used in the present work are as follows: (i) its type is Mamdani, (ii) AND operator is Min, (iii) OR operator is Max, (iv) implication used is Min, (v) aggregation used is Max, and (vi) the defuzzification used is Centroid (Figure B.9).

TABLE 9.14
Fuzzy Associated Memory for Output C_x with 49 Rules [36]

	e_x						
$\dot{e}_x$	**LN**	**MN**	**SN**	**ZE**	**SP**	**MP**	**LP**
LN	LN	LN	MN	MN	MN	SN	ZE
MN	LN	MN	MN	MN	SN	ZE	SP
SN	MN	MN	MN	SN	ZE	SP	MP
ZE	MN	MN	SN	ZE	SP	MP	MP
SP	MN	SN	ZE	SP	MP	MP	MP
MP	SN	ZE	SP	MP	MP	MP	LP
LP	ZE	SP	MP	MP	MP	LP	LP

Example 9.2 Target Tracking

The relevant MATLAB programs are **ExamplesSolSW/Example 9.2AKF; 9.2FKTT4Rules; 9.2FKTT49Rules**. The target data in x, y directions are generated using constant acceleration model with process noise increment [36]. With sampling interval $T=0.1$ s, a total of $N=100$ scans are generated. The data simulation proceeds with the following assumed parameter values and equations: (i) initial states of target $(x,\dot{x},\ddot{x})$ are (0 m, 100 m/s, 0 m/s²), respectively, and initial states of $y,\dot{y},\ddot{y}=$(0 m, −100 m/s, −10 m/s²), and (ii) process noise variance $Q=0.0001$. The target state equation is $X(k+1)=FX(k)+Gw(k)$ where, k is the scan number and w is the white Gaussian process noise with zero mean and covariance Q. The measurement equation is $Z_m(k)=HX(k)+v(k)$ with $H=\begin{bmatrix}1 & 0 & 0\end{bmatrix}$ and x and v is the measurement noise with zero mean and covariance $R=\sigma^2$ (σ =10 m). The system matrices are given as

$$F=\begin{bmatrix} 1 & T & T^2/2 & 0 & 0 & 0 \\ 0 & 1 & T & 0 & 0 & 0 \\ 0 & 0 & 1 & 0 & 0 & 0 \\ 0 & 0 & 0 & 1 & T & T^2/2 \\ 0 & 0 & 0 & 0 & 1 & T \\ 0 & 0 & 0 & 0 & 0 & 1 \end{bmatrix} \tag{9.75}$$

$$G=\begin{bmatrix}T^3/6 & T^2/2 & T & T^3/6 & T^2/2 & T\end{bmatrix} \tag{9.76}$$

$$H=\begin{bmatrix} 1 & 0 & 0 & 0 & 0 & 0 \\ 0 & 0 & 0 & 1 & 0 & 0 \end{bmatrix} \tag{9.77}$$

$$R=\begin{bmatrix} \sigma^2 & 0 \\ 0 & \sigma^2 \end{bmatrix} \tag{9.78}$$

The initial conditions, F, G, H, Q, and R for both the filters are kept same. The initial state vector $\hat{X}(0/0)$ is kept close to true initial states. The results for both the filters are compared in terms of true and estimated states, and states errors with bounds at every scan number. Every effort was made to tune the KF properly. The same FCV is used for *x*, *y* directions. The performances of both schemes are also compared in terms of root sum square position error (RSSPE) = $\sqrt{(x(k/k)-\hat{x}(k/k))^2+(y(k/k)-\hat{y}(k/k))^2}$. Figure 9.10 compares the RSSPE computed using true and estimated states for both the filters. Although the performance of the KF is satisfactory and acceptable, the FKF performs better than the KF. For the same target data (i.e., *x* & *y* axes), the performance of KF with FKF is compared for two cases: (i) when all the 49 rules are taken into consideration, and (ii) only when 4 rules are used (Table 9.15).

Figure 9.11 compares the RSSPE of FKF for these two cases. It is very much clear that filter with 49 rules shows better performance than the filter with 4 rules only, although the performance with 4 rules is quite acceptable. This indicates that in order to have a good FKF, just a sufficient number of rules are needed to get continuous and smooth inputs/outputs mapping. Too many rules are often not necessary.

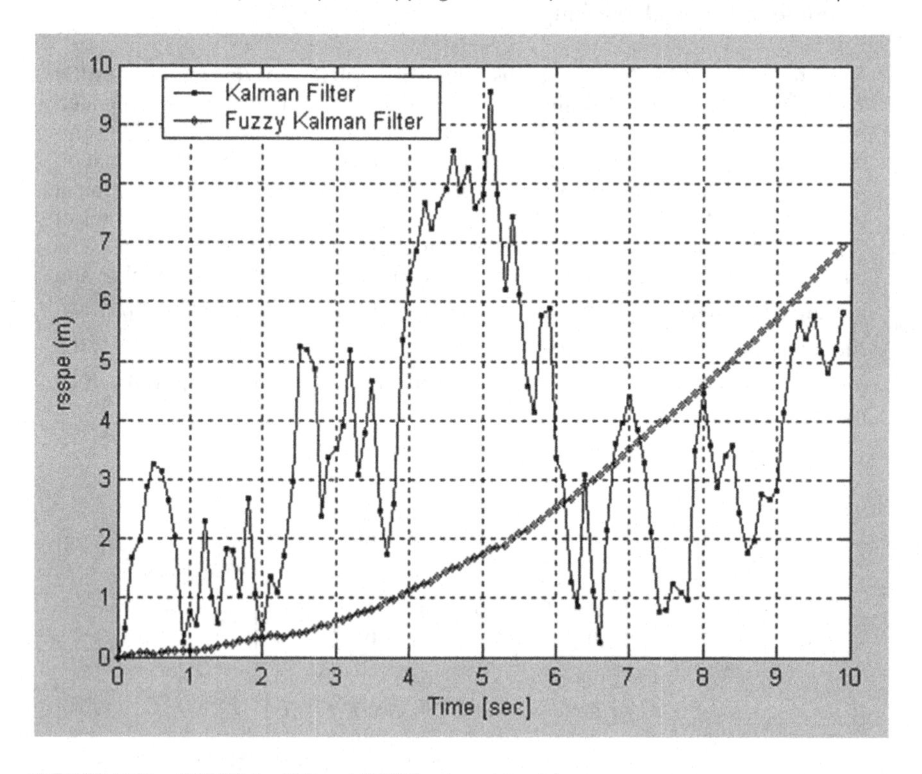

FIGURE 9.10 RSSPE for KF and FKF for target tracking.

TABLE 9.15
Four Rules for FKF

	e_x	
$\dot{e}_x$	**LN**	**LP**
LN	LN	ZE
LP	ZE	LP

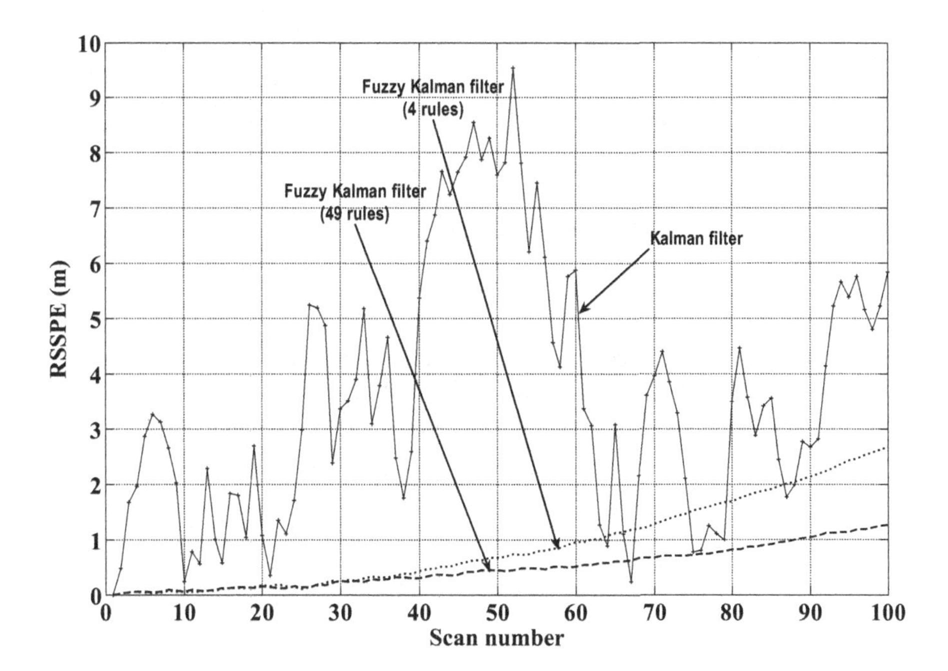

FIGURE 9.11 RSSPE for KF and FKF for 49 and 4 rules (Example 9.2)

9.8.1 Tracking of Maneuvering Target

For tracking a maneuvering target, it is essential to re-design the FCV to capture the various possible maneuver modes of the target. Re-designing of FCV involves the following: (i) proper selection of membership functions of inputs and output, (ii) tuning of selected membership functions, (iii) proper selection of Fuzzy operators (e.g. t-norm & s-norm) [37], and (iv) proper selection of Fuzzy implication, aggregation, and defuzzification methods. The MATLAB®-based functions, such as 'genfis1()' to create initial FCV and 'anfis()' to tune it, are used. These functions require training and checking data which are obtained from true and measured target positions. The target states are generated using a 3DOF kinematic model with process noise acceleration increments, Eqns. (9.90)–(9.93) and additional arbitrary accelerations. With sampling interval of 1 s, a total of 150 scans are generated. The data simulation proceeds with the following assumed parameter values: (i) initial states $(x,\dot{x},\ddot{x},y,\dot{y},\ddot{y})$ of target are (100, 30, 0, 100, 20, 0), respectively, (ii) process noise variance $Q = 0.1$. It is assumed that $Q_{xx} = Q_{yy} = Q$, and (iii) measurement noise variance $R = 25$. It is assumed that $R_{xx} = R_{yy} = R$. It is further assumed that the target has an additional acceleration of $(x_{\text{acc}}, y_{\text{acc}})$ at certain scans and an acceleration of $(-x_{\text{acc}}, -y_{\text{acc}})$ at certain other scans. Data simulation is carried out using the base state model with process noise vector w (which is a 2×1 vector) modified to include additional accelerations at the specified scan points, to induce a specific maneuver as follows:

$$\left.\begin{aligned} w(1) &= \text{guass}()*\sqrt{Q_{xx}} + x_{\text{acc}} \\ w(2) &= \text{guass}()*\sqrt{Q_{yy}} + y_{\text{acc}} \end{aligned}\right\} \quad \text{at certain scans} \tag{9.79}$$

$$\left.\begin{aligned} w(1) &= \text{guass}()*\sqrt{Q_{xx}} - x_{\text{acc}} \\ w(2) &= \text{guass}()*\sqrt{Q_{yy}} - y_{\text{acc}} \end{aligned}\right\} \text{ at certain other scans} \tag{9.80}$$

At the other scan points, the vector w is simply defined by

$$\left.\begin{aligned} w(1) &= \text{guass}()*\sqrt{Q_{xx}} \\ w(2) &= \text{guass}()*\sqrt{Q_{yy}} \end{aligned}\right\} \tag{9.81}$$

The function guass() uses a central-limit theorem to generate Gaussian random numbers with mean 0 and variance 1. The initial FCV for x-axis is created and tuned using inputs u_x^1, u_x^2 and output o_x obtained using the following equations:

$$u_x^1(k) = z_x(k) - x(k) \tag{9.82}$$

$$u_x^2(k) = \frac{u_x^1(k) - u_x^1(k-1)}{T} \tag{9.83}$$

$$o_x(k) = m * u_x^1(k) \tag{9.84}$$

Here, x, z_x are true and measured target x-positions, respectively. m is the unknown parameter that should be properly selected based on maneuver capability of a particular target of interest. For present case, m =2 (half of the total simulated points for training and remaining half as a checking data set). The same procedure is followed to obtain tuned FCV for y-axis. The trained FCVs are then plugged into FKF. In order to generate mild maneuver data, the accelerations are injected at scans 8 ($x_{\text{acc}} = 6\text{ m/s}^2$ and $y_{\text{acc}} = -6\text{ m/s}^2$) and 15 ($x_{\text{acc}} = -6\text{ m/s}^2$ and $y_{\text{acc}} = 6\text{ m/s}^2$) only. The performance of FKF is compared with KF and AKF. The equations of AKF are same as those of KF but with varying process noise covariance Q, estimated online using Maybeck's method [2]. The equations required to estimated Q are given by

$$Q(k) = G^{\#}\left[\hat{P}^{-} - F\hat{P}^{+}F^{T}\right](G^{\#})^{T} \tag{9.85}$$

$$\hat{P}^{-} = K(k)\hat{A}(k)*(H^{T})^{\#} \tag{9.86}$$

$$\hat{P}^{+} = \hat{P}^{-} - K(k)H\hat{P}^{-} \tag{9.87}$$

$$\hat{A}(k) = \frac{1}{WL}\sum_{j=k-WL+1}^{k} e(j)e(j)^{T};\ k \geq WL \tag{9.88}$$

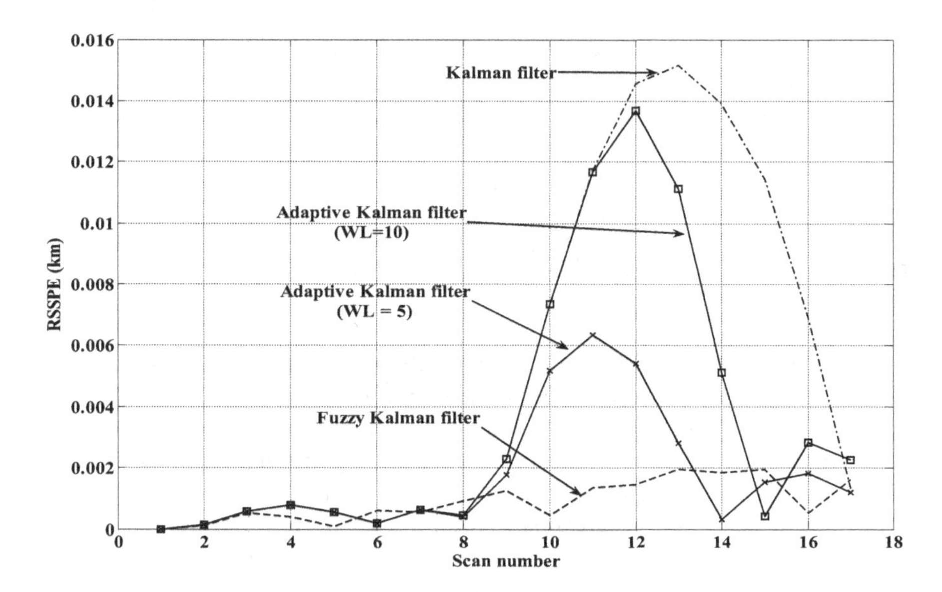

FIGURE 9.12 RSSPE for KF, FKF, and AKF for tracking of a maneuvering target (Example 9.2).

Here, # stands for pseudo-inverse, e is the innovation sequence vector, and WL (=5 for present case) is the window length. It is important to note that online value of Q can be made available to AKF only from WLth scan which means that accuracy of Q will depend upon its initial guess (for k =1 to WLth-1 scans) chosen by a filter designer. The AKF exhibited better tracking accuracy, especially during target maneuver, as compared to KF but still it had slightly degraded performance as compared to FKF. For non-maneuvering phases of target motion, AKF and KF perform almost similarly. During the maneuvering portion, it is found that the magnitude of online Q increases and so the Kalman gain, hence automatically more weight is assigned to measurement model which aids in convergence of the estimated states to true values at much faster rate than as seen for KF only. A sensitivity study of AKF w.r.t. different values of WL (5 and 10) was carried out, and its performance in terms of RSSPE is compared with KF and FKF (Figure 9.12). It is observed that RSSPE of AKF during maneuvering phase is high for $WL = 10$ as compared to for $WL = 5$. This could be due to non-availability of online Q and with an assumption that its initial guess is not sufficient at a time when actual maneuver starts, that is, at $k = 8$.

9.9 DERIVATIVE-FREE KALMAN FILTER FOR STATE ESTIMATION

A derivative-free Kalman filter (DFKF) [38] helps alleviate the problems associated with EKF, especially for nonlinear systems, and yields identical performance as compared to EKF when the assumption of local linearity is not violated. It does not require any linearization and uses deterministic sampling approach to capture the mean and covariance estimates with a minimal set of sample points or so-called sigma points. The emphasis is shifted from linearization of nonlinear systems to

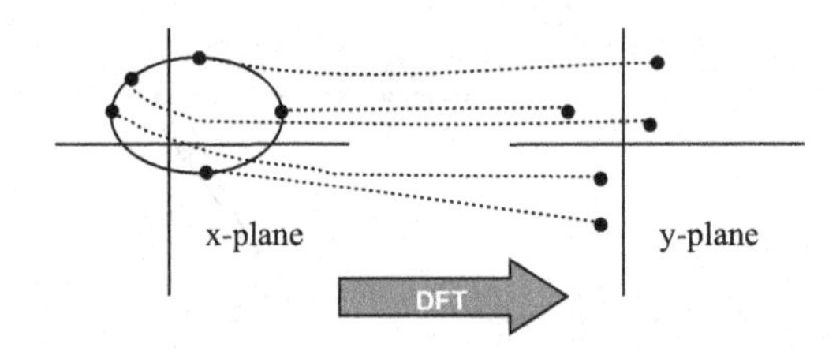

FIGURE 9.13 View of derivative-free transform-sigma point concept.

sampling approach of PDF. The fundamental difference is that in EKF, the non-linear models are linearized to parameterize the PDF in terms of its mean and covariance, whereas in DFKF, the PDF is parameterized through non-linear transformation of deterministically chosen sample points. The non-linear transformation is termed as derivative-free transformation (DFT) due to a fact that the transformation does not involve any differentiation expression. Figure 9.13 shows a pictorial representation of DFT. Consider propagation of a random variable x of dimension L ($L=2$) through a non-linear function $y=f(x)$. Assume that mean and covariance of sigma points, shown by black dots in left side of Figure 9.13, for random variable are $\bar{x}$ and P_x, respectively. These sigma points and their associated weights are deterministically created using the following equations [38,39]:

$$\left.\begin{aligned} \chi_0 &= \bar{x} \\ \chi_i &= \bar{x} + \left(\sqrt{(L+\lambda)P_x}\right)_i \qquad i=1,\ldots,L \\ \chi_i &= \bar{x} - \left(\sqrt{(L+\lambda)P_x}\right)_{i-L} \qquad i=L+1,\ldots,2L \end{aligned}\right\} \tag{9.89}$$

$$\left.\begin{aligned} W_0^{(m)} &= \frac{\lambda}{L+\lambda} \\ W_0^{(c)} &= \frac{\lambda}{L+\lambda} + (1-\alpha^2+\beta) \\ W_i^{(m)} &= W_i^{(c)} = \frac{1}{2(L+\lambda)} \qquad i=1,\ldots,2L \end{aligned}\right\} \tag{9.90}$$

The associated weights can be positive or negative, but to provide unbiased transformation, they must satisfy the condition $\sum_{i=1}^{2L} W_i^{(m \text{ or } c)} = 1$. For square root in Eqn. (9.104), it is proposed to use numerically efficient and stable method such as Cholesky decomposition. The scaling parameters used for the creation of sigma points and their associated weights are: (i) α which determines the spread of sigma points around $\bar{x}$, (ii) β to incorporate any prior knowledge about distribution of $\bar{x}$, $\lambda = \alpha^2(L+\kappa) - L$ and (iii) κ is the secondary tuning parameter. The sigma points are propagated through the non-linear function, $y_i = f(\chi_i)$, where $i = 0,\ldots,2L$, resulting in transformed sigma points. The mean and covariance of transformed points are formulated as

$$\bar{y} = \sum_{i=0}^{2L} W_i^{(m)} y_i \tag{9.91}$$

$$P_y = \sum_{i=0}^{2L} W_i^{(c)} \{y_i - \bar{y}\}\{y_i - \bar{y}\}^T \tag{9.92}$$

The DFKF is straightforward extension of the DFT for the recursive estimation problems, and the states of DFKF can be reconstructed by introducing the concept of augmented state vector that consists of actual system and process noise states each with n-dimension, and m-dimensional measurement noise state. The dimension of the augmented state vector becomes $n_a = n + n + m = 2n + m$. Although augmentation technique has additional sigma points, it implicitly incorporates the effects of noises at various stages. The steps required in implementation of DFKF at every sampling point can be specified as mentioned later.

DFKF Initialization

$$\left.\begin{aligned} &\hat{X}(0/0) = E[X(0/0)] \\ &\hat{P}(0/0) = E\left[\left(X(0/0) - \hat{X}(0/0)\right)\left(X(0/0) - \hat{X}(0/0)\right)^T\right] \end{aligned}\right\} \tag{9.93}$$

Augmented state and its error covariance are represented as

$$\left.\begin{aligned} &\hat{X}^a(0/0) = E\left[X^a(0/0)\right] = \begin{bmatrix} \hat{X}^T(0/0) & \underbrace{0,\ldots,0}_{n-\dim w} & \underbrace{0,\ldots,0}_{m-\dim v} \end{bmatrix}^T \\ &\hat{P}^a(0/0) = E\left[\left(X^a(0/0) - \hat{X}^a(0/0)\right)\left(X^a(0/0) - \hat{X}^a(0/0)\right)^T\right] \\ &\quad = \begin{bmatrix} \hat{P}(0/0) & 0 & 0 \\ 0 & Q & 0 \\ 0 & 0 & R \end{bmatrix}_{2n+m \text{ by } 2n+m} \end{aligned}\right\} \tag{9.94}$$

Sigma Points Computation

$$\left.\begin{aligned}
&\chi_0^a(k/k) = \hat{X}^a(k/k) \\
&\chi_i^a(k/k) = \hat{X}^a(k/k) + \left(\sqrt{(n_a+\lambda)\hat{P}^a(k/k)}\right)_i \qquad i = 1,\ldots,n_a \\
&\chi_i^a(k/k) = \hat{X}^a(k/k) - \left(\sqrt{(n_a+\lambda)\hat{P}^a(k/k)}\right)_{i-n_a} \qquad i = n_a+1,\ldots,2n_a
\end{aligned}\right\} \tag{9.95}$$

Here, $\chi^a = \begin{bmatrix} \underbrace{\chi}_{\text{state}} & \underbrace{\chi^w}_{\text{process noise}} & \underbrace{\chi^v}_{\text{meas. noise}} \end{bmatrix}$

State and Covariance Propagation

$$\left.\begin{aligned}
&\chi(k+1/k) = f\left(\chi(k/k), u(k), \chi^w(k/k), k\right) \\
&\tilde{X}(k+1/k) = \sum_{i=0}^{2n_a} W_i^{(m)} \chi_i(k+1/k) \\
&\tilde{P}(k+1/k) = \sum_{i=0}^{2n_a} W_i^{(c)} \left[\chi_i(k+1/k) - \tilde{X}(k+1/k)\right]\left[\chi_i(k+1/k) - \tilde{X}(k+1/k)\right]^T
\end{aligned}\right\} \tag{9.96}$$

$$\left.\begin{aligned}
&W_0^{(m)} = \frac{\lambda}{n_a+\lambda} \\
&W_0^{(c)} = \frac{\lambda}{n_a+\lambda} + (1-\alpha^2+\beta) \\
&W_i^{(m)} = W_i^{(c)} = \frac{1}{2(n_a+\lambda)} \qquad i = 1,\ldots,2n_a
\end{aligned}\right\} \tag{9.97}$$

State and Covariance Update

$$\left.\begin{aligned}
&y(k+1/k) = h\left(\chi(k/k), u(k), k\right) + \chi^v(k/k) \\
&\tilde{Z}(k+1/k) = \sum_{i=0}^{2n_a} W_i^{(m)} y_i(k+1/k)
\end{aligned}\right\} \tag{9.98}$$

$$\left.\begin{aligned}
S &= \sum_{i=0}^{2n_a} W_i^{(c)}\left[y_i(k+1/k)-\tilde{Z}(k+1/k)\right]\left[y_i(k+1/k)-\tilde{Z}(k+1/k)\right]^T \\
P_{xy} &= \sum_{i=0}^{2n_a} W_i^{(c)}\left[\chi_i(k+1/k)-\tilde{X}(k+1/k)\right]\left[y_i(k+1/k)-\tilde{Z}(k+1/k)\right]^T \\
K &= P_{xy}S^{-1} \quad \text{(filter gain)}
\end{aligned}\right\} \tag{9.99}$$

$$\left.\begin{aligned}
\hat{X}(k+1/k+1) &= \tilde{X}(k+1/k)+K\left(Z_m(k+1)-\tilde{Z}(k+1/k)\right) \\
\hat{P}(k+1/k+1) &= \tilde{P}(k+1/k)-KSK^T
\end{aligned}\right\} \tag{9.100}$$

Example 9.3 Kinematic Consistency for an Aircraft

The performance of the filter for kinematic consistency checking using realistic longitudinal short-period and lateral-directional (for flight condition: Mach = 0.5 and altitude = 4 km) data generated from a 6DOF simulation of an aircraft is evaluated and compared with that of the UD extended KF (UDEKF). The relevant MATLAB routines are **'ExamplesSolSW/Example 9.3DFKF'.** The basic kinematic models used in state estimation are mentioned later.

STATE OR PROCESS MODEL

$$\left.\begin{aligned}
\dot{u} &= -(q-\Delta q)w+(r-\Delta r)v-g\sin\theta+(a_x-\Delta a_x), \\
\dot{v} &= -(r-\Delta r)u+(p-\Delta p)w+g\cos\theta\sin\phi+a_y, \\
\dot{w} &= -(p-\Delta p)v+(q-\Delta q)u+g\cos\theta\cos\phi+(a_z-\Delta a_z), \\
\dot{\phi} &= (p-\Delta p)+(q-\Delta q)\sin\phi\tan\theta+(r-\Delta r)\cos\phi\tan\theta, \\
\dot{\theta} &= (q-\Delta q)\cos\phi-(r-\Delta r)\sin\phi, \\
\dot{h} &= u\sin\theta-v\cos\theta\sin\phi-w\cos\theta\cos\phi
\end{aligned}\right\} \tag{9.101}$$

Here, $\Delta a_x, \Delta a_z, \Delta p, \Delta q, \Delta r, K_\alpha, K_\theta$ are the bias terms, and p, q, r, a_x, a_y, a_z are the control inputs to the process model.

OBSERVATION OR MEASUREMENT MODEL

$$\left.\begin{aligned}
&Z_{t|m} = \begin{bmatrix} V_m & \alpha_m & \beta_m & \phi_m & \theta_m & h_m \end{bmatrix} \\
&V_m = \sqrt{u_n^2+v_n^2+w_n^2}, \quad \alpha_m = K_\alpha \tan^{-1}\left[\frac{w_n}{u_n}\right], \quad \beta_m = \sin^{-1}\left[\frac{v_n}{\sqrt{u_n^2+v_n^2+w_n^2}}\right], \\
&\phi_m = \phi+\Delta\phi, \qquad \theta_m = K_\theta\theta, \qquad h_m = h
\end{aligned}\right\} \tag{9.102}$$

Here, u_n, v_n, w_n are the velocity components along the three axes at the nose boom of the aircraft and are computed as follows:

$$\left.\begin{array}{l} u_n = u - (r - \Delta r)Y_n + (q - \Delta q)Z_n,\ v_n = v - (p - \Delta p)Z_n + (r - \Delta r)X_n, \\ w_n = w - (q - \Delta q)X_n + (p - \Delta p)Y_n \end{array}\right\} \quad (9.103)$$

Here, X_n, Y_n and Z_n are the offset distances from nose boom to CG, and their values are kept at 12.586, 0.011, and 0.14 (meters), respectively. The measurement noise with SNR of 10 is added only to the observables $V, \alpha, \beta, \phi, \theta, h$ and no noise is added to the rates and accelerations during the data generation. The additional information used in both the filters is

a. initial state, in SI units are, $\hat{X}^1(0/0)\,\&\,\hat{X}^2(0/0) =$

$$= \begin{bmatrix} u & v & w & \phi & \theta & h & \Delta a_x & \Delta a_z & \Delta p & \Delta q & \Delta r & K_\alpha & K_\theta \end{bmatrix}$$

$$= \begin{bmatrix} 167 & 0.001 & 17.305 & 0 & 0.10384 & 4000 & 0 & 0 & 0 & 0 & 0 & 1 & 1 \end{bmatrix}$$

b. sampling interval : $T = 0.025$ and process noise variance:

$$Q = 1.0e - 15 * \text{eye}(nx)$$

c. measurement noise variance : $R = E\left[(Z_m - Z_t)(Z_m - Z_t)^T\right]$, where Z_t is the noise free measurement from simulator and Z_m is the noisy measurement

d. initial state error covariance : eye(nx) for UDEKF and

$$\begin{bmatrix} \text{eye}(nx) & 0 & 0 \\ 0 & Q & 0 \\ 0 & 0 & R \end{bmatrix} \text{for DFKF.}$$

where, $nx = 13$ is the number of estimated states

The results are generated for 25 Monte Carlo simulations. Figure 9.14 shows the comparison of true, measured, and estimated observables such as $V, \alpha, \beta, \phi, \theta, h$. From the plots, it is clear that wherever (between 0–5 s or around 10 s) the non-linearity in measurement data is more severe, the performance of UDEKF is degraded as compared to DFKF. This can be further proved by comparing the measurement residuals with 1 sigma bounds (i.e., $\pm\sqrt{HPH' + R}$). It is observed from Figure 9.15 that the theoretical bounds are comparable (due to same initial conditions) for both filters, but in case of UDEKF, its residuals go out of bounds for more number of times as compared with DFKF.

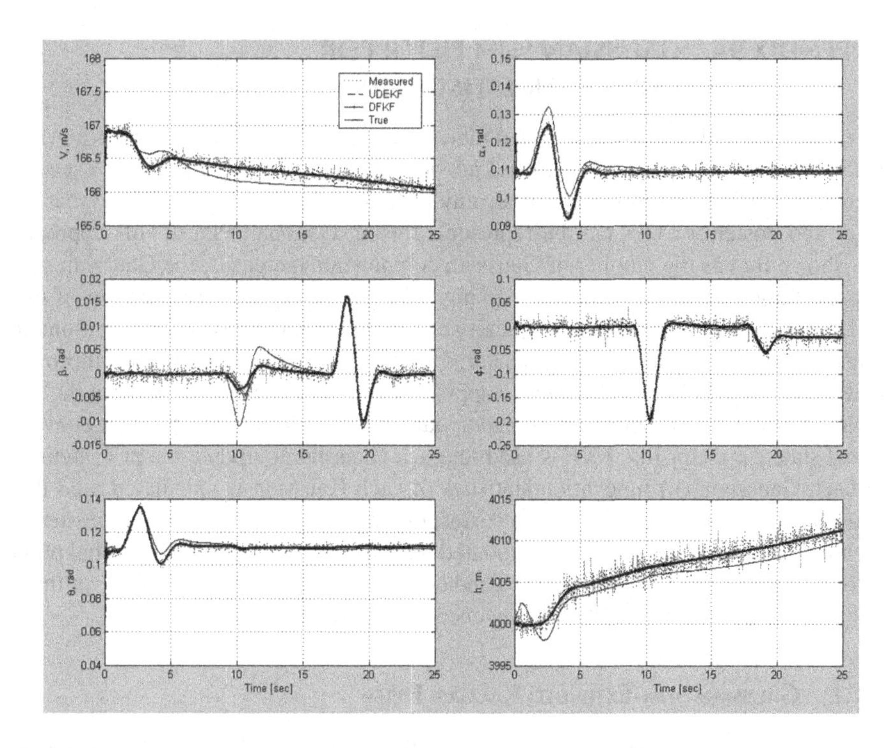

FIGURE 9.14 True, measured, and predicted observables for DFKF (Example 9.3).

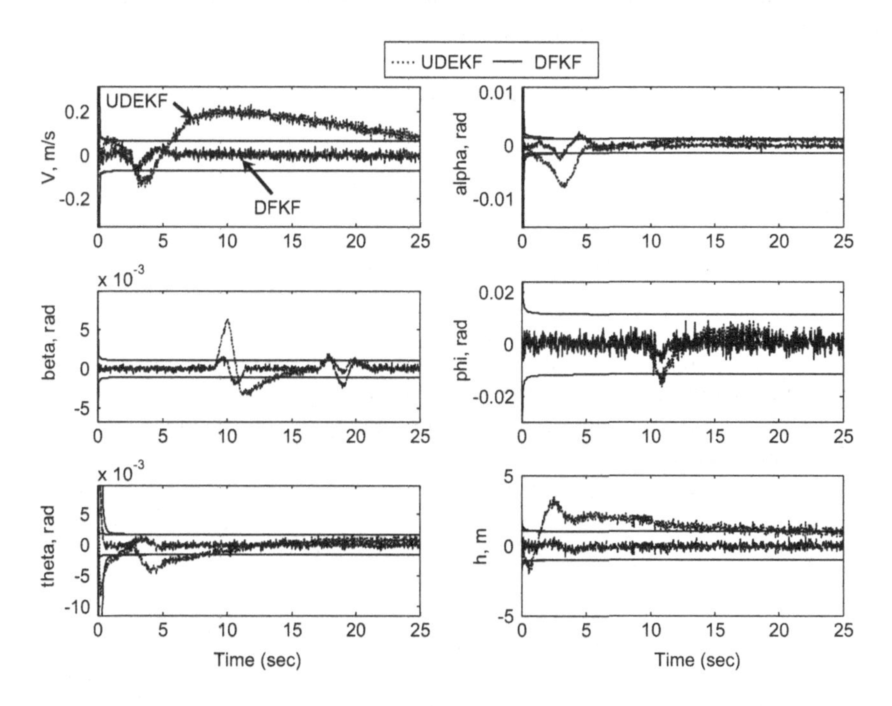

FIGURE 9.15 Innovations for DFKF.

APPENDIX 9A GAUSSIAN SUM FILTER FOR PARAMETER ESTIMATION

A variant to nonlinear filtering, Gaussian sum filter (GSF), can be used for APE, where the prior and posterior densities are approximated as weighted sum of several Gaussians [40]. A GSF can incorporate any Kalman-based nonlinear filter so that the prior and posterior PDFs can be represented using Gaussian PDFs. This approach has shown that as the number of Gaussian components increases, the Gaussian sum approximation converges uniformly to any PDF. However, the large number of (filter) components increases the complexity and computational time for estimation; an effective solution to this is also considered in this paper. The GSF and an improved GSF make use of sum of Gaussians to approximate the posterior PDF and use the following three steps to achieve the global update of the posterior density: (i) A conventional state estimator like EKF is used for each Gaussian to update the prior density of each Gaussian; (ii) innovation statistics of each Gaussian is calculated, and this is used to update the weights of the corresponding Gaussian; and (iii) the posterior moments of each Gaussian and the updated weights together provide the global point estimate. We study two types of GSFs and derive convergence results for one of these filters using the normalized Lyapunov energy functional (LEF).

9A.1 GAUSSIAN SUM-EXTENDED KALMAN FILTER

The main objective of the GSF is to approximate an arbitrary a posteriori density function and its moments. Consider a nonlinear discrete-time model corrupted by noise in both the states and the measurements:

$$x_{k+1} = f(x_k) + g_k \tag{9A.1}$$

$$y_{k+1} = h(x_{k+1}) + v_{k+1} \tag{9A.2}$$

$$Q_k = E[g_k g_k^T];\ R_{k+1} = E[v_{k+1} v_{k+1}^T] \tag{9A.3}$$

The noise processes, g_k and v_{k+1} (temporally uncorrelated with each other), are zero-mean random sequences with covariance matrices Q_k and R_{k+1}. For a stochastic dynamic system, the a posteriori (after incorporating the observations into the KF/EKF) normal/Gaussian PDF is expressed as

$$p(\hat{x}_k | z_k) = \sum_{i=1}^{N} w_k^i \mathrm{N}\ (\hat{x}_k; x_k^i, \hat{P}_k^i) \tag{9A.4}$$

Here, N is the total number of Gaussian densities used; w_k^i, x_k^i, and $\hat{P}_k^i$ represent, respectively, the components' conditional weight/s, mean/s, and covariance/s of the ith Gaussian with respect to the first k measurements. The usual condition on the weight is

$$\sum_{i=1}^{N} w_k^i = 1;\ w_k^i \geq 0 \tag{9A.5}$$

Since all the components of the mixture PDF are Gaussian, only estimates of their means and covariance/s need to be considered between k and $k+1$.

9A.1.1 Time Propagation Evolution

The prior (before the observations are incorporated into the EKF) mean $\tilde{x}_{k+1}$ and the prior covariance $\tilde{P}_{k+1}$ of the ith Gaussian component are calculated using the time update equations of the EKF.

$$\tilde{x}_{k+1}^i = f_k(\hat{x}_k^i) \tag{9A.6}$$

$$\tilde{P}_{k+1}^i = F_k \hat{P}_k^i F_k^T + Q_k \tag{9A.7}$$

The discrete recursion (9A.6) is performed by choosing proper initial condition of $x(.,.)$, where the Jacobian is given as

$$F_k = \frac{\partial f_k(\hat{x}_k^i)}{\partial \hat{x}_k^i} \tag{9A.8}$$

The state transition matrix F_k can be approximated using Taylor series expansion. In time update, the weights are not updated:

$$\tilde{w}_{k+1}^i = \hat{w}_k^i \tag{9A.9}$$

9A.1.2 Measurement Data Update

Following the standard KF, we have the data update recursions as

$$\hat{x}_{k+1}^i = \tilde{x}_{k+1}^i + K_{k+1}^i [y_{k+1} - h_{k+1}(\tilde{x}_{k+1}^i)] \tag{9A.10}$$

$$S_{k+1}^i = H_{k+1} \tilde{P}_{k+1}^i H_{k+1}^T + R_{k+1} \tag{9A.11}$$

$$K_{k+1}^i = \tilde{P}_{k+1}^i H^T{}_{k+1} (S_{k+1}^i)^{-1} \tag{9A.12}$$

$$\hat{P}_{k+1}^i = (I - K_{k+1}^i H_{k+1}) \tilde{P}_{k+1}^i \tag{9A.13}$$

$$H_{k+1} = \frac{\partial h_{k+1}(\tilde{x}_{k+1}^i)}{\partial \tilde{x}_{k+1}^i} \tag{9A.14}$$

In the subsequent analysis, the measurement vector/matrix H, the process noise covariance matrix Q, and the measurement noise covariance matrix R are represented without any time index/subscript for the sake of simplicity; also, the variation of H, since it is time-dependent, will be taken care in the program code for the implementation of the GSF.

The Gaussian sum estimate

Finally, the Gaussian sum weight and the state estimates are given as

$$\hat{w}_{k+1}^{i} = \frac{\tilde{w}_{k+1}^{i} \mathrm{N}(y_{k+1} - h_{k+1}(\tilde{x}_{k+1}^{i}), S_{k+1}^{i})}{\sum_{i=1}^{N} \tilde{w}_{k+1}^{i} \mathrm{N}(y_{k+1} - h_{k+1}(\tilde{x}_{k+1}^{i}), S_{k+1}^{i})} \tag{9A.15}$$

$$\hat{x}_{k+1} = \sum_{i=1}^{N} \hat{w}_{k+1}^{i} \hat{x}_{k+1}^{i} \tag{9A.16}$$

9A.2 Gaussian Sum Filter with Pruning

The required number of components in the GSF can be set depending on the application. However, as this number increases, the computational complexity and time also increase; hence, pruning of the number of filters is needed, and this can help in reducing the computational complexity, and possibly maintaining the estimation accuracy. In order to differentiate between less important weights and the important ones, an optimal threshold value needs to be evaluated; as in typical pruning methods, all the weights with the magnitudes below a specified threshold need not be considered for the update of the final state estimate. However, pruning of the component filters based on amplitude of the weights only may prune some weights which might be important for contribution to the final state estimate. We study an algorithm called GSF with pruning (GSFP) wherein the weights and weight changes considered together for pruning are studied. Pearson correlation coefficient is used to set the threshold, that is, between (−1 and +1). If the weight amplitude and the weight/change relationship of a weight w_k^i is unimportant, then the corresponding correlation coefficient would be closer to 0, and the corresponding filter components are considered not so valuable in the estimation process. The correlation coefficient to be calculated for each Gaussian component is given as

$$\mathrm{coeff}_W(w^i, \Delta w^i) = \frac{m(\sum w^i \Delta w^i) - \sum w^i \sum \Delta w^i}{\sqrt{(m\sum (w^i)^2 - (\sum w^i)^2)(m\sum \Delta(w^i)^2 - (\sum \Delta w^i)^2)}} \tag{9A.17}$$

Here, m is the number of time steps between consecutive pruning of filters, and Δw is the weight change magnitude:

$$\Delta w(k)_i = \left| w(k)_i - w(k-1)_i \right| \tag{9A.18}$$

The value of m and the threshold are decided by experimentation, that is, by trial-and-error/sensitivity analysis. When the initial state is not known and is taken very far away from the true value, the number of Gaussian components required is more to provide better performance. In such cases, overfitting might happen and pruning will remove such weights, which in turn would have increased/maintained the accuracy along with decreasing the computational cost. Therefore, pruning helps in improving the performance of the filter especially when the initial guess of the state estimate is not close to the true state.

Algorithm for GSFP: (i) Run N Gaussian sum EKF for m time steps, (ii) compute the Pearson coefficient, (9A.17), $\text{coeff}_W(w^i, \Delta w^i)$ related to magnitude of weight and change of weight, (iii) truncate filters which have, (iv) normalize the weights of the remaining filters $N = N - N_T$ where N_T is the number of truncated filters, and (v) go back to Step (i).

9A.3 Lyapunov Stability Analysis of GSF via Observer

First, an appropriate observer is proposed, and then, the conditions for the convergence of the GSEKF are established via the convergence of this observer. An observer for the nonlinear system of (9A.1) and (9A.2) is considered:

$$\hat{x}^i_{k+1} = f(\hat{x}^i_k) + K^i_{k+1}(y_{k+1} - \tilde{y}^i_{k+1}) = f(\hat{x}^i_k) + K^i_{k+1} r'^i_{k+1} \tag{9A.19}$$

$$\hat{y}^i_{k+1} = h(\hat{x}^i_{k+1}) \cong H\hat{x}^i_{k+1} \tag{9A.20}$$

The observer gain is taken as the gain of the GSEKF with associated covariance matrices:

$$K^i_{k+1} = \tilde{P}^i_{k+1} H^T (S^i_{k+1})^{-1} \tag{9A.21}$$

$$S^i_{k+1} = H\tilde{P}^i_{k+1} H^T + R \tag{9A.22}$$

The state errors are

$$\hat{e}^i_{k+1} = x_{k+1} - \hat{x}^i_{k+1} \tag{9A.23}$$

Since there are N filters, the weights of each filter have to be considered to show the combined final effect. The combined weights, a posterior error, a priori error, residual error, a posterior state covariance, and a priori state covariance are defined for the sake of clarity and completion for the time index 'k':

$$w_k = [w^1_k \; w^2_k \cdots w^N_k] \tag{9A.24}$$

$$\hat{e}_k = \sum_{i=1}^{N} w_k^i \hat{e}_k^i \tag{9A.25}$$

$$\tilde{e}_k = \sum_{i=1}^{N} w_k^i \tilde{e}_k^i \tag{9A.26}$$

$$r_k = \sum_{i=1}^{N} w_k^i r_k^i \tag{9A.27}$$

$$\hat{P}_k = \sum_{i=1}^{N} w_k^i \hat{P}_k^i \tag{9A.28}$$

$$\tilde{P}_k = \sum_{i=1}^{N} w_k^i \tilde{P}_k^i \tag{9A.29}$$

The observer error is obtained by subtracting (9A.19) from (9A.1):

$$\hat{e}_{k+1}^i = x_{k+1} - \hat{x}_{k+1}^i = f(x_k) - f(\hat{x}_k^i) - K_{k+1}^i(y_{k+1} - \tilde{y}_{k+1}^i) \tag{9A.30}$$

In fact, since the total estimate from the observer has to be considered for comparison with the actual states, we use a short-hand notation/symbol for sigma operation: $E\{.\} = \sum_{i=1}^{N} w_k^i(.)$, and with this notation, we rewrite (9A.30) as

$$\hat{e}_{k+1} = E\{\hat{e}_{k+1}^i\} = f(x_k) - E\{f(\hat{x}_k^i)\} - E\{K_{k+1}^i(y_{k+1} - \tilde{y}_{k+1}^i)\} \tag{9A.31}$$

$$\begin{aligned} &\hat{e}_{k+1} = F_x x_k + \phi_k - E\{K_{k+1}^i(y_{k+1} - \tilde{y}_{k+1}^i)\}; \\ &\text{wherein } \phi_k = -F_x x_k + f(x_k) - E\{f(\hat{x}_k^i)\} \end{aligned} \tag{9A.32}$$

The nonlinear function as 'phi' in (9A.32) is included for facilitating the convergence analysis; otherwise both (9A.31) and (9A.32) are exactly identical. In Eqn. (9A.32), one can substitute for $K(.)$ in (9A.33) from (9A.21) and (9A.22) to obtain, after the simplification of the measurement terms, that is, the residuals:

$$\hat{e}_{k+1} = F_x x_k + \phi_k - E\{\tilde{P}_{k+1}^i H^T (H\, \tilde{P}_{k+1}^i H^T + R)^{-1} H \tilde{e}_{k+1}^i)\} \tag{9A.33}$$

We would utilize Lyapunov–Krasovskii's approach to derive the conditions for the convergence of the observer error, (9A.33), for which we use the following development initially:

Let the error state in general be described as

$$e(k+1) = Ge(k) \tag{9A.34}$$

Let us define a feasible normalized LEF as

$$V(e(k)) = e^T(k)Y(k)e(k) \tag{9A.35}$$

In (9A.35), $Y(.)$ is the information matrix (as the inverse of the conventional covariance matrix P), it is a positive definite matrix, and we use the following expression as the (equivalent or analogous) derivative of the LEF:

$$\Delta\{V(e(k))\} = V(e(k+1)) - V(e(k)) \tag{9A.36}$$

Substituting for $V(.)$s in (9A.36), we obtain

$$\begin{aligned}\Delta\{V(e(k))\} &= V(e(k+1)) - V(e(k)) \\ &= e^T(k+1)Ye(k+1) - e^T(k)Ye(k) \\ &= e^T(k)G^TYGe(k) - e^T(k)Ye(k) \\ &= e^T(k)[G^TYG - Y]e(k)\end{aligned} \tag{9A.37}$$

Since the 'derivative' term of the error should be negative definite for the convergence of the observer error dynamics, we have

$$\Delta\{V(e(k))\} = -e^T(k)[-(G^TYG - Y)]e(k) \tag{9A.38}$$

Then for the convergence of the observers' error dynamics, we have the following condition:

$$-(G^TYG - Y) = Y - G^TYG \tag{9A.39}$$

which should be positive definite (PD). This means that in terms of the magnitudes of the involved terms, the following should hold true:

$$\left|Y - G^TYG\right| > 0 \tag{9A.40}$$

Since the magnitude of any matrix is measured by its norm, one can arrive at the final condition as

$$\left\|Y - G^TYG\right\| > 0 \Rightarrow \left\|G\right\|^2 < 1 \tag{9A.41}$$

Next, the error dynamics (9A.33) should be brought in the form of (9A.41), in fact of (9A.34), so that one can easily derive the conditions for the convergence, and in order to do this, it is necessary to define certain bounds:

1. The steady state solution, $P(\infty)$ of Eqn. (9A.7), must be in the range:

$$p_l I \le P(.) \le p_u I \tag{9A.42}$$

Here, $p_u, p_l > 0$ are +ve numbers, $P(.)$ is inherently symmetric/positive definite (PD) matrix; the constants specify the bounds.

2. The term $F_k x_k$ should also be bounded as

$$\|F_k x_k\| = \rho_1 \|e_k\|^2 \tag{9A.43}$$

The variable 'ρ_1' is the upper bound of the associated nonlinear function as a scalar factor of the state error norm.

3. The nonlinear function 'phi' in (9A.32) is bounded, because it is composed of the bounded functions of the dynamic system:

$$\|\phi(.)\| \le \rho_2 \|x_k - \hat{x}_k\|^2 = \rho_2 \|e_k\|^2 \tag{9A.44}$$

The variable 'ρ_2' is the upper bound of the associated nonlinear function as a scalar factor of the state error norm. Also, the following bounds on the norms of matrices H and R are needed to simplify the error dynamics:

$$\|H^T H\| \le h^2; \quad \|R\| = r; \quad \|e_k\|^2 \le \varepsilon^2; \tag{9A.45}$$

$$\|e_k\| \le \varepsilon = \text{some constant}$$

Next, the observer's error dynamics of (9A.33) should be brought in the form of (9A.41):

$$\begin{aligned} \hat{e}_{k+1} &= F_x x_k + \phi_k - E\{\tilde{P}^i_{k+1} H^T (H\, \tilde{P}^i_{k+1} H^T + R)^{-1} H\tilde{e}^i_{k+1})\} = G\hat{e}_k \\ \|\hat{e}_{k+1}\| &= \|G\| \|\hat{e}_k\| \end{aligned} \tag{9A.46}$$

It is better to recast (9A.46) in terms of the norms as follows:

$$\|\hat{e}_{k+1}\| \equiv \|F_x x_k\| + \|\phi_k\| - \left\|E\{\tilde{P}^i_{k+1} H^T (H\, \tilde{P}^i_{k+1} H^T + R)^{-1} H\tilde{e}^i_{k+1})\}\right\| \tag{9A.47}$$

Substituting various bounds defined earlier into (9A.47), one obtains

$$\|\hat{e}_{k+1}\| \equiv \rho_1 \|\hat{e}_k\|^2 + \rho_2 \|\hat{e}_k\|^2 - \left\|E\{\tilde{P}^i_{k+1} H^T (H\, \tilde{P}^i_{k+1} H^T + R)^{-1} H\tilde{e}^i_{k+1})\}\right\| \tag{9A.48}$$

After substituting other bounds from (9A.45), simplifying the resultant expression, and utilizing the definition of error norm, one obtains

$$\left\|\hat{e}_{k+1}\right\| \equiv \rho_1\left\|e_k\right\|^2 + \rho_2\left\|e_k\right\|^2 - \left(\frac{p_l h^2}{(p_l h^2 + r)}\varepsilon\right) \tag{9A.49}$$

$$\left\|\hat{e}_{k+1}\right\| \equiv \left\{\rho_1\varepsilon + \rho_2\varepsilon - \frac{p_l h^2}{(p_l h^2 + r)}\right\}\varepsilon = \left\|G\right\|\left\|e_k\right\| \tag{9A.50}$$

Equation (9A.50) is now in the required form of the second equation of (9A.46). From (9A.50), one obtains the following expression:

$$\left\|G\right\| \equiv \left\{\rho_1\varepsilon + \rho_2\varepsilon - \frac{p_l h^2}{(p_l h^2 + r)}\right\} \tag{9A.51}$$

Now applying the condition of (9A.41) and considering the positive square root of 1, one obtains

$$\begin{aligned}
&\left\|G\right\|^2 \equiv \left\{\rho_1\varepsilon + \rho_2\varepsilon - \frac{p_l h^2}{(p_l h^2 + r)}\right\}^2 < 1 \\
&\left\{\rho_1\varepsilon + \rho_2\varepsilon - \frac{p_l h^2}{(p_l h^2 + r)}\right\} < \mathrm{sqrt}(1) = \pm 1 \\
&\left\{\rho_1\varepsilon + \rho_2\varepsilon - \frac{p_l h^2}{(p_l h^2 + r)}\right\} < 1
\end{aligned} \tag{9A.52}$$

Thus, the convergence condition for the observer's error dynamics is obtained as

$$(\rho_1 + \rho_2)\varepsilon < 1 + \frac{p_l h^2}{(p_l h^2 + r)} \tag{9A.53}$$

All the coefficients appearing in (9A.53) are well defined as these are the positive bounding constants in Eqns. (9A.42)–(9A.45). The condition of (9A.53) means that if it is satisfied, then the time derivative, analogously the differential in (9A.38) of the LEF, would be negative definite, and the observer error dynamics would converge and the error would (asymptotically) tend to zero. This, in turn, also establishes that the GSEKF would also converge, since its gain and covariance matrix were used in the observer error dynamics. This is a novel way of establishing the asymptotic stability of the discrete time GSEKF via the use of the deterministic observer.

9A.4 Aircraft Parameter Estimation

The longitudinal motion of aircraft is represented using the following postulated aircraft model [1]:

$$\begin{aligned}
\dot{u} &= \frac{\bar{q}S}{m}C_X - qw - g\sin\theta \\
\dot{w} &= \frac{\bar{q}S}{mV}C_Z + qu + g\cos\theta \\
\dot{\theta} &= q \\
\dot{q} &= \frac{\bar{q}S\bar{c}}{I_{yy}}C_m
\end{aligned} \tag{9A.54}$$

In (9A.54), C_x, $C_{z,}$ and C_m are the non-dimensional aerodynamic coefficients expressed as

$$\begin{aligned}
C_Z &= C_{Z0} + C_{Z\alpha}\alpha + C_{Zq}\frac{q\bar{c}}{2V} + C_{Z\delta_e}\delta_e \\
C_X &= C_{X0} + C_{X\alpha}\alpha + C_{X_{\alpha^2}}\alpha^2 \\
C_m &= C_{m0} + C_{m\alpha}\alpha + C_{m_{\alpha^2}}\alpha^2 + C_{mq}\frac{q\bar{c}}{2V} + C_{m\delta_e}\delta_e
\end{aligned} \tag{9A.55}$$

Here, the state variables are true air velocity V, angle of attack α, pitch angle θ, and pitch rate q. The elevator deflection δ_e is the control input to aircraft dynamics. Other relevant parameters are mass of the aircraft 'm' and its geometrical parameters consisting of wing area S, wing chord $\bar{c}$, and moment of inertia I_{yy}. The dynamic pressure $\bar{q}$ is given by $\frac{1}{2}\rho V^2$, where ρ is the density of air and V is the true air speed, and all the variables have the SI metric units and have appropriate dimensions. The measurements vector is given by

$$\begin{aligned}
u &= u \\
w &= w \\
q &= q \\
\theta &= \theta \\
a_x &= \frac{\bar{q}S}{m}C_X + \frac{F_T}{m}\cos\sigma_T \equiv a_{x_m} \\
a_z &= \frac{\bar{q}S}{m}C_Z - \frac{F_T}{m}\sin\sigma_T \equiv a_{z_m} \\
\dot{q} &= \dot{q}
\end{aligned} \tag{9A.56}$$

In (9A.56), σ_T is the tilt angle of the engine. However, for the sake of simplicity, the engine effects are not simulated in the present case. It is presumed that each component of the measurements vector has some bias and/or random noise, but it is not shown in the model of (9A.56). The unknown parameter vector β that is to be estimated consists of the aerodynamic derivatives:

$$\beta = [C_{X_0}\ C_{X_\alpha}\ C_{X_{\alpha^2}}\ C_{Z_0}\ C_{Z_\alpha}\ C_{Z_{\delta_e}}\ C_{m_0}\ C_{m_\alpha}\ C_{m_{\alpha^2}}\ C_{m_q}\ C_{m_{\delta_e}}] \qquad (9A.57)$$

To evaluate the efficacy of GSEKF and GSFP, short-period data of a light transport aircraft (LTA) are simulated, and the non-dimensional longitudinal parameters are estimated. The 4-degrees-of-freedom longitudinal axis model is used for estimation; using Eqns. (9A.54)–(9A.56), the data are generated with a sampling interval of 0.03 s for duration of nearly 7 s by giving a doublet input to the elevator control surface. The initial states for $[u\ , w,\ q,\ \theta\ ]$ are appropriately. Random Gaussian noise with standard deviation 0.001 (as the process noise) is added to the states u, w, q and θ. Random noise with SNR=10 dB is added to these measurements, Eqn. (9A.56), to generate noisy data. A fourth-order RK integration method is used to obtain the longitudinal flight variables u, w, q and θ from Eqn. (9A.54). Two examples are considered to demonstrate the effectiveness of GSEKF over EKF and the performance improvement using GSFP; the metric parameter estimation error norm (PEEN) is evaluated for this and is defined as

$$\text{PEEN} = 100 * \frac{\text{norm}(\beta_{\text{true}} - \beta_{\text{est}})}{\text{norm}(\beta_{\text{true}})} \qquad (9A.58)$$

where β_{true} is the vector of true parameters, and β_{est} is the vector of estimated parameters.

The parameter estimation results for the two examples are shown in Table 9A.1, and the comparison of PEENs and the computational times are depicted in Table 9A.2. *The MATLAB scripts for these examples are not provided due to propriety reasons.*

Example 9A.1

In this example, the initial states/the parameters for the EKF are taken 10% away from their true values (used for generating the simulated data). The estimated values of the parameters are compared with the true values as in 2nd and 3rd columns of Table 9A.1. It is seen that the estimates are fairly close to the true values even when there is some noise (SNR = 10) in the measurement data, and the PEEN related to this is 5.63% which is reasonably small. In the case of the GSEKF, the number of Gaussian components required for effective filtering is less, as the initial state is close to the true value used for simulation; only 5 filters have been used for GSEKF and GSFP. The estimated parameters in Table 9A.1 are those of average of last ten data points. The GSEKF offers somewhat better estimation than EKF. When 3 Gaussian components were used, PEEN of GSEKF and GSFP was 3.228, though the value is higher than that of with 5 components, and the estimation accuracy is still better than EKF. When the number of components was increased to 6, the PEEN of GSEKF and GSFP remained the same as that obtained using 5 components.

Example 9A.2

In this example, the initial state is chosen far (~100%) from the true values. More number of Gaussian components are required by GSF/GSFP, since the initial states are not accurately known and are presumed to be far from the true values; 30 filters have been used for GSEKF and GSFP. Comparison is done by increasing and decreasing the number of filters, and the results are presented in Tables 9A.1 and 9A.2. Even in this case, the results are fairly satisfactory, and definitely better than those of the EKF.

EFFECTS OF PRUNING

Calculating $\text{coeff}_W(w_i, \Delta w_i)$ will help to locate the filters that do not contribute to the estimation process, and the pruning helps to negate overfitting and still maintains the accuracy. The computational load of GSEKF is more than that of GSFP. The pruning of filters that do not contribute significantly to the estimation process decreases the computational load of GSFP compared to GSEKF. To better visualize the effect of change in the number of filters, further study was performed with 25 and 35 Gaussian components. It was observed that when 25 components were used, the performance accuracy decreased as compared to 30 Gaussian components. There was no performance improvement when the number of filters was increased from 30 to 35. This implies that 30 Gaussian components are sufficient for this case. The comparative CPU time requirements for the examples 9A.1 and 9A.2 are shown in Table 9A.2. The execution time for EKF is the least. In the case of 3 Gaussian components, the execution time by GSFP is marginally more than time taken by GSEKF as the number of components that get pruned is very less. As the number of Gaussian components increases, the computational load would significantly reduce due to pruning. The improved GSEKF gives better performance than GSEKF when the number of filters required for estimation is more, especially when the initial states are not known very accurately a priori.

TABLE 9A.1
Estimated Parameters for GSEKF/GSFP Examples: 9A.1 and 9A.2

Parameter	True Values	EUD(~EKF) SNR=10	EUD(~EKF) SNR=10[c]	GSEKF-9A.1	GSFP 9A.1	GSEKF 9A.2	GSFP 9A.2
C_{X_0}[a]	−0.0540	*−0.0592*[b]	−0.0882	−0.054	−0.054	−0.053	−0.054
C_{X_α}	0.2330	*0.2543*	0.4937	0.234	0.234	0.231	0.233
$C_{X_{\alpha^2}}$	3.6089	*3.7058*	3.2183	3.653	3.653	3.633	3.647
C_{Z_0}	−0.1200	*−0.1249*	−0.1012	−0.122	−0.122	−0.123	−0.123
C_{Z_α}	−5.6800	*−5.7247*	−5.7807	−5.667	−5.667	−5.644	−5.654
$C_{Z_{\delta_e}}$	−0.4070	*−0.5049*	−0.4138	−0.402	−0.402	−0.395	−0.399
C_{m_0}	0.0550	*0.0576*	0.0606	0.054	0.054	0.054	−0.056
C_{m_α}	−0.7290	*−0.7092*	−0.654	−0.715	−0.715	−0.707	−0.708
$C_{m_{\alpha^2}}$	−1.7150	*−1.7843*	−1.9417	−1.687	−1.687	−1.667	−1.671
C_{m_q}	−16.300	*−15.3075*	−13.992	−16.069	−16.069	−15.755	−15.95
$C_{m_{\delta_e}}$	−1.9400	*−1.8873*	−1.8097	−1.9037	−1.9037	−1.882	−1.867
PEEN (%)		*5.6329*	13.3006	1.345	1.345	3.0966	2.065

[a] The aerodynamic derivatives are expressed in the form of the usual text format for convenience.
[b] Estimates in italics, Column 3, are from Ref. [21].
[c] The initial values are far away from true values.

TABLE 9A.2
Comparison of PEENs and Computational Times

	Number of Filters	PEEN in %		Comp. Times	
		GSEKF	GSFP	GSEKF	GSFP
Example 9A.1	3	3.228	3.228	1.05	1.17
	5	1.345	1.345	1.52	1.45
	6	1.345	1.345	1.56	1.22
Example 9A.2	25	3.539	3.246	4.88	2.12
	30	3.097	2.068	5.62	2.81
	35	3.097	2.065	6.033	2.86

APPENDIX 9B GAUSSIAN SUM INFORMATION FILTER FOR PARAMETER ESTIMATION

Here, Gaussian sum-extended information filter (GSEIF) is studied, and its performance is evaluated and compared with conventional EIF using MATLAB implementations with realistic simulated data for APE. The asymptotic stability procedure is similar to that of the GSEKF, and hence, it is not repeated here.

9B.1 Gaussian Sum-Extended Information Filter

Here, GSF has been incorporated with extended information filter; the prior and posterior PDFs can be represented using Gaussian PDFs:

$$x_{k+1} = f(x_k) + g_k \tag{9B.1}$$

$$z_{k+1} = h(x_{k+1}) + v_{k+1} \tag{9B.2}$$

$$Q_k = E[g_k g_k^T]; \; R_{k+1} = E[v_{k+1} v_{k+1}^T] \tag{9B.3}$$

The extended information filter (EIF) has two phases: prediction phase and correction phase. The prediction phase uses the posteriori estimate at previous time step to produce a priori estimate at current time step. In the correction phase, measurement at current time step is taken to correct the a priori estimate so that it will be more accurate, and it will result into a new posteriori estimate. The prediction phase of EIF uses the posteriori estimate at previous time step to produce a prior estimate at current time step, and in the correction phase, measurements (observables) at current time are taken to correct the a priori estimate to yield a new posteriori estimate.

9B.1.1 Time Propagation Evolution

The predicted information state $y(.)$ and information matrix $Y(.)$ are defined as

$$\tilde{y}_{k+1}^{j} = \hat{Y}_k^j f_k(\hat{x}_k^j) \tag{9B.4}$$

$$\tilde{Y}_{k+1}^{j} = (F_k(\hat{Y}_k^j)^{-1} F_k^T + Q_k)^{-1} \tag{9B.5}$$

$$F_k = \frac{\partial f_k(\hat{x}_k^j)}{\partial \hat{x}_k^j} \tag{9B.6}$$

Here, $Y(.) = P^{-1}(.)$, (P is the KF covariance matrix), $Y(.)$ can also be called as information Gramian.

9B.1.2 Measurement Data Update

The information pair (information state y, or i, and the information matrix Y or I as intermediate computations are termed for the sake of convenience) associated with an observation is given by

$$i_k^j = H_k^T R_k^{-1}(e_k^j + H_k \tilde{x}_k^j) \tag{9B.7}$$

$$I_k = H_k^T R_k^{-1} H_k \tag{9B.8}$$

Innovation sequence (in the covariance domain) is

$$e_k^j = z_k - h(\tilde{x}_k^j) \tag{9B.9}$$

$$H_{k+1} = \frac{\partial h_{k+1}(\tilde{x}_{k+1}^j)}{\partial \tilde{x}_{k+1}^j} \tag{9B.10}$$

The final estimated information state (y) and information matrix (Y) are given as

$$\hat{y}_k^j = \tilde{y}_k^j + i_k^j;\ \hat{Y}_k^j = \tilde{Y}_k^j + I_k^j \tag{9B.11}$$

The estimated covariance state x can be calculated using the estimated information state and the information matrix as follows:

$$\hat{x}_k^j = (\hat{Y}_k^j)^{-1} \hat{y}_k^j \tag{9B.12}$$

This state is the one that is normally appreciated in our analysis of the results and even further use in control and forecasting applications in/for real systems and situations; however, the use of information concept is highly motivating and useful in decentralized and distributed sensor data fusion algorithms and applications.

The Gaussian sum estimate

The final weight estimate is given by

$$\hat{w}_{k+1}^{j} = \frac{\hat{w}_{k+1}^{j}(z_{k+1} - h_{k+1}(f_k(\hat{x}_k^j)), S_{k+1}^j)}{\sum_{j=1}^{N} \hat{w}_{k+1}^{j}(z_{k+1} - h_{k+1}(f_k(\hat{x}_k^j)), S_{k+1}^j)} \tag{9B.13}$$

$$S_{k+1}^j = H_k(\tilde{Y}_{k+1}^j)^{-1} H_k^T + R_k \tag{9B.14}$$

The final state estimate at each time step is given as

$$\hat{x}_{k+1} = \sum_{j=1}^{N} \hat{w}_{k+1}^j \hat{x}_{k+1}^j \tag{9B.15}$$

9B.2 Aircraft Parameter Estimation

For the performance evaluation of the GSEIF, the two examples as in Section 9A.4 are taken here with the same simulated conditions.

Example 9B.1

The estimated values of the parameters are compared with the true values as in 2nd column of Table 9B.1. It is seen that the estimates are fairly close to the true values even when there is some noise (SNR = 10) in the measurement data, and the PEEN related to this is 4.12% which is reasonably small. In the case of the GSEIF, the number of Gaussian components required for effective filtering is less, as the initial state is close to the true value used for simulation; only 5 filters have been used for GSEIF. The GSEIF offers better estimation than EIF. When 3 Gaussian components were used, PEEN of GSEIF was 3.23%, though the value is higher than that of with 5 components with PEEN of 1.35%, and the estimation accuracy is still better than EIF. When the number of components was increased to 6, the PEEN of GSEIF remained the same as that obtained using 5 components. This shows that 5 Gaussian components are sufficient for estimation in this case, and increasing the number of Gaussian components beyond 5 will not enhance the performance.

Example 9B.2

Here, 30 Gaussian components have been used for GSEIF. Comparison is done by increasing and decreasing the number of components, and the results are presented in Table 9B.2. As GSEIF uses a number of Gaussian components for estimating, the results are fairly satisfactory and definitely better than those of the EIF.

The comparative CPU time requirements for the examples 9B.1 and 9B.2 are shown in Table 9B.2. The computational time of GSEIF depends upon the number of filters used. As the number of Gaussian components increases, the computational load would also increase.

The computational time increases as the number of components in GSEIF increases. The simulation results obtained prove that GSEIF gives better estimation results than EIF.

TABLE 9B.1
Estimated Parameters for EIF/GSEIF Examples 9B.1 and 9B.2

Parameter	True Values	EIF-Ex1	GSEIF-Ex1	EIF-Ex2	GSEIF-Ex2
C_{X_0}[a]	−0.0540	−0.0617	−0.0556	−0.0733	−0.0547
C_{X_α}	0.2330	0.2557	0.2329	0.3696	0.2300
$C_{X_{\alpha^2}}$	3.6089	3.7092	3.6474	3.4706	3.6288
C_{Z_0}	−0.1200	−0.1239	−0.1209	−0.1404	−0.1216
C_{Z_α}	−5.6800	−5.6988	−5.6594	−5.6538	−5.6389
$C_{Z_{\delta_e}}$	−0.4070	−0.4450	−0.4006	−0.4244	−0.3941
C_{m_0}	0.0550	0.0511	0.0532	0.0901	0.0532
C_{m_α}	−0.7290	−0.7147	−0.7134	−0.9724	−0.7060
$C_{m_{\alpha^2}}$	−1.7150	−1.6432	−1.6828	−0.9506	−1.6644
C_{m_q}	−16.300	−17.036	−16.0190	−13.8587	−15.7331
$C_{m_{\delta_e}}$	−1.9400	−1.8931	−1.8999	−1.7300	−1.8800
PEEN (%)		4.119	1.3517	14.593	3.1053

[a] The aerodynamic derivatives are expressed in the form of the text for convenience.

TABLE 9B.2
Comparison of PEENs and Computational Times

	Number of Filters	PEEN in %		Comp. Times in Seconds	
		EIF	GSEIF	EIF	GSEIF
Example 9B.1	3		3.2256		0.96875
	5	4.119	1.3517	0.3281	1.5156
	6		1.3517		1.875
Example 9B.2	20		4.6785		3.5346
	25	14.593	3.1053	0.3281	3.9844
	30		3.1053		4.3594

APPENDIX 9C APE USING ANNS

The application of artificial neural networks, ANNs, in parameter estimation has been receiving a great deal of attention in the recent days; however, their usage for APE is relatively less.

9C.1 APE WITH FEED-FORWARD NEURAL NETWORKS

Here, we present the APE results obtained by MATLAB implementations with realistically simulated data. The following postulated aircraft model is used to depict aircraft longitudinal motion:

$$\begin{aligned} \dot{u} &= \frac{\bar{q}S}{m}C_X - qw - g\sin\theta \\ \dot{w} &= \frac{\bar{q}S}{mV}C_Z + qu + g\cos\theta \\ \dot{\theta} &= q \\ q &= \frac{\bar{q}S\bar{c}}{I_{yy}}C_m \end{aligned} \tag{9C.1}$$

$$\begin{aligned} C_Z &= C_{Z_0} + C_{Z_\alpha}\alpha + C_{Z_q}\frac{q\bar{c}}{2V} + C_{Z_{\delta_e}}\delta_e \\ C_X &= C_{X_0} + C_{X_\alpha}\alpha + C_{X\alpha^2}\alpha^2 \\ C_m &= C_{m0} + C_{m\alpha}\alpha + C_{m_{\alpha^2}}\alpha^2 + C_{m_q}\frac{q\bar{c}}{2V} + C_{m\delta_e}\delta_e \end{aligned} \tag{9C.2}$$

The unknown parameter vector β that is to be estimated consists of the aerodynamic derivatives:

$$\beta = [C_{X_0}\ C_{X\alpha}\ C_{X_{\alpha^2}}\ C_{Z_0}\ C_{Z\alpha}\ C_{Z_{\delta_e}}\ C_{m0}\ C_{m\alpha}\ C_{m_{\alpha^2}}\ C_{m_q}\ C_{m\delta_e}] \tag{9C.3}$$

To evaluate the efficacy of FFNN-BPN and FFNN-BPN-RLS training algorithms, the short-period data of a LTA are simulated, and the non-dimensional longitudinal parameters are estimated. Using Eqns. (9C.1) and (9C.2), the data are generated with a sampling interval of 0.03 seconds for duration of nearly 7 seconds by giving a doublet input to the elevator control surface. A fourth-order RK integration method is used to obtain the longitudinal flight variables u, w, q, and θ from Eqn. (9C.1).

In order to train the network, simulation is carried out to generate time histories for the variables α, α^2, q, δ_e and coefficients C_x, C_m and C_z. The data α, α^2, q, δ_e are given to the NW as inputs, and C_x, C_m and C_z are presented as outputs. The NW is then trained using the algorithms FFNN-BPN and FFNN-BPN-RLS to obtain the updated weights which can be used to estimate the parameters. Once the network is trained, delta method is used to estimate the parameters.

The parameter estimation results for the examples taken are shown in Table 9C.1. The tuning parameters for the training algorithms are given in Table 9C.2, and the comparison of PEENs and the computational times is depicted in Table 9C.3. In FFNN-RLS, computation of Kalman gains is needed. However, this method which uses an improved form of backpropagation using Kalman gain (and covariance) requires lesser number of training iterations to converge. Hence, the overall computational time is seen to be about 60% lesser in FFNN-BPN-RLS as compared to FFNN-BPN.

TABLE 9C.1
Estimated Parameters for FFNN-BPN/FFNN-RLS

Parameter	True Values	FFNN-BPN	FFNN-BPN-RLS
C_{X_0}[a]	−0.0540	−0.0560	−0.0515
C_{X_α}	0.2330	0.2919	0.2159
$C_{X_{\alpha^2}}$	3.6089	3.5830	3.6416
C_{Z_0}	−0.1200	−0.1226	−0.1209
C_{Z_α}	−5.6800	−5.6799	−5.6687
$C_{Z_{\delta_e}}$	−0.4070	−0.4231	−0.3944
C_{m_0}	0.0550	0.0479	0.0481
C_{m_α}	−0.7290	−0.6803	−0.6819
$C_{m_{\alpha^2}}$	−1.7150	−1.8535	−1.8182
C_{m_q}	−16.300	−16.5652	−16.0477
$C_{m_{\delta_e}}$	−1.9400	−1.9669	−1.9374
PEEN (%)		1.886	1.693

[a] The aerodynamic derivatives are expressed in the form of the text for convenience.

TABLE 9C.2
Tuning Parameters for FFNN-BPN/FFNN-RLS

Tuning Parameters	FFNN-BPN	FFNN-RLS
No. of hidden layers	1	1
No. of nodes in hidden layer	6	6
Sigmoid slope parameter λ_1	0.8	0.8
Sigmoid slope parameter λ_2	0.75	0.75
Learning parameter	0.2	0.2
Momentum parameter	0.4	-
No. of training iterations	1000	200

TABLE 9C.3
Comparison of PEENs and Computational Times

	PEEN in %	Comp. Times in seconds
FFNN-BPN	1.886	3.912173
FFNN-BPN-RLS	1.693	1.129653

9C.2 APE with Recurrent Neural Networks

To estimate the parameters of an aircraft, two approaches of RNN are used: (i) the weight-bias method, and (ii) the gradient-based method, for which the simulated flight data are generated using MATLAB implementations for the same example as in Eqns. (9C.1) and (9C.2). In RNN-WB method, the weights and biases are

TABLE 9C.4
Estimated Parameters for RNN-WB/RNN-Gradient

Parameter	True Values	RNN-WB	RNN-Gradient
C_{X_0}	−0.0540	−0.05399	−0.0560
C_{X_α}	0.2330	0.2330	0.2502
$C_{X_{\alpha^2}}$	3.6089	3.6087	3.5722
C_{Z_0}	−0.1200	−0.1200	−0.1200
C_{Z_α}	−5.6800	−5.6797	−5.6800
$C_{Z_{\delta_e}}$	−0.4070	−0.4068	−0.4070
C_{m_0}	0.0550	0.0619	0.0553
C_{m_α}	−0.7290	−0.7905	−0.7298
$C_{m_{\alpha^2}}$	−1.7150	−1.5754	−1.7114
C_{m_q}	−16.300	−16.1216	−16.2090
$C_{m_{\delta_e}}$	−1.9400	−1.9319	−1.9358
PEEN (%)		1.31	1.142

[a] The aerodynamic derivatives are expressed in the form of the text for convenience.

TABLE 9C.5
Tuning Parameters

Tuning Parameters	RNN-WB	RNN-Gradient
Sigmoid slope parameter λ	0.5	0.5
Learning parameter μ	–	0.01
Sigmoid magnitude parameter ρ	40	1
No. of training iterations	15,000	6000

calculated once in the beginning using the data simulated data. The initial parameter values are then chosen randomly, and iterations are performed to estimate the parameters. The estimated values of the parameters from both methods are compared with the true values as in 2nd column of Table 9C.4. It is seen that the estimates are fairly close to the true values even when there is some noise (SNR = 10) in the data, and the PEEN related to this is 1.31% for RNN-WB with 15,000 iterations and 1.14% for RNN-Gradient with only 6000 iterations. The parameters used for the NWs are given in Table 9C.5. Proper choice of λ and ρ is necessary for good convergence of the estimated values to the true values. The value of λ is chosen less than 1.

The comparative CPU time requirements for the two methods are shown in Table 9C.6. RNN-Gradient method requires lesser number of training iterations compared to RNN-WB method to converge to the true values of the parameters. Hence, the overall computational time is seen to be about 20% lesser in RNN-Gradient as compared to RNN-WB. Though RNN-Gradient method converges faster, for the sake of comparison, both the methods were run for same number of iterations.

TABLE 9C.6
Comparison of PEENs and Computation

	PEEN in %	Comp. Time (s)
RNN-WB	1.31	2.787585
RNN-Gradient	1.142	2.125303

EPILOGUE

More than one dozen parameter estimation methods are treated with several example cases and many exercises in Ref. [1]. Various aspects of SID are treated in Ref. [41]. The maximum-likelihood/output error approach is found to be very popular for APE. In Ref. [42], a new compact formula for computing the uncertainty bounds for the OEM/MLE estimates is given. It utilizes the information matrix, output sensitivity, the covariance of the measurement noise, and the autocorrelation of the residuals. The KF algorithms have wide variety of applications: state estimation, parameter estimation, sensor data fusion, and fault detection. Numerically reliable algorithms are treated in Ref. [8,15]. The approaches for state/parameter estimation using fuzzy logic-based methods and derivative-free KF are relatively new, and these would find increasingly more application for aerospace vehicle data analysis. Also, the APE with artificial neural networks is gaining increasing popularity. The Lyapunov stability method is re-invented for deriving the conditions for asymptotic convergence of several new ramifications of the existing estimation algorithms (Gaussian sum filters) via equivalent and yet simpler observers that use the same gain/covariance/Gramian information (formulae) from these ramified estimators.

EXERCISES

9.1 Why for a given data set a 'long' (in terms of the order of the model) AR model would be required compared to LS or ARMA model for the same data set?
9.2 Why the $C_{m\alpha}$ and C_{n_ϕ} are identical for the case study of the AGARD SDM?
9.3 There are two distinct parts in the expression of FPE (Table 9.1). Explain the significance of each wrt the order 'n'.
9.4 What is the significance of the fudge factor in assessing the uncertainty of the parameter estimates?
9.5 Compare features of method for the determination of neutral and maneuver points by classical and parameter estimation methods.
9.6 Given $y(k) = b_0u(k) + b_1u(k-1) - a_1y(k-1)$, give interpretation of this recursion and obtain TF form and comment.
9.7 Give significance of the elements of the matrix W and vector b in RNN parameter estimation scheme.
9.8 What is the significance of using (residual) error time derivative in the FCV?
9.9 Why the results of FKF with 4 rules are almost similar to those with 49 rules?
9.10 Comparing the concepts of EKF and DFKF, tell what is really emphasized by the differences between the two approaches?
9.11 Can you guess what types of 'errors' can be defined for SID problems before one can specify a cost function for optimization?
9.12 Why the choice of adequate model structure is important in SID problem?

9.13 Compare the Bayes theorem of Appendix B and MLE, then from the probabilistic point of view what is the fundamental difference between the two?

9.14 Generate simulated data using the state and aerodynamic models of the aircraft dynamics (Chapters 3–5):

$$\dot{V} = -\frac{\bar{q}S}{m}C_D - g\sin(\theta - \alpha)$$

$$\dot{\alpha} = -\frac{\bar{q}S}{mV}C_L + \frac{g}{V}\cos(\theta - \alpha) + q$$

$$\dot{q} = \frac{\bar{q}S\bar{c}}{I_y}C_m$$

$$\dot{\theta} = q$$

The aerodynamic model is taken as

$$C_D = C_{D0} + C_{D\alpha}\alpha + C_{D\delta_e}\delta_e$$

$$C_L = C_{L0} + C_{L\alpha}\alpha + C_{L\delta_e}\delta_e$$

$$C_m = C_{m_0} + C_{m\alpha}\alpha + C_{m_q}\frac{q\bar{c}}{2V} + C_{m\delta_e}\delta_e$$

Time histories of variables $\dot{V},\dot{\alpha},\dot{q},\dot{\theta},V,\alpha,q,\theta,\delta_e$, and coefficients C_D,C_L, and C_m should be generated using the given equations with sinusoidal input $\delta_e = A\sin(\theta)$; $A = 1, \theta = 0 : \pi/8 : n\pi$, and $n = 25$. For the simulation, true values of aerodynamic coefficients are given as given later.

Parameters	True Values
C_{D_0}	0.046
C_{D_α}	0.543
$C_{D\delta_e}$	0.138
C_{L_0}	0.403
C_{L_α}	3.057
$C_{L\delta_e}$	1.354
C_{m_0}	0.010
C_{m_α}	−0.119
C_{m_q}	−1.650
$C_{m\delta_e}$	−0.571

The other parameters related to simulated aircraft are $\bar{c} = 10$ m, $S = 23.0$ m^2, $m = 7000$ kg, $I_y = 50000$ kg-m^2, $V = 100$ m/s, $\bar{q} = 5000$ kg/m-s^2, and $g = 9.81$ m/s^2. The initial value of α, q, and θ can be taken as 0.1 rad, 0.0 rad/s, and 0.1 rad, respectively. You can generate 200 data samples by simulation for analysis. For a given set of the true values, do the following:

i. Generate the time histories of variables $\dot{V},\dot{\alpha},\dot{q},\dot{\theta},V,\alpha,q,\theta,\delta_e$, and coefficients C_D,C_L, and C_m with sinusoidal input data and SNR=100 (almost no noise!), using MATLAB.

ii. Use the ANFIS to predict the aerodynamic coefficients. You will have to write the MATLAB code based on the theory of ANFIS discussed in Section 9.9.1.

iii. Use delta method to estimate the aerodynamics derivatives (ref. Section 9.8.1).

If the code is written properly, then you should be able to get the results given in the solution manual.

9.15 What is the fundamental relation between the information matrix and the covariance matrix? What is the significance of this relationship from the accuracy of data/estimates?

9.16 How would you transform the value of $C_{n\beta}$ estimated from flight data to a reference value?

9.17 A parametric relation between the measured data and the parameter to be estimated is given as $z = x\beta + n$. Determine the corresponding covariance relations.

9.18 If we attempt to include $C_{m\dot{\alpha}}$ term in the longitudinal short-period model for identification/estimation purpose, what difficulty is encountered?

9.19 In LAM analysis via data partitioning method, what is the main requirement for length of the data/number of data points from the point of view of consistency of the estimates?

9.20 What is model error?

9.21 What really the Gaussians (probability density) in Kalman and maximum-likelihood represent?

9.22 Kalman filter is obtained by parameterization of a Gaussian distribution. What is this parameterization?

9.23 In essence, the KF is an implementation of the Bayes filter (see App B). Explain this.

9.24 If in a KF, some consecutive measurements are not available, then what would be the status of the state covariance matrix? What is the measure of this growth?

9.25 If a variable with Gaussian PDF is passed thru' a nonlinear function, will it retain its Gaussianness? How would you compute the PDF of the transformed values?

9.26 Although EKF is applicable to nonlinear systems, the PDF involved in EKF is (still) a multivariate Gaussian. Why?

9.27 Will, in principle, the DFKF be more accurate than EKF? If so why?

9.28 Are the sigma points determined stochastically or deterministically?

REFERENCES

1. Raol, J.R., Girija, G., and Jatinder, S. Modelling and Parameter Estimation of Dynamic Systems, IEE Control Engg. Series Book, Vol. No. 65, IEE, London, August 2004.
2. Maybeck, P. S. *Stochastic Models, Estimation and Control*, Vol. 1. Academic Press, New York, 1979.
3. Balakrishna, S., Raol, J. R, and Rajamurthy, M. S. Contributions of congruent pitch motion cue to human activity in manual control. *Automatica*, 19(6), 749–754, 1983.
4. Girija, G., and Raol, J. R. PC based flight path reconstruction using UD factorization filtering algorithms. *Defense Science Journal*, 43, 429–447, 1993.
5. Singh, J., Application of time varying filter to aircraft data in turbulence. *Journal of Institution of Engineers (India)*, Aerospace, AS/1, 80, 7–17, 1999.
6. Kamali, C. Real time parameter estimation techniques for aircraft fault tolerant control systems, Doctoral Thesis, Visvesvaraya Technological University, Belgaum, Karnataka, April, 2007.

7. Parameswaran, V., Jategaonkar, R. V., and Press, M. Five-hole flow angle probe calibration from dynamic and tower flyby maneuvers. *Journal of Aircraft*, 42(1), 80–86, 2005.
8. Raol, J. R., and Girija, G. Sensor data fusion algorithms using square-root information filtering. *IEE Proceedings-Radar, Sonar Navigation*, 149(2), 89–96, 2002.
9. Maine, R. E., and Iliff, K.W. Application of parameter estimation to aircraft stability and control – the output error approach. NASA RP-1168, 1986.
10. Jategaonkar, R. V., Raol, J. R., and Balakrishna, S. Determination of model order for dynamical systems. *IEEE Transactions on Systems, Man and Cybernetics*, SMC-12, 56–62, 1982.
11. Middleton, R. H., and Goodwin, G. C. *Digital Estimation and Control: A Unified Approach*. Prentice Hall, Inc., Hoboken, NJ, 1990.
12. Haest, M., Bastin, G., Gevers, M., and Wertz, V. ESPION: An expert system for system identification. *Automatica*, 26(1), 85–95, 1990.
13. Plaetschke, E., and Schulz, G. Practical input signal design. AGARD LS No. 104, 1979.
14. Gelb, A. (Ed.) *Applied Optimal Estimation*. MIT Press, Cambridge, MA, 1974.
15. Bierman, G. J. *Factorization Methods for Discrete Sequential Estimation*. Academic Press, New York, 1977.
16. Girija, G., Parameswaran, V., Raol, J. R., and Srinathkuamr, S. Estimation of aircraft performance and stability characteristics by dynamic maneuvers using maximum likelihood estimation techniques. *Preprints of 9th IFAC/IFORS Symposium on Identification and System Parameter Estimation*, Budapest, Hungary, July 8–12, 1991.
17. Singh, J., and Raol, J. R. Improved estimation of lateral-directional derivatives of an augmented aircraft using filter error method. *Aeronautical Journal*, 14(1035), 209–214, 2000.
18. Parameswaran, V., Raol, J. R., and Madhuranath P. Mathematical model building from flight data for simulation validation of a modified aircraft. *Journal of Aeronautical Society of India*, 46(4), 186–195, 1994.
19. Srinathkumar, S., Parameswaran, V., and Raol, J. R. Flight test determination of neutral and maneuver point of aircraft. *AIAA Atmospheric Flight Mechanics Conference*, Baltimore, USA, August 7–9, 1995.
20. Saraswathi, L., and Prabhu, M. Estimation of stability and control derivatives from flight data of a trainer aircraft. *Journal of Institution of Engineers*, IE(I)-AS, 79, 15–18, 1998.
21. Shanthakumar, N., and Janardhana, R. Estimation of stability and control derivatives of light canard research aircraft. *Defence Science Journal*, 54(3), 277–292, 2004.
22. Singh, J., and Girija, G. Identification of helicopter rigid body dynamics from flight data. *Defence Science Journal*, 48(1), 69–86, 1998.
23. Raol, J. R., and Girija, G. Estimation of aerodynamic derivatives of projectiles from aeroballistics range data using maximum likelihood method. *Journal of Institution of Engineers*, IE(I) AS, 71, 17–20, 1990.
24. Balakrishna, S., Niranjana, T., Rajamurthy M. S., Srinathkumar, S., Rajan, S. R., and Singh, S. K. Estimation of aerodynamic derivatives using dynamic wind tunnel simulation technique. DLR-Mitt. 93-14, *Proceedings of DLR-NAL System Identification Conference* at NAL, November 24–25, 1993.
25. Iliff, K. W. Maximum likelihood estimates of lift and drag characteristics obtained from dynamic aircraft manoeuvres. *Atmospheric Flight Mechanics Conference*, Arlington, TX, pp. 137–150, 1976.
26. Knaus, A. A technique to determine lift and drag polars in flight. *Journal of Aircraft*, 20(7), 587–592, 1983.
27. Girija, G., Basappa, Raol, J. R., and Madhuranath, P. Evaluation of methods for determination of drag polars unstable/augmented aircraft. *Proceedings of 38th Aerospace Sciences Meeting and Exhibit*, Reno, NV, USA, Paper No. AIAA 2000-0501, January 2000.

28. Parameswaran, V., Girija, G., and Raol, J. R. Estimation of parameters from large amplitude maneuvers with partitioned data for aircraft. *AIAA Guidance, Navigation, and Control Conference and Exhibit*, Austin, Texas, USA. Paper No. AIAA-2003-xxxx, 11–17 August 2003.
29. Batterson, J. G., and Klein, V. Partitioning of flight data for aerodynamic modeling of aircraft at high angle of attack. *AIAA Journal of Aircraft*, 26(2), 334–340, 1989.
30. Basappa, Singh, J, and Raol, J. R. Parameter estimation using large amplitude data from real time simulator of an aircraft. *IEEE/CCECE06 Conference*, Ottawa, Canada, May 7–10, 2006.
31. Morelli, E. A. Global nonlinear aerodynamic modeling using multivariate orthogonal functions. *Journal of Aircraft*, 32(2), 270–277, 1995.
32. Trankle, T. L., and Bachner, S. D. Identification of a full subsonic envelope nonlinear aerodynamic model of the F-14 Aircraft. AIAA Paper 93-3634, 1993.
33. Shaik, I., and Singh, J. Aerodynamic global modeling using multivariate orthogonal functions and splines. *Accepted for Presentation at the 20th National Convention of Aerospace Engineers*, NCASE, Trivandrum, October 29–30, 2006.
34. Klein, V. Determination of airplane model structure from flight data using splines and stepwise regression. NASA TP 2126, 1983.
35. Klein, L. A. *Sensor and Data Fusion: A Tool for Information Assessment and Decision Making*. SPIE Press, Washington, D.C., 2004.
36. Kashyap, S. K., and Raol, J. R. Fuzzy logic applications for filtering and fusion for target tracking. *Defense Science Journal*,58(1), 120–135, 2008.
37. Kashyap, S. K., and Raol, J. R. Unification and interpretation of fuzzy set operations. *Proceedings of CCECE-CCGEI, IEEE Canadian Conference on Electrical and Computer Engineering*, Ottawa, Canada, May 2006.
38. Simon, J. J., and Uhlmann, J. K. Unscented filtering and nonlinear Estimation. *Proceeding of the IEEE*, 92(3), 401–422, 2004.
39. Kashyap, S. K., and Raol, J. R. Evaluation of derivative free Kalman filter and fusion in non linear estimation. *Proceedings of CCECE-CCGEI, IEEE Canadian Conference on Electrical and Computer Engineering*, Ottawa, Canada, May 2006.
40. Sara, M. G., Selvi, S. S., and Raol, J. R. Aircraft parameter estimation using Gaussian sum filter with Lyapunov stability analysis. *Journal of Aerospace Sciences & Technologies*, Bangalore, India, 73(3), 151–162, 2021.
41. Ljung, L. *System Identification: Theory for the User*. Prentice-Hall, Englewood Cliffs, NJ, 1987.
42. Morelli, E. A., and Klein, V. Determining the accuracy of aerodynamic model parameters from flight test data. Paper No. AIAA-95–3499, 1995.

10 Aircraft Handling Qualities Analysis

10.1 INTRODUCTION

The human pilot would be generally very successful in flying an airplane if she/he blends maximum performance and adequate handling qualities (HQs and FQs, flying qualities, for a civilian/transport/passenger aircraft) [1]. For an efficient flight operation, satisfactory HQs are essential; and Cooper and Harper [2] state that the *handling qualities are those characteristics of an aircraft that govern the ease and precision with which a pilot is able to perform the tasks required in support of an aircraft role.* In the opinion of the pilot, the HQs depend on aircraft dynamics, control system performance, cockpit environment, outside view, and instrument display [1]. In early years of HQ research, the pilots' evaluations in various types of existing aircrafts, variable stability research aircraft/in-flight-simulation (VSRA/IFS), and ground-based simulators have had helped in the development of HQ criteria. Certain government agencies would usually and often compulsorily demand compliance with certain HQ requirements for military as well transport aircraft. The purpose of the military requirement is to ensure a certain level of mission performance and also safe operation of a new aircraft, whereas for the civil aircraft, it is the safe operation (rather than mission effectiveness). Modern aircraft development requires comprehensive evaluation of HQs for different controller modes, loadings, and operational missions at various points in the flight envelop; however, the flight testing would be nearly impossible and/or would consume/require lot of time/effort if testing is done at all these conditions; hence, it becomes necessary to supplement the test results obtained by the pilots at several critical conditions, with those obtained from analytical evaluation of the HQs, using mathematical models of the aircraft and the human pilot to describe the performance on computers. The main objective of a good aircraft design and control system is to provide an aircraft-control system with good HQ throughout the flight envelope.

10.2 PILOT OPINION RATING

The pilots are provided with rating scales; while flying various configurations, the pilot would give a rating number on this scale and comments on workload experienced by her. This rating scale is known as Cooper-Harper rating scale (CHRS) [2] and is given in Figure 10.1. The CHRS is also called HQ rating scale (HQRS). The HQ evaluation pilot, after performing a maneuver, arrives at a numerical number of the scale (1–10) based on the series of the decisions made by her for a given flight operation/testing of the aircraft, wherefore the judgments are made in the context of the defined mission. The rating scale is a guide to the quality of the aircraft and its

DOI: 10.1201/9781003293514-11

a/c characteristics	demand on the pilot in required task/operation	PR
Excellent, Highly desirable	pilot compensation is not a factor for desired performance	1
Good, negligible deficiencies	pilot compensation is not a factor for desired perfomance	2
Fair, some mildly unpleasant deficiencies	minimal pilot compensation required for desired performance	3
Minor but annoying deficiencies	desired performance requires moderate pilot compensation	4
Moderately objectionable deficiencies	adequate performance requires considerable pilot compensation	5
Very objectionable but tolerable deficiencies	adequate performance requires extensive pilot compensation	6
Major deficiencies	adequate performance not attainable with maximum tolerable pilot compensation. Controllability is not in question	7
Major deficiencies	considerable pilot compensation is required for control	8
Major deficiencies	intense pilot compensation is reqired to reatain control	9
Major deficiencies	control will be lost during some portion of required operation	10

FIGURE 10.1 Pilot/HQ rating scale [1].

overall performance. Once trained, the pilot can easily use the scale that is not necessarily linear. In the evaluation of the HQ by the test pilot and the aviation authorities, it is also important to have the idea of the pilot's control activity/behavior for operating the aircraft, which can be ascertained by knowing the mathematical model of a human operator (say pilot's) (HOMM).

10.3 HUMAN OPERATOR MODELING

Human pilot while flying and controlling an aeroplane must be using some strategy for this flight operation; hence, the pilot is regarded as one of the controllers in the overall man-machine/aircraft interaction system. In the pilot-aircraft system (PAS) to evaluate the HQs, the HOMM would play an important role. Manual control tasks that a pilot is called upon to perform are compensatory, pursuit, precognitive, and preview. In a compensatory task, only the system error is displayed or presented to the pilot. In a precognitive task, the pilot experiences the pure open-loop programmed control like situation; here, the overall input to the system is a sinusoidal (signal) that is very quickly learned by the pilot.

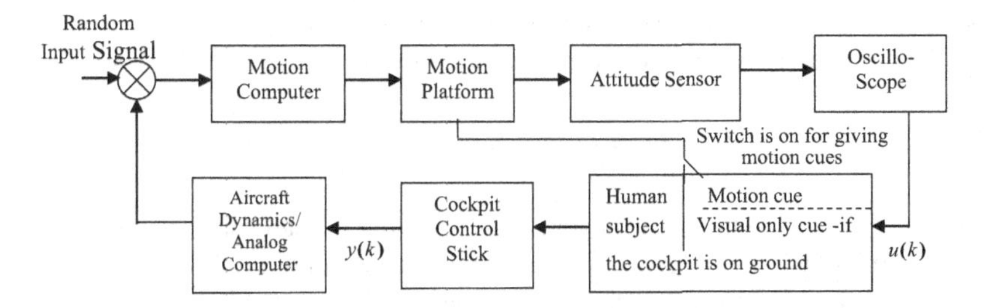

FIGURE 10.2 Manual compensatory tracking experiment in motion-based flight simulator [4].

10.3.1 Motion Plus Visual and Only Visual Cue Experiments

Human operator's control theoretic mathematical models in manual control experiments with compensatory tracking tasks in motion-based flight research simulator were obtained during 1986–1980 at NAL (CSIR). The simulator facility had three degrees of freedom: pitch, roll and heave motion [3,4]. The I/O data were generated while the human operator performed a manual control task, Figure 10.2. The input data to the pilot was in the form of a visual sensory stimulus as derived from the horizontal line on an oscilloscope that represented a tracking error. The actual input signal for the modeling exercise was taken from the equivalent electrical signal to the display device. As such, the tracking input was a band-limited random signal common to both the experiments. The output signal was derived from the movement of the stick (that was fixed with a potentiometer) used by the operator in performing the task. The idea was that first, the operator performs the task while he is in the cockpit that was mounted on the motion-based platform of the simulator. The tracking error was presented to him, and he had to continuously try to keep the error at minimum. When his performance had reached a steady state, the data were recorded and subsequently sampled at 20 samples/s. The training performance criterion was the integrated square error, measured by a device that used analog operational amplifier, for the duration of the task. The human operator as a subject had to perform the same task repeatedly to reach his steady-state performance in the terms of this criterion. Subsequently, for the same task, the cockpit was unfastened from the moving base and put on the ground, and only visual cue was made available to the operator. However, since the motion platform still will be in action, its dynamics are common to both the experiments. The time-series/transfer function (TF) modeling was carried out for two such human operators, while they tracked independently two types of dynamics simulated on an analog computer. Various model selection criteria were used, and 3rd/4th-order least squares (LS) time-series models in discrete-time domain (i.e., pulse transfer functions, 600 data points were used) were found adequate. Then, Bode diagrams were obtained from these pulse TFs, and further analysis and interpretations were carried out. A human operator's mathematical model (in the control theoretic sense) in such a task can be defined by the LS model structure:

$$A(q^{-1})y(k) = B(q^{-1})u(k) + e(k);\; y(k) = \left\{\frac{1}{A(q^{-1})}\right\}\{B(q^{-1})\}u(k) + \frac{1}{A(q^{-1})}e(k) \quad (10.1)$$

As can be seen from eqn. (10.1) that the operator's response naturally gets separated into the numerator and denominator contributions, $(q^{-1}) \rightarrow (z) \rightarrow (j\omega)$, as [3]:

$$H_{sp}(j\omega) = B(j\omega);\ H_{EN}(j\omega) = 1/A(j\omega); \tag{10.2}$$

Here, H_{sp}, the numerator part, is correlated with the human sensory and prediction part and the denominator H_{EN} with the equalizing and the neuromuscular parts. A visual input is considered as relatively unpredictable task, and if the motion cue is added, in addition to the visual cues, it will elicit the lead response from the operator, since the motion cue is a tactile cue and gives an advanced information, like an anticipatory feel, and this can be deciphered with the help of the model of eqn. (10.2). This would show up in the sensory and prediction part H_{sp}. The phase improvement, or phase 'lead', generated by the operator during the congruent motion cues over the visual cues, can be attributed to the functioning of the 'predictor operator' in the human pilot. Thus, a simple LS time-series model can be obtained to isolate the contributions of motion cues, and any other cues from body sensors to have a better understanding of human operator's behavior in manual control tracking tasks.

Example 10.1

From the experiments just described, the discrete time transfer functions (DFTF were obtained for two subjects (human operators) from the time-series data using the LS model. These TF were converted to continuous time transfer function (CTTF) by using complex curve-fitting approach [3]. One pair of these TFs (human operator's model) are given for a subject who performed manual control tracking task with motion and visual cues and only visual cue:

$$\text{CTTF}_{mv} = \frac{1.1(1+0.28s)}{1+\left(\frac{2\times 0.58}{13}\right)s+\left(\frac{s}{13}\right)^2};\ \text{and CTTF}_v = \frac{0.72(1+0.24s)}{1+\left(\frac{2\times 0.45}{9.13}\right)s+\left(\frac{s}{9.13}\right)^2} \tag{10.3}$$

Obtain the Bode diagrams of these control theoretic human operator's models.

Solution 10.1

The TFs are put into simple and standard forms for use in the MATLAB program:

$$\text{CTTF}_{mv} = \frac{52.05(s+3.57)}{s^2+15.08s+169};\ \text{and CTTF}_v = \frac{14.398(s+4.17)}{s^2+8.22s+83.36} \tag{10.4}$$

Then, we have nummv=[52.05 185.82]; denmv=[1 15.08 169]; bode(nummv, denmv); hold; numv=[14.398 60.035]; denv=[1 8.22 83.36]. This will obtain the Bode plots for both the CTTFs, see Figure 10.3. We observe that there is some phase lead due to the presence of the motion cue. We can see that the motion cue elicits the lead response in the human operator. The motion cue is considered as congruent because it is aiding piloting task.

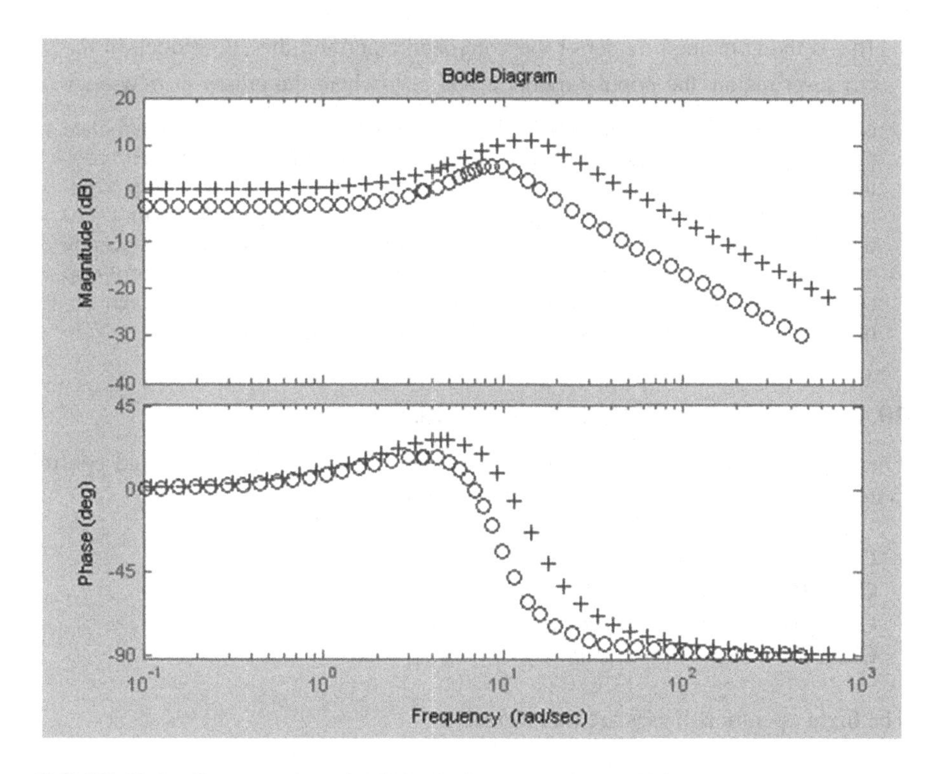

FIGURE 10.3 Human operator's TF Bode diagram (M&V +; V 0).

The well-known human operator/controller models are [1,5]: (i) the crossover model and (ii) the optimal control model. These models are valid under the following conditions: (i) forcing function is a random input signal, (ii) aircraft dynamics are assumed to be linear, and (iii) human operators fully attentive to the compensatory tracking task. Human operator even from control theoretic point of view is a nonlinear complex dynamic system. However, a reasonably good approximation can be made by using quasilinearization looking for an adequate linear correlation between input stimulus and output responses, i.e., neuromuscular. Using this approach, the controller behavior of the human pilot can be represented by: (i) random input quasilinear describing function and (ii) a remnant, the pilot output that is not linearly correlated with the input. A general HOTF model would be:

$$H_p(s) = \frac{K_p(s+1/T_{\text{lead}})e^{-j\omega\tau}}{(s+1/T_{\text{neurom}})(s+1/T_{\text{lag}})} \tag{10.5}$$

It is found that, in general, the HO's frequency response closely resembles an acceptable feedback control system. The crossover model [5] is also very useful and popular:

$$H_pH_c(s) = H_p\left(K, K/s, \frac{K}{s^2}, \frac{K}{s-2}, \ldots, \frac{K}{s(s-1/\tau)}\right) \tag{10.6}$$

This is the combined model of the controlled plant and the HO model. In the crossover region, the popular model is $\frac{1}{s}\omega_c e^{-s\tau}$, where the crossover frequency is 4.7 rad/s for an integrator plant and 3.25 rad/s for K/s^2 plant, and the effective time delay is 0.18 and 0.33 for the respective aircraft dynamics. Thus, the HO uses the simplified adaptation rule: equalization such that the PAS open-loop transfer (describing) function is much greater than 1.0 at low frequencies with a –20 dB/decade slope in the region of crossover frequency. This crossover frequency ranges from 3 to 8 rad/s. Intuitively, the crossover model satisfies the open-loop control system TF: large magnitude at low frequency and small gain at high frequency, and also large phase margin at the crossover frequency.

10.4 HANDLING QUALITIES CRITERIA

The HQC depend on several factors [1]: type of aircraft, flight phase and control system-aircraft characteristics. The aeroplanes are classified as:

Class I *Small, light*
Class II *Medium weight, low-to-medium maneuverability*
Class III *Large, heavy, low-to-medium maneuverability*
Class IV *High maneuverability*

The flight operation phases are categorized as:

Phase A *nonterminal flight that requires rapid maneuvering, precision tracking, or precise flight path control*
Phase B *nonterminal flight that is normally accomplished using gradual, moderate maneuvers and without precision tracking, although accurate flight path control may be required.*
Phase C *terminal flight that is normally accomplished using gradual maneuvers and usually requires accurate flight path control. This comprises of take-off, approach, and landing.*

Flying qualities level of acceptability related to the ability to complete the mission are also predefined as:

Level 1: *flying qualities clearly adequate for the flight phase of the mission*
Level 2: *flying qualities adequate to accomplish the mission flight phase, but there would exist an increase in pilot workload or degradation in mission effectiveness, or both.*
Level 3: *flying qualities such that the aeroplane can be controlled safely, but pilot work load is excessive or mission effectiveness is inadequate or both are possible. Category flight phase A can be terminated safely, and Category flight phase B and C can be completed.*

Primarily, the HQC would be evaluated by using mathematical models of the aeroplane as well as of the pilot-control systems. The quantitative requirements of the specifications are given in terms of parameters of a linear model of the aircraft. In case of control system nonlinearities/higher-order response of the flight control-aircraft combination, an equivalent mathematical model would be defined. For all the types of models, the requirements in terms of frequency, damping ratio, and phase angles would apply. Some of these criteria are described next [6–26]. In fact, Ref. [8] describes several HQ criteria for fixed-wing aircraft and their rationale, algorithms (formulae, etc.), criteria requirements, and guide for application. However, the source [27] described an integrated HQSW package for rotorcraft and fixed-wing aircraft, with more emphasis on the former.

10.4.1 Longitudinal HQ Criteria

Evaluation of aircraft's longitudinal HQ is very important, and there are certain very important and basic aspects of the HQ/FQ requirements (as per MIL-F-8785C specifications) that are very briefly stated here: (i) for Levels 1 and 2, there shall be no tendency for the airspeed to diverge aperiodically when the aircraft is disturbed from the trim the controls fixed and free. This requirement will be considered satisfied if the changes in the pitch control force and position with respect to airspeed are smooth and the local gradients are stable; (ii) when the aircraft is accelerated/decelerated rapidly (through the speed range), the magnitude and the rate of the associated trim change shall not be so great as to cause difficulty in maintaining the desired load factor by the normal piloting technique; (iii) for the Phugoid stability, the damping ratio could be 0.04 for Level 1, 0.0 for Level 2, and for Level 3, the long period time constant at least 55 s; (iv) the flight path angle (FPA) change with respect to change in airspeed shall be –ve or less +ve than: (a) 0.06 deg./knot for Level 1, (b) 0.15 deg./knot for Level 2, and (c) 0.24 deg./knot for Level 3; (v) for Levels 1 and 2, oscillations in the normal acceleration at the pilot's station should be $< +/-0.05\,g$.; and (vi) in a steady turning flight, there should not be objectionable nonlinearity in the variation of the pitch controller force with the steady-state normal acceleration.

More specific HQ criteria are described in next few sections.

10.4.1.1 LOTF (Lower-Order Equivalent TF)

An adequate LOTF is required to assess the longitudinal HQs of an aircraft for SP performance, and this is done in terms of the equivalent parameters [1,9]: $\zeta_{sp}, \omega_{sp}, 1/T_\theta, \tau_\theta$. This is because the aircraft with feedback controllers (called filters with gains and time constants), and other subsystem dynamics becomes very higher-order dynamics system (HODS). Thus, it would be very convenient to reduce this higher-order aircraft/control dynamic system to an equivalent lower-order system and judge the adequacy of the parameters mentioned above in terms of evaluation specifications. The HQ evaluation then becomes easy in similar manner as for the conventional aircraft, more so if the responses are of the conventional type. If the responses are not of the same type as the conventional aircraft, then this approach is not preferable. The LOTFs are obtained in the following forms:

$$\frac{q(s)}{\delta_e(s)} = \frac{K_\theta(s + 1/T_\theta)e^{-\tau_\theta s}}{s^2 + 2\zeta_{sp}\omega_{sp}s + \omega_{sp}^2}; \text{ and } \frac{n_z(s)}{\delta_e(s)} = \frac{K_n e^{-\tau_n s}}{s^2 + 2\zeta_{sp}\omega_{sp}s + \omega_{sp}^2} \tag{10.7}$$

The frequency range for fitting the LOTF to the HODS is taken as 0.1–10 rad/s. The procedure of obtaining the LOTF is explained in Section 10.10. The parameters of the LOTF are better used in conjunction with another criterion known as CAP discussed next.

10.4.1.2 CAP-Control Anticipation Parameter

The idea behind CAP is that the initial pitch attitude response of the aircraft is useful in ascertaining the ultimate response of the flight path [10,11]. The initial pitch acceleration is sensed by the pilot. The load factor parameter being related to the flight path curve is useful in defining this criterion.

CAP1: This criterion is defined as

$\frac{\ddot{\theta}_0}{\Delta n_{zs}}$(rad/s^2/g); an approximation of this is given as (for LOTF) $\frac{\omega_{sp}^2}{n/\alpha}$.

Here, the numerator is the initial pitch acceleration, and the denominator is the steady-state change in the load factor.

CAP2: is defined as $\frac{\ddot{\theta}_{\text{max HODS}}}{n_{zs}}$(rad/s^2/g), based on the maximum pitch acceleration.

CAP3: is defined as $\frac{q_{ss}/t_{ss}}{n_{zs}}$ based on the quasi-steady-state pitch rate.

CAP4: is defined as $\frac{\omega_{sp}^2}{n/\alpha} \cdot \frac{\dot{q}_{\text{max HODS}}}{(\dot{q}_{\text{max LOTF}})_{t=\tau}}$ based on the peak value of the pitch acceleration for step command from the pilot. The CAP4 is approximately equal to the CAP1. The CAP values where applicable are computed from the parameters of the LOTF or the HODS.

The CAP and the equivalent short-period damping ζ_{sp} should be within the HQ-level bounds. Table 10.1 gives the minimum values for the Category C flight phase.

The product $\omega_{sp} T_\theta$ represents the phase lag at ω_{sp}. It is also an equivalent time delay between the pitch attitude response and the flight path response. If the two responses are adequately separated, then the pilot gets a proper cue for controlling the outer slow path. This non-dimensional equivalent parameter should fall within the specified HQ bounds. Other limits are: For category A flight phase→Level 1, the $\omega_{sp} \geq 1.0$, and for Level 2, the $\omega_{sp} \geq 0.6$. For Category C, Table 10.2 gives the HQ specified requirements.

The time delay τ_θ max limits are specified as: for Level 1 as 0.1 s, for Level 2 as 0.2 s, and for Level 3 as 0.25 s. The equivalent time constant is important in the

TABLE 10.1
HQ Specified Minimum Values [8]

Class of the Aircraft	Level 1		Level 2		Level 3	
	ω_{sp}	n_α	ω_{sp}	n_α	ζ_{sp}	T_θ
I,II-C,IV	0.87	2.7	0.6	1.8	0.05	6
II-L,III	0.70	2.0	0.4	1.0	0.05	6

TABLE 10.2
HQ Specified Minimum Values for Time Constant [8]

Level	Class	Minimum ($1/T_\theta$)
1	I,II-C,IV	0.38
	II-L,III	0.28
2	I,II-C,III	0.24
	II-L,III	0.14

attitude dynamics as well as for the flight path. Additionally, the short-period frequency and the acceleration parameters are also evaluated. It is interesting to note here that the CAP1 approximation is related to the static margin. The ω_{sp}, n_α boundaries for Categories of flights and the Levels are also specified in MIL-F-8785C. The pitch acceleration, an important flight mechanics parameter, is related to pitch control derivative as:

$$\ddot{\theta} = \frac{\overline{q}S\overline{c}}{I_{yy}} C_{m\delta_e} (\delta_{e\,\max} - \delta_e) \tag{10.8}$$

At HAOA, a high value of the nose-up pitch acceleration is encountered and adequate control recovery power is required to overcome this pitch acceleration.

10.4.1.3 Bandwidth Criterion [12]

The HQs of an airplane, in a closed loop, can also be determined from its stability margins. The BW is taken as the lesser of the gain (6dB) and phase (45 deg.) margins' frequencies. A phase delay parameter characterizes the phase roll-off, and it is very similar to the equivalent time delay parameter. The phase delay is defined as:

$$T_{ph} = \frac{-(\phi_{2\omega_{180}} + 180)}{57.3 \times 2\omega_{180}} \tag{10.9}$$

Here, ω_{180} is the frequency at which the phase is −180 deg.; $\phi_{2\omega_{180}}$ is the phase at twice the ω_{180} frequency. These parameters are computed from the Bode diagram of the pitch attitude TF. The phase delay v/s BW HQ-level bounds are specified. The BW is also an important control design criterion. Higher BW would be preferable, but it might cause increased 'dropback'. The Unified BW criterion (UBC) considers the flight path dynamics, and it is useful for the landing task [13]. It considers pitch attitude BW, phase delay, flight path BW, pitch attitude dropback, and the pitch rate overshoot. HQ-level boundaries are defined for these parameters. The UBC also checks for the dropback.

10.4.1.4 Neal-Smith Criterion

This criterion is used for the assessment in terms of the pilot workload [14]. The aircraft is flown in closed loop as shown in Figure 10.4. The first block is the pilot

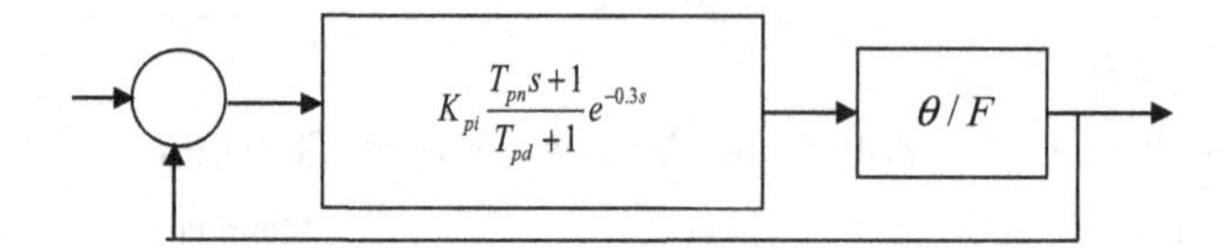

FIGURE 10.4 Pitch compensation pilot/aircraft system.

model. It is assumed that the pilot is performing the closed-loop pitch attitude control task, and hence, the pilot's compensation is a measure of the pilot's workload. The HQ-level boundaries are specified in terms of the closed-loop response (resonance) amplitude and the pilot's compensation. The pilot model parameters are tuned to obtain minimum closed-loop resonance. The droop allowed is up to –3dB. The lead/lag compensation used by pilot (determined from the pilot's control theoretical model) is used in the HQ-level boundaries. The point of this criterion is that it evaluates the closed-loop performance.

10.4.1.5 Closed-Loop Criterion

This criterion uses an additional cascade TF with the pilot's model [15]: $\frac{5s+1}{s}$. This term signifies integration at low frequency. The two forms used are given as:

$$Y_{P1} = K_{pi}\frac{T_{pn}+1}{T_{pd}+1}e^{-0.25s},\ Y_{P2} = K_{pi}\frac{T_{pn}+1}{T_{pd}+1}\frac{(5s+1)}{s}e^{-0.25s} \tag{10.10}$$

There are no HQ boundaries given. The additional term is necessitated due to the use of 3DOF equations for computation. It might be necessary for the pilot to perform this integration to avoid the droop at frequencies lower than the BW frequency. The closed droop should not be more than –3dB for Levels 1 and 2. The closed-loop resonance should not be greater than 3dB for Level 1 and 9dB for Level 2 (over the range 0–10rad/s).

10.4.1.6 Pitch Rate Response

The aircraft/control system's state responses to a unit step can be obtained [16]. Also, its response to turbulence can be obtained, and RMS values can be computed. All the four longitudinal states are plotted. From the pitch rate response, several time domain specifications can be studied. These specifications are standards from the conventional control system requirements: rise time, transient peak ratio, etc. The pitch rate peak ratio is computed as (max value–steady state value)/(steady state value–min value). The peak ratio signified the damping of the system. A tangent line (extending from the steady state-horizontal line to the time axis) is drawn at the point of maximum slope. The effective time delay is computed as: time of the intersection of the tangent at maximum slope with the steady-state line–time of the intersection of the tangent with the time axis. The limits are given in Tables 10.3 and 10.4. The criteria are applicable to conventional and many pitch augmentation systems.

TABLE 10.3
HQ Specified Values for Transient Response Parameters [8]

Level	1	2	3
Peak ratio (max)	0.3	0.6	0.85
Effective time delay	0.12	0.17	0.21
Damping ratio	0.36	0.16	0.052

TABLE 10.4
HQ Specified Values for Rise Time RT[a] [8]

	Terminal Flight Phase		Nonterminal Flight Phase	
Level	Min RT	Max RT	Min RT	Max RT
1	9	200	9	500
2	3.2	645	3.2	1600

[a] All the RT entries should be divided by the true velocity (in ft/s).

10.4.1.7 C* (C-star) Criterion

It is a combination of the pitch rate (at low speeds) and the normal acceleration (at high speeds) responses [17,18]. It is assumed that the pilot responds to a combination of both types of responses. The criterion is used by the control system designers. The C* boundaries are specified for various categories of the flight phases. The C* is given as:

$$C^*(t) = \text{normal acceleration in } g\text{'s} + 12.4q(\text{rad/s})$$

The value of 12.4 is arrived at as crossover velocity (121 m/s) divided by 'g'.

10.4.1.8 Gibson's Criterion

This criterion is based on both time and frequency responses of the aircraft [19,20]. The pitch attitude after following the stick command remains constant when the input is removed. The attitude dropback is defined as the condition where the attitude goes back to its previous value after the stick input is removed. The criterion is applicable for linear systems and mainly to maneuvering type of aircraft, since the criterion does not consider flight path control. Since the criterion is generally used for flight control development and optimization, the HQ-level boundaries are not really defined. The pitch attitude dropback should not be negative for Category A and C flight phases. For Category A, the frequency response specifications are: $\frac{\omega_{\text{sp min}}}{\sqrt{n/\alpha}} = 0.55; \frac{\omega_{\text{sp max}}}{\sqrt{n/\alpha}} = 1.95.$

For approach and landing, some boundaries are specified. The frequency at −120 deg. phase (lag) is required to be within 0.25 and 0.5 Hz. The pitch attitude frequency response should have gain attenuation greater than 0.1 deg./dB at −180 deg. phase. The phase rate should be lower than 100 deg./Hz. Many other aspects of Gibson's criterion are discussed in Refs. [16,17].

10.4.2 Lateral-Directional HQ Criteria

Due to the importance of the flight at HAOA, L-D criteria are also being utilized [8].

10.4.2.1 LOTF (Lower-Order Equivalent TF)

The LOTFs are computed as follows [21,22]:

$$\frac{\phi(s)}{\delta_a(s)} = \frac{Ke^{-\tau s}}{s(s+1/T_R)};\ \frac{\beta(s)}{\delta_r(s)} = \frac{Ke^{-\tau s}}{s^2+2\zeta_d\omega_d s+\omega_d^2};$$

$$\frac{\phi(s)}{\delta_a(s)} = \frac{L_{\delta a}(s^2+2\zeta_\phi\omega_\phi s+\omega_\phi^2)e^{-\tau s}}{(s\pm 1/T_s)(s+1/T_r)(s^2+2\zeta_d\omega_d s+\omega_d^2)}; \tag{10.11}$$

$$\frac{\beta(s)}{\delta_r(s)} = \frac{N_{\delta a}(s+1/T_{\beta 1})(s+1/T_{\beta 2})(s+1/T_{\beta 3})e^{-\tau s}}{(s\pm 1/T_s)(s+1/T_r)(s^2+2\zeta_d\omega_d s+\omega_d^2)}$$

The values of the gains and time delay could be different for each TF. Poles and zeros of the open-loop TF are computed. The roll angle and sideslip TF could be fitted independently to the given HODS frequency responses. From these LOTF models, several modal characteristics are computed and are dealt with in the description of other criteria [6,8].

10.4.2.2 Role Angle/Side Slip Mode Ratio

The ratio is computed at the DR frequency. This modal characteristic helps in deciding the damping factor (damping ratio and frequency product). It signifies the relative roll oscillations wrt the sideslip excursions. This latter aspect is very important in turn coordination.

10.4.2.3 Lateral/Directional Modes

Various such modes are evaluated. These are discussed next.

Roll mode: The requirements are specified in Table 10.5. There is a good correlation between roll mode time constant and the pilot rating.

Spiral stability: The time to double the roll angle should not be less than the specified values (see Table 10.6) following a disturbance up to 20 deg. in roll angle. This would reduce the work load of the pilot.

Coupled bank-spiral oscillations: The aircraft should not have a coupled bank-spiral mode for flight phase of Category A. It should not have coupled bank-spiral oscillation for Category C. For B and C categories, the roll-spiral damping coefficient $\zeta_{rs}\omega_{rs}$ should not be greater than the specified values. The idea is that the roll control effectiveness should not be sacrificed. Also, the combined time delay due to various components in the aircraft-control loop should be limited so that the pilot's tracking is not degraded (Table 10.6).

TABLE 10.5
HQ Specified Values for Max Roll Time Constant [8]

Category of Flight Phase	Class of the Aircraft	Level 1	Level 2	Level 3
A	I,IV	1.0	1.4	
	II,III	1.4	3.0	
B	All	1.4	3.0	10
C	I,II-C,IV	1.0	1.4	
	II-L,III	1.4	3.0	

TABLE 10.6
L-D Modes Specified Values [8]

	Flight Phase Category	Level 1	Level 2	Level 3
Time to double the roll angle should be greater than	A, C	12	8	4
	B	20	8	4
$\zeta_{rs}\omega_{rs}$	B, C	0.5	0.3	0.15
Equivalent roll time delay should be less than	–	0.1	0.2	0.3

TABLE 10.7
DR Mode Parameters Specified Values [8]

Level	Category	Class of Aircraft	DR Damping	Damping* DR Frequency	DR Frequency (rad/s)
1	A	I,IV	0.19	0.35	1.0
		II,III	0.19	0.35	0.4
	B	All	0.08	0.15	0.4
	C	I,II-C,IV	0.08	0.15	1.0
		II-L,III	0.08	0.10	0.4
2	All	All	0.02	0.05	0.4
3	All	All	0	–	0.4

Dynamic L-D response parameters: The minimum required values are specified in Table 10.7. The damping limits the DR oscillations.

10.4.2.4 Roll Rate and Bank Angle Oscillations

The assessment is done in terms of normalized roll rate and bank angle oscillations for a step roll command input [23–26]. The roll rate oscillations should be kept to an absolute minimum. The requirements are based on the linear model of the aircraft. The roll rate and bank angle oscillations boundaries are defined. The roll rate at the first minimum, following a step input, after the first peak should have the same

TABLE 10.8
Time to Attain the Specified Change in Roll Angle for Class I and II Aircraft [8]

Class	Level	Category A		Category B		Category C	
		45 deg.	60 deg.	45 deg.	60 deg.	25 deg.	30 deg.
I	1	–	1.3	–	1.7	–	1.3
	2	–	1.7	–	2.5	–	1.8
	3	–	2.6	–	3.4	–	2.6
II-L	1	1.4	–	1.9	–	–	1.8
	2	1.9	–	2.8	–	–	2.5
	3	2.8	–	3.8	–	–	3.6
II-C	1	1.4	–	1.9	–	1.0	–
	2	1.9	–	2.8	–	1.5	–
	3	2.8	–	3.8	–	2.0	–

polarity. It should not be less than the specified percentage of the roll rate of the first peak: For Level 1 Category A & C 60%, for Category B 25%; and for Level 2 Category A & C 5%, for category B 0%. For bank angle, HQ-level boundaries are specified in Ref. [8].

10.4.2.5 Roll Performance

This is indicated by the time to achieve the specified roll angle change [6,8]. Certain requirements for various classes of aircraft are often specified, see Table 10.8.

The pilot should be able to make certain necessary maneuvers and changes in the roll attitude. As far as possible, the side slip should be zero.

10.4.2.6 Sideslip Excursions

While making turn, coordination between yawing and banking is required in order that the aircraft does not 'skid'. The pilot should be able to make precise changes in the heading. There should be minimum yaw coupling in roll entries/exits. The sideslip excursions should be minimal. The pilot usually performs a coordinated turn. The required yaw control for coordination of turn entries and recoveries should not be objectionable to the pilot. The limits on the amount of the sideslip permitted after a small-step roll command (up to the magnitude that causes a 60 deg. roll angle within 2 s) are specified. The maximum sideslip excursions (during coordination in turn entry/exit) permitted are: 6 deg. for Level 1/Category A; 10 deg. for Level 1/Category B & C; and 15 deg. for Level 2/all categories, for right roll command (this causes adverse beta). For left roll command (which produces the right-side slip/pro verse), the limits are: 2 deg. for Level 1/Category A; 3 deg. for Level 1/Category B & C; and 4 deg. for Level 2/all categories [6,8].

Unit step and pulse responses and the turbulence response of the aircraft are studied as is done in the case of the longitudinal mode. The idea is that these responses should be acceptable from control point of view and the RMS value of the turbulence should be acceptable.

10.5 EVALUATION OF HQ CRITERIA

The HQ criteria are also evaluated by the pilot in a fixed base or motion-based simulators or finally in an actual airplane by performing a flight operation/maneuvers as specified by the design engineer/analyst. The HQ evaluation pilot should have several qualities [1]: (i) motivation, (ii) communication skills, (iii) confidence in her own capability and courage to give a frank opinion of the performance, and (iv) objectivity, piloting experience. Minimum three pilots are required for such exercises. It must be hastened to add that the pilot opinion (rating) evaluation is also a part of this exercise. The HQ criteria for fixed-wing aircraft applicable to fighter aircraft are detailed in Ref. [8].

10.5.1 HQ for Large Transport Aircraft (LTA)

The LTA here is under Class III. It is important to comprehensively study and evaluate the HQ of this class of aircraft. HQ criteria based on the dynamic characteristics of the aircraft only are [1,28]: (i) LOTF of the aircraft, (ii) open-loop time response – pitch rate response criterion, (iii) steady manipulator forces in maneuvering flight, (iv) dynamic manipulator forces in maneuvering flight, (v) compatibility of steady manipulator forces and pitch acceleration sensitivity, (vi) C-star time history envelope, (vii) large supersonic aircraft criterion, (viii) shuttle pitch rate time history envelope, (ix) dropback criterion, and (x) rise time and settling time criterion.

HQ criteria based on PAS are: (i) pilot-aircraft closed-loop dynamic performance, (ii) pilot-aircraft 'inferred' closed-loop dynamic performance, and (iii) modified Neal-Smith criterion. The HQLTASW [28] is evaluated with the NLR's database for the Fokker F28/Mk6000 aircraft. This aircraft is a medium-weight, twin-engined jet transport aircraft. The rate-command/attitude hold control system was selected for pitch and roll control. The longitudinal HQ evaluation for a transport aircraft is primarily important. This is because the main purpose of a transport aircraft is takeoff, cruise, and landing. The LOTF were used for the HQ evaluation studies. The NLR evaluated HQ of this aircraft by using the ground-based and in-flight simulation experiments, as well as various analytical criteria. A typical configuration, E5 [1], has a complete pitch rate TF as:

$$\frac{q(s)}{q_c(s)} = \frac{19.2(s+0.0831)(s+0.706)(s+0.870)}{(s+0.0775)(s+0.838)(s+10)(s^2+2(0.699)(1.2)s+1.44)} \quad (10.12)$$

Some overall results of the HQ evaluation using the HQLTASW and compared with NLR's predictions are shown in Table 10.9 (with entries as the HQ levels) for this configuration.

The HQ levels specified by the CAP criterion v/s short-period damping ratio for the configuration E-5 (and also other configurations studied in Ref. [1]) are depicted in Figure 10.5 [28]. The chosen configuration E5 meets the Level 1 HQ. The HQ-level requirements of short-period frequency and load factor (wrt alpha) are depicted in Figure 10.6. The E5 configuration meets the Level 1 specification as can be seen

TABLE 10.9
Comparison of HQ Evaluation Results for a Large Transport Aircraft-E5 [28]

	BW	N-S	APhR	ω_{sp} v/s n_α	CAP v/s ζ_{sp}	TPR	Closed Loop	Ground Based Sim.(3 Pilots)	IFS (3 Pilots)
NLR	–	1	–	1	1	1	–	1,1,1	1,2,2
HQLTASW (NAL)	2	1	1	1	1	1	1	–	–

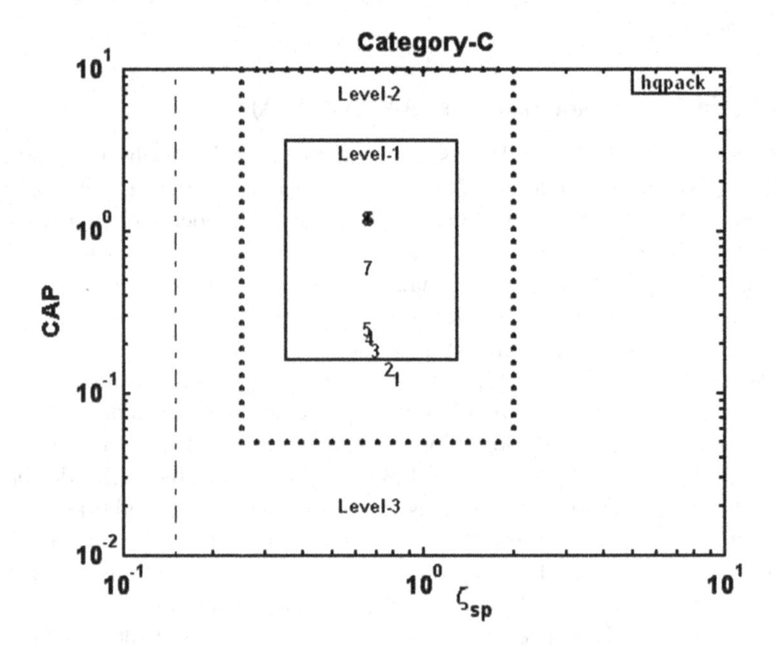

FIGURE 10.5 CAP criterion evaluation for E5 [8].

from the figure. The pitch rate history in the normalized form should lie within boundaries specified in Figure 10.7. The configuration E5 falls within the specified boundaries. It might be necessary to modify certain criteria (like BW/N-S) for their applicability to large transport aircraft.

10.5.2 Rotorcraft Handling Qualities

The Boeing/Sikorsky RAH-66 Comanche military helicopters are designed to meet the HQ specifications of ADS-33D [29]. The ADS-33D (aeronautical design standards) contains the requirements for the flying and ground handling qualities for rotorcraft [30]. These requirements should be complied with, by the process of analysis, simulation, and flight tests at various stages of the design and development of a helicopter. Some of these are: (i) Prior to the critical design review – analytical checks should be computed/made using the available math models, and (ii) prior to first flight –

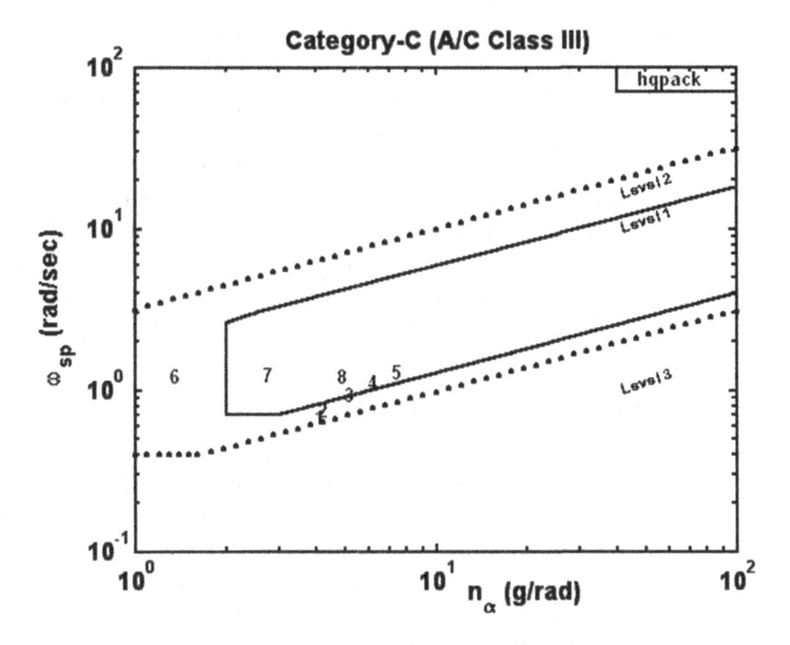

FIGURE 10.6 Short-period frequency v/s N-alpha criterion evaluation for E5 [8].

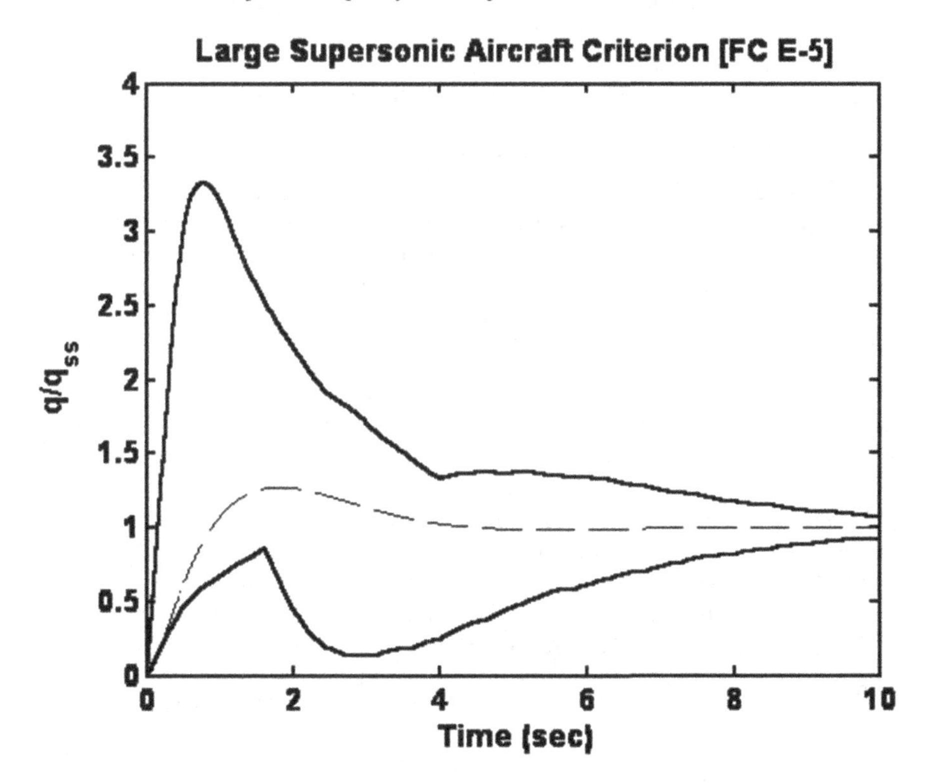

FIGURE 10.7 Pitch rate time history criterion evaluation for E5 [8].

analytical checks should be made using full nonlinear math models, which include the feel system and SCAS elements. Also, the assessment should be done in flight simulators by the test pilots, and (iii) after the first flight – except for the maneuvers which are hazardous or impractical, the flight test verification of all maneuvers should be carried out. The ADS-33D defines the flying qualities in terms of Levels 1, 2, and 3 as well as UCE of 1, 2, and 3 (usable cue environment). The UCE depend on the assessment of the displays and vision aids, which is provided to the pilot. The UCE level is determined from VCR (visual cue rating) scale for at least 3 pilots. The ADS-33D requirements are specified in terms of small-, medium-, and large-amplitude changes. In essence, the ADS-33D HQ assessment relies upon the objective analysis.

10.5.3 Handling Qualities Analysis Tool (HAT)

The toolbox is an integrated MATLAB/GUI-based SW for analytical evaluation of the HQs of aircraft and rotorcraft [27]. The HQ criteria for Military helicopters specified in ADS-33 and the HQ and PIO prediction criteria specified for fixed-wing aircraft in MIL-STD-1797A are covered in this toolbox. The input information could be in the form of time histories, state space models, TF, or frequency response data. It is important that an adequate theoretical/analytical evaluation of HQ is carried out at the design and development stage of an airplane, before thorough evaluation in flight. For this purpose, mathematical models are used and experimental evaluations can be made on flight simulators. Table 10.10 gives a brief overview of the certain possible HQ criteria to be evaluated for a helicopter.

A brief description of some of the tabulated criteria is given next.

10.5.3.1 HLSR-Pitch Axis Response (PAR) Criteria

Bandwidth (BW) criterion: It addresses the small-amplitude and short-term pitch attitude changes to control inputs. The ω_{BW} and phase delay τ_p are determined from the frequency response of the pitch attitude to longitudinal cyclic frequency sweep input. For rate command, ω_{BW} should be lesser of $\omega_{\text{BW}g}$ and $\omega_{\text{BW}p}$. For attitude command, $\omega_{\text{BW}} = \omega_{\text{BW}p}$. Phase delay criterion: $\tau_p = \dfrac{-(\phi_{2\omega_{180}} - 180)}{57.3(2\omega_{180})}$. ω_{180} is the frequency at which the phase angle is -180 deg., $\omega_{\text{BW}g}$ is the frequency at which the gain margin is 6 dB, and $\omega_{\text{BW}p}$ is the frequency at which the phase margin is 45 deg. The above quantities are computed for given test conditions and are placed on the pitch BW HQ-level templates to see which criteria meet the levels of HQ.

Dynamic stability (DySt) criterion: These are the natural frequency and damping ratio of the oscillatory mode especially of Phugoid and Dutch roll. In hover, these modes are coupled. These parameters are determined from the responses and compared against the pitch axis dynamic stability HQ-level boundaries.

Attitude quickness (AQ) criterion: The attitude quickness measures the rotorcraft agility. The metrics are: $\dfrac{q_{\text{peak}}}{\Delta\theta_{\text{peak}}}$ and the minimum attitude change. These parameters are compared on the ADS-33 HQ-level boundaries for the target acquisition and tracking task.

TABLE 10.10
Possible HQ Criteria Matrix for Helicopters

Hover and Low-Speed Requirements (HLSR)							Forward Flight Requirements (FFR)						
Direct Axis (Response Criteria)				Interaxis Coupling			Direct Axis				Interaxis Coupling		
Pitch	Roll	Yaw	Heave	Yaw due to Coll	Pitch due to Roll	Roll due to Pitch	Pitch	Roll	Yaw	Heave	Pitch due to Coll	Pitch due to Roll	Roll due to Pitch
BW	BW	BW	HR	AA	AA	AA	BW	BW	BW	Flight path control			
DySt	DySt	DySt	TR				DySt	DySt	DySt				
AQ	AQ	AQ					Pitch Control power	AQ					
LA	LA	LA						LA	LA				
								SpSt					

Large-amplitude (LA) criterion: ADS-33 specifies minimum achievable angular or attitude change, for example: for moderate agility mission test element, Level 1 limit is +/−13 deg./s., and +20/−30 deg. attitude.

10.5.3.2 HLSR-Roll Axis Response (RAR) Criteria

Bandwidth (BW) criterion: is similar to the PAR BW criterion.

Dynamic stability (DySt) criterion: addresses the oscillatory roll modes following a lateral doublet in hover. These oscillations are dominated by DR mode. The limits on the natural frequency and damping of these oscillations are similar to the pitch axis limits.

Attitude quickness (AQ) criterion: is for the bank angle changes of moderate amplitude between 10 and 60 degs.

Large-amplitude (LA) criterion: It measures the absolute control power in terms of the maximum achievable roll rate (for rate command system). It measures the maximum achievable roll attitude change for attitude command system. A typical limit for moderate agility MTE is +/−50 deg./s for Level 1 HQ.

10.5.3.3 HLSR-Yaw Axis Response (YAR) Criteria

Bandwidth (BW) criterion: is similar to the PAR BW criterion; however, the minimum limits imposed are higher than those for the pitch axis.

Dynamic stability (DySt) criterion: Here, more relaxed limits are on the damping ratio compared to pitch- and roll-axis oscillations. For Level 1, the minimum value of 0.19 is specified for yaw damping ratio for all oscillations with a frequency of more than 0.5 rad/s.

Attitude quickness (AQ) criterion: The parameters are: the heading quickness $\frac{r_{\text{peak}}}{\Delta\psi_{\text{peak}}}$ and the minimum heading change $\Delta\psi_{\min}$ required.

Large-amplitude (LA) criterion: The large-amplitude heading change is for the yaw control power in terms of the maximum achievable yaw rate: the limit for moderate agility MTE is +/−22 deg./s for Level 1 HQ.

10.5.3.4 HLSR-Heave Axis Response (HAR) Criteria

Height response (HR) criterion: is to evaluate the rotorcraft's dynamic behavior following a step collective input in hover. While applying the collective input, the pitch, roll, and heading excursions should be maintained essentially constant. According to ADS-33, the vertical rate response should have a quantitative first-order appearance for at least 5 s. The first-order TF can be fitted to this 5-s response:

$\frac{\dot{h}}{\delta_c} = \frac{Ke^{-\tau s}}{1+Ts}$, where T is the equivalent time constant and τ is the equivalent vertical axis time delay. This criterion can also be evaluated in the frequency domain. For HQ Level 1, the max T is 5 s and max delay is 0.2 s, and for HQ Level 2, the max T is infinity and the delay is 0.3 s.

Torque response (TR) criterion: The torque displayed to the pilot is used as a measure of the maximum allowable power that can be commanded without exceeding the engine or transmission limits.

Some case study results of BW criterion for the BO 105 helicopter data are given in Table 10.11.

TABLE 10.11
BW Criteria Evaluation for BO 105 Helicopter [27]

	Condition	ω_{180}	ω_{BWp}	ω_{BWg}	ω_{BW}	τ_p	HQ Levels		
							Task A	Task F	Task D
Pitch axis	Hover	6.298	2.8	4.12	2.8	0.078	1	1	1
Roll axis	Hover	12.27	8.09	4.87	4.87	0.042	1	1	1
Yaw axis	Hover	2.235	0.246	1.39	0.246	0.074	3	–	–
Pitch axis	Forward flight	5.862	2.56	3.49	2.56	0.081	1	1	1
Roll axis	Forward flight	12.06	5.47	6.56	5.47	0.048	1	1	1
Yaw axis	Forward flight	6.39	3.27	5.09	3.27	0.015	2	–	–

Task A – Target acquisition and tracking; Task B – all mission task elements, UCE=1, and fully attended; Task D – all mission task elements, UCE>1 and/or divided attention.

10.6 HQ ASPECTS FOR UAVs

The unmanned aerial vehicles (UAVs) are finding increasing applications in military and other civilian operations, and the study of their dynamics is gaining importance in recent times. For the remotely piloted vehicles, several aspects are important: remote interaction of the test pilot with the vehicle, ground station interaction/software, flight displays, and data link time delays. The UAVs are used for conducting specialized and complex mission. For the remote test pilot of the UAV, the 'seat-of-the pants' cues are not available. The data link/transmission time delays can have adverse effect on the system's HQ. The small UAVs exhibit higher natural frequencies compared to large aircraft. So even if these UAVs are rated (by test pilots) to have good HQs, they would fall outside the conventionally specified HQ boundaries for short period. A dynamic scaling method can be used [31] to adjust the military standard for short-period mode. One can use the dynamic ('Froude') scaling to provide common ratios between inertia-to-gravity and aerodynamic-to-gravity forces for vehicles with different geometrical dimensions (large- to small-scale vehicles). The following relation is handy, with N as the scaling factor:

$$\omega_{\text{small}} = \sqrt{N}\,\omega_{\text{large}} \tag{10.13}$$

Table 10.12 shows the comparison of a few parameters of Cessna-182 aircraft and the StablEyes UAV. The StablEyes (BYU Captsone 2004) has span of 0.61 m, mac of 0.15 m, cruise velocity of 15 m/s, and the average wing sweep of 8 deg.

Despite the differences in the size and dimension of these two aircraft, there is a good similarity between the non-dimensional derivatives. The Cessna-182 has HQ Level 1. However, the UAV had HQ Level of 2. As rated by the pilot, the UAV had good HQ rating. When the scaling of $N=80$ was used, and the HQ boundaries were adjusted, the UAV fell within the adjusted HQ Level 1 [31].

TABLE 10.12
Comparison of a Few Parameters of a Light Aircraft and a UAV [31]

	Mass (kg)	I_{xx}, I_{yy}, I_{zz} kg m^2	ω_{sp}	ζ_{sp}	ω_{DR}	ζ_{DR}	C_{m_q}	C_{l_β}	C_{n_β}
Cessna-182	1200	700, 1000, 1450	6.2	0.89	2	0.3	−12.4	−0.092	0.0587
StablEyes	0.445	0.002, 0.004, 0.005	14.9	1.0	9	0.16	−15.8	−0.096	0.119

10.7 PILOT-AIRCRAFT INTERACTIONS

Pilot-induced oscillations (PIOs) are oscillations of the PAS occurring inadvertently. For high-performance aircraft with fully powered control systems, PIOs could occur rather frequently [5]. The pilot can be regarded as an adaptive controller (of the vehicle) whose capabilities exceed those of the most sophisticated unmanned control systems. The performance of the pilot and the system can be predicted under certain circumstances. The man-machine system is thus amenable to mathematical analysis. This is also due to the fact that the system adjustments adopted by the pilot for uncoupled multiloop feedback systems are consistent with those of good feedback systems. For more complicated multiloop systems, the pilot model and analysis methods have a continuing practical usefulness. The mathematical models used in PAS analysis could be transfer function type. The inputs and outputs of the nonlinear elements (friction, breakout, nonlinear gearing) of interest in PIO can be approximated by a pair of sine waves. Prior to a PIO, the pilot adopts a quasi-stationary set of feedbacks and equalizations (phase lead/lag, gains) that amount to the performance of a good control system (stability, small error). After a PIO, the airframe motion changes to a nearly sinusoidal motion. For pre-PIO adaptation, the simplified rule is: $|Y_o| = |Y_p Y_c| \cong \left|\frac{\omega_c}{s}\right|$ in the vicinity of the crossover frequency of 1.0–2.0 rad/s [5]. In case of the actual PIO phase, terminal phase of pilot adaptation is synchronous or precognitive behavior, and the pilot essentially duplicates the sinusoidal with no phase lag and $Y_p \cong K_p$. However, in most PIO cases, the pilot has adapted the pre-PIO strategies. Prior to PIO, the pilot uses visual and motion cues to perform the basic flying task. When the PIO limit cycle occurs, the dominant input is the acceleration felt by the pilot. The pilot outputs are the forces and displacements applied to the control stick/pedals. The pilot then has both force and displacement control loops within her neuromuscular actuator system.

While adaptation to some flight operation/phase, something causes a sudden change in the situation dynamics [5]: (i) change in the pilot's organization/adjustment of the system, (ii) initiation of a large steady maneuver, or (iii) a damper failure. Subsequently, the oscillations build up and are sustained for a few cycles – a limit cycle exists. Soon, the pilot realizes this or feels and starts controlling and pumps in the stick commands.

10.7.1 Longitudinal PIO Criteria

Some analytical criteria are used for the prediction of PIO tendencies of an aircraft. Some of these criteria (for the prediction of Category I PIO) from the open literature are given here [32–45].

10.7.1.1 Ralph Smith Criterion

The longitudinal PIO are classified into three types [32–35]: Type I PIO is expected to occur when the pilot 'switches' control tracking pitch attitude to tracking of the pilot-felt normal acceleration. Type II PIO is expected to occur as a result of an abrupt turbulence or non-tracking abrupt maneuver. Type III PIO tendency is caused by the pitch attitude tracking only. The main idea is that the aggressive tracking behavior (of pilot/aircraft) should not result in pilot-aircraft closed-loop instability. The conditions for the R-S criterion (Type I & II) are:

$$\left|\frac{a_{z_p}}{\theta(j\omega_R)}\right| > 0.012 \text{ (in g/deg./s) and } \phi_m = 180 + \phi(j\omega_R) - 14.3\omega_R \le 0 \quad (10.14)$$

Here, $\phi(j\omega_R)$ is the phase angle of the normal acceleration TF $\left(\frac{a_z}{F_s(j\omega_R)}\right)$ and 14.3 ω_R is the phase lag as a result of pilot delay of 0.25 s. The condition for Type III PIO is that the phase angle of pitch attitude TF (wrt force input at the crossover frequency) would be less than −180 deg. The ω_R is the resonance frequency of the PSD of the normal acceleration due to the closed-loop control of pitch angle. In case of Type II PIO, the resonance frequency is from the PSD of the pilot-aircraft open-loop system. The pilot model for the normal acceleration is simple gain with a pure time delay of 0.25 s, and for the theta loop, it is lead/lag model with 0.3-s time delay, the time constants of which are 'adjusted' as per the crossover pilot model. The R-H criterion requires an explicit pilot model.

10.7.1.2 Smith-Geddes Criterion

This criterion is based on phase margin [36,37]. The theta loop should have sufficient phase margin in order to avoid the PIO sensitivity. The pilot model is not required. The flight is PIO sensitive if the phase angle of $\frac{\theta}{F_s(j\omega_c)}$ TF (at the open-loop gain crossover frequency) is < -165 deg. It is PIO prone if the angle is < -180 deg. Also, for the normal acceleration TF, if $\phi_m = 180 + \phi(j\omega_c) - 14.3\omega_c \le 0$, then the flight is PIO prone.

10.7.1.3 Phase Rate Criterion [19,38,39]

The excessive phase lags are encountered in modern high-performance aircraft. The APhR (average phase rate) is the pitch attitude phase per unit frequency, and the range of frequency is up to $2\omega_{180}$. This would capture the higher-order effects. The APhR is computed as:

$$\text{APhR} = \frac{-(180 + \phi_{2\omega_{180}})}{\omega_{180}} \quad (10.15)$$

The aircraft is PIO prone if APhR is greater than 100 deg./Hz. The criterion is applicable to unaugmented/augmented aircraft.

10.7.1.4 Loop Separation Parameter

During landing, the flight path and theta loops could be treated as independent of each other since, during the flare, the pilot would concentrate rather on flight path control [40]. This means that the pilot models could be different for each loop and also the dominant resonance frequencies. If LSP=pilot-aircraft closed-loop system resonance frequency of (pitch attitude – flight path) is less than 2 rad/s, then the aircraft is prone to PIO. Appropriate pilot models are required.

10.7.1.5 N-S Time Domain Criterion

The criterion uses the pilot compensation model defined as [14,41–43]:

$$\frac{\delta_e(s)}{\theta_e(s)} = \frac{400K_p(T_{\text{lead}}s+1)e^{-0.23s}}{s^2+28s+400} \tag{10.16}$$

This model includes the second-order lag filter and equivalent time delay of 300 ms. The pitch attitude control task is monitored for 5 s. The theta-error (of the closed-loop control) rms value is minimized. The rms value of the second derivative of the theta-error is computed. If this is greater than 100, the prediction is that the aircraft is PIO prone.

10.7.1.6 Bandwidth PIO Criterion

The criterion is specified in terms of the PIO rating boundaries in terms of BW and phase delay of the theta TF [34,44,45]. Unified BW criterion uses the flight path BW and the time domain theta dropback also. The parameters are obtained from the Bode diagrams of the theta, pitch rate, and the flight path. Also, step response of the pitch attitude is used. The unified BW PIO criterion boundaries are defined. Table 10.13 specifies the PIO conditions.

10.7.1.7 Lateral PIO Criteria

Due to the augmentation of the roll control, some low-frequency oscillations in roll could be possible. This could be mainly due to coupled roll/spiral modes and control surface rate saturation. The too high or too low roll response could lead to lateral PIO tendencies.

10.7.1.8 Ralph-Smith

It can be used for the prediction of roll attitude PIO [33]. If the phase of the roll angle to roll control force TF (at the crossover frequency) is less than −180, then the aircraft is prone to roll PIO.

10.7.1.9 Phase Rate

The criterion is used as a guide in control law design [46]. The requirement is the same as for the APhR for the longitudinal PIO.

TABLE 10.13
PIO Conditions for BW Criteria

Condition	Remarks
Theta BW > 1 and Phase delay < 0.14 (s)	Not susceptible to PIO
Theta BW > 1, Phase delay < 0.14 and pitch rate overshoot (in frequency domain) > 9 dB	Possibility of the pitch bobble
Theta BW < 1 and pitch overshoot > 12 dB	PIO susceptibility
If Phase delay ≥ 0.14	Fighter aircraft are always PIO susceptible
If Phase delay ≥ 0.14 and flight path BW < 0.55 rad/s;	Transport aircraft susceptible
If Phase delay ≥ 0.2	Always susceptible to PIO

10.8 MODEL ORDER REDUCTION FOR EVALUATIONS OF HQ

The inherently unstable/augmented aircraft systems are often highly augmented leading to very high-order dynamic systems (HOTF) [47]. Mathematical description of aircraft and flight control system includes the rigid and elastic body dynamics, actuator model, sensor models, and flight control filters, and this would result in models of very high order. By LOTF (lower-order T.F.) approximation, one can express the characteristics of an aircraft/control system in terms of flight mechanics models in order to appropriately evaluate the HQs. For this model, order reduction techniques for the generation of equivalent LOTF are required. Generation of LOTF for analytical evaluation of HQs is a very important part of the overall FCS development process for a modern high-performance aircraft. The direct methods for evaluating HQs are piloted flight simulation and flight testing. The HQ criteria are often specified in terms of 2nd-order T.F. constants with some additional parameters, like time delay. Generation of LOTF having a specified structure and order becomes important for highly augmented aircraft. Due to increased control system integration, sophistication, and coupling, it becomes difficult to compute LOTF as additional modes in the region of pilot crossover are introduced. There are several methods for obtaining LOTF models for HQ evaluation: (i) frequency response matching (amplitude and phase) over a selected range using numerical search routines, (ii) system identification methods based on I/O data collected from flight tests and piloted simulation, and (iii) balanced model reduction technique based on model decomposition of the HOTF models into high, mid, and low bands and Hankel reduction of each of the 3 subsystems (6–8). In case of frequency response matching technique, the cost normally used is:

$$J = \frac{1}{N}\sum_{j=1}^{N}(\Delta G_i^2 + W\Delta P_j^2) \tag{10.17}$$

Here, N = number of discrete frequencies ω_j, ΔG is the amplitude difference, in dB, between the actual (HOTF data) and (LOTF) model-predicted response, and ΔP is the phase difference in degrees, and W is the weighting factor, usually chosen as

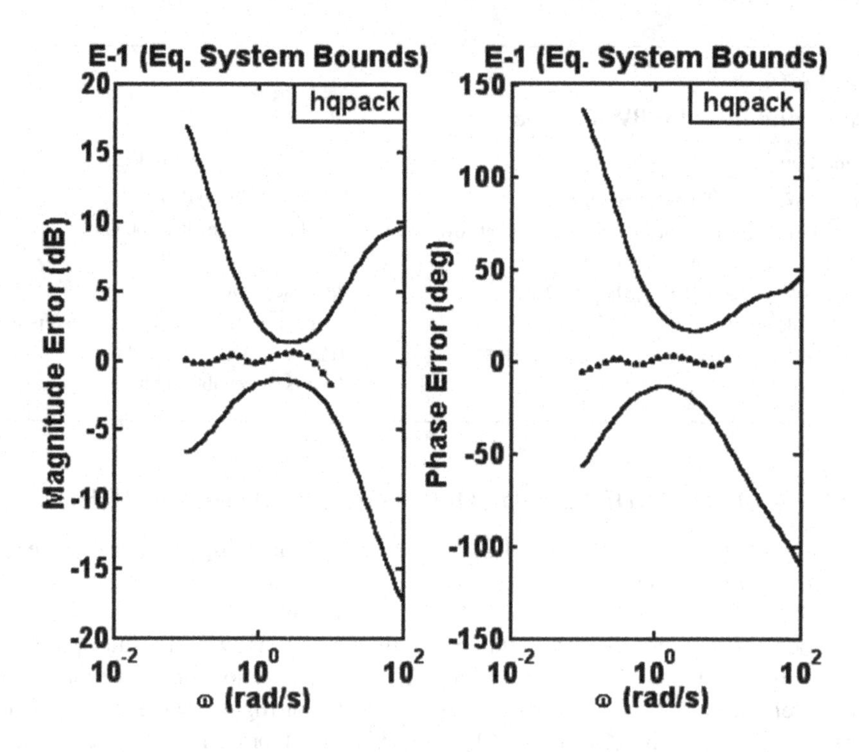

FIGURE 10.8 LOTF estimation error bounds a configuration [8].

0.018. The frequency range should be chosen to include the essential modes of the aircraft, the usual choice being 0.1–10 rad/s. For the frequency range, 20 points per decade can be used for model fitting. The structure of the LOTF model should be akin to the aircraft modes to be evaluated for HQ. Figure 10.8 shows the error bounds obtained for fitting a LOTF to a HODS of a transport aircraft.

EPILOGUE

Only a brief introduction of HQ analysis and criteria has been given here. The subject is actually very vast and would need a separate and special treatment, earliest being [7]. Also, the HQ requirements and PAIs for UAVs and other special types of aircraft and even new flight phases, like coordinated operation of multiple UAVs flown by a remote pilot, are current topics of research and investigations. A historical perspective and various aspects of pilot evaluation and handling qualities are highlighted in 1984-Wright Brothers Lectureship in Aeronautics article in Refs. [47], and [48]. Human pilot modeling using the model error approach is dealt with in [49]. Due to the present-day FBW control technology adopted in many aircraft systems, the issues of PIO/PAI are significantly reduced, since mainly the onboard computer controls the aircraft and its missions rather than the pilot who is acting in a supervisory or monitoring mode most of the time.

EXERCISES

10.1 Show how the CAP1 approximation is related to the aircraft's static margin.
10.2 From control/system point of view, what does a lower limit on ω_{sp} signify?
10.3 What do lower limit and upper limit on short-period damping ratios signify?
10.4 Obtain the step responses for the human operators' models of Example 10.1.

REFERENCES

1. Mooij, H. A. *Criteria for Low Speed Longitudinal Handling Qualities (of Transport Aircraft with Closed-Loop Flight Control Systems).* Martinus Nijhoff Publishers, the Netherlands, 1984.
2. Cooper, G. E., and Harper, R. P. The use of pilot rating in the evaluation of aircraft handling qualities. AGARD-R-567, 1969 and NASA TN D-5153, April 1969.
3. Balakrishna, S., Raol, J. R., and Rajamurthy, M. S. Contributions of congruent pitch motion cue to human activity in manual control. *Automatica*, 19(6), 749–754, 1983.
4. Raol, J. R., Girija, G., and Jatinder, S., *Modelling and Parameter Estimation of Dynamic Systems*, IEE Control Engg. Series Book, Vol. No. 65, IEE, London, August 2004.
5. Ashkenas, I. L., Jex, H. R., and McRuer, D.T. Pilot-induced oscillations: Their cause and analysis. Norair Report NOR-64-143; Systems Technology, Inc., report STI TR-239-2, Hawthorne, California, USA, 1964.
6. Anon. Flying qualities of piloted aircraft. USAF Military Standard MIL-STD-1797A. January 1990.
7. Hodgkinson, J. Aircraft handling qualities. AIAA Education Series, Virginia 1999.
8. Shaik, I., and Chetty, S. HQPACK User's Guide, Vol. 1, 2. PC based SW package in MATLAB ver 5.2 for the prediction of handling qualities and pilot induced oscillation tendencies of aircraft. NAL Project document, PD-FC-9810, December 1998.
9. Fuller, S. G., and Potts, D. W. (eds.) Design criteria for the future of flight controls. AFWAL-TR-82-3064, July 1982.
10. Bhirle, W. A handling qualities theory for precise flight path control. AFFDL-TR-65-198, June 1966.
11. Ashkenas, I. L. Summary and interpretation of recent longitudinal flying qualities results. AIAA Conference Paper 69-898, August 1969.
12. Hoh, R. H., Mitchel, D. G., and Hodgkinson, J. Bandwidth- a criterion for highly augmented aircraft. AGARD-CP-333, June 1982.
13. Mitchel, D. G., Hoh, R. H., Aponso, B. L., and Klyde, D. H. Proposed incorporation of mission oriented requirements into MIL-STD-1797A. WL-TR-94-3162, October 1994.
14. Neal, T. P., and Smith, R. E. An in-flight investigation to develop control system design criteria for fighter airplane. AFFDL-TR-70-74, I & II, December 1974.
15. Crombie, R. B., and Moorhouse, D. J. Flying qualities design criteria. *Proceedings of AFFDL Flying Qualities Symposium*, Wright-Patterson Airforce Base, Ohio, October 1979. AFWAL-TR-80-3067, May 1980.
16. Chalk, C. R. Calspan recommendations for SCR flying qualities design criteria. NASA-CR-159236, April 1980.
17. Tobie, H. N. A new longitudinal handling qualities criterion. The Boeing Company document, 1964.
18. Smith, R. E. Effects of control system dynamics of fighter approach and landing longitudinal flying qualities. AFFDL-TR-78-122, March 1978.
19. Gibson, J. C. Piloted handling qualities design criteria for high order flight control systems. AGARD-CP-333, June 1982.
20. Gibson, J. C. Handling qualities of unstable aircraft. ICAS-86-5.3.4, 1986.

21. Bishoff, D. E., and Palmer, R. E. Investigation of lower order lateral-directional transfer function models for augmented aircraft. AIAA-CP-82-1610, 1982.
22. Palmer, R. E., and Bischoff, D. E. Experience in determining lateral-directional equivalent system models. *Presented at 'Design Criteria for the Future of Flight Controls' Workshop*, Wright-Patterson Air Force Base, Ohio, 2–5 March 1982.
23. Hall, G. W., and Boothe, E. M. An in-flight investigation of lateral-directional dynamics for the landing approach. AFFDL-TR-70-145, October 1971.
24. Wasserman, R., Kokhart, F. F., and Ledder, H. J. In-flight investigation of an unaugmented class III airplane in hte landing approach task. Phase I: Lateral-directional study. AFFDL-TR-71-164, 1, January 1972.
25. Chalk, C. R., Neal, T. P., et al. Background information and user's guide for MIL-F-8785B (ASG). Military specification – flying qualities of piloted airplanes. AFFDL-TR-69-72, August 1969.
26. Chalk, C. R., DiFranco, D. A., et al. Revisions to MIL-F-8785B(ASG) proposed by Cornell Aeronautical laboratory (Contract F33615-71-C-1254). AFFDL-TR-72-41, April 1973.
27. Shaik, I., Wolfgang, V. G., and Hamers, M. HAT- A handling qualities analysis tool box. NAL Project document FC 0210, Nov. 2002 (also as IB 111-2002/27, DLR).
28. Shaik, I. Evaluation of NLR handling qualities database for a large transport aircraft using HQPACK. Project Document FC 0519, National Aerospace Laboratories, Bangalore, December 2005.
29. Shanthakumaran, P. An industry perspective of future helicopter handling qualities. *Proceedings of International Seminar on Futuristic Aircraft Technologies*, (DRDO/Aero. Soc. of India) 3–5 December 1996.
30. Ockier, C. J., Evaluation of the ADS-33D handling qualities criteria using the BO 105 helicopter. FB 98-07, DLR, Germany, January 1998.
31. Foster, T. M., and Bowman, W. J. Dynamic stability and handling qualities of small unmanned-aerial-vehicles. *43rd AIAA Aerospace Sciences Meeting and Exhibit*, Reno, Nevada, 10–13 January 2005.
32. Smith, R. H. A theory for longitudinal short period pilot induced oscillations. AFFDL-TR-77-57, June 1977.
33. Smith, R. H. Notes on lateral-directional pilot induced oscillations. AFWAL-TR-81-3090, May 1981.
34. Bjorkman, E. A. Flight test evaluation of techniques to predict longitudinal pilot induced oscillations. AFIT thesis, AFIT/GAF/AA/86-J-1, December 1986.
35. Radford, R. C., Smith, R. E., and Bailey, R. E. Landing flying qualities evaluation criteria for augmented aircraft. NASA-CR-163097.
36. Smith, R. H., and Geddes, N. D. Handling qualities requirements for advanced aircraft design criteria for fighter airplanes. AFFDL-TR-78-154, August 1979.
37. Smith, R. H. The Smith-Geddes criteria. *Presented at the SAE Aerospace Control and Guidance Systems Committee Meeting*, Reno, NV, March 1993.
38. Gibson, J. C. The prevention of PIO by design. AGARD-CP-560, January 1995.
39. Gibson, J. C. Looking for the simple PIO model. AGARD-CP-335. February 1995.
40. Martz, J. J., Biezad, D. J., DiDomenico, E. D. Loop separation parameter: A new metric for landing flying qualities. *Journal of Guidance and Control*, 11(6), 535–541, 1988.
41. Onstott, E. D., and Faulker, W. H. Prediction, evaluation and specification of closed loop multi-axis flying qualities. AFFDL-TR-78-3, February 1978.
42. Bailey, R. E., and Bidlack, T. J. Unified pilot induced oscillation theory Vol. IV: Time domain Neal-Smith criterion. WL-TR-96-3031. Vol. 4, December 1995.
43. Bailey, R. E., and Bidlack, T. J. A quantitative criterion for pilot induced oscillations: Time domain Neal-Smith criterion. AIAA Conference Paper AIAA-96-3434.

44. Klyde, D. H., Mcruer, D. T., and Myers, T. T. Unified pilot induced oscillation theory. WL-TR-96-3028, December 1995.
45. Mitchell, D. G., and Hoh, R. H. Development of a unified method to predict PIO. AIAA Conference Paper AIAA-96-3435-CP, 1996.
46. Gibson, J. C. Prevention of PIO by design. AGARD-CP-560, January 1995.
47. Chetty, S., and Marchand, M. Frequency domain identification of lower order equivalent system. In *System Identification;* edrs. Srinathkumar, S., and Doherr, K.-F.; DLR (German Aerospace Research), Germany, Report DLR-mitt., 93–14, November–December 1993.
48. Harper, R. P., and Cooper, G. E. Handling qualities and pilot evaluation. *Journal of Guidance*, 9(5), 515–528, 1986.
49. Mook, D. J., and Bailey, R. E. Pilot control identification using minimum model error estimation. *AIAA Atmospheric Flight Mechanics Conference*, Hilton Head Island. AIAA Paper 92-4421, August 10–12, 1992.

11 Aeroservoelastic Concepts

11.1 INTRODUCTION

Aeroservoelasticity (ASE) is defined by the interplay between the airframe structure, aircraft control system, and aerodynamic forces acting on the aircraft. These interactions are of major concern to designers and quite complex to model. The reason why designers are so concerned about ASE is the adverse effect it can have on aircraft stability and control characteristics. Significant modifications to the control law algorithms may be required to make the aircraft flyable in its operational flight envelope.

Modern fighter aircraft today are designed to be lightweight with a highly flexible structure to make them super maneuverable. The use of active control technology has proved beneficial in such cases to achieve the aircraft desired performance and mission capability.

11.1.1 MODELING PROCEDURES

A lot of effort has gone into modeling the unsteady aerodynamics resulting from aircraft motion to given inputs or disturbances. The majority of analysts make use of frequency domain analysis to predict and validate unsteady aerodynamic effects. The recent trend, however, has been to use a linear, time-invariant state–space form to model ASE effects [1].

11.1.1.1 Minimum State for Approximating Unsteady Aerodynamics

Software codes and tools have been developed to determine rational functions that can be used to transform aeroelastic equations from the frequency domain to the time domain. The flip side of using rational functions is the significant increase in the dimension of the state vector. Among the various methods used to determine rational functions, the Minimum-State method provides the lowest-order model to solve ASE equations [2]. The accuracy of results from these models can be improved by judiciously weighing the significant aerodynamic terms and imposing appropriate constraints. Such techniques not only improve the fidelity of the model but also help reduce the computational runtime.

11.1.1.2 Unsteady Aerodynamic Corrections Factor Methodology

Recent years have witnessed a quantum jump in the use of computational fluid dynamic (CFD) codes for predicting aircraft unsteady aerodynamics and conducting aeroelastic studies. Despite the boom in digital computing and high-end workstations, the large computational requirements of CFD desist its use for ASE on a routine basis. An alternative method to predict aerodynamic forces with reasonable accuracy, at least in the subsonic and supersonic flight regimes, is to use linear

DOI: 10.1201/9781003293514-12

theory, and a more reasonable and less-expensive approach to solve unsteady aerodynamic problems is to use a linearized lifting surface approach based on small disturbance theory. Correction factors derived from experiments and CFD can be applied to correct the magnitude of computed pressure, thereby improving the overall accuracy of ASE predictions. The correction factors can be obtained by either of these three methods by (i) matching the pressure values derived from CFD with analytically computed values, (ii) matching one or more airfoil section properties, and (iii) matching the total forces and moments.

11.1.2 Analysis Methods

Flight through atmospheric gusts can lead to increased structural loads and deformation that can be detrimental to aircraft performance. Certification for gust loads is, therefore, a very critical aspect of airworthiness. Power spectral density methods are often used to determine the aircraft response to gusts. An alternative method is to use matched filter theory that can provide maximized and time-correlated gust loads.

11.1.2.1 Matched Filter Theory (MFT)

In MFT, to analytically compute the maximum dynamic response in the aircraft structure to input gust, an impulse is first used to excite the system and obtain output load responses. One of the selected output responses is then normalized and used as input to the known dynamic model to get the final time-correlated responses and the maximum dynamic response of the system. MFT is a fast and efficient tool to estimate gust loads. This can help aircraft designers to ensure that there are no structural failures caused by critical gust loads.

11.1.3 Synthesis Methodology

One can use also some synthesis methods.

11.1.3.1 Integrated Structure Control Law Design Methodology

A subsystem, despite being optimally designed to give a good performance, can lead to an overall unsatisfactory performance of the aircraft when it starts to dynamically interact with other subsystems. For example, adverse dynamic coupling between the flight control system and aeroelastic characteristics has been observed in the past for aircraft like F16 and F18. Integrated structural control law design methodology, within the framework of multidisciplinary design and optimization approach, has been developed and used to improve the overall system performance.

11.1.3.2 Design Using Constrained Optimization with Singular Value Constraints

Integrating aeroelastic equations into the mathematical formulation for robust control law design will generally lead to state–space model of a high order. Balanced truncation and residualization techniques are often used to reduce the math model of the controller to a lower order. Singular value analysis is applied to determine the significant states to be retained in the lower-order model. Constrained optimization can be used to ensure that the controller robustness is not compromised. Factors like control surface deflection and rate limits can be used to ensure robustness.

11.1.4 Validation of Methods through Experimentation

Although enhanced modeling and analytical procedures have enabled a more accurate prediction of ASE characteristics, validation through experimentation is important before the results can be used for aircraft performance evaluation. Results from the wind tunnel testing of actively controlled aeroelastic wind tunnel models can be used to validate the predictions from ASE design methodologies. Such experimental testing can provide useful insight into flutter suppression and rolling maneuver load alleviation.

11.2 FLIGHT DYNAMICS OF A FLEXIBLE AIRCRAFT

Here, we consider some aspects of flight dynamics pertaining to flexible aircraft [3]. Recent years have witnessed an increased use of Multidisciplinary Optimization (MDO) in aircraft design and development. Several advanced analysis codes are used to achieve high levels of fidelity in design. MDO techniques allow the designers to address the strong coupling between aerodynamics and aircraft structures while attempting to reduce the overall weight and minimize drag. The objective, of course, is to achieve optimal configurations that enhance the aircraft's flying performance.

The traditional approach considers the conceptual aircraft configuration design, weight optimization, structural analysis, and control design in a sequential manner. This approach will not consider the influence of control design on aircraft configuration. Studies have shown that ASE should be considered at the early design stage. One way to minimize the structural weight is to design the control system and structure simultaneously. Including aerodynamics in this multidisciplinary approach will further help in designing an aircraft with better performance.

11.2.1 ASE Formulation

Since an integrated approach is seen to be more effective than considering the control design and structural analysis separately, numerical models for elastic aircraft are required for analysis and control design. Such models include nonlinear EOMs for flexible aircraft. These formulations either use the mean axis method or the quasi-coordinate method. By adopting the mean axis body reference frame, one can avoid rigid elastic coupling terms from appearing in the equations of motion.

11.2.1.1 Kinetic Energy

Figure 11.1 shows the inertial frame and body reference frame. The absolute position and velocity of an infinitesimal mass element of the deformed wing are given by

$$\vec{R} = \vec{R}_0 + \left(\vec{r} + \vec{d}_E\right) = \vec{R}_0 + \vec{p} \tag{11.1}$$

$$\dot{\vec{R}} = \dot{\vec{R}}_0 + \frac{\delta p}{\delta t} + \vec{\omega} \times \vec{p} \tag{11.2}$$

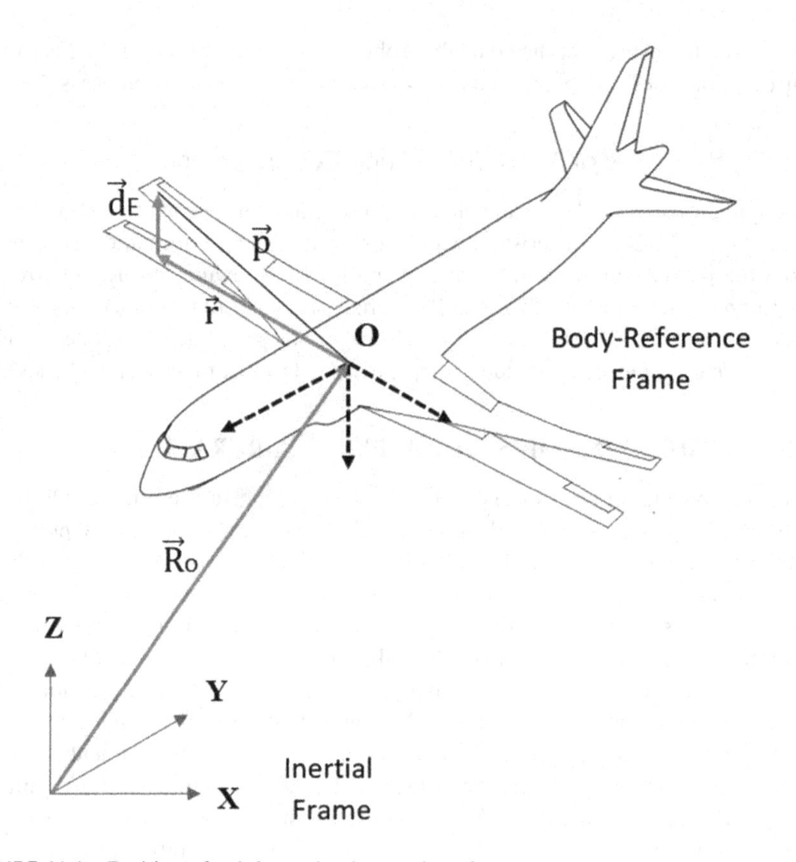

FIGURE 11.1 Position of a deformed point on aircraft.

where $\vec{R}_0$ is the inertial position of the origin of body reference coordinate frame, $\vec{d}_E$ is the total displacement or elastic deformation of an arbitrary point on the aircraft, $\vec{r}$ is the position of the point relative to the aircraft C.G., and $\vec{\omega}$ is the angular velocity at which the local body reference frame rotates relative to the inertial frame.

The kinetic energy of deformable aircraft can be expressed as:

$$K_E = \frac{1}{2}\int_V \frac{d\vec{R}}{dt}\frac{d\vec{R}}{dt}\rho\, dV \tag{11.3}$$

Applying mean axis constraints and locating the origin of the body reference axes at the C.G. eqn. (11.2) can be substituted into eqn. (11.3) to express the kinetic energy of aircraft in the following form.

$$K_E = \frac{1}{2}\int_V \left[\left(\frac{d\vec{R}_0}{dt}\cdot\frac{d\vec{R}_0}{dt}\right) + \left(\frac{\delta\vec{p}}{\delta t}\cdot\frac{\delta\vec{p}}{\delta t}\right) + (\vec{\omega}\times\vec{p})\cdot(\vec{\omega}\times\vec{p})\right]\rho\, dV \tag{11.4}$$

where the mean axis constraints define the relative linear and angular momenta to be zero, i.e.,

$$\int_V \frac{\delta \vec{p}}{\delta t} \rho dV = 0 \tag{11.5}$$

$$\int_V \vec{p} \times \frac{\delta \vec{p}}{\delta t} \rho dV = 0 \tag{11.6}$$

The kinetic energy in eqn. (11.4) can equivalently be expressed as in eqn. (11.7).

$$K_E = \frac{1}{2} M \left(\frac{d\vec{R}_0}{dt} \cdot \frac{d\vec{R}_0}{dt} \right) + \frac{1}{2} \int_V \frac{\delta \vec{p}}{\delta t} \frac{\delta \vec{p}}{\delta t} \rho dV + \frac{1}{2} \vec{\omega}^T [J] \vec{\omega} \tag{11.7}$$

where $[J]$ is the inertia tensor and M the mass of the aircraft. The inertia tensor can be assumed to be constant for small displacements.

For a body undergoing elastic deformation, $\vec{d}_E$ can be expressed in terms of the generalized displacement coordinates $\eta_i(t)$ and the vibration mode shapes Φ_i, where index i denotes the ith vibration mode.

$$\vec{d}_E(\vec{r}, t) = \sum_{i=1}^{N} \vec{\Phi}_i(\vec{r}, t)\, \eta_i(t) \tag{11.8}$$

The generalized mass of the ith mode is defined by

$$M_i = \int_V \vec{\Phi}_i \vec{\Phi}_i \rho dV \tag{11.9}$$

Due to orthogonality of the free vibration modes, we have

$$\int_V \vec{\Phi}_i \vec{\Phi}_j \rho dV \equiv 0, i \neq j \tag{11.10}$$

Further, since $\vec{r}$ does not vary with time, the second term of eqn. (11.7) can be expressed as:

$$\frac{1}{2} \int_V \frac{\delta \vec{p}}{\delta t} \frac{\delta \vec{p}}{\delta t} \rho dV = \frac{1}{2} \int_V \frac{\delta \vec{d}_E}{\delta t} \frac{\delta \vec{d}_E}{\delta t} \rho dV = \frac{1}{2} \sum_{i=1}^{N} M_i \dot{\eta}_i^2 \tag{11.11}$$

The expression for K_E in eqn. (11.7) now becomes

$$K_E = \frac{1}{2} M \left(\frac{d\vec{R}_0}{dt} \cdot \frac{d\vec{R}_0}{dt} \right) + \frac{1}{2} \sum_{i=1}^{N} M_i \dot{\eta}_i^2 + \frac{1}{2} \vec{\omega}^T [J] \vec{\omega} \tag{11.12}$$

11.2.1.2 Potential Energy

The potential energy P_E comprises the gravitational and elastic strain energies:

$$P_E = -\int_V \dot{g}\left(\vec{R}_0 + \vec{p}\right)\rho\, dV - \frac{1}{2}\int_V \frac{\delta^2 \vec{d}_E}{\delta t^2}\cdot \vec{d}_E\, \rho\, dV \tag{11.13}$$

Substituting eqn. (11.8) into eqn. (11.13), P_E can be expressed as:

$$P_E = -Mg\,\vec{R}_0 + \frac{1}{2}\sum_{i=1}^{N} M_i\,\omega_i^2\,\eta_i^2 \tag{11.14}$$

where ω_i is the in vacuo frequency for the ith mode.

11.2.1.3 Generalized Forces and Moments

Using Lagrange's expression from eqn. (2.7.18) of Chapter 2, the generalized force Q_i is expressed as:

$$Q_i = \frac{d}{dt}\left(\frac{\partial K_E}{\partial \dot{q}_i}\right) - \frac{\partial K_E}{\partial q_i} + \frac{\partial P_E\left(q_i, t\right)}{\partial q_i},\quad i = 1,\ldots,N \tag{11.15}$$

where q_i denotes the generalized coordinates. For the problem at hand, the generalized coordinates include the variables $x, y, z, \phi, \theta, \psi, \eta_i$, and their rates.

Equation (2.7.7) of Chapter 2 states that the generalized forces in terms of virtual work can be expressed as:

$$Q_i = \frac{\partial y}{\partial q_i}\left(\delta W\right) \tag{11.16}$$

If $\vec{F}$ and $\vec{M}$ are the total forces and moments acting on the aircraft, relative to body reference axes, then

$$\vec{F} = X\hat{i} + Y\hat{j} + Z\hat{k} \text{ and } \vec{M} = L\hat{i} + M\hat{j} + N\hat{k} \tag{11.17}$$

Here, X, Y, Z are the generalized forces and L, M, N are the generalized moments that need to be determined.

Vectors $\vec{R}_0$ and $\vec{R}_0$ that define the inertial position and velocity of the origin of body reference coordinate frame in Figure 11.1 can be defined in terms of the generalized coordinates as follows:

$$\vec{R}_0 = x\hat{i} + y\hat{j} + z\hat{k} \text{ and } \vec{R}_0 = u\hat{i} + v\hat{j} + w\hat{k} \tag{11.18}$$

The velocity components u, v, w can be expressed in terms of x, y, z and angular rates p, q, r as follows:

$$u = \dot{x} + qz - ry \tag{11.19a}$$

$$v = \dot{y} - pz + rx \tag{11.19b}$$

$$w = \dot{z} + py - qx \tag{11.19c}$$

The vector $\vec{\omega}$, defining the inertial angular velocity of the body reference coordinate frame, can be expressed as:

$$\vec{\omega} = p\hat{i} + q\hat{j} + r\hat{k} \tag{11.20}$$

The angular rates can be expressed in terms of the Euler angles ϕ, θ, ψ as follows:

$$p = \dot{\phi} - \dot{\psi}\sin\theta \tag{11.21a}$$

$$q = \dot{\psi}\sin\phi\cos\theta + \dot{\theta}\cos\phi \tag{11.21b}$$

$$r = \dot{\psi}\cos\phi\cos\theta - \dot{\theta}\sin\phi \tag{11.21c}$$

Substituting the expressions of u, v, w and p, q, r from eqns. (11.19) and (11.21) into eqn. (11.12) and applying Lagrange's equations defined in eqns. (11.15) and (11.16) result in a set of equations of motion for elastic airplanes. One form of the translational and rotational equations of motion, to describe the flight dynamics of a flexible aircraft, is provided next.

11.2.1.4 Equations for Elastic Airplane

The translational equations for elastic airplanes can be expressed as:

$$M\left[\dot{u} - rv + qw + g\sin\theta\right] = X \tag{11.22a}$$

$$M\left[\dot{v} + ru - pw - g\sin\phi\cos\theta\right] = Y \tag{11.22b}$$

$$M\left[\dot{w} + pv - qu - g\cos\phi\cos\theta\right] = Z \tag{11.22c}$$

The rotational equations for elastic airplanes can be expressed as:

$$\left[I_{XX}\dot{p} - (I_{XY}\dot{q} + I_{XZ}\dot{r}) + (I_{ZZ} - I_{YY})qr + (I_{XY}r + I_{XZ}q)p + (r^2 - q^2)I_{YZ}\right] = L \tag{11.23a}$$

$$\left[I_{YY}\dot{q} - (I_{XY}\dot{p} + I_{YZ}\dot{r}) + (I_{XX} - I_{ZZ})pr + (I_{YZ}p - I_{XY}r)q + (p^2 - r^2)I_{XZ}\right] = M \tag{11.23b}$$

$$\left[I_{ZZ}\dot{r} - (I_{XZ}\dot{p} + I_{YZ}\dot{q}) + (I_{YY} - I_{XX})pq + (I_{XZ}q - I_{YZ}p)r + (q^2 - p^2)I_{XY}\right] = N \tag{11.23c}$$

The generalized force Q_i for the elastic degrees of freedom can be expressed as follows:

$$\left[\ddot{\eta}_i + 2\xi_i\omega_i\dot{\eta}_i + \omega_i^2\eta_i\right] = \frac{Q_i}{m_i} \quad i = 1,2,\ldots,N \tag{11.24}$$

where m_i, ξ_i, and ω_i represent the generalized modal mass, modal damping, and in vacuo frequency of the ith vibration mode. The terms η_i, $\dot{\eta}_i$, and $\ddot{\eta}_i$ are the generalized displacement, rate, and acceleration of the ith vibration mode.

The generalized forces and moments can be expressed in terms of their nondimensional coefficients as follows:

$$X = \bar{q}SC_X,\ Y = \bar{q}SC_Y,\ Z = \bar{q}SC_Z \tag{11.25}$$

$$L = \bar{q}SbC_l,\ M = \bar{q}S\bar{c}C_m,\ N = \bar{q}SbC_n \tag{11.26}$$

$$Q_i = C_{Q_i}\bar{q}Sl \tag{11.27}$$

where C_X, C_Y, C_Z are the force coefficients in the longitudinal, lateral, and vertical axes; C_l, C_m and C_n are the rolling, pitching, and yawing moment coefficients, respectively; $\bar{q}$ and S are the dynamic pressure and wing surface area; b represents the wing span and $\bar{c}$ the mean aerodynamic chord. Term 'l' denotes the characteristic length equal to $\bar{c}$ for symmetric modes and b for asymmetric modes. Equation (11.23) may be further simplified by assuming $I_{XY} = I_{YZ} \approx 0$, due to the symmetry about X-axis for a fixed-wing aircraft.

For aeroelastic analysis (being carried out) from flight test data, the nondimensional generalized force and moment coefficients are expressed in terms of the rigid and flexible stability and control derivatives and parameter estimation techniques like, equation error, output error, or filter error, are used to identify derivative models. Parameter estimation for flexible aircraft is discussed in the next section.

11.3 PARAMETER ESTIMATION FOR FLEXIBLE AIRCRAFT

Design of flight control system of highly flexible aircraft can be very complex and would require accurate aeroelastic models. Over the last few decades, parameter estimation techniques have been routinely applied to extract rigid body models from aircraft flight test data. Such models are, however, not applicable where the rigid body and elastic modes exist close to each other.

Some applications of system identification techniques to flexible aircraft using extended dynamic models have been reported in the literature [4] and [5]. The attempts to estimate flexible aircraft stability and control derivatives, however, are a few because of the challenges arising from the complexity of instrumenting and flight testing the aircraft and the requirement of a significantly higher-order mathematical model. Further, in addition to the traditional approach of using maximum likelihood estimation techniques, an analysis may require the finite element method (FEM) to estimate aeroelastic derivatives.

Among the traditional parameter estimation techniques, the equation error method is the simplest of all that; it is non-iterative and not sensitive to the initial guess values of derivatives. The disadvantage, however, is that it requires accurate measurements of aeroelastic forces and moments along with complete information on rigid body and structural state variables. If an adequate number of sensors can be used to derive information on structural states, then regression analysis can be used to determine the forces, moments, and state variables. Further, the application of the approach in the frequency domain can be used for real-time estimation.

11.3.1 Flexible Aircraft Model

A lot of effort has been put in over the last decade to build integrated models that include both rigid and elastic degrees of freedom. The aim has been to develop a mathematical model for flexible aircraft and then introduce simplifications depending on their applicability to the problem being addressed. In this section, we summarize the extended aerodynamic models used for the analysis of the flight dynamics of a flexible aircraft. For the linear elastic theory to be valid, the elastic deformations are assumed to be small. Further, frequencies and mode shapes of the structural modes are assumed to be known (mostly from finite element analysis).

11.3.1.1 Structural Deformation Model

Theoretically, an infinite number of vibrational modes may be existing in a flexible/elastic body; the modes with increasing frequencies would be having reducing magnitudes, and thus after a certain number of modes, the higher numbers would not be contributing much to the overall amplitude of the 'flexibility' of the body and hence can be ignored; hence in practice a finite set of N modes is considered for analysis. The generalized modal displacement $\eta_i(t)$ is governed by the second-order equation given by eqn. (11.24). The structural deformation model in eqn. (11.24) assumes elastic deformation and vibrations to be small to be within the linear elastic range. Modal parameters m_i, ξ_i, ω_i, and Φ_i can be obtained from FEM or ground vibration tests (GVT). These have useful orthogonal properties for a flight dynamic analysis.

11.3.1.2 Extended Aerodynamic Model

The extended aerodynamic model for parameter estimation, in addition to the standard rigid body stability and control derivatives, shall include the additional set of derivatives to model the flexibility effects. The nondimensional force coefficients can be expressed in terms of the stability and control derivatives as follows:

$$C_X = C_{X0} + C_{X\alpha}\alpha + C_{Xq}\frac{q\bar{c}}{2U} + C_{X\delta_e}\delta_e + \sum_{i=1}^{N}\left[C_{X\eta_i}\eta_i + C_{X\dot{\eta}_i}\frac{\dot{\eta}_i\bar{c}}{2U}\right] \quad (11.28a)$$

$$C_Y = C_{Y0} + C_{Y\beta}\beta + C_{Yp}\frac{pb}{2U} + C_{Yr}\frac{rb}{2U} + C_{Y\delta_a}\delta_a + C_{Y\delta_r}\delta_r + \sum_{i=1}^{N}\left[C_{Y\eta_i}\eta_i + C_{Y\dot{\eta}_i}\frac{\dot{\eta}_i b}{2U}\right] \tag{11.28b}$$

$$C_Z = C_{Z0} + C_{Z\alpha}\alpha + C_{Zq}\frac{q\bar{c}}{2U} + C_{Z\delta_e}\delta_e + \sum_{i=1}^{N}\left[C_{Z\eta_i}\eta_i + C_{Z\dot{\eta}_i}\frac{\dot{\eta}_i\bar{c}}{2U}\right] \tag{11.28c}$$

Sometimes, it may be more convenient to use the polar coordinate velocity form of equations (α, β, V), that requires expressions for the lift and drag coefficients C_L and C_D. For such cases, we express the extended models for C_L, C_D as follows:

$$C_L = C_{L0} + C_{L\alpha}\alpha + C_{Lq}\frac{q\bar{c}}{2U} + C_{L\delta_e}\delta_e + \sum_{i=1}^{N}\left[C_{L\eta_i}\eta_i + C_{L\dot{\eta}_i}\frac{\dot{\eta}_i\bar{c}}{2U}\right] \tag{11.29a}$$

$$C_D = C_{D0} + C_{D\alpha}\alpha + C_{Dq}\frac{q\bar{c}}{2U} + C_{D\delta_e}\delta_e + \sum_{i=1}^{N}\left[C_{D\eta_i}\eta_i + C_{D\dot{\eta}_i}\frac{\dot{\eta}_i\bar{c}}{2U}\right] \tag{11.29b}$$

Likewise, the aeroelastic moment coefficients can be expressed as:

$$C_l = C_{l0} + C_{l\beta}\beta + C_{lp}\frac{pb}{2U} + C_{lr}\frac{rb}{2U} + C_{L\delta_a}\delta_a + C_{L\delta_r}\delta_r + \sum_{i=1}^{N}\left[C_{l\eta_i}\eta_i + C_{l\dot{\eta}_i}\frac{\dot{\eta}_i b}{2U}\right] \tag{11.30a}$$

$$C_m = C_{m0} + C_{m\alpha}\alpha + C_{m\dot{\alpha}}\frac{\dot{\alpha}c}{2U} + C_{mq}\frac{q\bar{c}}{2U} + C_{m\delta_e}\delta_e + \sum_{i=1}^{N}\left[C_{m\eta_i}\eta_i + C_{m\dot{\eta}_i}\frac{\dot{\eta}_i\bar{c}}{2U}\right] \tag{11.30b}$$

$$C_n = C_{n0} + C_{n\beta}\beta + C_{np}\frac{pb}{2U} + C_{nr}\frac{rb}{2U} + C_{n\delta_a}\delta_a + C_{n\delta_r}\delta_r + \sum_{i=1}^{N}\left[C_{n\eta_i}\eta_i + C_{n\dot{\eta}_i}\frac{\dot{\eta}_i b}{2U}\right] \tag{11.30c}$$

Equations (11.30) assume the conventional decoupling of the longitudinal and lateral directional coefficients. The aerodynamic model is similar to that used for rigid body aircraft flight data analysis, except that the additional terms from modal displacement and model rates are included in the model.

The nondimensional generalized force coefficient C_{Q_i} can likewise be expanded in terms of the coupled rigid-flexible and all-flexible derivatives:

$$C_{Q_i} = \begin{bmatrix} C_{Qi0}^{\eta_i} + C_{Qi\alpha}^{\eta_i}\alpha + C_{Qi\beta}^{\eta_i}\beta + C_{Qip}^{\eta_i}\frac{pb}{2U} + C_{Qiq}^{\eta_i}\frac{q\bar{c}}{2U} + C_{Qir}^{\eta_i}\frac{rb}{2U} \\ +C_{Qi\delta_e}^{\eta_i}\delta_e + C_{Qi\delta_a}^{\eta_i}\delta_a + C_{Qi\delta_r}^{\eta_i}\delta_r + \left\{ \sum_{j=1}^{N} C_{\eta_j}^{\eta_i}\eta_j + \sum_{j=1}^{N} C_{\dot{\eta}_j}^{\eta_i}\dot{\eta}_j \right\} \end{bmatrix} \tag{11.31}$$

11.3.2 Estimation of Modeling Variables

Estimation of stability and control derivatives from eqns. (11.28) to (11.31) would require information on the modal states η_i and $\dot{\eta}_i$. The simplest approach is to apply least square estimators to compute the modal displacement from strain and attitude data, while the air data and gyroscopic measurements can be used to compute the modal rates. Linear and angular acceleration data can be used to determine the modal accelerations. A Kalman filter is generally applied to obtain improved estimates of the modal states. Once the modal states are determined, it becomes possible to remove the effect of modal contributions from measured data to obtain other state variables. The nondimensional generalized force and moment coefficients can be computed once the complete set of state vectors, including the modal states, is available.

11.3.2.1 Estimating Modal Displacements

Modal displacements can be computed from the measured strain data and attitude data using the output model given in eqn. (11.32)

$$\epsilon^j = \Psi_i^j\ \eta_i,\ \ j = 1,\ldots,\ n_\varepsilon \tag{11.32}$$

In eqn. (11.32), n_ε is the number of strain sensors used to gather onboard data, and Ψ_i^j denotes the strain modes that can be obtained using FEM and defines the strain measured from the jth sensor for the ith vibration mode. It may be noted that any change in aircraft inertia would alter the values of Ψ_i^j. However, the inertia may be assumed constant for shorter durations of a flight. Further, one needs to judiciously select the number of strain sensors to be put onboard aircraft and the number of vibration modes to be considered. A large number of sensors or vibration modes could increase the computational burden while too few sensors or vibration modes may degrade the accuracy of the estimated modal displacements.

The Euler angles can be estimated using eqns. (11.32) and (11.33), and the data obtained from attitude sensors are:

$$\phi^j = \phi + v_{i\phi}^j\eta_i \tag{11.33a}$$

$$\theta^j = \theta + v_{i\theta}^j\eta_i \tag{11.33b}$$

$$\psi^j = \psi + v_{i\psi}^j\eta_i \tag{11.33c}$$

where $j = 1,\ldots,n_a$ indicates the attitude sensor index. The terms $v^j_{i\phi}$, $v^j_{i\theta}$, and $v^j_{i\psi}$ are the gradients of the displacement mode shapes $\Phi_i(\vec{r}, t)$ obtained through FEM. The angles ϕ,θ and ψ are defined in the body reference mean axis system and can be obtained from the measured attitude data ϕ^j,θ^j, and ψ^j using eqn. (11.33).

The vector of unknown parameters now, therefore, includes the ϕ,θ and ψ along with the modal displacements:

$$\Theta = \begin{bmatrix} \phi & \theta & \psi & \eta_1 & \eta_2 & \ldots & \eta_N \end{bmatrix}^T \tag{11.34}$$

11.3.2.2 Estimating Modal Rates

The observed angular rates p^j, q^j, and r^j affected by the structural deformation are:

$$p^j = p + \dot{v}^j_{i\phi}, \text{ where } \dot{v}^j_{i\phi} = \sum_{i=1}^{N} v^j_{i\phi}\dot{\eta}_i \text{ and } j = 1,\ldots,n_g \tag{11.35a}$$

$$q^j = q + \dot{v}^j_{i\theta}, \text{ where } \dot{v}^j_{i\theta} = \sum_{i=1}^{N} v^j_{i\theta}\dot{\eta}_i \tag{11.35b}$$

$$r^j = r + \dot{v}^j_{i\psi}, \text{ where } \dot{v}^j_{i\psi} = \sum_{i=1}^{N} v^j_{i\psi}\dot{\eta}_i \tag{11.35c}$$

The above eqn. (11.35) can be used to compute the aircraft angular rates p,q, and r.

11.3.2.3 Estimating Airflow Angles

Knowing p, q, r, the airflow angles α and β can be estimated from the sensor outputs using:

$$\alpha^j = \alpha + \frac{py^j - qx^j + \sum_{i=1}^{N} \Phi^j_{iz}\dot{\eta}_i}{V} \tag{11.36a}$$

$$\beta^j = \beta + \frac{rx^j - pz^j + \sum_{i=1}^{N} \Phi^j_{iy}\dot{\eta}_i}{V} \tag{11.36b}$$

where the terms x, y, and z indicate the location at which the flow angles are measured away from the aircraft C. G. and the terms Φ^j_{iz} and Φ^j_{iy} represent the displacement mode shapes at the location of rate gyros and flow vane sensors.

Application of least squares method to the set of eqns. (11.35) and (11.36) yields the vector of unknown parameters:

$$\Theta = \begin{bmatrix} p & q & r & \alpha & \beta & \dot{\eta}_1 & \dot{\eta}_2 & \ldots & \dot{\eta}_N \end{bmatrix}^T \tag{11.37}$$

11.3.2.4 Estimating Modal Accelerations

Accelerometer sensors are very sensitive to aircraft motion and deformation. Using outputs of the accelerometer sensors located at a distance of x_a, y_a, and z_a from the C. G., the acceleration at the C. G. can be computed using the following equations:

$$ga_x^j = \frac{\overline{q}S}{m}C_X - x_a\left(q^2 + r^2\right) + y_a\left(pq - \dot{r}\right) + z_a\left(pq - \dot{r}\right) + \sum_{i=1}^{N}\Phi_{ix}^j\ddot{\eta}_i \quad (11.38a)$$

$$ga_y^j = \frac{\overline{q}S}{m}C_Y - x_a\left(pq + \dot{r}\right) - y_a\left(p^2 + r^2\right) + z_a\left(qr - \dot{p}\right) + \sum_{i=1}^{N}\Phi_{iy}^j\ddot{\eta}_i \quad (11.38b)$$

$$ga_z^j = \frac{\overline{q}S}{m}C_Z - x_a\left(pr - \dot{q}\right) + y_a\left(qr + \dot{p}\right) - z_a\left(p^2 + q^2\right) + \sum_{i=1}^{N}\Phi_{iz}^j\ddot{\eta}_i \quad (11.38c)$$

where g is the acceleration due to gravity. The vector of unknown parameters in eqn. (11.38) is given by

$$\Theta = \begin{bmatrix} C_X & C_Y & C_Z & \ddot{\eta}_1 & \ddot{\eta}_2 & \ldots & \ddot{\eta}_N \end{bmatrix}^T \quad (11.39)$$

Estimation may require several sets of accelerometer outputs to adequately represent the modal accelerations.

11.3.2.5 Kalman Filtering to Improve the Modal State Estimation

Least square estimation is generally applied to obtain the estimates of modal displacement, rates, and accelerations. Improved estimation can be achieved by utilizing Kalman filtering, which may either be carried out in real-time or offline mode post flight data gathering. Since the process and measurement noise covariance matrices Q and R will be different at each time step, a time-varying filter Kalman filter would be more appropriate for estimation.

11.3.2.6 Practical Aspects

It is well known that errors in measurements can adversely affect the accuracy of the estimated parameters. In the present situation, errors in modal frequencies and damping ratios can lead to biased estimates of aeroelastic stability and control derivatives. For example, one needs to first obtain $\dot{q}$ to compute C_m. The modal contribution $\dot{\eta}$ needs to be subtracted from the pitch rate gyro reading and the resulting pitch rate q is then differentiated to obtain C_m. Any error in $\dot{\eta}$ will lead to biased estimates of pitching moment derivatives. Similarly, inaccurate FEM can lead to errors in displacement mode shapes Φ which, in turn, can affect the accuracy of modal state estimation. Extra care, therefore, is required to instrument the flight vehicle and design experiments for better accuracy in estimation results.

11.4 X-56A AIRCRAFT AND FLIGHT TESTS

A successful application of the system identification technique to estimate the aeroelastic flight dynamic model from X-56A flight test data was reported in Ref. [5]. An overview of the practical aspects of instrumenting the aircraft, design of flight experiments, and test results, discussed in Ref. [5] is presented here.

11.4.1 AIRCRAFT

X-56A is a tailless flying wing configuration having swept back wings with a high aspect ratio. The wings have a span of 28 ft and are equipped with winglets. The aircraft is controlled through ten control surfaces, two along the trailing edge of the center body and four each along the trailing edge of the left and right wings. The aircraft was remotely flown from a ground control station to gather flight test data for aeroelastic analysis.

11.4.2 INSTRUMENTATION

The X-56A aircraft was equipped with string potentiometers to measure the control surface, air data vanes on the nose boom to measure the angle of attack and sideslip angle, and a set of 18 accelerometers and rate gyros to measure the aircraft acceleration and rates in all the three axes. An embedded GPS/INS system was installed close to the C. G. Differential GPS provided measurements of position and inertial velocity. Strain data was gathered through 24 strain gauges and a fiber-optic strain sensor. Time skews were estimated and removed from the measured data to avoid phase lags before the parameter estimation.

11.4.3 FLIGHT TESTING

Aircraft was trimmed in straight and level flight and multi-sine inputs were applied to all the control surfaces simultaneously to excite the aircraft's rigid body and structural modes. Since the inputs are mutually orthogonal harmonic multiples, simultaneous application to different control surfaces does not lead to correlation in input signals; otherwise, a filter error method might be required. Application of these inputs generated small amplitude responses about the trim condition so as not to violate the assumptions for linearity. The inputs were designed to be sufficiently different to excite distinct responses for each set of inputs.

11.4.4 FLIGHT TEST RESULTS

In vacuo frequencies, damping ratios and mode shapes describing the free vibrations of the structure were obtained from the FEM model that was tuned to the data generated from GVT. The main structural modes of interest were the first symmetric wing bending (SWB) and symmetric wing torsion modes. The SWB mode was observed to be approximately twice the frequency of the rigid body short-period mode.

Measurements from accelerometers and control surface deflections from the left and right wings were averaged to reduce the number of outputs and inputs. Data compatibility checks were used to estimate the systematic errors like scale factor in the angle of attack. Time delays in the control input surface deflections were estimated along with the aeroelastic stability and control derivatives using the output error estimation method in the frequency domain. The model structure was determined according to the principle of parsimony and the adequacy of the model to be fitted to the data. This helped to achieve quick convergence and reduced error bounds on the parameter estimates because overfitting of data with any higher-order model was avoided.

Typical response fit to Fourier transforms of the model outputs and measured signals of α, q, θ and a_z, reproduced from Ref. [5], is shown in Figure 11.2; the superscripts *EGI* indicates the data from the accelerometer located in the embedded GPS Inertial system, *cf* and *ca* indicate the readings from the forward and aft location of sensors at center fuselage, and *mf*, *of* and *ma*, *oa* define the average accelerometer readings from the left and right wings for sensors at the forward and aft locations at the wing midspan and outboard. The rigid body short-period mode at lower frequencies and the structural wing bending and torsion modes at higher frequencies are clearly seen and well captured by the estimated model. The estimated control derivatives and angle of attack derivatives had good agreement with the numerical predictions and results from wind tunnel tests. Figure 11.3 shows model validation

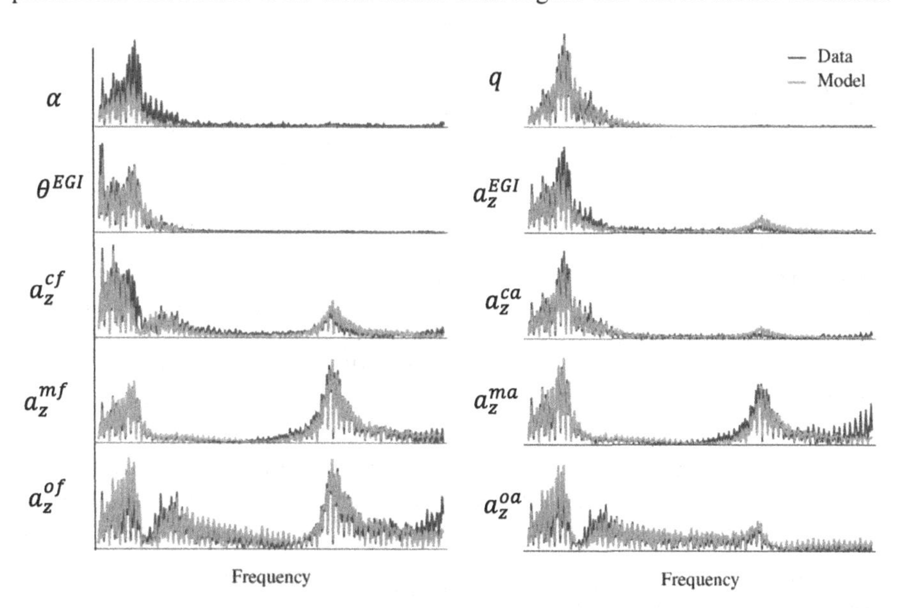

FIGURE 11.2 Frequency response match between the model estimated and measured data signals. (Grauer, J. A., and Boucher, M. J. Identification of aeroelastic models for the X-56A longitudinal dynamics using multi-sine inputs and output error in the frequency domain. *Aerospace*, 6, 24, 2019; doi:10.3390/aerospace6020024; accessed June, 2022; No special permission required as per https://www.mdpi.com/openaccess.)

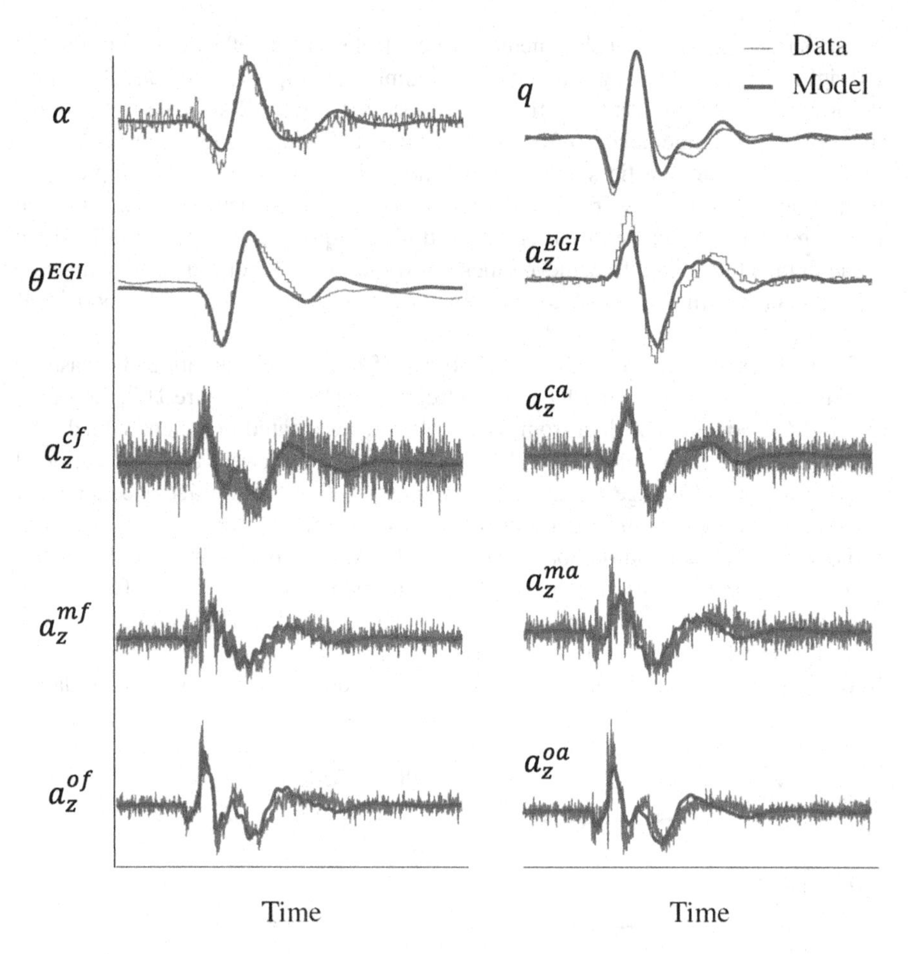

FIGURE 11.3 Model validation with complementary flight data. (Grauer, J. A., and Boucher, M. J. Identification of aeroelastic models for the X-56A longitudinal dynamics using multi-sine inputs and output error in the frequency domain. *Aerospace*, 6, 24, 2019; doi:10.3390/aerospace6020024; accessed June, 2022; No special permission required as per https://www.mdpi.com/openaccess.)

with complementary flight data generated at a different altitude with a different set of control inputs [5]. The close agreement between the measured and model responses bears testimony to the good predictive capabilities of the estimated model.

EPILOGUE

We have given here only a brief description of the major concepts of aeroservoelastic aircraft, since in itself, it is a big topic that covers modeling, system identification, parameter estimation, design optimization, control synthesis, and fault identification and management and would cover several aspects of elasticity of the material, structural deformation, vibrational modes and their determination by experimental and analytical methods, and optimal designs of structures to withstand the forces that prevent the deformations, and even the morphology of the wings, etc. Reference [6] deals with several aspects of modeling and control for aeroservoelastic aircraft.

EXERCISES

11.1 What is the aeroelastic effect?
11.2 How does a flutter occur?
11.3 Why the study of aeroservoelasticity is important in the flight mechanics analysis?
11.4 Why optimization is important in the study of aeroservoelasticity?
11.5 What is the role of modeling in the study of aeroservoelasticity?

REFERENCES

1. Noll, T. E. Aeroservoelasticity, NASA TM 102620, March 1990.
2. Tiffany, S., and Karpel, M. Aeroservoelastic modeling and applications using minimum-state approximations of the unsteady aerodynamics. *Presented at the 30th AIAA Structures, Structural Dynamics, and Materials Conference*, Alabama, April 1989; also published as NASA TM-101574, April 1989.
3. Wazak, M. R., and Schmidt, D. K. Flight dynamics of aeroelastic vehicles. *Journal of Aircraft*, 25(6), 563–571, 1988.
4. Grauer, J. A., and Boucher, M. J. Real-time parameter estimation for flexible aircraft. *AIAA Aviation Forum, Atmospheric Flight Mechanics Conference*, Atlanta, Georgia, June 25–29, 2018.
5. Grauer, J. A., and Boucher, M. J. Identification of aeroelastic models for the X-56A longitudinal dynamics using multi-sine inputs and output error in the frequency domain. *Aerospace*, 6, 24, 2019; doi:10.3390/aerospace6020024, accessed June 2022; https://www.mdpi.com/openaccess.
6. Tiwari, A. *Aeroservoelasticity: Modelling and Control*. Springer, New York, 2015.

Appendix A
Atmospheric Disturbance Models

The Earth's atmosphere as such is highly dynamic and in a constant state of flux; the temperature and pressure gradients, and atmospheric gust and turbulence are very important considerations in the study of an aircraft flight. The pressure and temperature of the atmosphere depend on altitude, location on the globe (longitude and latitude), and the time of the day; the seasonal and solar sunspot activities also affect the atmosphere.

A.1 ATMOSPHERE

Due to the spinning of the Earth and the movement of the air, there are significant Coriolis effects. A chunk/parcel of air is rotating because of the rotation of the Earth as a whole. The Earth's atmosphere is divided into five portions: (i) troposphere, its height depends on the latitude of the place, and varied from (up to) 9 to 16 kms; (ii) stratosphere, depends on latitude and varied from 15 to 30 kms; (iii) ionosphere, here the air molecules are ionized due to ultrasonic radiation and cosmic rays, extends up to ~100 kms; (iv) exosphere up to 300–500 kms, the air molecules are ionized, the continuum principle breaks down, reflection of EM wave would occur; and (v) mesosphere up to ~ 1000 kms, λ, the mean free path is 1000 km, whereas this path is 1 cm. between stratosphere and ionosphere. The variation of the temperature and pressure with altitude depends on the latitude of the place and the season: (i) tropical is latitude +/−300; (ii) temperate is latitude 300–600; and (iii) arctic is >600. The standard atmosphere is valid for 300–600; sea-level temperature is 150; sea-level pressure is 760 mm of Hg; i.e., 1013.25 millibars; and 1 Bar = 106 dynes/cm^2. The lapse rate of the temperature (linear drop) is 6.250 C/km in the temperate temperature region (drops from 300 to −700C from sea level to the altitude of 16 kms). In the Arctic region, the temperature lapse rate is 3.50 C/km from 00 to up to 7.5 kms.

A standard atmosphere (SATM) is defined in order to relate the flight tests, the WT results, and general airplane design and performance to a common reference. This STAM gives mean values of pressure, temperature, density, and other properties as function of altitude. These values are obtained from the experimental balloons, and sounding rocket measurements combined with a mathematical model of the atmosphere. This SATM, in common use over the past decades, has been the 1959ARDC Model Atmosphere (The US Airforce's Air Research and Development Command).

A.2 TURBULENCE MODELS

The movement of the air in the atmosphere creates gusts. The velocity of air may vary in a random fashion in both space and time. The two popular mathematical model forms that explain the random behavior of gust are the Dryden and the von

Karman model. The power spectral density for the turbulence velocity given by the Dryden model (Example 6.3 of Chapter 6) is as follows:

$$
\begin{aligned}
\Phi_{ug}(\omega) &= \frac{2\sigma_u^2 L_u}{\pi V}\frac{1}{\left[1+(L_u\Omega)^2\right]} \\
\Phi_{vg}(\omega) &= \frac{\sigma_v^2 L_v}{\pi V}\frac{1+3(L_v\Omega)^2}{\left[1+(L_v\Omega)^2\right]^2} \\
\Phi_{wg}(\omega) &= \frac{\sigma_w^2 L_w}{\pi V}\frac{1+3(L_w\Omega)^2}{\left[1+(L_w\Omega)^2\right]^2}
\end{aligned}
\tag{A.1}
$$

Here, $\Omega = \frac{\omega}{V} = \frac{2\pi}{\lambda}$, Ω is the spatial frequency, ω is the circular or temporal frequency, V is the aircraft speed, λ is the wavelength, L_u, L_v, L_w are the scale of the turbulence, and $\sigma_u, \sigma_v, \sigma_w$ are the turbulence intensities. The subscripts u, v, w represent the gust velocity components. The von Karman's spectral representation of turbulence is given by [1]:

$$
\begin{aligned}
\Phi_{ug}(\Omega) &= \frac{2\sigma_u^2 L_u}{\pi}\frac{1}{\left[1+(1.339L_u\Omega)^2\right]^{5/6}} \\
\Phi_{vg}(\Omega) &= \frac{2\sigma_v^2 L_v}{\pi}\frac{1+8/3(1.339L_v\Omega)^2}{\left[1+(L_v\Omega)^2\right]^{11/6}} \\
\Phi_{vg}(\Omega) &= \frac{2\sigma_w^2 L_w}{\pi}\frac{1+8/3(1.339L_w\Omega)^2}{\left[1+(1.339L_w\Omega)^2\right]^{11/6}}
\end{aligned}
\tag{A.2}
$$

The above spectral representation given by von Karman is widely used to simulate turbulence effects.

A.3 LONGITUDINAL MODEL WITH TURBULENCE

The rigid body airplane EOM were discussed in Chapter 3, and the aerodynamic models for the force and moment coefficients, in terms of Taylor's series, were discussed in Chapter 4. Here, we present the longitudinal aerodynamic model assuming one-dimensional gust (only w_g, i.e., up and down gust), since w_g is mainly responsible for normal accelerations. The upward gust is assumed positive as it produces positive increment in the AOA (Figure A.1).

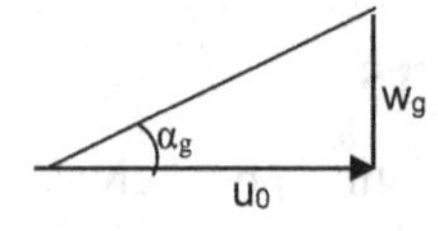

FIGURE A.1 Upward gust.

TABLE A1
The Standard Atmosphere Data-in Metric Units

Height (m)	Temperature (deg. C)	Pressure (hPa)	Density (kg/m³)
0000	15.0	1013	1.2
1000	8.5	900	1.1
2000	2.0	800	1.0
3000	−4.5	700	0.91
4000	−11.0	620	0.82
5000	−17.5	540	0.74
6000	−24.0	470	0.66
7000	−30.5	410	0.59
8000	−37.0	360	0.53
9000	−43.5	310	0.47
10000	−50.0	260	0.41
11000	−56.5	230	0.36
12000	−56.5	190	0.31
13000	−56.5	170	0.27
14000	−56.5	140	0.23
15000	56.5	120	0.19
16000	−56.5	100	0.17
17000	−56.5	90	0.14
18000	−56.5	75	0.12
19000	−56.5	65	0.10
20000	−56.5	55	0.088
21000	−55.5	47	0.075
22000	−54.5	40	0.064
23000	53.5	−34	0.054
24000	−52.5	29	0.046
25000	−51.5	25	0.039
26000	−50.5	22	0.034
27000	−49.5	18	0.029
28000	48.5	16	0.025
29000	−47.5	14	0.021
30000	−46.5	12	0.018
31000	−45.5	10	0.015
32000	−44.5	8.7	0.013
33000	−41.7	7.5	0.011
34000	−38.9	6.5	0.0096
35000	−36.1	5.6	0.0082

The angle of attack and pitch rate due to gust are given by [2]:

$$\alpha_g = \frac{w_g}{u_0} \text{ and } q_g = -\dot{\alpha}_g \tag{A.3}$$

The aerodynamic model equations for vertical force and pitching moment coefficient can therefore be expressed as:

$$C_z = C_{z\alpha}(\alpha + \alpha_g) + C_{z_q}(q - \dot{\alpha}_g)\frac{c}{2u_0} + C_{z\delta_e}\delta_e \tag{A.4}$$

$$C_m = C_{m\alpha}(\alpha + \alpha_g) + C_{m_q}(q - \dot{\alpha}_g)\frac{c}{2u_0} + C_{m\dot{\alpha}}(\dot{\alpha} + \dot{\alpha}_g)\frac{c}{2u_0} + C_{m\delta_e}\delta_e \tag{A.5}$$

A.4 TABLE OF ATMOSPHERIC DATA

The variations in pressure and temperature with altitude in a standard atmosphere are also available in the form of Tables [1,3]. The standard atmosphere is thought of based on the average pressure, temperature, and air density at various altitudes [4]. This information is very useful for engineering calculations for aircraft. It shows what pressures and temperatures are to be expected at various altitudes. The standard atmosphere is based on mathematical formulas that relate the temperature and pressure as altitude is gained or reduced. The results are close to averages of balloon and airplane measurements at these altitudes.

A table using U.S. units – altitude in feet, temperatures in Fahrenheit, and pressures in inches of mercury, is given in Ref. [4]. It gives density in slugs per cubic foot because it uses the American system. People often use pounds per cubic foot as a measure of density in the U.S.; however, pounds are a measure of force, not mass. Slugs are the correct measure of mass, and you need to multiply slugs by 32.2 for a rough value in pounds.

REFERENCES

1. Nelson, R. C. *Flight Stability and Automatic Control*, 2nd Edn. McGraw-Hill International Editions, New York, 1998.
2. Klein, V. Maximum likelihood method for estimating airplane stability and control parameters from flight data in frequency domain, NASA-TP-1637, May 1980.
3. Anderson, J. D. Jr. *Introduction to Flight*, 3rd Edn, McGraw-Hill, New York, 1989.
4. Williams, J. www.usatoday.com/weather/wstdatmo.htm (USATODAT.com), accessed in 2007.

Appendix B
Artificial Neural Network-Based Modeling

Computationally neural nets represent a radically new approach to problem solving. The methodology represented can be contrasted with the traditional approach to artificial intelligence (AI). Whereas the origins of AI lay in applying conventional serial processing techniques to high-level cognitive processing like concept formation, semantics, symbolic processing, etc. (in a top-down approach), the neural nets are designed to take the opposite: the bottom-up approach. The idea is to have a human-like reasoning emerge on the macro-scale. The approach itself is inspired by such basic skills of the human brain as its ability to continue functioning with noisy and/or incomplete information, its robustness or fault tolerance, its adaptability to changing environments by learning, etc. The artificial neural networks/nets (ANNs) attempt to mimic and exploit the parallel processing capability of the human brain in order to deal with precisely the kinds of problems that the human brain itself is well adapted for [1]. The ANNs are new paradigms for computations and modeling of dynamic systems. The ANNs are themselves modeled based on the massively parallel (neuro-) biological structures found in human brains [1]. ANN simulates a highly interconnected parallel computational structure with relatively simple individual processing elements called neurons. The human brain has 10–500 billion neurons, has about 1000 main modules with 500 neural networks (NWs), and each such NW has 100,000 neurons [1]. The axon of each neuron connects to several hundred or thousand other neurons. Neuron is a basic building block of a nervous system. Comparison of a biological neuron and ANN is shown in Table B.1 and Figure B.1.

Table B.1
Biological and Artificial Neuronal Systems

Artificial Neuronal System	Biological Neuronal System
• the data enter the input layer of the ANN • one hidden layer is between input and output layers; one layer is normally sufficient • output layer that produces the NW's output • responses can be linear or non-linear • weights are the strength of the connection between nodes	• dendrites are the input branches – tree of fibers, and connect to a set of other neurons – the receptive surfaces for input signals • soma cell body wherein all the logical functions of the neurons are performed • axon nerve fiber – output channel, the signals are converted into nerve pulses (spikes) to target cells • synapses are the specialized contacts on a neuron, and interface some axons to the spines of the input dendrites, can enhance/dampen the neuron excitation

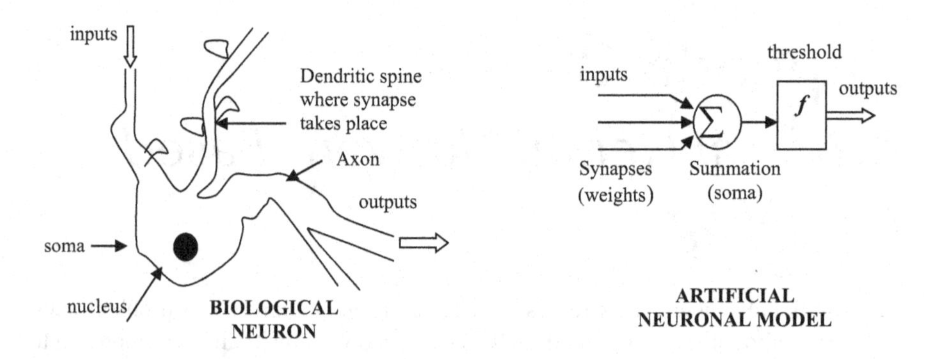

FIGURE B.1 A comparison of biological neuron and ANN models [4].

Some equivalence of the neuron process to an electronic circuit is discussed next. The voltage amplifier simulates a cell body. The wires represent the input structure (dendrites) and output structure (axon). The variable resistors model the synaptic weights and the sigmoid function is the saturation characteristic of the amplifier. It is apparent from the foregoing that, for ANN, a part of the behavior of real neurons is used and ANN can be regarded as multi-input non-linear device with weighted connections. The cell body is non-linear limiting function or activation step (hard) limiter (for logic circuit simulation/modeling) or S-shaped sigmoid function for general modeling of non-linear systems.

ANNs have found successful applications in image processing, pattern recognition, non-linear curve fitting/mapping, flight data analysis, adaptive control, system identification, and parameter estimation. In many such cases, the ANNs are used for prediction of the phenomena that they have learnt earlier by way of training from the known samples of data. They have good ability to learn adaptively from the data. The specialized features that motivate the strong interest in ANNs are as follows: (i) The neural network (NW) expansion is a basis (recall Fourier expansion/orthogonal functions as basis for the expression of time-series data/periodic signals), (ii) the NW structure does extrapolation in adaptively chosen directions, (iii) the NW structure uses adaptive bases functions – the shape and location of these functions are adjusted by the observed data, (iv) NW's approximation capability is good, (v) the NW structure is repetitive in nature and has advantage in HW/SW implementations, and (vi) this repetitive structure provides resilience to failures. The regularization feature is a useful tool to effectively include only those basis functions that are essential for the approximation and is achieved implicitly by stopping rules and explicitly by putting penalty for parameter deviation.

The conventional computers based on Von Neumann's architecture cannot match the human brain's capacity for several tasks like (i) speech processing, (ii) image processing, (iii) pattern recognition, (iv) heuristic reasoning, and (v) universal problem solving. The main reason of this capability of the brain is that each biological neuron is connected to about 10,000 other neurons, and this gives it a massively parallel computing capability. The brain effectively solves certain problems that have two main characteristics: (i) these problems are generally ill-defined, and (ii) they

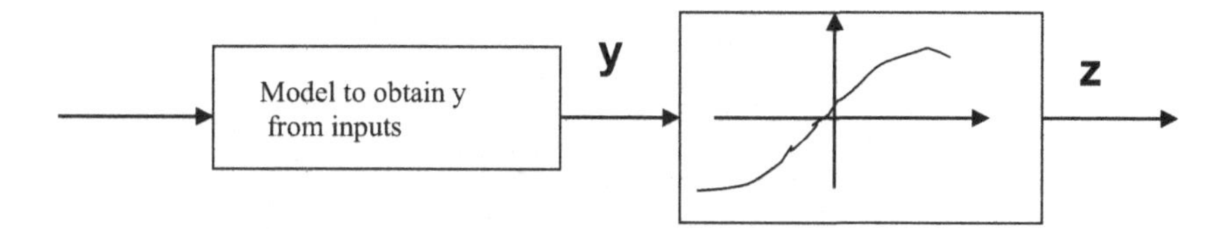

FIGURE B.2 Sigmoid neuron model.

usually require a very large amount of processing. The primary similarity between biological nervous system and ANN is that each system typically consists of a large number of simple elements that learn, are collectively able to solve complicated and ambiguous problems. It seems that a 'huge' amount of 'simplicity' rules the roost.

The computations based on ANNs can be regarded as some '6th-generation computing' as a kind of extension of massively parallel processing (MPP) of the then existing '5th-generation computing' [2]. The MPP super computers give much throughput per money value than conventional computers for the algorithms within the domain of programming constraint of such computers. The ANN can be considered as any general-purpose algorithm within the constraints of such 6th-generation hardware. The brain itself lives within essentially the same sort of constraints. ANN applications for large and more complex problems become more interesting. One can develop new generations of ANN design providing new learning-based capabilities in control and system identification. These new designs could be characterized as extensions as an aid to existing control theory, and new general control designs compatible with ANN implementation. One can exploit the power of new ANN chips or optical HW which extend the ideas of MPP, and reduced instruction sets down to a deeper level permitting orders of magnitudes more through put per money value.

There are mainly two types of neurons used in ANNs: sigmoid and binary. In sigmoid neuron, the output is a real number (Figure B.2): $z = 1/(1 + \exp(-y)) = (1 + \tanh(y))/2$. The y could be a simple linear combination of the input components, or it could be the result of some more complicated operations: $y =$ summation of (weights * x) – bias → $y =$ weighted linear sum of the components of the input vector and the threshold. In binary neuron, the non-linear function, see Figure B.3, is:

$$z = \text{saturation}(y) = \{1 \text{ if } y >= 0 \text{ and } \{0 \text{ if } y < 0 \qquad \text{(B.1)}$$

Here, the y could be a simple linear combination of the input components or it could be the results of some more complicated operations. The binary neuron has

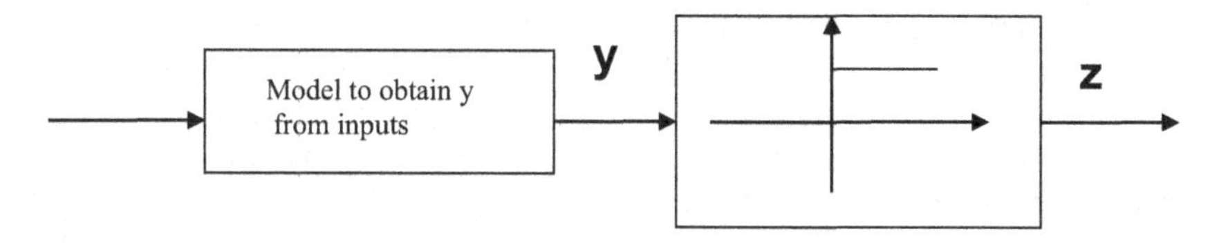

FIGURE B.3 Binary neuron model.

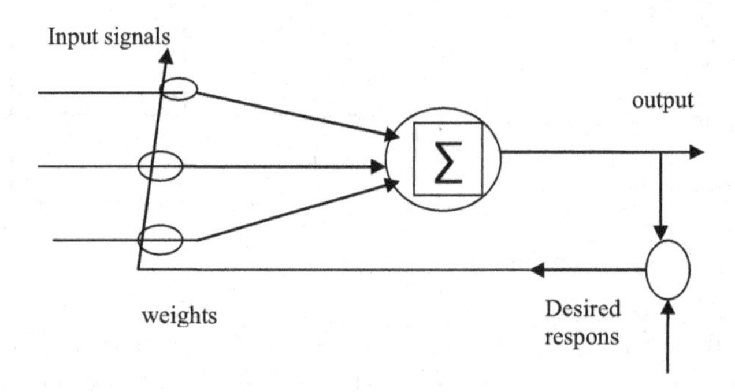

FIGURE B.4 Basic building block of ANN.

saturation/hard non-linear function. This type of binary neuron is called a perceptron. One can have a higher-order perceptron where the input y is a second-order polynomial (as a function of input data). A basic building block of most ANNs and adaptive circuits is the adaptive linear combiner (ALC) shown in Figure B.4. The inputs are weighted by a set of coefficients, and the ALC outputs linear combination of inputs. For training, the measured input patterns/signals and desired responses are presented to the ALC. The weight adaptation is done by LMS algorithm (least mean square, Widrow-Hoff delta rule), which minimizes the sum of squares of the linear errors over the training set [3]. More useful ANNs have multi-layer perceptrons (MLPN) with input layer, one or at most two hidden layers, and output layer with sigmoid (non-binary) neuron in each layer. In this network (NW), the I/O data are real-valued signals. The output of MLPN is a differential function of the NW parameters. This NW is capable of approximating arbitrary non-linear mapping using given set of examples. For a fully connected MLPN, 2–3 layers are generally sufficient. The number of hidden-layer nodes should be much less than the number of training samples. One important property of the NW is the generalization property that is a measure of how well the NW performs on actual problem once training is completed. The generalization is influenced by number of data samples, the complexity of the underlying problem, and the NW size.

B.1 FEEDFORWARD NEURAL NETWORK (FFNN)-BASED MODELING

The FFNN (feed forward neural network) is a non-cyclic type ANN with a layered topology and no feedback. FFNN is an information processing system of a large number of simple processing elements. FFNN can be thought as a non-linear black box (model structure but not in the conventional sense of polynomial or transfer function model), the parameters (weights) of which can be estimated by conventional optimization methods. A typical FFNN topology is shown in Figure B.5. The FFNNs are suitable for system identification, time-series modeling, parameter estimation, sensor failure detection, and related applications [4,5]. The chosen FFNN is first trained using the training set data, and then, it is used for prediction using another input set, which belongs to the same class of the data. This is the validation set [4].

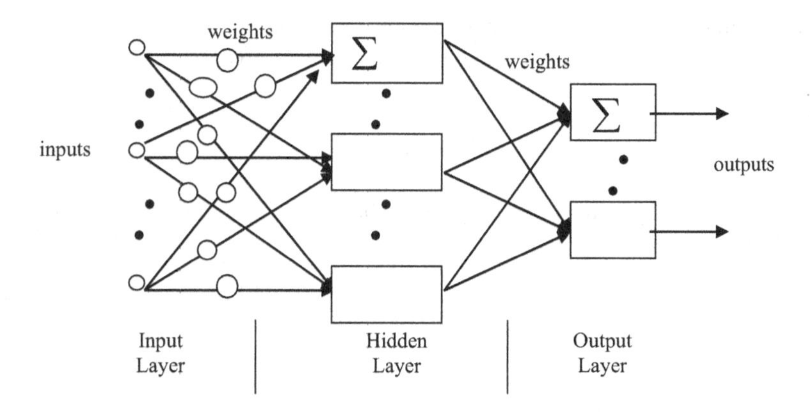

FIGURE B.5 A FFNN topology.

The process is similar to the one used as cross-validation in system identification literature.

B.1.1 Backpropagation (BPN) Training Algorithm

The backpropagation algorithm is a learning rule for multi-layered neural networks [2], credited to Rumelhart and McClelland. The algorithm gives a prescription for adjusting the initially randomized set of synaptic weights (existing between all pairs of neurons in each successive layer of the network) so as to minimize the difference between the network's output of each input fact and the output with which the given input is known (or desired) to be associated. The backpropagation rule takes its name from the way in which the calculated error at the output layer is propagated backwards from the output layer to the n-th hidden layer to the n_1-th hidden layer, and so on. Because this learning process requires us to 'know' the correct pairing of input-output facts beforehand, this type of weight adjustment is called supervised learning. The FFNN training algorithm is described using the matrix/vector notation for easy implementation in PC MATLAB. Alternatively, the NN tool box of the MATLAB can be used.

The FFNN has the following variables: (i) u_0 as input to (input layer of) the network, (ii) n_i as the number of input neurons (of the input layer) equal to the number of inputs u_0, (iii) n_h as the number of neurons of the hidden layer, (iv) n_o as number of output neurons (of the output layer) equal to the number of outputs z, (v) $W_1 = n_h \times n_i$ as the weight matrix between input and hidden layers, (vi) $W_{10} = n_h \times 1$ as the bias weight vector, (vii) $W_2 = n_o \times n_h$ as the weight matrix between hidden and output layers, (viii) $W_{20} = n_o \times 1$ as the bias weight vector, and (ix) μ as the learning rate or step size. The algorithm is based on steepest descent optimization method [4]. The signal computation is done using the following equations, since u_0 and initial guesstimates of the weights are known.

$$y_1 = W_1 u_0 + W_{10}; \; u_1 = f(y_1) \tag{B.2}$$

Here, y_1 and u_1 are the vector of intermediate values and the input to the hidden layer, respectively. The $f(y_1)$ is a sigmoid activation function given by:

$$f(y_i) = \frac{1 - e^{-\lambda y_i}}{1 + e^{-\lambda y_i}} \tag{B.3}$$

Here, λ is a scaling factor to be defined by the user. The signal between the hidden and output layers is computed as:

$$y_2 = W_2 u_1 + W_{20};\ u_2 = f(y_2) \tag{B.4}$$

A quadratic function is defined as:

$$E = \frac{1}{2}(z - u_2)(z - u_2)^T \tag{B.5}$$

Equation (B.5) signifies the square of the errors between the NW output and desired output. The u_2 is the signal at the output layer and z is the desired output. The following result from the optimization theory is used to derive a training algorithm:

$$\frac{dW}{dt} = -\mu(t)\frac{\partial E(W)}{\partial W} \tag{B.6}$$

The expression for the gradient is obtained as:

$$\frac{\partial E}{\partial W_2} = -f'(y_2)(z - u_2)u_1^T \tag{B.7}$$

Here, u_1 is the gradient of y_2 with respect to W_2. The derivative f' of the node activation function is given as:

$$f'(y_i) = \frac{2\lambda_i e^{-\lambda y_i}}{\left(1 + e^{-\lambda y_i}\right)^2} \tag{B.8}$$

The modified error of the output layer is expressed as:

$$e_{2b} = f'(y_2)(z - u_2) \tag{B.9}$$

Finally, the recursive weight update rule for the output layer is given as:

$$W_2(i+1) = W_2(i) + \mu e_{2b} u_1^T + \Omega[W_2(i) - W_2(i-1)] \tag{B.10}$$

The Ω is the momentum factor used for smoothing out the (large) weight changes and to accelerate the convergence of the algorithm. The backpropagation of the error and the update rule for W_1 are given as:

$$e_{1b} = f'(y_1)W_2^T e_{2b};\ W_1(i+1) = W_1(i) + \mu e_{1b} u_0^T + \Omega[W_1(i) - W_1(i-1)]; \tag{B.11}$$

The data are presented to the network in a sequential manner again and again but with initial weights as the outputs from the previous cycle until the convergence is reached. The entire process is recursive-iterative.

B.1.2 FFNN-BPN Training with Least-Squares Algorithm

The FFNN with backpropagation using recursive least-square (FFNN-RLS) uses an enhanced form of the backpropagation technique. The conventional approach minimizes the mean-squared error with respect to the weights. Here, the algorithm minimizes the mean-squared error between the expected output and actual output with respect to the summation outputs and Kalman filter gain and covariance are computed in each layer to update the weights [1,4]. The first two steps, I and II, in the algorithms are the same as FFNN-BPN. The updates for Kalman gain $K(i)$ and covariance matrix $P(i)$ for each layer are given as:

$$K_1 = P_1 u_0 (f_1 + u_0 P_1 u_0)^{-1}; \quad P_1 = \frac{P_1 - K_1 u_0 P_1}{f_1}$$
$$K_2 = P_2 u_1 (f_2 + u_1 P_2 u_1)^{-1}; \quad P_2 = \frac{P_2 - K_2 u_1 P_2}{f_2} \tag{B.12}$$

The modified output error e_{2b} and the backpropagation inner layer error e_{1b} are computed as per FFNN-BPN scheme. The weight updates formula for the output layer is given as:

$$W_2(i+1) = W_2(i) + (d - y_2) K_2^T \tag{B.13}$$

Here, d is the summation of output and is calculated by using inverse function as:

$$d = \frac{1}{\lambda} \ln \frac{1+\dot{x}}{1-\dot{x}} \tag{B.14}$$

Here, λ is the sigmoid slope parameter. The weight updates formula for the output layer is given as:

$$W_1(i+1) = W_1(i) + \mu e_{1b} K_1^T \tag{B.15}$$

The updated weights are used and the algorithm is repeated till desired convergence is achieved.

B.1.3 ANN-Based Parameter Estimation

We have seen that a FFNN can be used for non-linear mapping of the input-output data. This suggests that it should be possible to use it for system identification/parameter estimation. The FFNN works with a black-box model structure, which cannot be physically interpreted, and hence, the parameters, i.e., the network weights, have no interpretation in terms of the parameters of the actual system. The parameter

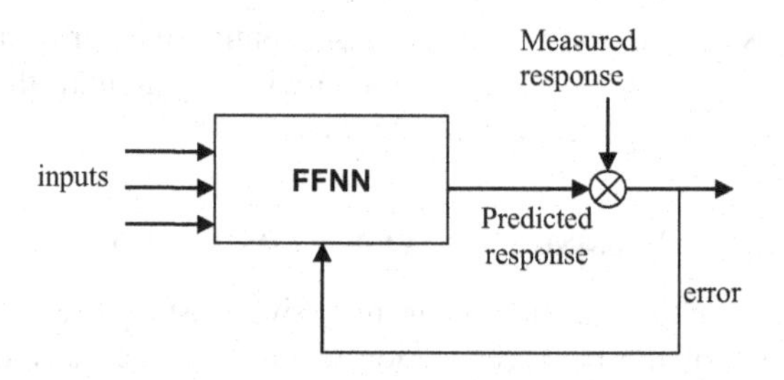

FIGURE B.6 A scheme for parameter estimation using FFNN.

estimation using FFNN is done in two steps: (i) the FFNN is presented with the measured data and trained to reproduce the clean/predicted responses that are compared with the actual system's responses in the sense of minimization of the output error as depicted in Figure B.6, and (ii) subsequently, the predicted responses are perturbed in turn for each parameter that is to be estimated. This obtains a changed predicted response. Next, assume that $z = \beta x$ and the FFNN is trained to produce clean z. Then, the trained FFNN is used to produce $z + \Delta z$ and $z - \Delta z$ by changing x to $x + \Delta x$ and $x - \Delta x$, and the parameter β is obtained as:

$$\beta = \frac{z^+ - z^-}{x^+ - x^-} \tag{B.16}$$

This is called a Delta method. The estimates are obtained by averaging these respective parameter time histories, by removing some initial and some final points.

B.2 RECURRENT NEURAL NETWORKS (RNNs)-PARAMETER ESTIMATION

The RNNs are NWs with feedback of an output to the internal states [6]. The RNNs are more suitable for the problem of parameter estimation of linear dynamic systems, as compared to FFNNs. RNNs are dynamic NWs that are amenable to explicit parameter estimation in state-space models [4]. Basically, RNNs are a type of the Hopfield NWs (HNN) [6]. One type of RNN is shown in Figure B.7, the dynamics of which are given by [7]:

$$\dot{x}_i(t) = -x_i(t)R^{-1} + \sum_{j=1}^{n} w_{ij}\beta_j(t) + b_i; \quad j = 1,\ldots,n \tag{B.17}$$

Here, (i) x is the internal state of the neurons, (ii) β is the output state, (iii) w is a neuron weight, (iv) b is the bias input to neurons, (v) f is the sigmoid non-linearity, $\beta = f(x)$, (vi) R is the neuron impedance, and (vii) n is the dimension of neuron state. The FFNNs and RNNs can be used for parameter estimation [4]. The RNNs can be used for parameter estimation of linear dynamic systems [4], Chapter 9, as well as non-linear time-varying systems [8]. They can be also used for trajectory prediction/matching [9]. Many variants of RNNs and their interrelationships have been reported [7].

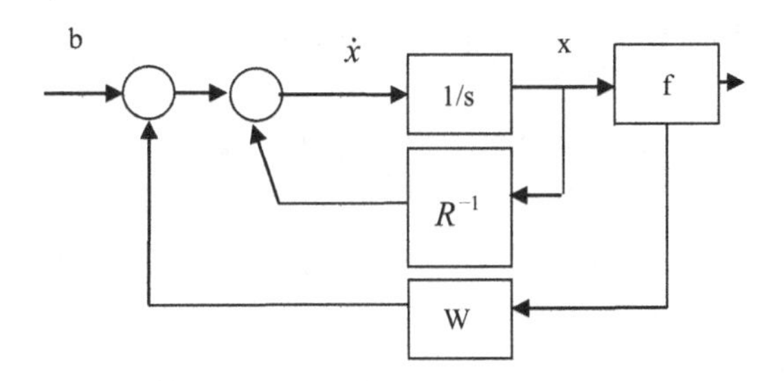

FIGURE B.7 A schematic of an RNN structure.

Also, the trajectory matching algorithms for these variants are given in Ref. [9]; these algorithms can be used for training RNNs for non-linear model fitting and related applications, in a similar manner as done using FFNNs.

B.2.1 Weight-Bias Estimation Method

The idea is to estimate the parameters of a dynamic system:

$$\dot{x} = Ax + Bu;\ x(0) = x_0 \tag{B.18}$$

The $n \times 1$ vector $\beta = \{A, B\}$ represents the parameter vector to be estimated. The dynamics are affected by the (sigmoid) non-linear function f: $\beta_i = f(x_i)$. The equation error cost function is given as:

$$E(\beta) = 1/2\sum(\dot{x} - Ax - Bu)^T(\dot{x} - Ax - Bu) \tag{B.19}$$

Using the result from optimization theory, we get:

$$\frac{d\beta}{dt} = -\frac{\partial E(\beta)}{\partial \beta} = -\frac{1}{2}\frac{\partial\left\{\left[\sum_{k=1}^{N} e^T(k)\, e(k)\right]\right\}}{\partial \beta} \tag{B.20}$$

The β as a parameter vector contains the elements of A and B, and we obtain expressions $\partial E/\partial A$ and $\partial E/\partial B$ for A and B vectors, with $\sum(.) = \sum_{k=1}^{N}(.)$, after simplification:

$$\begin{aligned} \frac{\partial E}{\partial A} &= A\sum x\, x^T + B\sum u\, x^T - \sum \dot{x}\, x^T \\ \frac{\partial E}{\partial B} &= A\sum x\, u + B\sum u^2 - \sum \dot{x}\, u \end{aligned} \tag{B.21}$$

In terms of individual elements of the matrices, we obtain:

$$\begin{aligned}
\frac{\partial E}{\partial a_{11}} &= a_{11}\sum x_1^2 + \cdots + b_1\sum x_1 u - \sum \dot{x}_1 x_1 \\
\frac{\partial E}{\partial a_{12}} &= a_{11}\sum x_1 x_2 + \cdots + b_1\sum u x_2 - \sum \dot{x}_1 x_2 \\
\frac{\partial E}{\partial a_{21}} &= a_{21}\sum x_1^2 + \cdots + b_2\sum u x_1 - \sum \dot{x}_2 x_1 \\
\frac{\partial E}{\partial a_{22}} &= a_{21}\sum x_1 x_2 + \cdots + b_2\sum u x_2 - \sum \dot{x}_2 x_2
\end{aligned} \tag{B.22}$$

$$\begin{aligned}
\frac{\partial E}{\partial b_1} &= a_{11}\sum x_1 u + \cdots + b_1\sum u^2 - \sum \dot{x}_1 u \\
\frac{\partial E}{\partial b_2} &= a_{21}\sum x_1 u + \cdots + b_2\sum u^2 - \sum \dot{x}_2 u
\end{aligned} \tag{B.23}$$

Assuming that the impedance R is very high, the dynamics of RNN are given as:

$$\dot{x}_i = \sum_{j=1}^{n} w_{ij}\beta_j + b_i \tag{B.24}$$

The energy functional of the RNN is given as [10]:

$$E = -1/2\sum_i\sum_j W_{ij}\beta_i\beta_j - \sum_i b_i\beta_i;\ \frac{\partial E}{\partial \beta_i} = -\sum_{j=1}^{n} w_{ij}\beta_j - b_i \tag{B.25}$$

$$\frac{\partial E}{\partial \beta_i} = -\left[\sum_{j=1}^{n} w_{ij}\beta_j + b_i\right] = -\dot{x}_i \tag{B.26}$$

$$\dot{x}_i = -\frac{\partial E}{\partial \beta_i} \text{ and since } \beta_i = f(x_i), \text{ and } \dot{x}_i = (f^{-1})'\,\dot{\beta}_i \tag{B.27}$$

We have $(f^{-1})'\,\dot{\beta}_i = -\frac{\partial E}{\partial \beta_i}$; hence,

$$\dot{\beta}_i = -\frac{1}{(f^{-1})'(\beta_i)}\frac{\partial E}{\partial \beta_i} = \frac{1}{(f^{-1})'(\beta_i)}\left[\sum_{j=1}^{n} w_{ij}\beta_j + b_i\right] \tag{B.28}$$

Finally, the expressions for the weight matrix W and the bias vector b are obtained as:

$$W = -\begin{bmatrix} \sum x_1^2 & \sum x_2x_1 & 0 & 0 & \sum ux_1 & 0 \\ \sum x_1x_2 & \sum x_2^2 & 0 & 0 & \sum ux_2 & 0 \\ 0 & 0 & \sum x_1^2 & \sum x_2x_1 & 0 & \sum ux_1 \\ 0 & 0 & \sum x_1x_2 & \sum x_2^2 & 0 & \sum ux_2 \\ \sum x_1u & \sum x_2u & 0 & 0 & \sum u^2 & 0 \\ 0 & 0 & \sum x_1u & \sum x_2u & 0 & \sum u^2 \end{bmatrix} \quad \text{(B.29)}$$

$$b = -\begin{bmatrix} \sum \dot{x}_1x_1 \\ \sum \dot{x}_1x_2 \\ \sum \dot{x}_2x_1 \\ \sum \dot{x}_2x_2 \\ \sum \dot{x}_1u \\ \sum \dot{x}_2u \end{bmatrix} \quad \text{(B.30)}$$

The algorithm for parameter estimation of a dynamic system is given as: (i) since the measurements of x, $\dot{x}$, and u are available for a certain time interval T, compute W matrix, and bias vector b, (ii) initial values of β_i are chosen randomly, and (iii) solve the following differential equation. Since $\beta_i = f(x_i)$ and the sigmoid non-linearity are known, by differentiating and simplifying, we get:

$$\frac{d\beta_i}{dt} = \frac{\lambda(\rho^2 - \beta_i^2)}{2\rho}\left[\sum_{j=1}^{n} w_{ij}\beta_j + b_i\right] \quad \text{(B.31)}$$

$$f(x_i) = \rho\left(\frac{1-e^{-\lambda x_i}}{1+e^{-\lambda x_i}}\right) \quad \text{(B.32)}$$

Integration of eqn. (B.31) would yield the solution to parameter estimation problem posed in equation error/RNN structure. Proper tuning of λ and ρ is very essential.

Often, λ is chosen as a small number, i.e., less than 1.0. The value of ρ is chosen such that when x_i (of RNN) approaches $\pm\infty$, the f approaches $\pm\rho$.

B.2.2 Gradient-Based Estimation Method

In the RNN-gradient-based method, the non-linearity directly operates on the equation error. The non-linear function does not enter into the neuron dynamic equation. However, it does affect the error by way of quantizing them and thereby reducing the effect of measurement outliers [9]. The cost function is the same as in RNN-WB method as given in eqn. (B.19). We assume that the state and its derivative data are available for the times: $k = 1, 2, \ldots, m$. The cost function is rewritten as:

$$E(\beta) = 1/2\sum(\dot{x}(k) - Ax(k) - Bu(k))^T(\dot{x}(k) - Ax(k) - Bu(k)); \sum \equiv \sum_{k=1}^{m} \quad \text{(B.33)}$$

$$\begin{aligned} \frac{\partial E}{\partial A} &= \sum(\dot{x}(k) - Ax(k) - Bu(k))(-x(k))^T = \sum e(k)(-x(k))^T \\ \frac{\partial E}{\partial B} &= \sum(\dot{x}(k) - Ax(k) - Bu(k))(-u(k))^T = \sum e(k)(-u(k))^T \end{aligned} \quad \text{(B.34)}$$

In the component form of the states and the parameters, we have:

$$e_i(k) = \dot{x}_i(k) - \sum_{p=1}^{n} a_{ip}x_p(k) - b_i u(k) \quad \text{(B.35)}$$

$$\frac{\partial E}{\partial a_{ij}} = \sum -e_i(k)(x_j(k)); \quad \frac{\partial E}{\partial b_{ij}} = \sum -e_i(k)(u_j(k)) \quad \text{(B.36)}$$

$$\frac{\partial v_s}{\partial t} = -\sum_{r=1}^{(n^2+n)} \mu_{sr}\frac{\partial E}{\partial v_r} \quad \text{(B.37)}$$

Here, $v(.)$ is a composite parameter vector that has the elements of matrices A and B in a proper order. The parameter variation with respect to time is proportional to –ve of the gradient of the error cost function with respect to the parameters, and this continuous time formula or the parameter update rule can be converted into discrete recursion for the implementation of a digital computer. The recursions are performed with the updated parameters until convergence.

REFERENCES

1. Eerhart, R. C, and Dobbins, R. W. *Neural Network PC Tools–A Practical Guide*. Academic Press, Inc., New York, 1993.
2. Werbos, P. J. Back propagation through time: what it does and how to do it. *Proceedings of the IEEE*, 78(10), 1550–1560, 1990.
3. Widrow, B., and Lehr, M. A. Thirty years of adaptive neural networks: perceptron, madaline and back propagation. *Proceedings of the IEEE*, 78(9), 1415–1442, 1990.
4. Raol, J. R., Girija, G., and Singh, J. *Modelling and Parameter Estimation of Dynamic Systems*. IET/IEE Control Series, Vol. 65, IET, London, 2004.
5. Raol, J. R. Aerodynamic modelling and sensor failure detection using feedforward neural networks. *Journal of Aeronautical Society of India*, 47(4), 193–197, 1995.
6. Hopfield, J. J, and Tank, D.W. Computing with neural circuits-A model. *Science*, 233, 625–633, 1986.
7. Hush, D. R., and Horne, B. G. Progress in supervised neural networks–What is new since Lippmann. *IEEE Signal Process Magazine*, 10, 8–39, 1993.
8. Zhenning, Hu., and Balakrishnan, S. N. Parameter estimation in nonlinear systems using Hopfield neural networks. *Journal of Aircraft*, 42(1), 41–53, 2005.
9. Raol, J. R. Parameter estimation of state space models by recurrent neural networks. *IEE Proceedings-Control Theory and Applications*, 142(2), 114–118, 1995.
10. Chu, S. R., and Tenorio, M. Neural networks for system identification. *IEEE Control System Magazine*, 10, 31–35, 1990.

Appendix C
Fuzzy Logic-Based Modeling

The random phenomena that represent uncertainty are modeled by probability theory, which is based on crisp (binary) logic. Our interest is to model uncertainty that abounds in Nature, and need to model uncertainty in science and engineering problems. One way is to have crisp logic:

The crisp or Boolean characteristic function
fA(x) = 1 if input x is in set A
= 0 if input x is not in set A

In classical set theory, a set consists of a finite/infinite number of elements that belong to some specified set called the Universe of discourse. In the crisp logic, we have answers like: yes or no; 0 or 1; −1 or 1; off or on. Examples are (i) a person is in the room or not in the room, (ii) an event A has occurred or not occurred, and (iii) the bulb/light is on or off. However, the real-life experiences tell us that some extension of the crisp logic is definitely needed. Events or occurrences leading to fuzzy logic (FL) are as follows: (i) a bulb/light could be dim, (ii) a day could be bright with a certain degree of brightness, (iii) a day could be cloudy with a certain degree, and (iv) weather could be warm or cold. Thus, FL allows for a degree of uncertainty and gradation; truth (1) and falsity (0) become the extremes of a continuous spectrum of uncertainty. This leads to multi-valued logic and the fuzzy set theory [1]. Fuzziness is the theory of sets, and the characteristic function is generalized to take an infinite number of values between 0 and 1: e.g., triangular form.
Fuzzy systems can model any continuous function or system. The quality of approximation depends on the quality of rules that can be formed by the experts. Fuzzy engineering is a function approximation (FA) with fuzzy systems. This approximation does not depend on words, cognitive theory, or linguistic paradigm. It rests on mathematics of FA and statistical learning theory (SLT). Much of the mathematics is well known, and as such, there is no magic in fuzzy systems. Fuzzy system is natural way to turn speech and measured action into functions that approximate the hard tasks. The words are just a tool/ladder we climb down on to get the task of FA. Fuzzy language is a means to the end of computing and not the goal. Basic unit of fuzzy FA is the 'If then' rule: 'If the wash water is dirty then add more detergent'. Thus, fuzzy system is a set of 'If then' rules that maps input sets like 'dirty wash water' to output sets like 'more detergent'. The overlapping rules define polynomials/richer functions. A set of possible rules is given below [2]:

Rule 1: if the air is cold then set the motor speed to stop
Rule 2: if the air is cool then set the motor speed to slow
Rule 3: if the air is just right then set the motor speed to medium

Rule 4: if the air is warm then set the motor speed to fast
Rule 5: if the air is hot then set the motor speed to blast

This gives the first-cut fuzzy system. More rules can be guessed and added by the experts and by learning new rules adaptively from training data sets. The ANNs can be used to learn the rules from the data. Much of the fuzzy engineering deals with tuning these rules and adding new rules or pruning old rules.

C.1 ADDITIVE FUZZY SYSTEM

Each input partially fires all the rules in parallel, and the system acts as an associative processor as it computes the output $F(x)$. The system then combines the partially fired, 'then' part fuzzy sets in a sum and converts this sum to a scalar or vector output. Thus, a match-and-sum fuzzy approximation can be viewed as a generalized AI expert system or as a (neural like) fuzzy associative memory. The AFSs (adaptive fuzzy systems) have proven to be universal approximators for rules that use fuzzy sets of any shape and are computationally simple. A feedback fuzzy system (rules feedback to one another and to themselves) can model a dynamic system: $dx/dt = F(x)$. The core of every fuzzy controller is the inference engine (FIE), the computation mechanism with which decision can be inferred even though the knowledge may be incomplete. This mechanism gives the linguistic controller, the power to reason by being able to extrapolate knowledge and search for rules, which only partially fit any given situation for which a rule does not exist. The FIE performs an exhaustive search of the rules in the knowledge base (rule base) to determine the degree of fit for each rule for a given set of causes (see Figure C.1). The input and output are crisp variables.

The several rules contribute to the final result to varying degrees. A degree of fit of unit implies that only one rule has fired and only one rule contributes to the final decision; a degree of fit of zero implies that none of the rules contribute to the final decision. Knowledge necessary to control a plant is usually expressed as a set of linguistic rules of the form: 'If' (cause) 'then' (effect). These are the rules with which new operators are trained to control a plant, and they constitute the knowledge base of the system. All the rules necessary to control a plant might not be elicited, or known. It is therefore essential to use some technique capable of inferring the control action from available rules. Fuzzy logic base control has several merits: (i) for a complex system, mathematical model is hard to obtain, (ii) fuzzy control is

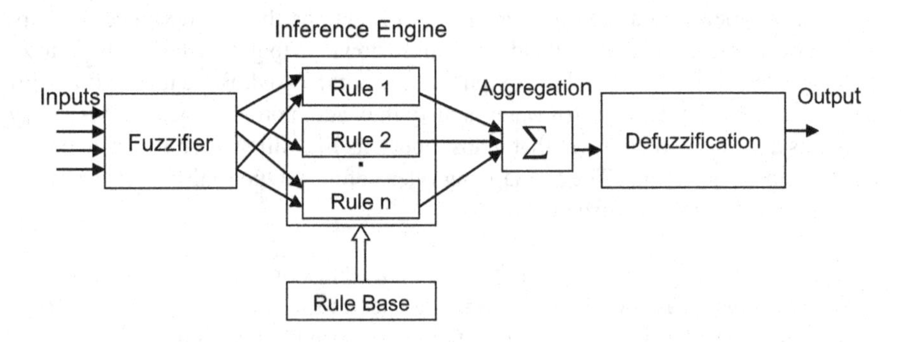

FIGURE C.1 Gross schematic of a FIS.

model-free approach, (iii) human experts can provide linguistic descriptions about the system and control instructions, and (iv) fuzzy controllers provide a systematic way to incorporate the knowledge of human experts. The assumptions in fuzzy logic control system are as follows: (i) plant is observable and controllable, (ii) expert linguistic rules are available or formulated based on engineering common sense, intuition, or an analytical model, (iii) a solution exists, (iv) look for a good enough solution (approximate reasoning in the sense of probably approximately correct solution, e.g., algorithm) and not necessarily the optimum one, and (v) desire to design a controller to the best of our knowledge and within an acceptable precision range. Figure C.2 depicts a macro-level computational schematic of the fuzzy inference

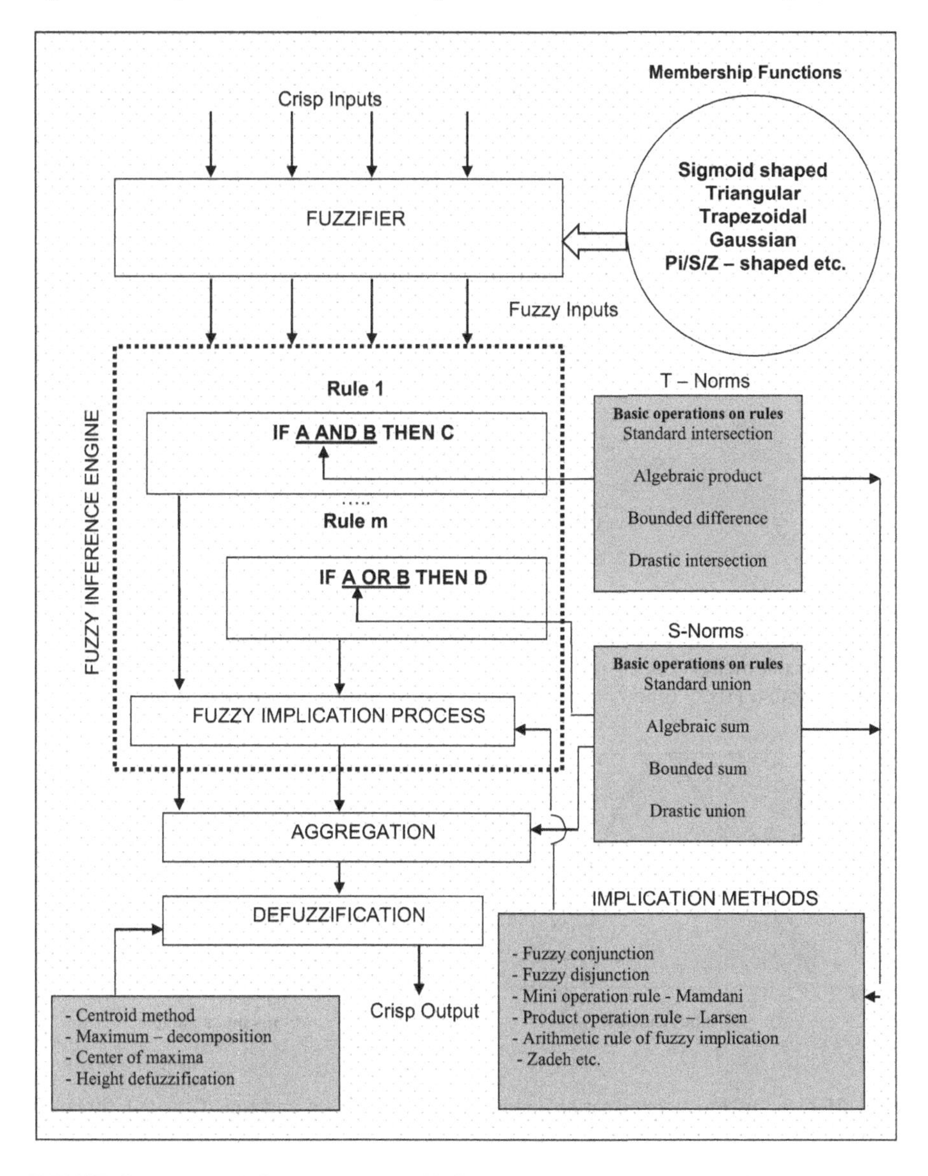

FIGURE C.2 Fuzzy inference system with fuzzy approximation computation process.

process: fuzzification (using membership functions), basic operations (on the defined rules), fuzzy implication process (using implication methods), aggregation, and finally defuzzification (using defuzzification methods) [2,3]. Example C.1 illustrates the definition of fuzzy membership functions (GMF-Gaussian MF, TMF-Triangular MF) as seen in Figure C.3, definition of fuzzy operators, numerical computations to obtain the fuzzy counter part of the classical intersection and union, and comparison with other such operations. The example requires careful study, but tells us many simple things about the fuzzy operators, which are routinely used in modeling and control of dynamic systems.°

Example C.1

The fuzzy operators corresponding to the well-known Boolean operators *AND* and *OR* are, respectively, defined as *min* and *max* [2,3]:

$$\mu_{A\cap B}(u) = \min\left[\mu_A(u), \mu_B(u)\right] \quad \text{(intersection)} \tag{C.1}$$

$$\mu_{A\cup B}(u) = \max\left[\mu_A(u), \mu_B(u)\right] \quad \text{(union)} \tag{C.2}$$

Another way to define *AND* and *OR* operators has been proposed by Zadeh [2,3]:

$$\mu_{A\cap B}(u) = \mu_A(u)\mu_B(u) \tag{C.3}$$

$$\mu_{A\cup B}(u) = \mu_A(u) + \mu_B(u) - \mu_A(u)\mu_B(u) \tag{C.4}$$

Define fuzzy sets A and B by Gassian and Trapezoidal membership functions ($\mu_A(u)$, $\mu_B(u)$), respectively (see GMF, TMF in Figure C.3a). Obtain the fuzzy intersection (element-wise) by putting the $\mu_A(u)$ and $\mu_B(u)$ in eqn. (C.1) or eqn. (C.3). Also, obtain Fuzzy union (element-wise) by using eqn. (C.2) or eqn. (C.4).

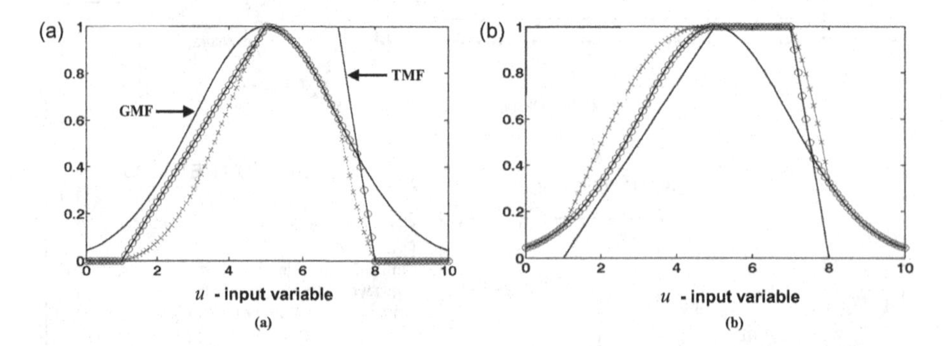

FIGURE C.3 (a) Fuzzy intersection operations; {eqn. (C.3) –x- and eqn. (C.1) -0-}. (b) Fuzzy union operations {eqn. (C.4) –x- and eqn. (C.2) -0-}.

Solution C.1

The results are shown in Figure C.3. Let us denote $a = \mu_A(u)$, and $b = \mu_B(u)$ to be membership grades (within interval [0,1]). Now assume that for some value of input u, $a < b$, therefore $\mu_{A \cap B}(u)$ using eqn. (C.1) will be

$$\mu_{A \cap B}(u) = \min(a,b) = a \tag{C.5}$$

Also, it is known that $b < 1$, multiplying this expression with a gives $ab < a$. Using eqn. (C.3) in the left-hand side of $ab < a$ results in $ab < \min(a,b)$ which shows that eqn. (C.3) <eqn. (C.1) (in the sense of the numerical values), i.e., the area covered by x-line (representing eqn. (C.3)) is lesser than that covered by 0-line (representing the eqn. (C.1). The same equality holds for $a > b$. Membership function obtained by eqn. (C.4) expands compared to one obtained by eqn. (C.2) (Figure C.3b). This can be easily proved by numerical example. Let us assume $a = 0.1$ and $b = 0.5$ for some value of input variable u. By putting them in eqn. (C.4) and eqn. (C.2), we get $\mu_{A \cup B}(u)$ equal to 0.55 and 0.5 (one can easily verify this), respectively, which means that eqn. (C.4)>eqn. (C.2), i.e., the area covered by x-line (representing eqn. (C.4)) is greater than that covered by 0-lines (representing the eqn. (C.2)). It is apparent from the above definitions that, unlike in crisp logic/set theory, the fuzzy operators can have seemingly different definitions. However, these would be mathematically consistent and would be intuitively valid and appealing [1].

C.2 ANFIS FOR PARAMETER ESTIMATION

An ANFIS (adaptive neuro-fuzzy inference system) of MATLAB for parameter estimation utilizes the rule-based procedure to represent the system behavior in the absence of the precise model of the system and uses the I/O data to determine the membership function's parameters (constants). It consists of FIS whose membership function's parameters are tuned using either a backpropagation algorithm or in combination with LS method. These parameters will change through the learning process. The computation of these parameters is facilitated by a gradient vector, which provides a measure of how well the FIS is modeling the I/O data for a given set of parameters. Once the gradient vector is obtained, any optimization routine can be applied in order to adjust the parameters so as to reduce the error-measure defined by the sum of the squared difference of the actual and desired outputs. The membership function is adaptively tuned/determined by using ANN and the I/O data of the given system. The FIS is shown in Figure C.1. The process of system parameter estimation using ANFIS is depicted in Figure C.4. The steps involved in the process are depicted in Figure C.5. Consider a fuzzy system having rule base: (i) If u1 is A1 and u2 is B1, then y1 = c11u1 + c12u2 + c10; if u1 is A2 and u2 is B2, then y2 = c21u1 + c22u2 + c20. Here, u1, u2 are crisp/non-fuzzy inputs, and y is the desired output. Let the membership functions of fuzzy sets Ai, Bi, i = 1,2 be μ_A, μ_B. The 'pi' is the product operator to combine the AND process of sets A and B, and N is the normalization. The Cs are the output membership function parameters. The steps are given as follows: (i) each neuron '*i*' in layer 1 is adaptive with a parametric activation function. Its output is the grade of membership function to which the

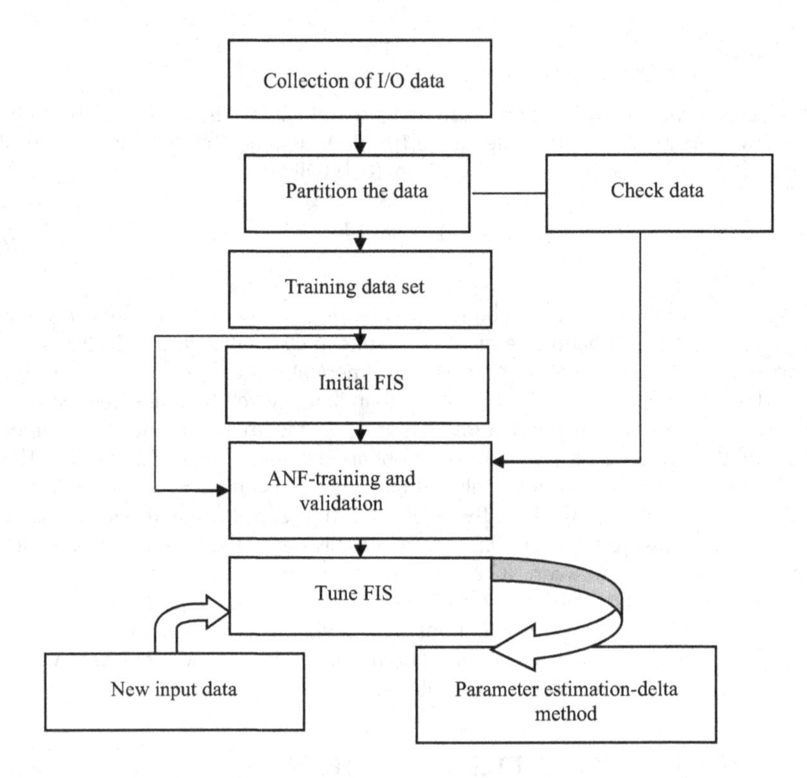

FIGURE C.4 ANFIS scheme for parameter estimation.

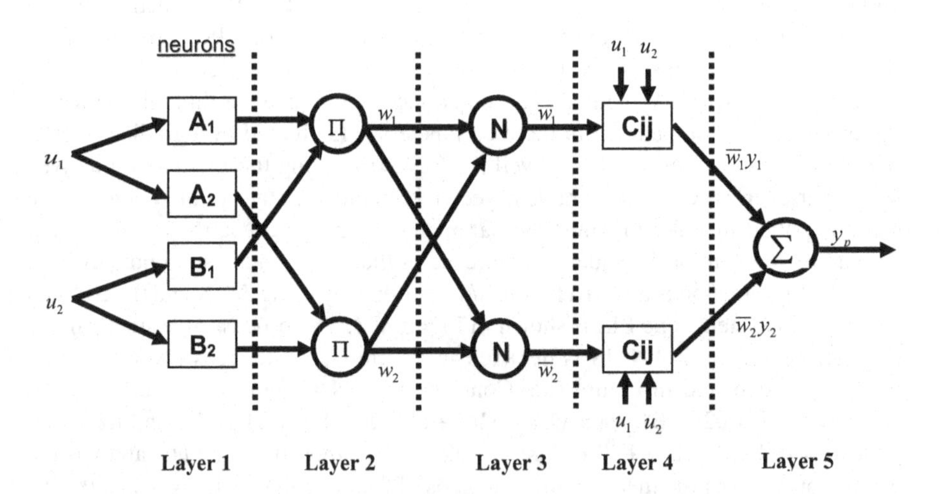

FIGURE C.5 Steps involved in the ANFIS for parameter estimation (MATLAB Fuzzy Tool Box).

given input satisfies the membership function, i.e., μ_A, μ_B. A generalized membership function $\mu(u) = \dfrac{1}{1+\left|\dfrac{u-c}{a}\right|^{2b}}$ is used, and the parameters (a, b, c) are premise parameters; (b) every node in layer 2 is fixed node, whose output (w_1) is the product Π of all incoming signals: $w_1 = \mu_{A_i}(u_1)\mu_{B_i}(u_2),\ i=1,2$; (iii) output of layer 3 for each node is the ratio of the ith rule's firing strength relative to the sum of all rules' firing strengths: $\bar{w}_i = \dfrac{w_i}{w_1+w_2}$; (iv) every node in layer 4 is an adaptive node with a node output: $\bar{w}_i y_i = \bar{w}_i(c_{i1}u_1 + c_{i2}u_2 + c_{i0}),\ i=1,2$. Here, c as consequent parameters, and finally (v) every node in layer 5 is a fixed node which sums all incoming signals $y_p = \bar{w}_1 y_1 + \bar{w}_2 y_2$, where y_p is the predicted output. When the premise parameters are fixed, the overall output is a linear combination of the consequent parameters. The output (linear in the consequent parameters) can be written as:

$$\begin{aligned} y_p &= \bar{w}_1 y_1 + \bar{w}_2 y_2 = \bar{w}_1(c_{11}u_1 + c_{12}u_2 + c_{10}) + \bar{w}_2(c_{21}u_1 + c_{22}u_2 + c_{20}) \\ &= (\bar{w}_1 u_1)c_{11} + (\bar{w}_1 u_2)c_{12} + \bar{w}_1 c_{10} + (\bar{w}_2 u_1)c_{21} + (\bar{w}_2 u_2)c_{22} + \bar{w}_2 c_{20} \end{aligned} \quad \text{(C.6)}$$

A hybrid algorithm adjusts the consequent parameters in a forward pass and the premise parameters in the backward pass. In the forward pass, the network inputs propagate forward until layer 4, where the consequent parameters are identified by the LS method. In a backward pass, the error signals propagate backward and the premise parameters are updated by a gradient descent method. The steps involved are as follows: (i) generation of initial FIS by using **INITFIS = genfis1(TRNDATA)**. The 'TRNDATA' is a matrix with $N+1$ columns where the first N columns contain data for each FIS, and the last column contains the output data. INITFIS is a single output FIS, ii) training of FIS: **[FIS, ERROR, STEPSIZE, CHKFIS, CHKERROR] = anfis(TRNDATA, INITFIS, TRNOPT, DISPOPT, CHKDATA).** Here, vector TRNOPT is used to specify training options, vector DISPOPT is to specify display options during training, CHKDATA is to prevent overfitting of the training data set, and CHKFIS is the final tuned fuzzy inference system.

Example C.2

Generate simulated data using the equation $y = a + bx_1 + cx_2$, with a = 1, b = 2, and c = 1. Estimate the parameters a, b, c using the ANFIS and the delta method.

Solution C.2

The simulated data is generated and is partitioned as training and check data sets. Then, the training data is used to obtain tuned FIS, which is validated with the help of the check data set. The tuned FIS is, in turn, used to predict the system output

TABLE C.1
Estimates by ANFIS

Parameters	True Values	Estimated Values	
		No Noise	SNR = 10
a	1	0.9999	0.9999
b	2	2.0000	2.0000
c	1	0.9999	0.9999

for new input data of the same class and parameters estimated using the delta method. The MATLAB routines given in: **'ExampSolSW/ExampleC.2ANFISPEstm'**. The results are shown in Table C.1.

REFERENCES

1. Kosko, B. *Neural Networks and Fuzzy Systems-A Dynamic Systems Approach to Machine Intelligence.* Prentice Hall, Englewood Cliffs, NJ, 1992.
2. King, R. E. *Computational Intelligence in Control Engineering.* Marcel Dekker, Inc., New York, 1999.
3. Kashyap, S. K., and Raol, J. R. Unification and interpretations of fuzzy set operations. NAL PD FC 0502, FMCD, NAL, Bangalore, March 2005.

Appendix D
Statistics and Probability

Several definitions and concepts from statistics, probability theory, and linear algebra are collected in nearly alphabetical order [1–5]. All of which might have not been used in this book; however, they would be very useful in general for aerospace science and engineering applications.

D.1 AUTOCORRELATION FUNCTION

If $x(t)$ is a random signal, then it is given as $R_{xx}(\tau) = E\{x(t)x(t+\tau)\}$; here, τ is the 'time lag', and E is the expectation operator. For stationary process, R_{xx} is not dependent on 't' and has a maximum value for $\tau = 0$. This is then the variance of the signal x. With increasing time if R_{xx} shrinks, then it means that the nearby values of the signal x are not correlated, and hence not dependent on each other. The autocorrelation of the white noise process is an impulse function. The autocorrelation of discrete time residuals is given as: $R_{rr}(\tau) = \frac{1}{N-\tau}\sum_{k=1}^{N-\tau} r(k)r(k+\tau)$; $\tau = 0,\ldots,\tau_{\max}$ (are discrete time lag).

D.2 BIAS IN AN ESTIMATE

It is given as $(\beta) = \beta - E(\hat{\beta})$, the difference between the true value of the parameter β and expected value of its estimate. The estimates would be biased if the noise were not zero mean. The idea is that for large amount of data used for the estimation of a parameter, an estimate is expected to center closely on the true value, and the estimate is called unbiased if $E\{\hat{\beta} - \beta\} = 0$. The bias should be very small.

D.3 BAYES THEOREM AND STATISTICS

It defines the conditional probability, the probability of the event A, given the event B: $P(A/B)\,P(B) = P(B/A)\,P(A)$. Here, $P(A)$ and $P(B)$ are the unconditional, or a priori, probabilities of event A and B, respectively. This is a fundamental theorem in probability theory. Bayes Theorem allows new information to be used to update the conditional probability of an event. It refers to repeatable measurements (as is done in the frequency-based interpretation of probability), and the interpretation of data can be described by Bayes Theorem. In that case, A is a hypothesis and B is the experimental data. The meanings of various terms are as follows: (i) $P(A|B)$ is the degree of belief in the hypothesis A, after the experiment which produced data B, (ii) $P(A)$ is the prior probability of A being true, (iii) $P(B|A)$ is the ordinary likelihood function used also by non-Bayesian believers, and $P(B)$ is the prior probability of obtaining data B.

D.4 CENTRAL LIMIT PROPERTY

Let a collection of random variables, which are distributed individually according to some different probability distributions, be represented as $z = x_1 + x_2 + \cdots + x_n$; then the central limit theorem states that the random variable z is approximately Gaussian (normally) distributed, if $n \to \infty$ and z has finite expectation and variance. This property allows a general assumption that noise processes are Gaussian, since we can say that these processes have arisen due to the sum of individual processes with different distributions.

D.5 CHI-SQUARE DISTRIBUTION AND TEST

The random variable χ^2 given by $\chi^2 = x_1^2 + x_2^2 + \cdots + x_n^2$, where x_i are the normally distributed variables with zero mean and unit variance, has the **pdf** (probability density function) with n degrees of freedom: $p(\chi^2) = 2^{-n/2}\Gamma(n/2)^{-1}(\chi^2)^{\frac{n}{2}-1}\exp(-\chi^2/2)$. $\Gamma(n/2)$ is the Euler's gamma function, and $E(\chi^2) = n$; $\sigma^2(\chi^2) = 2n$. In the limit, the χ^2 distribution approaches the Gaussian distribution with mean n and variance $2n$. Once the probability density function is numerically computed from the random data, the χ^2 test is used to determine if the computed probability density function is Gaussian.

For normally distributed and mutually uncorrelated variables x_i, with mean m_i and with variance σ_i, the normalized sum of squares is formed as: $s = \sum_{i=1}^{n} \frac{(x_i - m_i)^2}{\sigma_i^2}$. Then, s obeys the χ^2 distribution with n DOF. The χ^2 test is used for hypothesis testing.

D.6 CONFIDENCE INTERVAL AND LEVELS

A confidence interval for a parameter is an interval constructed from empirical data, and the probability that the interval contains the true value of the parameter can be specified. The confidence level of the interval is the chance that this interval (that will result once data are collected) will contain the parameter. In estimation result, requirement of high confidence in the estimated parameters or states is imperative. This information is available from the estimation-process results. A statistical approach is used to define the confidence interval within which the true parameters/states are assumed to lie with 95% of confidence. This signifies a high value of probability with which the truth lies within the upper and lower confidence intervals. If $P\{\, l < \beta < u \,\} = \alpha$, then α is the probability that β lies in the interval (l, u). The probability that the true value, β, is between l (the lower bound) and u (the upper bound) is 'α'. As the interval becomes smaller and smaller, the estimated value $\hat{\beta}$ is regarded more confidently as the true parameter.

D.7 CONSISTENCY OF ESTIMATES

An estimator is asymptotically unbiased if the bias approaches zero with the number of data tending to infinity. It is reasonable that as the number of data increases, the estimate tends to the true value; this is property called 'consistency'. It is a stronger property than asymptotic unbiasedness, because it has to be satisfied for every single realization of estimates. All the consistent estimates are unbiased asymptotically. This convergence is required to be with probability 1 (one):

$$\lim_{N\to\infty} P\left\{ \left|\hat{\beta}(z_1,z_2,\ldots,z_n)-\beta\right| < \delta \right\} = 1 \quad \forall\delta > 0$$

The probability that the error in estimates (w.r.t the true values) is less than a certain small positive value is one.

D.8 CORRELATION COEFFICIENT AND COVARIANCE

It gives the degree of correlation between two random variables. It is given as $\rho_{ij} = \frac{\text{cov}(x_i,x_j)}{\sigma_{x_i}\,\sigma_{x_j}}$ $-1 \le \rho_{ij} \le 1$. $\rho_{ij} = 0$ for independent variables x_i and x_j, and for definitely correlated processes $\rho = 1_{ij}$. If a variable d is dependent on many x_i, then correlation coefficient for each of x_i can be utilized to determine the degree of this correlation with d as:

$$\rho(d,x_i) = \frac{\sum_{k=1}^{N}(d(k)-\underline{d})\,(x_i(k)-\underline{x}_i)}{\sqrt{\sum_{k=1}^{N}(d(k)-\underline{d})^2}\,\sqrt{\sum_{k=1}^{N}(x_i(k)-\underline{x}_i)^2}}$$

The 'under bar' represents the mean of the variable. If $|\rho(d,x_i)|$ approaches unity, then d can be considered linearly related to particular x_i. The covariance between two variables is defined as $\text{cov}(x_i,x_j) = E\{[x_i - E(x_i)][x_j - E(x_j)]\}$.

By definition, the covariance matrix should be symmetric and positive semi-definite. It gives theoretical prediction of the state-error variance. If the parameters are used as variables in the definition, it gives the parameter estimation error covariance matrix. The square roots of the diagonal elements of this matrix give standard deviations of the errors in estimation of states or parameters as the case may be. It is emphasized that the inverse of the covariance matrix is the indication of the information content in the signals about the parameters or states. Large covariance signifies higher uncertainty and low information and low confidence in the estimation results.

D.9 DISTRIBUTION FUNCTION

The distribution of a set of numerical values shows how they are distributed over the real numbers, and it is completely characterized by the empirical distribution function. The probability distribution of a random variable is defined by its probability distribution function (PDF). For each real value of x, the cumulative distribution function of a set of numerical data is the fraction of observations that are less than or equal to x and a plot of the empirical distribution function is an uneven set of stairs, with the width of the stairs being the spacing between adjacent data. The height of the stairs depends on how many data points have exactly the same value. The distribution function is zero for small enough (negative) values of x, and is unity for large enough values of x and increases monotonically. If $y > x$, the empirical distribution function evaluated at y is at least as large as the empirical distribution function evaluated at x.

D.10 DISJOINT OR MUTUALLY EXCLUSIVE EVENTS

Two events are independent if the occurrence of one event gives no information about the occurrence of the other event. The two events, A and B, are independent if the probability that they both occur is equal to the product of the probabilities of the two individual events: $P(A, B) = P(A)\ P(B)$. Two events are disjoint or mutually exclusive if the occurrence of one is incompatible with the occurrence of the other, i.e., if they cannot both happen at once (if they have no outcome in common). Equivalently, two events are disjoint if their intersection is the empty set.

D.11 EXPECTATION

The expected value of a random variable is the long-term average of its values. For a discrete random variable (one that has a countable number of possible values), the expected value is the weighted average of its possible values, and the weight assigned to each possible value is the chance/probability that the random variable takes that value. The mathematical expectation $E(x) = \sum_{i=1}^{n} x_i P(x = x_i)$ and $E(x) = \int_{-\infty}^{\infty} x\ p(x) dx$ with P as the probability distribution of variables x and p the **pdf** of variable x. It is a weighted mean, and the weights are individual probabilities. The expected value of sum of two variables is the sum of their expected values $E(X+Y) = E(X) + E(Y)$; similarly, $E(a \times X) = a \times E(X)$.

D.12 EFFICIENCY OF AN ESTIMATOR

If $\hat{\beta}_1$ and $\hat{\beta}_2$ are the unbiased estimates of the parameter vector β, then compare these estimates in terms of error covariance matrices: $E\left\{(\beta - \hat{\beta}_1)(\beta - \hat{\beta}_1)^T\right\} \leq E\left\{(\beta - \hat{\beta}_2)(\beta - \hat{\beta}_2)^T\right\}$. The estimator $\hat{\beta}_1$ is said to be superior to $\hat{\beta}_2$ if the inequality is satisfied. If it is satisfied for any other unbiased estimator, then we get the efficient estimator. The mean square error and the variance are identical for unbiased estimators, and such optimal estimators are minimum variance unbiased estimators. The efficiency of an estimator can be defined in terms of Cramer-Rao inequality (Chapter 9). It gives a theoretical limit to the achievable accuracy, irrespective of the estimator used: $E\left\{[\hat{\beta}(z)-\beta][\hat{\beta}(z)-\beta]^T\right\} \geq I_m^{-1}(\beta)$, the matrix I_m is the Fisher information matrix, and its inverse is a theoretical covariance limit. If it is assumed that the estimator is unbiased, then such an estimator with equality valid is called an efficient estimator. The Cramer-Rao inequality means that for an unbiased estimator, the variance of parameter estimates cannot be lower than its theoretical bound $I_m^{-1}(\beta)$. The Cramer-Rao bounds define uncertainty levels around the estimates obtained by using the maximum-likelihood/output error method (Chapter 9).

D.13 F-DISTRIBUTION AND F TEST

For x_1 and x_2, the normally distributed random variables with arbitrary means and variances as σ_1^2 and σ_2^2, the following variances are defined: $s_1^2 = \frac{1}{N_1 - 1}\sum_{i=1}^{N_1}(x_{1i} - \bar{x}_1)^2$ and $s_2^2 = \frac{1}{N_2 - 1}\sum_{i=1}^{N_2}(x_{2i} - \bar{x}_2)^2$.

With s_1^2 and s_2^2 the unbiased estimates of the variances, and x_{1i} and x_{2i} the samples from the Gaussian distribution, we have $x_1^2 = \frac{(N_1 - 1)s_1^2}{\sigma_{x1}^2}$ and $x_2^2 = \frac{(N_2 - 1)s_2^2}{\sigma_{x2}^2}$ as the χ^2 distributed variables with DOF $h_1 = N_1 - 1$ and $h_2 = N_2 - 1$. The ratio defined as $F = \left(\frac{h_2}{h_1}\right)\left(\frac{x_1^2}{x_2^2}\right) = \frac{s_1^2\sigma_{x2}^2}{s_2^2\sigma_{x1}^2}$ is described by F-distribution with (h_1, h_2) degrees of freedom. The F-distribution is used in F-test, which provides a measure for the probability that the two independent samples of variables of sizes n_1 and n_2 have the same variance. The ratio $t = \frac{s_1^2}{s_2^2}$ follows F-distribution with h_1 and h_2 DOF. The test hypotheses are formulated and tested for making decision on the (unknown!) truth:

$$H_1(\sigma_1^2 > \sigma_2^2) \; : \; t > F_{1-\alpha}$$

$$H_2(\sigma_1^2 < \sigma_2^2) \; : \; t < F_{\alpha}$$

at the level of $1-\alpha$ or α. The F-test is useful in determining a proper order or structure in time-series and transfer function models (Chapter 9).

D.14 IDENTIFIABILITY

One must be able to identify the coefficients/parameters of the postulated mathematical model (Chapters 2 and 9), from given I/O data of a system under experiment (with some statistical assumptions on the noise processes, which contaminate the measurements). The identifiability refers to this aspect. The input to the system should be persistently exciting. The spectrum of the input signal should be broader than the bandwidth of the system that generates the data.

D.15 INFORMATION MEASURE

The concept of information measure is of a technical nature and does not directly equate to the usual emotive meaning of information. The entropy, somewhat directly related to dispersion, covariance, or uncertainty, of a random variable x (with probability density $p(x)$) is defined as:

$$H_x = -E_x\{\log p(x)\}, \; E \text{ is the expectation operator.}$$

For Gaussian m vector x, it is given as $H_x = 1/2m(\log 2\text{pi} + 1) + 1/2\log|P|$; pi = π, with P as the covariance matrix of x. Entropy is thought of as a measure of disorder or lack of information. Let $H_\beta = -E_\beta[\log p(\beta)]$, the entropy prior to collecting data z and $p(\beta)$ the prior density function of β. When data is collected, we have $H_{\beta/z} = -E_{\beta/z}\{\log p(\beta|z)\}$. Then, the measure of the average amount of information provided by the experiment with data z and parameter β is given by: $I = H_\beta - E_z\{H_{\beta/z}\}$. This is 'mean information' in z about β. The entropy is the dispersion or covariance of the density function, and hence the uncertainty. Thus, the information is seen as the difference between the prior uncertainty and the 'expected' posterior uncertainty; due to collection of data z, the uncertainty is reduced and hence the information is gained. The information is non-negative measure, and it is zero if $p(z,\beta) = p(z)\cdot p(\beta)$, i.e., if the data is independent of the parameters.

D.16 MEAN-SQUARED ERROR (MSE)

The mean-squared error of an estimator of a parameter is the expected value of the square of the difference between the estimator and the parameter. It is given as: $\text{MSE}(\hat{\beta}) = E\{(\hat{\beta} - \beta)^2\}$. It measures how far the estimator is off from the true value, on the average in repeated experiments.

D.17 MEDIAN AND MODE

The median is a middle value: the smallest number such that at least half the numbers are no greater than it. If the values have an odd number of entries, the median is the middle entry after sorting the values in an increasing order. If the values have an even number of entries, the median is the smaller of the two middle numbers after sorting. The mode is the most common or frequent value. There could be more than one mode. Mode is a relative maximum. In estimation, data affected by random noise are used and the estimate of the parameter vector is some measure or quantity related to the probability distribution – it could be mode, median, or mean of the distribution. The mode defines the value of x for which the probability of observing the random variable is a maximum. Thus, the mode signifies the argument that gives the maximum of the probability distribution.

D.18 MONTE-CARLO SIMULATION

In a dynamic system simulation, one can study the effect of random noise on parameter/state estimates to evaluate the performance of the estimator. We first get one set of estimated parameters, then change the seed number for random number generator, and add these random numbers to measurements as noise. We get estimates of the parameters with the new data. We can formulate a number of such data sets with different seeds and obtain parameters to establish the variability of the estimates across different realizations of the data. We next obtain the mean value and the variance of the parameter estimates using all the individual estimates from all these realizations. The mean of the estimates should converge to the true values. The approach can be used for any type of system. Depending upon the problem's complexity, 500 simulation runs or as small as 20 runs could be used to generate average results.

D.19 PROBABILITY AND RELATIVE FREQUENCY

Relative frequency is the value calculated by dividing the number of times an event occurs by the total number of times an experiment has been carried out. The probability of an event is then thought of as its limiting value when the experiment is carried out several times, and the relative frequency of any particular event will settle down to this value. Probability is expressed on a scale from 0 to 1; a rare event has a probability close to 0, and a very common event has a probability of 1.

If X is a continuous random variable, there is a function $p(x)$ (**pdf**) such that for every pair of numbers $a <= b$, $P(a <= X <= b)$ = (area under p between a and b). The probability density function of a random variable with a standard normal/Gaussian distribution is the normal curve. Only continuous random variables have probability density functions.

The probability distribution of a random variable specifies the chance that the variable takes a value in any subset of the real numbers, and the probability distribution of a random variable is completely characterized by the cumulative PDF. The probability distribution of a discrete random variable can be characterized by the chance that the random variable takes each of its possible values. The probability distribution of a continuous random variable is characterized by its pdf.

The Gaussian pdf is given by: $p(x) = \frac{1}{\sqrt{2\pi}\,\sigma} \exp\left(-\frac{(x-m)^2}{2\sigma^2} \right)$, with m as the mean and σ^2 as the variance of the distribution. For the measured random variables, given the state x (or parameters), the **pdf** is given by:

$p(z|x) = \frac{1}{(2\pi)^{n/2}|R|^{1/2}} \exp\left(-\frac{1}{2}(z - Hx)^T R^{-1}(z - Hx) \right)$, with R as the covariance matrix of measurement noise.

REFERENCES

1. Hsia, T. C. *System Identification – Least Squares Methods.* Lexington Books, Lexington, Massachusetts, Toronto, 1977.
2. Sorenson, H. W. *Parameter Estimation – Principles and Problems.* Marcel Dekker, New York and Basel, 1980.
3. Drakos, N. "untitled", Computer based learning unit. University of Leeds, 1996. (Internet Site: rkb.home.cern.ch/rk6/AN16pp/node165.html)
4. Gelb, A. (Ed.) *Applied Optimal Estimation.* M.I.T. Press, Cambridge, MA, 1974.
5. Papoulis, A. *Probability, Random Variables and Stochastic Processes*, 2nd Edn. McGraw Hill, Singapore, 1984.

Appendix E
Signal and Systems Concepts

Several concepts from systems theory are collected in nearly alphabetical order [1,2]. All of which might have not been used in this book; however, they would be very useful in general for aerospace science and engineering applications.

E.1 DATA ACQUISITION AND SAMPLING OF SIGNALS

This is an important ingredient of the total flight testing, simulation, and estimation exercises. For successful data analysis, a good set of data should be generated, gathered, and recorded and/or telemetered. The data acquisition personnel and data analysts should keep several aspects in mind: (i) how the data got to the analysis program from the sensors and how the data were filtered; (ii) how the data were digitized, time-tagged, and recorded; (iii) for extracting more information from data, as much information as available from the related sources should be recorded/gathered. This requires a systems approach for analyzing the data. This is to say that one should look at the entire system – from input to output, because the connections and interactions between various components are of varied nature and important; and (iv) one should safeguard against overreliance on misleading simulation, unforeseen circumstances, and neglect of the synergism of data system problems.

In most of the flight mechanics modeling and analysis exercises, the digitized data are required. It is important to see that the sampling rate is properly selected in order that there is no loss of useful information as well as it does not increase burden of collecting too much data. It is based on the Nyquist theory, which applies to many different fields where data is captured. In general terms, it states the minimum number of resolution elements required to properly describe or sample a signal. In order to reconstruct (interpolate) a signal from a sequence of samples, sufficient samples must be recorded to capture the peaks and trough of the original waveform. For example, when a digital recording uses a sampling rate of 50 kHz, the Nyquist frequency is 25 kHz. If a signal being sampled contains frequency components that are above the Nyquist limit, aliasing will be introduced in the digital representation of the signal unless those frequencies are filtered out prior to digital encoding. The Nyquist sampling theorem states that a sample with a regular sample interval of T sec (a sample rate of $1/T$ samples/s) can contain no information at a frequency higher than $1/(2T)$ Hz. This limit frequency is called the Nyquist frequency, the half-sample frequency, or the folding frequency. Frequency limits of 12.5 or 25 Hz (25 or 50 samples/s) are sufficiently enough to include all useful aircraft stability and control information on the modes/data. The higher frequencies in the original continuous time signal could contain: (i) structural resonance modes, higher than the rigid body dynamic modes;

(ii) ac (altering current) power frequencies; (iii) engine vibration modal responses; and (iv) thermal noise and other nuisance data. These high-frequency data shift to an apparent lower frequency, and this phenomenon is called aliasing or frequency folding (Nyquist folding). The high-frequency noise/data contaminate the low-frequency stability and control data. After sampling, there is no way to remove the effect of this aliasing, and hence, the data should be adequately treated before sampling. The effective method is to apply pre-sample filtering to the data. This means filter the signal (+noise, or unwanted higher-frequency data) to remove the high-frequency/unwanted data/responses before the actual sampling is performed. To avoid aliasing effect, (i) pre-filter the data before digitization, so that unwanted signals in the higher band that would otherwise alias the low frequencies would be eliminated; however, the pre-filtering would introduce lags in signal, so the time lag should be accounted for; (ii) sample the signal at a very high rate, so that the folding frequency is moved farther away and the high-frequency signals (noise, etc.) would alias the frequencies, which are near the new folding frequency, which is farther away from the system/signal frequency of interest. This assures that the useful low-frequency band signals are not affected, as can be seen from Figure E.1.

However, it is prudent to sample the aircraft responses at 100 or 200 samples/s and then digitally filter the data and thin it to 25 or 50 samples/s. The pre-sample filter requirements are: (i) a low-pass filter at 40% of the Nyquist frequency can be used, (ii) for systems with high sampling rates, a 1st-order filter would be adequate, and (iii) for low sample rate systems higher-order filters may be necessary.

In the digitized signal, the resolution is exactly one count, and if the resolution magnitude is much smaller than the noise level of the digitized signal, the resolution problem is a non-issue. A good resolution is 1/100 of the maneuver size and a low resolution is acceptable for Euler angles, say, 1/10 of the maneuver size. The time tagging of data is very important. The time errors, in data measurements, would affect the accuracy of estimated derivatives depending upon the severity of these errors. The 'time tagging' refers to the information about the time of each measurement. Any time lag due to analog pre-filtering or sensors should be taken into account; otherwise, it will cause error in time tag. Some derivate estimates are

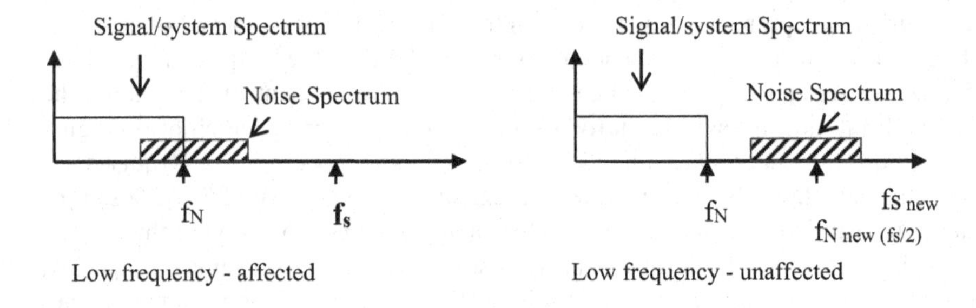

Figure E.1 Frequency aliasing.

extremely sensitive to time shifts in certain signals. Time errors > 0.01 s may cause problems, whereas accuracy of 0.02–0.05 s is usually tolerable for signals like altitude and airspeed.

E.2 DATA EDITING AND FILTERING

Before any post-flight data analysis for parameter estimation and related work that is carried out, it would be desirable to pre-process the data to remove spurious spikes and any high-frequency noise: (i) to edit the data to remove wild points and replace the missing points (often an average from two samples is used), (ii) to filter the edited/raw data to reduce the effect of noise, and (iii) to decimate the data to obtain the data at the required sampling rate for further post-processing; for this, the data should have been filtered at higher rate than required. Editing by using finite difference method and filtering using FFT (fast Fourier transform)/spectral analysis can be done. The editing is a process of removing the data spikes (wild points) and replacing them by suitable data points. Wild points occurring singly can be detected by using the slope of the data set or first finite difference. Any data point exhibiting higher than the prefixed slope is considered as wild point and eliminated. When wild points occur in groups, the surrounding points are considered to detect them. A finite difference array consisting of first-, second-, and third-difference (up to nth differences) is formed. The maximum allowable upper and lower limits for these differences are pre-specified, and if the array indicates any value greater than the limits, the points are treated as wild points and eliminated. The points are replaced by suitable points by interpolation considering surrounding points. If the editing limits are too high, the edited data will leave a large amount of wild points, and if the limit is too small, the edited data would appear distorted.

Filtering is the process of removing the noise components presented in the edited/raw data. These filters introduce the time-lag effects in the flight data, thereby compounding the problem of parameter estimation. Discrete Fourier transform (DFT) method allows processing from time domain to frequency domain and does not introduce the time lags. However, it is offline procedure. Based on the spectral analysis through DFT, one can know the frequency contained in the raw signal. The filtering/editing nowadays can be easily carried out using certain functions from MATLAB signal processing tool box. However, certain fundamental aspects are briefly given next.

The Fourier transform (FT) from time to frequency domain is given by:

$$H(f) = 1/N\left[\sum_{t=0}^{N-1} h(t)\exp(-j2\pi ft/N)\right] \tag{E.1}$$

Here, $h(t)$ is the time function of the signal, $H(t)$ the complex function in frequency domain, and N the number of samples.

The inverse FT (from frequency to time domain) is given by:

$$h(t) = 1/N \left[\sum_{f=0}^{N-1} H(f)\exp(+j2\pi ft/N) \right] \tag{E.2}$$

From above eqns. it is clear that the transformation eqns. are different because $\exp(-j2\pi ft/N)$ term is changed to $\exp(+j2\pi ft/N)$. But it can be shown that:

$$\begin{aligned} h(t) &= 1/N \left[\sum_{f=0}^{N-1} H^*(f)\exp(-j2\pi ft/N) \right] \\ &= 1/N \left[\sum_{f=0}^{N-1} H(f)\exp(+j2\pi ft/N) \right] \end{aligned} \tag{E.3}$$

Here, $H^*(f)$ is the conjugate of $H(f)$, and * is for the conjugate operation. By comparing eqns. (E.1) and (E.3), it can be seen that the same transformation routine can be used. After the inverse transformation, the conjugate is not necessary because the data in time domain is real. The signal to be filtered is first transformed into frequency domain using DFT. Using cut-off frequency, the Fourier coefficients of the unwanted frequencies are set to zero. Inverse FT time domain yields the filtered signal. For proper use of filtering method, selection of the cut-off frequency is crucial. In spectral analysis method, the power spectral density of the signal is plotted against frequency and the inspection of the plot would help distinguish the frequency contents of the signal. This information is used for selecting the proper cut-off frequency for the filter. The FT of the correlation functions is often used in analysis. The FT of the autocorrelation function (ACF):

$$\phi_{xx}(\omega) = -\int_{-\infty}^{\infty} \phi_{xx}(T)\cdot\exp(-j\omega T)\cdot dT \tag{E.4}$$

is called the power spectral density function of the random process $x(t)$. The 'power' term in here is used in a generalized sense and indicates the expected squared value of the members of the ensemble. $\phi_{xx}(\omega)$ is indeed the spectral distribution of power density for $x(t)$ in that integration of $\phi_{xx}(\omega)$ over frequencies in the band from ω_1 to ω_2 yields the mean-squared value of the process whose ACF consists only of those harmonic components of $\phi_{xx}(T)$ that lie between these frequencies. The mean-squared value of $x(t)$ itself is given by the integration of the power density spectrum for the random process over the full range of frequencies:

$$\phi_{xx}(T) = \frac{1}{2\pi}\int_{-\infty}^{\infty} \phi_{xx}(\omega)\cdot\exp(j\omega T)\cdot d\omega \tag{E.5}$$

$$E\{x^2\} = \phi_{xx}(0) = \frac{1}{2\pi}\int_{-\infty}^{\infty} \phi_{xx}(\omega)\cdot d\omega \tag{E.6}$$

E.3 FUDGE FACTOR/ACCURACY DETERMINATION

The uncertainty bounds for parameter estimates obtained from the Cramer-Rao bounds/output error method (OEM) are multiplied with a fudge factor to reflect correctly the uncertainty. This is because when the OEM (which does not handle the process noise) is used on flight data, which are often affected by the process noise (like atmospheric turbulence), the uncertainty bounds do not correctly reflect the effect of this noise on the uncertainty of the estimates. And hence, a fudge factor of about 3–5 can be used in practice. This number has been arrived (on the average) at by performing flight simulation-based parameter estimation exercises for a fighter aircraft for longitudinal and lateral-directional maneuvers and using the formula:

$$FF = \sqrt{\frac{\text{sampling frequency}}{(2\ \text{bandwidth of residuals})}}. \tag{E.7}$$

E.4 GENETIC ALGORITHMS

Genetic algorithms are heuristic/directed search methods and computational models of adaptation and evolution based on natural selection strategy of evolution of biological species. The search for beneficial adaptations to a continually changing environment in the nature (i.e., evolution) is fostered by the cumulative evolutionary knowledge that each species possesses of its forebears. This knowledge, which is encoded in the chromosomes of each member of a species, is passed from one generation to the next by a well-known mating process wherein the chromosomes of 'parents' produce 'offspring'. Thus, the genetic algorithms mimic and exploit the genetic dynamics underlying natural evolution to search for optimal and global solutions of general combinatorial optimization problems. The applications are the traveling salesman problem, VLSI circuit layout, gas pipeline control, the parametric design of an aircraft, learning in neural nets, models of international security, strategy formulation, and parameter estimation.

E.5 SIGNALS/INSTRUMENTATION REQUIRED FOR PARAMETER ESTIMATION FROM FLIGHT TEST DATA

Minimum Practical Set: time; control positions $\delta e, \delta a, \delta r$; angular rates p, q, r; linear accelerations a_x, a_y, a_z^2.

Other useful signals: α, β; Euler angles; angular accelerations $\dot{p}, \dot{q}, \dot{r}$; ρ (air density); engine parameters r.p.m, thrust; configuration parameters m, inertia; fuel, loading data (C.G. aft or fore of a datum.

Most important aspects are: for control surface positions, the direct measurements of surface hinge angles (rudder, aileron, and elevator, etc.) would be desirable. If these signals are taken at the actuator, command measurements require the

Table E.1
Permissible Error Budget

Parameter / Signal	Allowable Errors
Mass	1%
MOI, I_y	2%
$X_{C.G.}$	0.005 m
$\bar{q}$ $(1/2\ \rho\ V^2)$	2%
V	1%
α	0.06 deg.
θ	0.3 deg.
q	0.01 rad/s
$\dot{q}$	0.05 rad/s^2

knowledge of actuator. For the angular rates, the alignment of the sensors should be within 0.1 deg. For the linear accelerations, the mounting location should be known – ideal being at the C.G. For the flow angles, the vanes are normally used for which the vane location must be known. If the nose boom is used, then empirically the nose (boom) length should be 2.5–3 times the fuselage diameter. For pressure ports, three locations is a minimum to deduce the flow angle in one-axis. The aircraft velocities can be obtained by integrating accelerometer measurements. The angular are not required but useful for correcting linear accelerometer reading to the C.G. Engine parameters (engine thrust; pressure, temperatures at various places in the propulsion system; rotational speeds; fuel flows; engine control settings, etc.) are not useful for stability and control derivatives estimation but are useful for estimating drag coefficients. The positions of flaps, landing gear up or down, tank, and fuel weight are other useful parameters. Typical allowable measurement and installation errors for the identification of longitudinal motion of a small aircraft are given in Table E.1. Linear accelerations range should be at least 4 g, and the accuracy of 0.1 g might be tolerable. Body rates accuracy of +/−0.1 deg./s would be preferable. The control surface deflection measurements should be at least accurate to 0.1 deg. The flow angles accuracy should be better than at least 0.1 deg.

E.6 UNITS AND THEIR CONVERSION FACTORS

Quantity	Unit (Base Unit/ Preferred Multiple and Sub-Multiples)		Conversion Factors
Length	m(meter)	Km, mm	1 statute mile = 1609.344 m
Area	m^2	cm^2, mm^2	1 international nautical mile = 1852 m
Volume	m^3	mm^3	1 ft = 0.304 m
Time	s (second)	minutes, hours	1 in = 25.4 mm
Velocity	m/s	------	1 in^2 = 645.16 $(mm)^2$
Angle	rad (radian)	Degree, minute, second	1 UK gallon = 4.536 liters
Angular Velocitsssy	rad/s	------	Hours and minutes are not SI, but still used widely
Acceleration	m/s^2	------	1 ft/sec = 0.3048 m/s
Angular acceleration	rad/s^2	KHz	1 mph = 1.609344 km/h
Frequency	Hz (hertz)	g (gram)	1 international knot = 1.852 km/h
Mass	Kg	g/s	$1^\circ = (2\pi/360)$ rad
Mass flow (fuel flow)	Kg/s	------	1 rpm = $(\pi/30)$ rad/s
Volumetric flow rate	m^3/s	MN, KN	1 rad/s = 9.5493 rpm
Force (thrust)	N (Newtons)		1 ft/sec^2 = 0.3048 m/s^2
Torque/moment of force	Nm		1 m/s^2 = 0.102 g, (g is acceleration due to gravity)
			1 cycle /sec = 1 Hz
			1 metric tonne = 1000 kg
			1 ton = 1.01605 metric tonne
			1 lb = 0.45359237 kg
			1 lb/sec = 0.45359237 kg/s
			1 gal/min = 6.309×10^{-5} m^3/s
			1 gal/min = 1.26280 ml/s
			1 N = 1 kg m/s^2
			1 ton f = 9.96402 kN
			1 Ibf = 4.44822 N
			1 kgf = 9.80665 N
			1 dyne = 10^{-5} N
			1 Nm = 1 kg m^2/s^2

Torque/length	N.m/m	Mpa, KPa	1 lbf in/in = 0.2248 N m/m
Specific fuel consumption	mg/Ns	KJ	1 lb/hr/lbf = 28.3255 mg/Ns
Pressure (stress)	μg/j	MW, kW	1 lb/hr/hp = 168.966 g/J
Work	Pa (pascal)	k Nm	1Pa = 1 N/m^2 = 1 kg/m s^2
Energy	J (joule)	kJ /kg K	1lbf / in^2 = 6894.76 Pa
Quantity of heat	J		1lbf / in^2 = 0.070307 kg/cm^2
Power	J		1 tone/in^2 = 15.4443 MPa
Energy	W (Watt)		1 mm Hg = 133.322 Pa
Energy (thermo-chemical)	kWh		1 Torr = 1 mm Hg
Density	Cal		1 mm H_2O = 9.81302 Pa
Torque	Kg/m^3		1 bar = 100k Pa = 0.1 MPa
Viscosity	Nm		1 pieze = 1 KPa
Dynamic	Pa s		1 Atm = 1.01325 bars
kinematic	N-s/m^2		1J = 1 Nm = 1 kg m^2/s^2
Thermal	M^2/s		1 ft lbf = 1.35582 J
(conductivity)	W/m K		1 Btu = 1.05506 kJ
Specific heat (gas constant)	J / kg K		1 thermochemical calorie = 4.184 J
			1 W = 1 J / s = 1 N m/s = 1 kg m^2/s^3
			1 hp = 0.7457 kW
			1 kWh = 3.86 × 10^4 J
			1 Btu = 3.96832 calories
			1 lb/ft^3 = 16.0185 kg/m^3
			1 lb/gal = 99.7764 kg/m^3
			1 lbf / ft = 1.35582 Nm
			1 Pa s = 1 kg/m s
			1 centipoise = 1m N s/ms
			1 centistoke = 1 mm^2/s
			1 Btu in /ft^2 hr °F = 0.144131 W/m K
			1 ft / lbf / lb K = 2.98907 J /kg K

REFERENCES

1. Rabinder, L. R., and Rader, C. M. (Eds.) *Digital signal Processing. Vol. I and II*, IEEE Press. USA, 1975.
2. Maine, R. E., and Iliff, K. W., Application of parameter estimation to aircraft stability and control – the output error approach, NASA RP-1168, 1986.

Bibliography

Some additional list of books and papers is provided here.

1. Jategaonkar, R. V. *Flight Vehicle System Identification* – A Time Domain Methodology. Progress in Astronautics and Aeronautics, Vol. 216, published by AIAA Inc., Reston, VA, 2006.
2. Klein, V., and Morelli, E. A. *Aircraft System Identification – Theory and Practice*, AIAA Education Series, USA, 2006.
3. Tischler, M. B., and Remple, R. K. *Aircraft and Rotorcraft System Identification—Engineering Methods with Flight Test Examples*, AIAA Education Series, USA, 2006.
4. Anderson, J. D. *Introduction to Flight.* McGraw Hill, Boston, MA, 2005.
5. Pamadi, B. N. *Performance, Stability, Dynamics, and Control of Airplanes*, AIAA Education Series, 2004.
6. Howison, H. *Practical Applied Mathematics—Modelling, Analysis, Approximation. Cambridge University Press*, Cambride, UK, 2005.
7. Newcome, L. R. *Unmanned Aviation—A Brief History of Unmanned Air Vehicles*, Virginia, AIAA Education Series, 2004.
8. Mueller, T. J. *Introduction to the Design of Fixed Wing Micro Air Vehicles—Including 3 Case Studies*, Virgina, AIAA Education series, 2007.
9. Rotstein, H., Ingvalson, R., Keviczky, T., and Balas, G. J. Fault-detection design for uninhabited aerial vehicles. *Journal of Guidance, Control and Dynamics*, 29(5), 1051–1060, 2006.
10. Kermode, A. C. *Mechanics of Flight*, 11th Edn. Revised by Barnard, R. H., and Philpott, D. R. Prentice Hall, Hoboken, NJ, 2006.
11. Zipel, P. *Advanced Six Degrees of Freedom Aerospace Simulation and Analysis in C++*. AIAA, USA, 2006.
12. Haghighat, S., Martins, J. R., and Liu, H. H. Aeroservoelastic design optimization of a flexible wing. *Journal of Aircraft*, 49, 432–443. doi:10.2514/1.C031344, American Institute of Aeronautics and Astronautics, March 2012.
13. Ananthasayanam, M. R. Flight mechanics from a viewpoint of optimal parameters, variables, and approaches. AIAA-2001-4315, *AIAA Atmospheric Flight Mechanics Conference and Exhibit*, 6–9 August 2001, Montreal, Canada.
14. Ananthasayanam, M. R. Pattern of progress of civil transport airplanes during the twentieth century. AIAA 2003-5621, *AIAA Atmospheric Flight Mechanics Conference and Exhibit*, 11–14 August 2003, Austin, Texas, USA.
15. Ananthasayanam, M. R. Some useful conceptual aspects in flight mechanics. AIAA 2004-570, *42nd AIAA Aerospace Sciences Meeting and Exhibit*, 5–8 January 2004, Reno, Nevada, USA.
16. Ananthasayanam, M. R. The persistent occurrence of number three in the history of space mechanics. AIAA 2004-3335, *40th AIAA/ASME/SAE/ASEE Joint Propulsion Conference and Exhibit*, 11–14 July 2004, Fort Lauderdale, Florida, USA.
17. Ananthasayanam, M. R., and Ibrahim, K. Historical evolution of the military fighter airplanes around the twentieth century. AIAA 2005-326, *43rd AIAA Aerospace Sciences Meeting and Exhibit*, 10–13 January 2005, Reno, Nevada, USA.

18. Narasimha, R., and Ananthasayanam, M. R. Engine-out take-off and climb—An old airworthiness problem reviewed. *40th Anniversary United Nations, October, 1985, Official Magazine of International Civil Aviation, ICAO Bulletin*, pp. 18–21, USA.
19. Ananthasayanam, M. R. The influence of culture and environment on the development of flight mechanics. AIAA 2006-96, *44th AIAA Aerospace Sciences Meeting and Exhibit*, 9–12 January 2006, Reno, Nevada, USA.

Index

Printed in the United States
by Baker & Taylor Publisher Services